Engineering Aspects of Food Biotechnology

Contemporary Food Engineering

Series Editor

Professor Da-Wen Sun, Director
Food Refrigeration & Computerized Food Technology
National University of Ireland, Dublin
(University College Dublin)
Dublin, Ireland
http://www.ucd.ie/sun/

Engineering Aspects of Food Biotechnology

EDITED BY
José A. Teixeira • António A. Vicente

CRC Press
Taylor & Francis Group
Boca Raton London New York

CRC Press is an imprint of the
Taylor & Francis Group, an **informa** business

CRC Press
Taylor & Francis Group
6000 Broken Sound Parkway NW, Suite 300
Boca Raton, FL 33487-2742

First issued in paperback 2017

© 2014 by Taylor & Francis Group, LLC
CRC Press is an imprint of Taylor & Francis Group, an Informa business

No claim to original U.S. Government works

ISBN-13: 978-1-4398-9545-0 (hbk)
ISBN-13: 978-1-138-19976-7 (pbk)

Library of Congress Cataloging-in-Publication Data

Engineering aspects of food biotechnology / [edited by] José A. Teixeira, António A. Vicente.
    pages cm. -- (Contemporary food engineering)
Includes bibliographical references and index.
ISBN 978-1-4398-9545-0 (hardback : acid-free paper) 1. Food--Biotechnology. I. Teixeira, José A. (José António), 1957- editor of compilation. II. Vicente, António A., editor of compilation.

TP248.65.F66E54 2014
664--dc23

2013026451

**Visit the Taylor & Francis Web site at
http://www.taylorandfrancis.com**

**and the CRC Press Web site at
http://www.crcpress.com**

Contents

SECTION I   Use of Biotechnology in the Development of Food Processes and Products

SECTION II   Advanced Unit Operations in Food Biotechnology

SECTION III   Adding Value to Food Processing By-Products—The Role of Biotechnology

Series Preface

CONTEMPORARY FOOD ENGINEERING

Food engineering is the multidisciplinary field of applied physical sciences combined with the knowledge of product properties. Food engineers provide the technological knowledge transfer essential to the cost-effective production and commercialization of food products and services. In particular, food engineers develop and design processes and equipment to convert raw agricultural materials and ingredients into safe, convenient, and nutritious consumer food products. However, food engineering topics are continuously undergoing changes to meet diverse consumer demands, and the subject is being rapidly developed to reflect market needs.

In the development of food engineering, one of the many challenges is to employ modern tools and knowledge, such as computational materials science and nanotechnology, to develop new products and processes. Simultaneously, improving food quality, safety, and security continues to be a critical issue in food engineering study. New packaging materials and techniques are being developed to provide more protection to foods, and novel preservation technologies are emerging to enhance food security and defense. Additionally, process control and automation regularly appear among the top priorities identified in food engineering. Advanced monitoring and control systems are developed to facilitate automation and flexible food manufacturing. Furthermore, energy saving and minimization of environmental problems continue to be important food engineering issues, and significant progress is being made in waste management, efficient utilization of energy, and reduction of effluents and emissions in food production.

The *Contemporary Food Engineering Series*, consisting of edited books, attempts to address some of the recent developments in food engineering. The series covers advances in classical unit operations in engineering applied to food manufacturing as well as such topics as progress in the transport and storage of liquid and solid foods; heating, chilling, and freezing of foods; mass transfer in foods; chemical and biochemical aspects of food engineering and the use of kinetic analysis; dehydration, thermal processing, nonthermal processing, extrusion, liquid food concentration, membrane processes, and applications of membranes in food processing; shelf-life and electronic indicators in inventory management; sustainable technologies in food processing; and packaging, cleaning, and sanitation. These books are aimed at professional food scientists, academics researching food engineering problems, and graduate-level students.

The editors of these books are leading engineers and scientists from different parts of the world. All the editors were asked to present their books to address the market's needs and pinpoint cutting-edge technologies in food engineering.

All contributions are written by internationally renowned experts who have both academic and professional credentials. All authors have attempted to provide critical, comprehensive, and readily accessible information on the art and science of a relevant topic in each chapter, with reference lists for further information. Therefore, each book can serve as an essential reference source to students and researchers in universities and research institutions.

Da-Wen Sun
Series Editor

Preface

Food biotechnology has had a very significant and growing impact on the food industry's activity in recent years and that impact is clearly set to keep increasing.

Food biotechnology's typical developments and applications have occurred in the field of genetics and in enzyme- and cell-based biological processes, aiming at producing and improving food ingredients and foods themselves. While these developments and applications are usually well reported in terms of the underlying science, there is a clear lack of information on the engineering aspects of such biotechnology-based food processes. This book provides a comprehensive review of those aspects, from the development of food processes and products to the most important unit operations implied in food biotechnological processes, including food quality control and waste management.

The book is intended for food technologists and engineers working as food developers, processors, or food process controllers. Undergraduate and postgraduate students and researchers will also find the materials covered in this book an invaluable contribution to their education and work.

This book is divided into three sections: Use of Biotechnology in the Development of Food Processes and Products, Advanced Unit Operations in Food Biotechnology, and Adding Value to Food Processing By-Products—The Role of Biotechnology.

The first section of this book presents information on the production, using biotechnology-based processes, of ingredients that play an important role in the food industry. Because it's not possible to address all the ingredients used in the food industry and that are produced involving a biotechnological step, we focused this part of the book on ingredients that, on one hand, have an extended role in food processing and that have been associated with relevant challenges on their use and production (e.g., in the case of enzymes), and on the other, are gaining an increasing importance in the sector (this being the case of pre- and probiotics and biopolymers). These products were also selected keeping in mind that their utilization in the food industry addresses two relevant issues—consumers' awareness of the relation between nutrition and good health and the importance of the environmental sustainability in the food chain. Fermentation is also a key technology in food processing as many traditional food products involve a fermentation step; this being the case, fermentation was also considered a subject of relevance for the book. Finally, a chapter on the production of "*in vitro*" meat is included as a challenge on the use of biotechnology for food applications. The importance of this comes from the fact that meat consumption is expected to grow an extra 74% in 2050 to 470 million tons and this will have, apart from a relevant impact on meat supply, a rather dramatic impact on the environment.

The second section of the book deals with the application of relevant unit operations developed to extract/purify the ingredients of biotechnological origin intended for food applications. Food technology has been, for years, a traditional industry and only recently has several unit operations that have long been implemented in the pharmaceutical and biotechnology industries gained importance in food processing. These operations are expected to contribute to an improved efficiency in overall production processes, minimize operation costs, and make such processes more environment friendly. They will also allow for the extraction and purification of molecules that could not be obtained by conventional operations, extending the range of applications of food-based products. As in the previous section of the book, unit operations that are of relevance to the food industry with a high potential for increased application were considered. A chapter is also included in the application of process analytical technology as this is an important tool to satisfy the increasingly sophisticated and strict policies for quality control and monitoring of specific phases of the process being adopted in the food industry.

The extraction and purification of valuable molecules are also an important concern; thus, the third section of this book deals with the role of biotechnology in adding value to food processing by-products, including post-harvest losses. This is a key issue in the food industry as waste management is a growing concern that makes industrialists and scientists look at waste not as a problem but as an opportunity to create value. Biotechnological solutions are described, either for adding value to food processing by-products or for waste minimization. Once again, not all cases were taken into account and only a group of selected industries was considered due to their relevance to the food sector. The relevance of these industries comes from their economical impact, the amount of waste being produced, and/or the potential associated with its valorization. The implementation of waste management and waste valorization strategies will make a relevant case for the food industry for the development of the biorefinery concept.

Editors

José A. Teixeira, PhD, is a professor and currently the head of the Biological Engineering Department (DEB) of the Engineering School of Minho University. The Biological Engineering Department is responsible for scientific research and teaching in the areas of biological and chemical engineering. The DEB carries out research in food safety and processing, environmental biotechnology and sustainable development, industrial biotechnology and chemistry, and health biotechnology.

Professor Teixeira focused his research activities on two main topics—fermentation technology (multiphase bioreactors, in particular) and food technology. A more recent area of interest is food nanotechnology as well as the production of bioactive compounds for food and medical application. He has been the scientific coordinator of 21 research projects, including two Alfa networks. He has coauthored 210 peer-reviewed papers and is the coeditor of two books, *Reactores Biológicos-Fundamentos e Aplicações* (in Portuguese) and *Engineering Aspects of Milk and Dairy Products*.

António A. Vicente, PhD, graduated with a degree in food engineering in 1994 from the Portuguese Catholic University, in Porto, Portugal, and completed his PhD in chemical and biological engineering in 1998 from the University of Minho, in Braga, Portugal. He received his Habilitation in 2010 from the University of Minho.

From early in his career he has kept a close contact with the food industry and he is involved in several research projects, both national and international, together with industrial partners either as a participant or as a project leader.

His main research interests are food processing by ohmic heating/moderate electric fields (namely, the study of the effects of electric currents on biomolecules and cells) and fermentation technology (including design and operation of bioreactors).

Dr. Vicente has published over 120 research articles in international peer-reviewed journals and 22 chapters in international books. He is presently associate professor with Habilitation at the Biological Engineering Department of the University of Minho, in Braga, Portugal.

Series Editor

Born in Southern China, Professor Da-Wen Sun is a world authority in food engineering research and education; he is a member of the Royal Irish Academy (RIA), which is the highest academic honor in Ireland; he is also a member of Academia Europaea (The Academy of Europe). His main research activities include cooling, drying, and refrigeration processes and systems, quality and safety of food products, bioprocess simulation and optimization, and computer vision technology. His innovative studies on vacuum cooling of cooked meats, pizza quality inspection by computer vision, and edible films for shelf-life extension of fruit and vegetables have been widely reported in national and international media. Results of his work have been published in about 600 papers including more than 250 peer-reviewed journal papers (h-index = 35). He has also edited 12 authoritative books. According to Thomson Scientific's *Essential Science Indicators*[SM] updated as on July 1, 2010, based on data derived over a period of 10 years plus four months (January 1, 2000 to April 30, 2010) from ISI Web of Science, a total of 2554 scientists are among the top 1% of the most cited scientists in the category of Agriculture Sciences, and Professor Sun tops the list with his ranking of 31.

He received a first class BSc Honors and MSc in mechanical engineering, and a PhD in chemical engineering in China before working in various universities in Europe. He became the first Chinese national to be permanently employed in an Irish university when he was appointed college lecturer at the National University of Ireland, Dublin (University College Dublin [UCD]), in 1995, and was then continuously promoted in the shortest possible time to senior lecturer, associate professor, and full professor. Dr. Sun is now a professor of Food and Biosystems Engineering and the director of the Food Refrigeration and Computerised Food Technology Research Group at the UCD.

As a leading educator in food engineering, Professor Sun has significantly contributed to the field of food engineering. He has trained many PhD students, who have made their own contributions to the industry and academia. He has also delivered lectures on advances in food engineering on a regular basis in academic institutions internationally and delivered keynote speeches at international conferences. As a recognized authority in food engineering, he has been conferred adjunct/visiting/consulting professorships from 10 top universities in China, including Zhejiang University, Shanghai Jiaotong University, Harbin Institute of Technology, China Agricultural University, South China University of Technology, and Jiangnan University. In recognition of his significant contribution to food engineering worldwide and for his outstanding leadership in the field, the International Commission of Agricultural and

Biosystems Engineering (CIGR) awarded him the "CIGR Merit Award" in 2000, and again in 2006, the Institution of Mechanical Engineers based in the United Kingdom named him "Food Engineer of the Year 2004." In 2008, he was awarded the "CIGR Recognition Award" in honor of his distinguished achievements as the top 1% of Agricultural Engineering scientists in the world. In 2007, he was presented with the only "AFST(I) Fellow Award" in that year by the Association of Food Scientists and Technologists (India), and in 2010, he was presented with the "CIGR Fellow Award"; the title of Fellow is the highest honor in CIGR and is conferred to individuals who have made sustained, outstanding contributions worldwide.

He is a fellow of the Institution of Agricultural Engineers and a fellow of Engineers Ireland (the Institution of Engineers of Ireland). He has also received numerous awards for teaching and research excellence, including the President's Research Fellowship, and has twice received the President's Research Award of the UCD. He is the editor-in-chief of *Food and Bioprocess Technology—an International Journal* (Springer; 2010 Impact Factor = 3.576, ranked at the 4th position among 126 ISI-listed food science and technology journals), series editor of *Contemporary Food Engineering* book series (CRC Press/Taylor & Francis), former editor of *Journal of Food Engineering* (Elsevier), and editorial board member for *Journal of Food Engineering* (Elsevier), *Journal of Food Process Engineering* (Blackwell), *Sensing and Instrumentation for Food Quality and Safety* (Springer), and *Czech Journal of Food Sciences*. He is also a chartered engineer.

On May 28, 2010, he was awarded membership in the RIA, which is the highest honor that can be attained by scholars and scientists working in Ireland; at the 51st CIGR General Assembly held during the CIGR World Congress in Quebec City, Canada, on June 13–17, 2010, he was elected incoming president of CIGR and will become CIGR president in 2013–2014—the term of his CIGR presidency is six years, two years each for serving as incoming president, president, and past president. On September 20, 2011, he was elected to Academia Europaea (The Academy of Europe), which is functioning as the European Academy of Humanities, Letters and Sciences and is one of the most prestigious academies in the world; election to the Academia Europaea represents the highest academic distinction.

Contributors

Ana Torrado Agrasar
Department of Analytical and Food
  Chemistry
University of Vigo
Ourense, Spain

Vitor D. Alves
CEER-Biosystems Engineering
ISA/Technical University of Lisbon
Lisboa, Portugal

Monika Antošová
Department of Chemical
  and Biochemical Engineering
Slovak University of Technology
Bratislava, Slovakia

M. Gabriela Bernardo-Gil
Institute for Biotechnology and
  Bioengineering
Instituto Superior Técnico
Technical University of Lisbon
Lisboa, Portugal

Tomáš Brányik
Department of Biotechnology
Institute of Chemical Technology
  Prague
Prague, Czech Republic

Bas Brinkhof
Department of Farm Animal Health
Utrecht University
Utrecht, the Netherlands

Lorenzo Pastrana Castro
Department of Analytical and Food
  Chemistry
University of Vigo
Ourense, Spain

Isabel Coelhoso
Department of Chemistry
FCT/Universidade Nova de Lisboa
Caparica, Portugal

Jane Selia dos Reis Coimbra
Departamento de Tecnologia de
  Alimentos
Universidade Federal de Viçosa
Viçosa, MG-Brazil

Lucília Domingues
IBB-Institute for Biotechnology
  and Bioengineering
Centre for Biological Engineering
University of Minho
Braga, Portugal

António Ferreira
Laboratory for Process, Environmental
  and Energy Engineering
Faculdade de Engenharia da
  Universidade do Porto
Porto, Portugal

Filomena Freitas
Department of Chemistry
FCT/Universidade Nova de Lisboa
Caparica, Portugal

B. Giménez
Institute of Food Science,
  Technology and Nutrition
  (ICTAN–CSIC)
Madrid, Spain

M.C. Gómez-Guillén
Institute of Food Science,
  Technology and Nutrition
  (ICTAN–CSIC)
Madrid, Spain

Elisa Alonso González
Department of Analytical and Food
   Chemistry
University of Vigo
Ourense, Spain

Michal Gramblička
Department of Chemical and
   Biochemical Engineering
Slovak University of
   Technology
Bratislava, Slovakia

Nelson Pérez Guerra
Department of Analytical and Food
   Chemistry
University of Vigo
Ourense, Spain

Henk P. Haagsman
Department of Infectious Diseases
   and Immunology
Utrecht University
Utrecht, the Netherlands

Ernesto Hernandez
School of Chemical Engineering
   and Analytical Science
University of Manchester
Manchester, United Kingdom

João Sérgio Azevedo Lima da Silva
IBB-Institute for Biotechnology
   and BioEngineering
Universidade do Minho
Braga, Portugal

M.E. López-Caballero
Institute of Food Science, Technology
   and Nutrition (ICTAN–CSIC)
Madrid, Spain

Fernando Filipe Macieira da Silva
IBB-Institute for Biotechnology
   and BioEngineering
Universidade do Minho
Braga, Portugal

João Paulo Martins
Instituto Federal de Educação
Ciência e Tecnologia do Sul de Minas
   Gerais
Pouso Alegre, MG-Brazil

Rui Miguel da Costa Martins
CBMA-Molecular Biology and
   Environmental Research Center
Universidade do Minho
Braga, Portugal

Luiza Helena Meller da Silva
Instituto de Tecnologia
Universidade Federal do Pará
Belem, PA-Brazil

Nuno G.T. Meneses
IBB-Institute for Biotechnology
   and Bioengineering
Centre of Biological Engineering
Universidade do Minho
Braga, Portugal

P. Montero
Institute of Food Science, Technology
   and Nutrition (ICTAN–CSIC)
Madrid, Spain

Solange I. Mussatto
IBB-Institute for Biotechnology
   and Bioengineering
Centre of Biological Engineering
Universidade do Minho
Braga, Portugal

Carla Oliveira
IBB-Institute for Biotechnology
   and Bioengineering
Centre for Biological Engineering
University of Minho
Braga, Portugal

Severino S. Pandiella
School of Chemical Engineering
   and Analytical Science
University of Manchester
Manchester, United Kingdom

Petra Patáková
Department of Biotechnology
Institute of Chemical Technology Prague
Prague, Czech Republic

Leona Paulová
Department of Biotechnology
Institute of Chemical Technology
    Prague
Prague, Czech Republic

Milan Polakovič
Department of Chemical
    and Biochemical Engineering
Slovak University of Technology
Bratislava, Slovakia

Philippe Ramos
IBB-Institute for Biotechnology
    and Bioengineering
Centre for Biological Engineering
University of Minho
Braga, Portugal

Maria A.M. Reis
Department of Chemistry
FCT/Universidade Nova de Lisboa
Caparica, Portugal

Fernando Alberto Rocha
Laboratory for Process,
    Environmental and Energy
    Engineering
Faculdade de Engenharia da
    Universidade do Porto
Porto, Portugal

Lígia Rodrigues
IBB-Institute for Biotechnology
    and Bioengineering
Centre for Biological Engineering
University of Minho
Braga, Portugal

María Luisa Rúa Rodríguez
Department of Analytical and Food
    Chemistry
University of Vigo
Ourense, Spain

Bernard A.J. Roelen
Department of Farm Animal Health
and
Department of Equine Sciences
Utrecht University
Utrecht, the Netherlands

Štefan Schlosser
Institute of Chemical and
    Environmental Engineering
Slovak University of Technology
Bratislava, Slovakia

José A. Teixeira
IBB-Institute for Biotechnology
    and Bioengineering
Centre for Biological Engineering
University of Minho
Braga, Portugal

António A. Vicente
IBB-Institute for Biotechnology
    and BioEngineering
University of Minho
Braga, Portugal

Łukasz Wiśniewski
Department of Chemical and
    Biochemical Engineering
Slovak University of Technology
Bratislava, Slovakia

Katarzyna Wrzosek
Department of Chemical and
    Biochemical Engineering
Slovak University of Technology
Bratislava, Slovakia

Section I

Use of Biotechnology in the Development of Food Processes and Products

1 Biotechnology-Derived Enzymes for Food Applications

Philippe Ramos, António A. Vicente, and José A. Teixeira

CONTENTS

1.1 INTRODUCTION

The first applications of enzymes were in the production of cheese, sourdough, beer, wine, and vinegar, and in the manufacture of commodities such as leather, indigo, and linen (Olempska-Beer et al. 2006). In those times, all the enzymes were applied as a part of natural extracts obtained from plants, animals, or microorganisms.

3

In the last few decades, the use of selected microbial strains has made it possible to have enzymes in pure and well-characterized forms, and has also permitted the industrial-scale production of enzymes. Since then, enzyme applications have been widespread in food processing and there are several foods and food processing technologies where the use of enzymes plays a key role. The important benefits of enzymes in food processing include the following: they are very specific and controllable chemical catalysts; the production of undesirable by-products may be avoided/minimized; compared with the chemical methods of conversion, mild conditions of temperature, pH, and pressure can be used; enzymes are biodegradable; and the immobilization of enzymes on solid supports can provide technological advantages in processing and can also avoid the existence of the enzyme in the final product, which excludes the need of separation processes and possible allergenicity issues resulting from the contact with the enzymes (MacCabe et al. 2002). An important step in the industrial production of enzymes for use in food processing occurred in 1874, when Christian Hansen extracted rennin (chymosin) from calves' stomachs for use in cheese production (Nielsen et al. 1994). Nowadays, chymosin is produced from microorganisms that contain the bovine prochymosin gene introduced through recombinant deoxyribonucleic acid (rDNA) techniques. Bovine chymosin has been expressed in *Escherichia coli* K-12, which became the first recombinant enzyme approved for use in food by the U.S. Food and Drug Administration. This has also been an important step ahead in the food application of enzymes as, besides the fact that the use of a heterologous protein expressed in a recombinant strain in food processing has been allowed for the first time, a high-purity and a low-cost enzyme could be obtained without using the traditional/natural "sources" of enzymes.

1.2 ENZYME SOURCES

Enzymes can be obtained from different sources and although some of them have been discovered a long time ago, nowadays, the most important enzyme sources are microorganisms (Lambert and Meers 1983), with animal- and plant-derived enzymes contributing to 10% and 5% of the total enzyme market, respectively (Illanes 2008). Several factors contribute to the extensive use of microorganisms for enzyme production: they are metabolically vigorous, versatile and easy to grow on a large scale by submerged or solid-state fermentation, simple to manipulate and, their nutritional requirements are easy to provide, and their supply does not depend on seasonal fluctuations (Blanch and Clark 1997). These qualities make the production of industrial enzymes easier, cheaper, and independent of raw materials, side products, seasons, and harmful substances (Illanes 2008). However, it must be pointed out that in some cases, the production of toxic compounds may occur (Ibrahim 2008).

1.3 ENZYME CLASSES FOR FOOD APPLICATIONS

The classification of enzymes is based on the Nomenclature Committee of the International Union of Biochemistry and Molecular Biology (IUBMB) (Anonymous 1986), and has six families, according to the type of reaction catalyzed. The Enzyme Commission (EC) of the IUBMB assigned four numbers for each enzyme, which compose the enzyme classification. The first number represents the class, the second number represents the subclass within a class and also represents the chemical group where it acts, the third number represents a subgroup within a subclass (sub-subclass) and is associated with the chemical groups involved in the reaction, and the fourth number is a correlative number of identification within a sub-subclass. The six classes are: oxidoreductases, transferases, hydrolases, lyases, isomerases, and ligases. The five classes that are relevant for the food industry are briefly presented and characterized below.

1.3.1 OXIDOREDUCTASES

Oxidoreductases represent a large class of enzymes that catalyze biological oxidation/reduction reactions (May 1999). The oxidation/reduction processes are present in a large number of chemical and biochemical reactions, and because of that, this type of enzymes has a significant industrial importance. There are 22 subclasses of oxidoreductases with a wide range of applications, including synthesis of chiral compounds, such as chiral alcohols, aldehydes, and acids; preparation and modification of polymers, especially biodegradable or biocompatible polymers; biosensors for a variety of analytical and clinical applications; and degradation of organic pollutants (Hummel 1999).

Generally, oxidoreductases need a nonprotein chemical group, a cofactor, to catalyze reactions. In some cases, oxidoreductases have prosthetic groups to facilitate reactions but in other cases, they do not exist and they are not paired. In those cases, enzymes need to interact with other cofactors, which are organic compounds, such as nicotinamide adenine dinucleotide (NAD(H)), nicotinamide adenine dinucleotide phosphate (NADP(H)), and adenosine triphosphate (ATP) (Liu and Wang 2007). These coenzymes act as transport metabolites, transporting hydrogen, oxygen, or electrons in one way, and other atoms or molecules in the reverse way, between different parallel reactions. Coenzymes are indispensable for the reactions mentioned above, but they have a few drawbacks, which need to be solved. Cofactors are complex, unstable, and quite expensive and because of that their regeneration has been proposed (Wichmann 2005). Oxidoreductases are present in the central metabolic pathways of the cell and are strictly intracellular (Illanes 2008).

1.3.2 TRANSFERASES

Transferases are enzymes responsible for transferring a functional group from a donor to a suitable acceptor. Transferases need coenzymes to catalyze their reactions, and they are strictly intracellular. There are nine subgroups of transferases that can be differentiated according to the group that is moved. In cell metabolism, the most important groups are methyltransferases, acyltransferases, glycosyltransferases, transaminases, and phosphotransferases (Illanes 2008).

1.3.3 HYDROLASES

In general, all classes of enzymes have an application in the food and feed production area, but hydrolases are possibly the predominant one. Hydrolases are responsible for the catalysis of bond cleavage by the reaction with water. In most of the cases, hydrolases are present in digestion because they are responsible for breaking down nutrients into small units. Many hydrolases are commercially available and do not require cofactors to catalyze reactions, many are extracellular, and accept the addition of water-miscible solvents (Bornscheuer and Kazlauskas 2006). There are 12 subclasses of hydrolases that differ in the type of susceptible bond. The prominent enzymes of this class are esterases, proteases, and glycosidases. Hydrolases are also able to catalyze the reverse reactions of bond formation with water elimination, under proper conditions (Illanes 2008).

1.3.4 LYASES

Lyases are enzymes acting on the cleavage of chemical bonds that are not hydrolytic or oxidative. The enzymes of this class are divided into seven subclasses, corresponding to their type of susceptible bond: C–C, C–O, C–N, C–S, C–X (halides), P–O, and other bonds. Most of these enzymes are intracellular, some of them do not require coenzymes and perform different metabolic functions associated with cell catabolism and biosynthesis, by acting in a reverse mode (Illanes 2008).

1.3.5 ISOMERASES

Isomerases are enzymes that catalyze the reactions of conversion of a substrate into an isomer. There are six subclasses of isomerases, according to the type of isomer produced: *cis-trans*-isomerases, intramolecular oxidoreductases, intramolecular transferases (mutases), intramolecular lyases, racemases and epimerases, and other isomerases. Generally, isomerases are intracellular and some of them need cofactors (Illanes 2008).

1.4   ENZYME APPLICATIONS

Enzymes have different areas of application and are extensively used in the production of detergents, starch, fuels, foods, animal feed, beverages, textile products, fats and oils, organic compounds, leather, and personal-care products (Kirk et al. 2002). The food industry is one of the industrial sectors where the application of enzymes is more important, as shown in Table 1.1 that lists some of the most relevant enzymes and their applications. These will be further detailed.

1.4.1   GLUCOSE OXIDASE

Among the nonhydrolytic enzymes of fungal origin, glucose oxidase (GOD) (EC 1.1.3.4) has seen large-scale technological applications since the early 1950s (Fiedurekl et al. 1997). GOD (β-D-glucose: oxygen 1-oxireductase) catalyzes the oxidation of β-D-glucose into gluconic acid by utilizing molecular oxygen as an electron acceptor with a simultaneous production of hydrogen peroxide (H_2O_2) (Macris 1995). GOD has gained considerable commercial importance in the last few years due to its multitude of applications (chemical, pharmaceutical, food, beverage,

TABLE 1.1

Enzymes Used in Food and Feed Processing

Enzyme	Role	Class
Lipoxygenase	Bread whitening	Oxidoreductases
Catalase	Food preservation	
Glucose oxidase	Dough strengthening	
Laccases	Clarification and flavor enhancer of drinks	
Fructosyltransferase	Synthesis of fructose oligomers	Transferases
Cyclodextrin glycosyltransferase	Cyclodextrin production	
Transglutaminase	Dough and meat processing	
Galactosidase	Viscosity reduction	Hydrolases
Urease	Removal of urea in alcoholic beverages	
Amylase	Starch liquefaction and saccharification	
Glucoamylase	Saccharification	
Glucanase	Viscosity reduction	
Lactase	Lactose hydrolysis	
Invertase	Sucrose hydrolysis	
Proteases	Protease hydrolysis in milk, cheese, and so on	
Lipase	Cheese flavor, lipid digestion, and so on	
Peptidase	Hydrolysis of proteins	
Pectinase	Mash treatment, juice clarification	
Phytases	Release of phosphate, enhanced digestibility	
Xylanases	Viscosity reduction	
Pullulanase	Saccharification	
Acetolactate decarboxylase	Beer maturation	Lyases
Xylose (glucose) isomerase	Glucose isomerization to fructose	Isomerases

Source: Tramice, A et al. 2007. *Journal of Molecular Catalysis B: Enzymatic* 47: 21–27. doi:10.1016/j. molcatb.2007.03.005; Illanes, A. 2008. *Enzyme Biocatalysis.* Springer Science, Valparaíso, Chile.

clinical chemistry, biotechnology, and other industries). GOD has been applied in glucose removal from dried egg; improvement of color, flavor, and shelf life of food materials; and oxygen removal from fruit juices, canned beverages, and mayonnaise to prevent rancidity. GOD has been used in food and beverage as an additive to remove residual glucose and oxygen to prolong their shelf life (Bankar et al. 2009). H_2O_2 produced by the action of GOD has a bactericidal function and can be removed with catalase (CAT), which converts H_2O_2 into oxygen and water. GOD/CAT used in combination are applied to remove glucose during the manufacture of egg powder, preventing browning during dehydration caused by the Maillard reaction in baking and providing slight improvements to the crumb properties in bread and croissants (Rasiah et al. 2005). GOD can also be used to remove oxygen from the top of bottled beverages before they are sealed; the same applies in fish and canned foods. This process prevents color and flavor loss as well as improves color and flavor stabilization (Bankar et al. 2009). Another application of GOD/CAT is in the wine industry, where it can lower the alcohol content of wine though the removal of some of the glucose (by converting it into δ-glucono-1,5-lactone), which would otherwise be converted into ethanol. Pickering et al. (1999) reduced the fermentative alcohol potential by pretreating grape juice with the GOD/CAT enzyme system to convert the available glucose into gluconic acid (87% of glucose conversion was achieved with this system). The most common microbial sources for GOD production by fermentation are *Aspergillus, Penicillium,* and *Saccharomyces* species. Most of the commercially produced GOD is isolated from the mycelium of *Aspergillus niger,* grown principally for the production of gluconic acid or its salts such as sodium gluconate or calcium gluconate (Bankar et al. 2009). Some examples of microbial systems producing GOD are listed in Table 1.2.

As in many products obtained by fermentation, GOD recovery is a fundamental step. GOD is produced intracellularly and extracellularly or sometimes as a mycelia-associated enzyme. Hence, cells have to be disrupted for the complete release of GOD into the broth (Bankar et al. 2009). The release of the enzyme from the mycelium may be facilitated by applying mechanical and physical forces, for example, agitation or sonication. For the release of intracellular as well as cell-bound GOD into the liquid broth, various methods such as sonication, bead-milling, homogenization, and freeze–thawing have been applied. These steps are followed by the separation of GOD from cells and cell-debris, where centrifugation or filtration techniques have been used. The purification step of GOD has been typically performed with ammonium sulfate, uranyl acetate, and potassium hexacyanoferrate (Bankar et al. 2009).

1.4.2 LACCASE

Laccases (benzenediol:oxygen oxidoreductase; EC 1.10.3.2) are a group of oxidative enzymes that have had an important role in several industrial fields, including the

---

TABLE 1.2
Examples of Microbial Systems Producing GOD

Microbial Systems	Yield (U/mL)	Reference
Penicillium variable P16	5.52	Petruccioli et al. (1999)
Aspergillus niger GOD gene expressed in *Saccharomyces cerevisae*	125	Bankar et al. (2009)
A. niger (BTL)	7.5	Macris (1995)
Recombinant *Saccharomyces cerevisae*	3.43 U/mg dry cell mass	Kapat et al. (2001)
A. niger AM111	2.5	Fiedurek and Gromada (2000)
Penicillium pinophilum DSM 11428	1.9	Rando and Gi (1997)
A. niger ZBY-7	6	Lu et al. (1996)

pulp and paper, textile, and food industries (Couto et al. 2006). The reactions catalyzed by laccases proceed by the monoelectronic oxidation of a suitable substrate molecule (phenols and aromatic or aliphatic amines) to the corresponding reactive radical. The redox process takes place with the assistance of a cluster of four copper atoms that form the catalytic core of the enzyme; they also confer a typical blue color to these enzymes because of the intensive electronic absorption of the Cu–Cu linkages (Piontek et al. 2002). These enzymes catalyze the oxidation of *ortho-* and *para*-diphenols, aminophenols, polyphenols, polyamines, lignins, and aryl diamines, as well as some inorganic ions coupled to the reduction of molecular dioxygen to water (Couto et al. 2006). The food industry has been using laccases from different microbial sources, for example, *Trametes versicolor* (Aktas et al. 2001; Specht et al. 2001; Aktas and Tanyolac 2003), *T. villosa* (Uchida et al. 2001), *Myceliophthora thermophila*, *Polyporus pinsitus* (Micard and Thibault 1999), *Pycnoporus cinnabarinus* (Kuuva et al. 2003), *T. hirsute* (Selinheimo et al. 2006), and *T. versicolor* (Minussi et al. 2002). In the food industry, laccases can be applied to certain processes that enhance or modify the color/appearance of food or beverage. Laccases have been involved in the elimination of undesirable phenolics, responsible for browning, haze formation, and turbidity development in clear fruit juices, beer, and wine (Couto et al. 2006). They are also able to increase the maximum resistance and decrease the extensibility in both flour and gluten dough (Kuuva et al. 2003). Other aspects of the food industry have been found to be interesting for laccases application. Minussi et al. (2002) described the potential applications of laccase in bioremediation, beverage processing, ascorbic acid determination, sugar beet pectin gelation, and baking, and as a biosensor. Reducing the costs of laccase production by optimizing the fermentation medium is the main challenge for industrial applications (Rodríguez and Toca-herrera 2007).

1.4.3 XYLANASES

Hemicellulose hydrolysis is achieved by a spectrum of enzymes, which include endoxylanase (endo-1,4-β-xylanase, EC 3.2.1.8), β-xylosidase (xylan-1,4-β-xylosidase, EC 3.2.1.37), α-arabinofuranosidase (α-glucosiduronase, EC 3.2.1.139), α-arabinofuranosidase (α-L-arabinofuranosidase, EC 3.2.1.55), and acetylxylan esterase (EC 3.1.1.72) (Juturu and Wu 2011). Endoxylanases work on the homopolymeric backbone of 1,4-linked β-D-xylopyranose producing xylooligomers, a crucial step for the hydrolysis of xylan, the major component of hemicellulose (Ahmed et al. 2009). Xylanases have been used in food industries as ingredients in dough making for bread preparations. Their action enhances the specific volume, color, and crumb structure of bread (Kulkarni et al. 1999). The production of xylanases has been reported from several microbial sources: *Clostridium stercorarium* (Bérenger et al. 1985), *Thermomonospora curvata* (Stutzenberger and Bodine 1992), *Neurospora crassa* (Deshpande et al. 1986), *Aspergillus awamori*, *Penicillium purpurogenum* (Haltrich et al. 1997), *Trichoderma reesei* (Haapala et al. 1994), and *Bacillus licheniformis* (Bajaj and Manhas 2012).

1.4.4 Pectinases

Pectinases are divided into two main groups: acidic and alkaline pectinases. Alkaline pectinases have been used in degumming and retting of fiber crops and pretreatment of pectic wastewater from fruit juice industries. Acidic pectinases have been largely used in the food industry, more precisely in fruit juice and wine-making industries (Kashyap et al. 2001). In the production of clear juices, enzymes are added to increase juice extraction and to remove the suspended matter. This technique has been applied in apple, pear, and grape juices, and wine, strawberry, raspberry, and blackberry juice production (Alkorta et al. 1998; Alana et al. 1990; Kashyap et al. 2001). In cloudy juices, these enzymes are added due to the high levels of polygalacturonase activity that contribute to the stabilization of citrus juices, purees, and nectars. They have also been used for the production of unicellular products. These products are formed by the transformation of organized tissues into a suspension of intact cells. They are fit to be used as the base material for pulpy juices and nectars, as baby foods, and as ingredients for dairy products such as puddings and yogurt (Kashyap et al. 2001). Some of the microbial producers of these enzymes are *A. niger* CH4 (Acuña-Argüelles et al. 1995), *Penicillium frequentans, Sclerotium rolfsii* (Borin et al. 1996), *Clostridium thermosaccharolyticum* (Van Rijssel et al. 1993), *Bacillus* species NT-33 (Cao et al. 1992), *Amucola* species, and *Penicillium italicum* CECT 22941 (Alana et al. 1990).

1.4.5 Cyclodextrin Glycosyl Transferases

Cyclodextrins (CDs) are cyclic oligosaccharides composed of α-1,4-glycosidic-linked glucosyl residues produced from starch or starch derivatives using cyclodextrin glycosyl transferase (CGTase). They have been used in a wide range of applications in food, pharmaceutical, and chemical industries (Biwer 2002). CGTase (1,4-α-D-glucan-4-α-D-(1,4-α-D-glucano)-transferase, EC 2.4.1.19) produces nonreducing cyclic dextrins from starch, amylose, and other polysaccharides by catalyzing different transglycosylation steps: intermolecular coupling and disproportionation, reacting at the $\alpha(1-4)$-positions of oligosaccharides, and modification of the length of noncyclic dextrins (Kobayashi 1996). Intramolecular cyclization then leads to CD formation (Biwer 2002). CD is used in food formulations for flavor protection or flavor delivery. They form inclusion complexes with a variety of molecules, including fats, flavors, and colors. Most natural and artificial flavors are volatile oils or liquids and complexation with CDs provides a promising alternative to the conventional encapsulation technologies used for flavor protection (Valle 2004). The bacterial species producing CGTase for use in cyclodextrin production are *Bacillus macerans, B. megaterium, B. circulans, B. stearothermophilus, B. ohbensis, Klebsiella pneumoniae, K. oxytoca, Micrococcus luteus, M. varians,* and *Thermoanaerobacter* spp. (Biwer 2002).

1.4.6 Fructosyltransferase

Fructooligosaccharides (FOSs) possess extraordinary importance as functional food ingredients owing to their prebiotic properties (Simmering and Blaut 2001).

Fructosyltransferase (FT) (EC 2.4.1.9) is an enzyme responsible for the production of FOSs at the industrial scale (Ghazi et al. 2007). This enzyme acts on sucrose by cleaving the β-(2→1) linkage, releasing glucose, and then transferring the fructosyl group to an acceptor molecule. The fungal FT acceptor substrates include sucrose, 1-ketose, nystose, and raffinose (Chuankhayan et al. 2010). In most cases, FTs are produced by fungal sources. Some examples of these organisms are *A. niger* AS 0023 (Hocine et al. 2000), *Penicillium citrinum* (Sangeetha et al. 2005), *Aspergillus japonicus* (Chien et al. 2001), *Aspergillus oryzae* CFR 202 (Sangeetha et al. 2005), *Aureobasidium pullulans* CFR 77 (Sangeetha, et al. 2004), *B. macerans* EG-6 (Park et al. 2001), and *Zymomonas mobilis* (Bekers et al. 2002).

1.4.7 TRANSGLUTAMINASE

Transglutaminase (TGase; protein–glutamine γ-glutamyltransferase, EC 2.3.2.13) catalyzes an acyl-transfer reaction between the γ-carboxyamide group of peptide-bound glutamine residues (acyl donors) and a variety of primary amines (acyl acceptors), including the ϵ-amino group of lysine residues in certain proteins. Without amine substrates, TGase catalyzes the deamidation of glutamine residues during which water molecules are used as acyl acceptors. TGase is also able to modify proteins by means of amine incorporation, cross-linking, and deamidation (Motoki and Seguro 1998). TG has been used to catalyze the cross-linking of whey proteins, soy proteins, wheat proteins, beef myosin, casein, and crude actomyosin (which are refined from mechanically deboned meat), leading to their texturization (Motoki and Seguro 1998). Traditionally, this enzyme could only be obtained from animal tissues, most commonly from guinea pig liver. The rare source and the complicated downstream procedure resulted in an extremely high price for an enzyme of utmost importance in food processing. More recently, the production of TG in microorganisms has become the most usual mode and the obtained enzyme has been shown to improve the food flavor, appearance, and texture. TG can also increase the shelf life of certain foods and can reduce their allergenicity (Zhu and Tramper 2008). The microbial production of TG started with *Streptoverticillium* (Takehana et al. 1994; Washizu et al. 1994), but it was rapidly engineered in *E. coli* and *Streptomyces lividans* (Takehana et al. 1994). Later, other microbial sources were used, such as *Streptoverticillium mobaraense* (Meiying et al. 2002), *Corynebacterium glutamicum* (Kikuchi et al. 2003), and *Streptomyces hygroscopicus* (Cui et al. 2007) among others.

1.4.8 AMYLASES

Amylases are enzymes that are responsible for catalyzing the hydrolysis of glycosidic α-1,4-bonds in polysaccharides such as starch and glycogen (Brena et al. 1996). The α-amylase is the most used form and performs the hydrolysis of the internal α-1,4-glycosidic links in amylose and amylopectin. This action creates a less viscous solution with lower-molecular- weight polysaccharides. The β-amylase that forms (exoamylases) are responsible for the hydrolysis of starch, beginning at nonreducing ends. With this reaction, they produce β-maltose and β-limit dextrins

(Brena et al. 1996). Some of the microbial producers of amylases are *Aspergillus flavus* var. *columnaris* (Ammar 2004), *Cryptococcus* species S-2 (Iefuji et al. 1996), *Schwanniomyces alluvius* UCD-54-83 (Wilson and Ingledew 1982), *A. niger* (Hernández et al. 2006), *Lactobacillus plantarum,* and *L. fermentum* (Sanni et al. 2002). In baking processes, amylases have been used to reduce dough fermentation time, improve the properties of the crumb and the dough, as well as retention of aromas and moisture levels (Gerday et al. 2000). The other applications of this group of enzymes are starch liquefaction and saccharification (Maarel and Veen 2002), paper desizing, and detergent applications (Gupta et al. 2003).

1.4.9 PROTEASES

Proteases are enzymes that catalyze the rupture of peptide bonds. They can perform this cleavage at the end of the polypeptide chain (exopeptidases) or within the chain (endopeptidases) (Palma et al. 2002). Proteases have different applications in the food industry, such as tenderization of meat, reduction of dough fermentation time in baking processes, improving the properties of crumb and dough, and retention of aromas and moisture levels (Gerday et al. 2000). Currently, there are a vast range of microbial protease producers, with a very significant commercial importance, such as *B. licheniformis*, alkalophilic *Bacillus* sp., *Aspergillus* sp., and so on (Kumar and Takagi 1999).

1.4.10 β-GALACTOSIDASE

The enzyme β-galactosidase is used for the reduction of the amount of lactose in milk at a low temperature. This enzyme is able to transform one molecule of lactose into one molecule of glucose and another molecule of galactose. The importance of this enzyme is increasing due to the increase of the percentage of population that is lactose intolerant, estimated as being almost two-thirds of the global population (Gerday et al. 2000).

β-Galactosidase has also been applied in the synthesis of galactooligosaccharides (GOS) by making use of its transgalactosylation activity in the presence of high lactose concentrations.

The major microbial producers of this enzyme are *E. coli, Kappaphycus fragilis, A. oryzae,* and *A. niger* (Santos et al. 1998; Neri et al. 2009).

1.5 IMPORTANCE OF RECOMBINANT MICROORGANISMS IN ENZYME PRODUCTION

The conventional production of enzymes by extraction from natural sources or by "conventional" microbial systems presented, in most of the cases, relevant limitations that hindered their industrial production-specific or expensive conditions that were required for microbial development, a large spectrum of enzymes was produced or in other cases, undesirable metabolites were obtained. Because of this, the use of recombinant microorganisms became a potential solution for enzyme production

(Olempska-Beer et al. 2006). The application of genetic modification made possible the production of enzymes in host microbial sources and also the modification or production of new enzymes. The modification of enzymes is very important as they maximize their specific characteristics allowing better products to be obtained at a lower cost. New enzymes have a huge potential interest in all industrial processes, contributing to the solution of the existing problems and presenting better options to the existing production processes (Fernandes 2010). The primary/classical approach for enzyme production was to use what was provided by nature and, if needed, to apply classical mutagenesis and up-scale the production process (fermentation, purification, and formulation).

Nowadays, the development of enzymes is completely different, as it is possible to create a biological diversity through genetic modification. Following genetic modification, a screening procedure is applied to select the adequate expression system and the production process is up-scaled (fermentation, purification, and formulation) (Kirk et al. 2002). In these situations, screening is an important tool to discover new enzymes. Screening of a new enzyme demands the definition of the type of reaction wanted; this will allow the selection of the groups of (or single) microorganisms that are to be screened; the final step is the design of an assay that is sufficiently sensitive, convenient, and capable of being applied in as many microorganisms as possible (Ogawa and Shimizu 1999). The characteristics such as enzyme stability, catalytic mechanism, substrate specificity, surface activity, folding mechanism, cofactor dependency, pH and temperature optima, and kinetic parameters can be successfully and simultaneously modified. This knowledge is applied, for example, in the modification of α-amylases used as a sweetener (such as glucose or fructose syrups). In this case, the goal was the improvement of the heat stability of these enzymes for a better performance in all the different steps in the process of modification (Olempska-Beer et al. 2006).

The production of recombinant enzymes is based on an extensive research and follows several steps, such as development of the host strain, construction of the expression vector, transformation of the host strain, identification of the best recombinant strain, additional improvements, and characterization of the production strain (Olempska-Beer et al. 2006).

1.5.1 Host Strains

The first microorganisms used in the production of recombinant enzymes were *Bacillus subtilis*, *B. licheniformis*, *A. niger*, or *A. oryzae*. All these organisms were utilized because they have a history of safe sources of native enzymes, and are amenable for genetic manipulation, and growth efficiently under industrial conditions. In Table 1.3, a list of enzymes obtained from recombinant microorganisms is presented, all of them recognized as nonpathogenic (Olempska-Beer et al. 2006).

The production of different enzymes by the same strain of microorganism is based on the introduction of different genes, responsible for providing the required information. These genes, encoding for the enzyme desired, are introduced into host strains in expression vectors. An expression vector is a deoxyribonucleic acid

TABLE 1.3
Recombinant Microorganisms for Enzyme Production

Source Microorganism	Enzyme	Applications	Reference
Aspergillus niger	Phytase	Animal feed industry	Vats and Banerjee (2002)
	Chymosin	Cheese production	Bodie et al. (1994)
	Lipase	Sterification of edible oils	Illanes (2008)
Aspergillus oryzae	GOD	Additive to remove residual glucose and	Bankar et al. (2009)
	Laccase	oxygen	Couto et al. (2006)
	Lipase	Color appearance of food or beverage Sterification of edible oils	Illanes (2008)
Bacillus licheniformis	α-Amylase	Starch conversion	Maarel and Veen (2002)
Bacillus subtilis	α-Amylase	Starch conversion	Maarel and Veen (2002)
Escherichia coli K-12	Chymosin	Cheese production	Bodie et al. (1994)
Fusarium venenatum	Xylanase	Enhanced specific volume, color, and crumb structure of bread	Olempska-Beer et al. (2006)
Kluyveromyces marxianus var. *lactis*	Chymosin	Cheese production	Bodie et al. (1994)
Pseudomonas fluorescens Biovar I	α-Amylase	Starch conversion	Maarel and Veen (2002)
Trichoderma reesei	Pectin lyase	Degradation of pectin polymers	Olempska-Beer et al. (2006)

Source: Adapted from Olempska-Beer, Z S et al. 2006. *Regulatory Toxicology and Pharmacology: RTP* 45 (2) (July): 144–158. doi:10.1016/j.yrtph.2006.05.001. http://www.ncbi.nlm.nih.gov/pubmed/16769167.

(DNA) plasmid that transports the expression cassette. The expression cassette is formed by a promoter, the gene encoding the desired enzyme, and a terminator. The promoter is responsible for the beginning of the transcription process and the terminator for the end. The most used plasmids are pUB110, pUC18, and pUC19 because they are well-characterized and are commercially available (Olempska-Beer et al. 2006).

1.5.2 ENZYME PRODUCTION

Currently, microbial enzymes are produced by controlled fermentation in specific reactors. A diversity of parameters is monitored, such as temperature, pH, and aeration. The composition of the broth is also closely controlled and contains the ingredients required to maximize enzyme production. In most cases, the culture medium is composed of dextrose, corn steep liquor, soybean meal, starch, yeast extract, ammonia, urea, and minerals such as phosphates, chlorides, or carbonates (Olempska-Beer et al. 2006). The overall production process can be divided into four stages—enzyme production/synthesis, enzyme recovery, enzyme purification, and enzyme product formulation.

The first step corresponds to the fermentation process that usually occurs in reactors under batch, fed-batch, or continuous conditions. The batch processes are the most utilized, but present some drawbacks such as difficulties in operational parameter stabilization and in the recovery of enzymes from the fermentation broth; in batch processes, low productivity values are usually obtained, thus hindering process competitiveness.

After the fermentation step, it is necessary to recover the enzyme. The procedure to be applied depends on the enzyme location: intracellular and extracellular. In the first case, it is necessary to understand in which part of the microorganism's structure the enzyme is located. High-pressure homogenization is the process usually applied for cell disruption. In this case, cells are subjected to high pressure through a nozzle at low outlet temperature (e.g., −20°C). The other methods for cell disruption include homogenization, milling, sonication, decompression, freezing–thawing, dispersion in water, and thermolysis. There are also methods for cell permeabilization such as alkali treatment, solvent treatment, enzymatic lysis, and autolysis (Illanes 2008), which can be used to extract the enzymes from the cells. The recovery of extracellular enzymes requires a solid–liquid separation and the conventional methods are centrifugation or filtration. Centrifugation is more appropriate for unicellular microorganisms, such as yeasts or bacteria, but in these cases, flocculation is often used as a complementary operation because of the small size of the individual cells. Filtration is more appropriate for multicellular organisms with a filamentous morphology. Bearing in mind the additional difficulties posed by intracellular enzymes, one of the top priorities when genetically modifying an organism for enzyme production is to favor modifications that allow the extracellular secretion of the enzyme (Illanes 2008).

After enzyme recovery, purification is the next step. The purification of enzymes produced intracellularly is different from the enzymes produced and excreted by the organism. After cell disruption, the medium contains not only the enzyme but also other proteins, nucleic acids, and other cell constituents, which make the purification procedure more complex and will typically involve several unit operations that reduce the process efficiency and increase the overall production cost. Currently, two techniques are mostly used to remove impurities that are present in the medium, such as nuclear treatment and precipitation with different agents.

Enzyme formulation is the final step and requires that the broth with purified enzymes be adapted to the final intended use (Illanes 2008). For logistic reasons, a lyophilization step may be required.

The utilization of enzymes in food processing requires that they are obtained at a low cost and, for this, their reutilization is desirable. This can be achieved by immobilization as, by this method, enzymes can be easily separated from the reaction medium. This can be achieved by mechanical containment or by attachment to a solid matrix. The use of an enzyme immobilization procedure allows not only the recycle of the enzymes but also the extension of its use life. Different materials have been used in enzyme immobilization, such as agar, alginate, carrageenan, polyacrylamide, and others (Sassolas et al. 2012).

REFERENCES

Acuña-Argüelles, M E, M Gutiérrez-Rojas, G Viniegra-González, and E Favela-Torres. 1995. Production and properties of three pectinolytic activities produced by *Aspergillus niger* in submerged and solid-state fermentation. *Applied Microbiology and Biotechnology* 43 (5) (October): 808–814. http://www.ncbi.nlm.nih.gov/pubmed/7576547.

Ahmed, S, S Riaz, and A Jamil. 2009. Molecular cloning of fungal xylanases: An overview. *Applied Microbiology Biotechnology* 84: 19–35. doi:10.1007/s00253–009-2079-4.

Aktas, N, N Kolankaya, A Tas, and A Tanyolac. 2001. Reaction kinetics for laccase-catalyzed polymerization of 1-naphthol. *Bioresource Technology* 80: 29–36.

Aktas, N, and A Tanyolac. 2003. Reaction conditions for laccase catalyzed polymerization of catechol. *Bioresource Technology* 87: 209–214.

Alana, A, I Alkorta, J B Dominguez, M J Llama, and J L Serra. 1990. Pectin lyase activity in a *Penicillium italicum* strain. *Applied and Environmental Microbiology* 56 (12): 3755–3759.

Alkorta, I, C Garbisu, M J Llama, and J L Serra. 1998. Industrial applications of pectic enzymes: A review. *Process Biochemistry* 33 (1) (January): 21–28. doi:10.1016/S0032-9592(97)00046-0. http://dx.doi.org/10.1016/S0032-9592(97)00046-0.

Ammar, M S. 2004. Purification and characterization of α-amylase isolated from *Aspergillus falvus* var. *Columnaris* abstract: Introduction: Materials and methods: Enzyme purification. *Association of University Bulletin Environmental Research* 7 (1): 93–100.

Anonymous. 1986. Nomenclature committee of IUB (NC-IUB) and IUPAC-IUB joint commission on biochemical nomenclature (JCBN). *Bioscience Reports* 6 (1): 121–125.

Bajaj, B K, and K Manhas. 2012. Production and characterization of xylanase from *Bacillus licheniformis* P11(C) with potential for fruit juice and bakery industry. *Biocatalysis and Agricultural Biotechnology* 1 (4) (October): 330–337. doi:10.1016/j.bcab.2012.07.003. http://linkinghub.elsevier.com/retrieve/pii/S1878818112001053.

Bankar, S B, M V Bule, R S Singhal, and L Ananthanarayan. 2009. Glucose oxidase—An overview. *Biotechnology Advances* 27 (4): 489–501. doi:10.1016/j.biotechadv.2009.04.003. http://dx.doi.org/10.1016/j.biotechadv.2009.04.003.

Bekers, M, J Laukevics, D Upite, E Kaminska, A Vigants, U Viesturs, L Pankova, and A Danilevics. 2002. Fructooligosaccharide and levan producing activity of *Zymomonas mobilis* extracellular levansucrase. *Process Biochemistry* 38: 701–706.

Bérenger, J-F, C Frixon, J Bigliardi, and N Creuzet. 1985. Production, purification, and properties of thermostable xylanase from *Clostridium stercorarium*. *Canadian Journal of Microbiology* 31 (7) (July 10): 635–643. doi:10.1139/m85-120. http://www.nrcresearchpress.com/doi/abs/10.1139/m85-120.

Biwer, A, G Antranikian, and E Heinzle. 2002. Mini-review: Enzymatic production of cyclodextrins. *Applied and Environmental Microbiology* 59: 609–617. doi:10.1007/s00253-002-1057-x.

Blanch, H W, and D S Clark. 1997. *Biochemical Engineering (Google eBook)*. CRC Press, United Kingdom. http://books.google.com/books?id=ST_p2AOApZsC&pgis=1.

Bodie, E A, G L Armstrong, and N S Dunn-Coleman. 1994. Strain improvement of chymosin-producing strains of *Aspergillus niger* var. *Awamori* using parasexual recombination. *Enzyme and Microbial Technology* 16 (5) (May): 376–382. doi:10.1016/0141-0229(94)90151-1. http://dx.doi.org/10.1016/0141-0229(94)90151-1.

Borin de, M F, S Said, and M J V. Fonseca. 1996. Purification and biochemical characterization of an extracellular endopolygalacturonase from *Penicillium frequentans*. *Journal of Agricultural and Food Chemistry* 44(6) (June 1): 39–47. http://agris.fao.org/agris-search/search/display.do?f=1997/US/US97096.xml;US9628895.

Bornscheuer, U T, and R J Kazlauskas. 2006. *Hydrolases in Organic Synthesis: Regio- and Stereoselective Biotransformations (Google eBook)*. John Wiley & Sons, Weinheim. http://books.google.com/books?id=JGjh-iNO07QC&pgis=1.

Brena, B M, C Pazos, L Franco-Fraguas, and F Batista-Viera. 1996. Chromatographic methods for amylases. *Journal of Chromatography* 684 (1–2) (September 20): 217–237. http://www.ncbi.nlm.nih.gov/pubmed/8906475.

Cao, J, L Zheng, and S Chen. 1992. Screening of pectinase producer from alkalophilic bacteria and study on its potential application in degumming of ramie. *Enzyme and Microbial Technology* 14 (12) (December): 1013–1016. doi:10.1016/0141-0229(92)90087-5. http://dx.doi.org/10.1016/0141-0229(92)90087-5.

Chien, C-S, W-C Lee, and T-J Lin. 2001. Immobilization of *Aspergillus japonicus* by entrapping cells in gluten for production of fructooligosaccharides. *Enzyme and Microbial Technology* 29: 252–257.

Chuankhayan, P, C-Y Hsieh, Y-C Huang, Y-Y Hsieh, H-H Guan, Y-C Hsieh, Y-C Tien, C-D Chen, C-M Chiang, and C-J Chen. 2010. Crystal structures of *Aspergillus japonicus* fructosyltransferase complex with donor/acceptor substrates reveal complete subsites in the active site. *The Journal of Biological Chemistry* 285 (30): 23251–23264. doi:10.1074/jbc.M110.113027.

Couto, S R, J Luis, and T Herrera. 2006. Industrial and biotechnological applications of laccases: A review. *Biotechnology Advances* 24: 500–513. doi:10.1016/j.biotechadv.2006.04.003.

Cui, L, G Du, D Zhang, H Liu, and J Chen. 2007. Food chemistry purification and characterization of transglutaminase from a newly isolated *Streptomyces hygroscopicus*. *Food Chemistry* 105: 612–618. doi:10.1016/j.foodchem.2007.04.020.

Deshpande, V, A Lachke, C Mishra, S Keskar, and M Rao. 1986. Mode of action and properties of xylanase and beta-xylosidase from *Neurospora crassa*. *Biotechnology and Bioengineering* 28 (12) (December): 1832–1837. doi:10.1002/bit.260281210. http://www.ncbi.nlm.nih.gov/pubmed/18555300.

Fernandes, P. 2010. Enzymes in food processing: A condensed overview on strategies for better biocatalysts. *Enzyme Research* 2010: 19. doi:10.4061/2010/862537.

Fiedurek, J, and A Gromada. 2000. Production of catalase and glucose oxidase by *Aspergillus niger* using unconventional oxygenation of culture. *Journal of Applied Microbiology* 89: 85–89.

Fiedurekl, J, A Gromada, and M Curie-sktodowska. 1997. Screening and mutagenesis of molds for improvement of the simultaneous production of catalase and glucose oxidase. *Enzyme and Microbial Technology* 20: 344–347.

Gerday, C, M Aittaleb, M Bentahir, J-P Chessa, P Claverie, T Collins, S D Amico et al. 2000. Cold-adapted enzymes: From fundamentals to biotechnology. *Trends in Biotechnology* 18 (March): 103–107.

Ghazi, I, L Fernandez-arrojo, H Garcia-arellano, M Ferrer, A Ballesteros, and F J Plou. 2007. Purification and kinetic characterization of a fructosyltransferase from *Aspergillus aculeatus*. *Journal of Biotechnology* 128: 204–211. doi:10.1016/j.jbiotec.2006.09.017.

Gupta, R, P Gigras, H Mohapatra, V K Goswami, and B Chauhan. 2003. Microbial α-amylases: A biotechnological perspective. *Process Biochemistry* 38 (11) (June): 1599–1616. doi:10.1016/S0032-9592(03)00053-0. http://linkinghub.elsevier.com/retrieve/pii/S0032959203000530.

Haapala, R, S Linko, E Parkkinen, and P Suominen. 1994. Production of endo-1,4-glucanase and xylanase by *Trichoderma reesei* immobilized on polyurethane foam. *Biotechnology Techniques* 8 (6) (June): 401–406. doi:10.1007/BF00154311. http://www.springerlink.com/content/l2032834m6m1p822/.

Haltrich, D, B Nidetzky, K D Kulbe, and S Zupan. 1997. Production of fungal xylanases. *Bioresource Technology* 58 (1996): 137–161.

Hernández, M S, M R Rodríguez, N P Guerra, and R P Rosés. 2006. Amylase production by *Aspergillus niger* in submerged cultivation on two wastes from food industries. *Journal of Food Engineering* 73 (1) (March): 93–100. doi:10.1016/j.jfoodeng.2005.01.009. http://linkinghub.elsevier.com/retrieve/pii/S0260877405000439.

Hocine, L L, Z Wang, B Jiang, and S Xu. 2000. Purification and partial characterization of fructosyltransferase and invertase from *Aspergillus niger. Journal of Biotechnology* 81: 73–84.

Hummel, W. 1999. Large-scale applications of NAD (P)-dependent oxidoreductases: Recent developments. *Trends in Biotechnology* 7799 (1996): 24–29.

Ibrahim, C O. 2008. Development of applications of industrial enzymes from Malaysian indigenous microbial sources. *Bioresource Technology* 99: 4572–4582. doi:10.1016/j.biortech.2007.07.040.

Iefuji, H, M Chino, M Kato, and Y Iimura. 1996. Raw-starch-digesting and thermostable. *Biochemical Journal* 996: 989–996.

Illanes, A. 2008. *Enzyme Biocatalysis*. Springer Science, Valparaíso, Chile.

Juturu, V, and J C Wu. 2011. Microbial xylanases: Engineering, production and industrial applications. *Biotechnology Advances* 30: 1–9. doi:10.1016/j.biotechadv.2011.11.006. http://dx.doi.org/10.1016/j.biotechadv.2011.11.006.

Kapat, A, J Jung, and Y Park. 2001. Enhancement of glucose oxidase production in batch cultivation of recombinant *Saccharomyces cerevisiae*: Optimization of oxygen transfer condition. *Journal of Applied Microbiology* 90: 216–222.

Kashyap, D R, P K Vohra, S Chopra, and R Tewari. 2001. Applications of pectinases in the commercial sector: A review. *Bioresource Technology* 77: 215–227.

Kikuchi, Y, M Date, K-I Yokoyama, and Y Umezawa. 2003. Secretion of active-form *Streptoverticillium mobaraense* transglutaminase by *Corynebacterium glutamicum*: Processing of the pro-transglutaminase by a cosecreted subtilisin-like protease from *Streptomyces albogriseolus* secretion of active-form streptovertic. *Applied and Environmental Microbiology* 69 (1): 358–366. doi:10.1128/AEM.69.1.358.

Kirk, O, T V Borchert, and C C Fuglsang. 2002. Industrial enzyme applications. *Current Opinion in Biotechnology* 13: 345–351. doi:10.1016/S0958-1669(02)00328-2.

Kobayashi, S. 1996. *Enzymes for Carbohydrate Engineering*, vol. 12. Elsevier, Amsterdam. doi:10.1016/S0921-0423(96)80360-1.http://dx.doi.org/10.1016/S0921-0423(96)80360-1.

Kulkarni, N, A Shendye, and M Rao. 1999. Molecular and biotechnological aspects of xylanases. *FEMS Microbiology Reviews* 23 (February): 411–456.

Kumar, C G, and H Takagi. 1999. Microbial alkaline proteases: From a bioindustrial viewpoint. *Biotechnology Advances* 17: 561–594.

Kuuva, T, R Lantto, T Reinikainen, J Buchert, and K Autio. 2003. Rheological properties of laccase-induced sugar beet pectin gels. *Food Hydrocolloids* 17: 679–684. doi:10.1016/S0268-005X(03)00034-1.

Lambert, P W, J L Meers, and D J Best. 1983. The production of industrial enzymes. *Biological Sciences* 300: 263–282.

Liu, W, and P Wang. 2007. Cofactor regeneration for sustainable enzymatic biosynthesis. *Biotechnology Advances* 25: 369–384. doi:10.1016/j.biotechadv.2007.03.002.

Lu, T, X Peng, H Yang, and L Ji. 1996. The production of glucose oxidase using the waste myceliums of. *Enzyme and Microbial Technology* 19: 339–342.

Maarel, M, J E C van der, and B van der Veen. 2002. Properties and applications of starch-converting enzymes of the α-amylase family. *Journal of Biotechnology* 94: 137–155.

MacCabe, A P, M Orejas, E N Tamayo, A Villanueva, and D Ramón. 2002. Improving extracellular production of food-use enzymes from *Aspergillus nidulans. Journal of Biotechnology* 96 (1) (June 13): 43–54. http://www.ncbi.nlm.nih.gov/pubmed/12142142.

Macris, H B J. 1995. Factors regulating production of glucose oxidase by *Aspergillus niger. Enzyme and Microbial Technology* 17 (94): 530–534.

May, S W. 1999. Applications of oxidoreductases. *Current Opinion in Biotechnology* 10: 370–375.

Meiying, Z, D Guocheng, and C Jian. 2002. pH control strategy of batch microbial transglutaminase production with *Streptoverticillium mobaraense. Enzyme and Microbial Technology* 31: 477–481.

Micard, V, and J Thibault. 1999. Oxidative gelation of sugar-beet pectins: Use of laccases and hydration properties of the cross-linked pectins. *Carbohydrate Polymers* 39: 265–273.

Minussi, R C, M Pastore, and N Dura. 2002. Potential applications of laccase in the food industry. *Trends in Food Science and Technology* 13: 205–216.

Motoki, M, and K Seguro. 1998. Transglutaminase and its use for food processing. *Trends in Food Science and Technology* 9: 204–210.

Neri, D F M, V M Balcão, R S Costa, I CAP Rocha, E M F C Ferreira, D P M Torres, L R M Rodrigues, L B Carvalho, and J A Teixeira. 2009. Galacto-oligosaccharides production during lactose hydrolysis by free *Aspergillus oryzae* β-galactosidase and immobilized on magnetic polysiloxane–polyvinyl alcohol. *Food Chemistry* 115 (1) (July): 92–99. doi:10.1016/j.foodchem.2008.11.068. http://linkinghub.elsevier.com/retrieve/pii/S030881460801412X.

Nielsen, P H, H Malmos, T Damhus, B Diderichsen, H K Nielsen, M Simonsen, H E Schiv et al. 1994. Enzyme applications (industrial). In *Kirk–Othmer Encyclopedia of Chemical Technology*, pp. 567–620, John Wiley & Sons, New York.

Ogawa, J, and S Shimizu. 1999. Microbial enzymes: New industrial applications. *Trends in Biotechnology* 7799 (98): 13–20.

Olempska-Beer, Z S, R I Merker, M D Ditto, and M J DiNovi. 2006. Food-processing enzymes from recombinant microorganisms—A review. *Regulatory Toxicology and Pharmacology: RTP* 45 (2) (July): 144–158. doi:10.1016/j.yrtph.2006.05.001. http://www.ncbi.nlm.nih.gov/pubmed/16769167.

Palma, J M, L M Sandalio, F J Corpas, M C Romero-Puertas, I McCarthy, and L A del Río. 2002. Plant proteases, protein degradation, and oxidative stress: Role of peroxisomes. *Plant Physiology and Biochemistry* 40 (6–8) (June): 521–530. doi:10.1016/S0981-9428(02)01404-3. http://linkinghub.elsevier.com/retrieve/pii/S0981942802014043.

Park, J-P, T-K Oh, and J-W Yun. 2001. Purification and characterization of a novel transfructosylating enzyme from *Bacillus macerans* EG-6. *Process Biochemistry* 37: 471–476.

Petruccioli, M, U F Federici, and U C Bucke. 1999. Enhancement of glucose oxidase production by *Penicillium Ariabile* P16. *Enzyme and Microbial Technology* 24: 397–401.

Pickering, G J, D A Heatherbell, and M F Barnes. 1999. Optimising glucose conversion in the production of reduced alcohol wine using glucose oxidase. *Food Research International* 31 (10): 658–692.

Piontek, K, M Antorini, and T Choinowski. 2002. Crystal structure of a laccase from the fungus *Trametes versicolor* at 1. 90-Å resolution containing a full complement of coppers. *The Journal of Biological Chemistry* 277 (40): 37663–37669. doi:10.1074/jbc.M204571200.

Rando, D, and G F Kohring Gi. 1997. Production, purification and characterization of glucose oxidase from a newly isolated strain of *Penicillium pinophilum*. *Applied Microbiology Biotechnology* 48: 34–40.

Rasiah, I A, K H Sutton, F L Low, H Lin, and J A Gerrard. 2005. Food chemistry crosslinking of wheat dough proteins by glucose oxidase and the resulting effects on bread and croissants. *Food Chemistry* 89: 325–332. doi:10.1016/j.foodchem.2004.02.052.

Rodríguez, S, and J L Toca-herrera. 2007. Laccase production at reactor scale by filamentous fungi. *Biotechnology Advances* 25: 558–569. doi:10.1016/j.biotechadv.2007.07.002.

Sangeetha, P T, M N Ramesh, and S G Prapulla. 2004. Production of fructo-oligosaccharides by fructosyl transferase from *Aspergillus oryzae* CFR 202 and *Aureobasidium pullulans* CFR 77. *Process Biochemistry* 39: 753–758. doi:10.1016/S0032-9592(03)00186-9.

Sangeetha, P T, M N Ramesh, and S G Prapulla. 2005. Recent trends in the microbial production, analysis and application of fructooligosaccharides. *Trends in Food Science and Technology* 16: 442–457. doi:10.1016/j.tifs.2005.05.003.

Sanni, A I, J Morlon-Guyot, and J P Guyot. 2002. New efficient amylase-producing strains of *Lactobacillus plantarum* and *L. fermentum* isolated from different Nigerian traditional fermented foods. *International Journal of Food Microbiology* 72 (1–2) (January 30): 53–62. http://www.ncbi.nlm.nih.gov/pubmed/11843413.

Santos, A, M Ladero, and F Garci. 1998. Kinetic modeling of lactose hydrolysis by a β-galactosidase from *Kluyveromices fragilis*. *Enzyme and Microbial Technology* 0229 (97): 558–567.

Sassolas, A, L J Blum, and B D Leca-bouvier. 2012. Immobilization strategies to develop enzymatic biosensors. *Biotechnology Advances* 30 (3): 489–511. doi:10.1016/j.biotechadv.2011.09.003. http://dx.doi.org/10.1016/j.biotechadv.2011.09.003.

Selinheimo, E, K Kruus, J Buchert, A Hopia, and K Autio. 2006. Effects of laccase, xylanase and their combination on the rheological properties of wheat doughs. *Journal of Cereal Science* 43: 152–159. doi:10.1016/j.jcs.2005.08.007.

Simmering, R, and M Blaut. 2001. Pro- and prebiotics ± the tasty guardian angels? *Applied Microbiology Biotechnology* 55: 19–28. doi:10.1007/s002530000512.

Specht, M, A Hetzheim, F Schauer, D Greifswald, È Hamburg, and D Hamburg. 2001. Synthesis of substituted imidazoles and dimerization products using cells and laccase from *Trametes versicolor*. *Tetrahedron* 57: 7693–7699.

Stutzenberger, F J, and A B Bodine. 1992. Xylanase production by *Thermomonospora curvata*. *Journal of Applied Microbiology* 72 (6) (June): 504–511. doi:10.1111/j.1365-2672.1992. tb01867.x. http://doi.wiley.com/10.1111/j.1365–2672.1992.tb01867.x.

Takehana, S, K Washizu, K Ando, S Koikeda, K Takeuchi, H Matsui, M Motoki, and H Takagi. 1994. Chemical synthesis of the gene for microbial transglutaminase from *Streptoverticillium* and its expression in *Escherichia coli*. *Bioscience, Biotechnology, and Biochemistry* 58 (1) (January): 88–92. http://www.ncbi.nlm.nih.gov/pubmed/7765335.

Tramice, A, E Pagnotta, I Romano, A Gambacorta, and A Trincone. 2007. Transglycosylation reactions using glycosyl hydrolases from *Thermotoga neapolitana*, a marine hydrogen-producing bacterium. *Journal of Molecular Catalysis B: Enzymatic* 47: 21–27. doi: 10.1016/j.molcatb.2007.03.005.

Uchida, H, T Fukuda, H Miyamoto, T Kawabata, M Suzuki, and T Uwajima. 2001. Polymerization of biphenol A by purified laccase from *Trametes villosa*. *Biochemical and Biophysical Research Communications* 287: 355–358. doi:10.1006/bbrc.2001.5593.

Valle, E M, and M Del. 2004. Cyclodextrins and their uses: A review. *Process Biochemistry* 39: 1033–1046. doi:10.1016/S0032-9592(03)00258-9.

Van Rijssel, M, G J Gerwig, and T A Hansen. 1993. Isolation and characterization of an extracellular glycosylated protein complex from *Clostridium thermosaccharolyticum* with pectin methylesterase and polygalacturonate hydrolase activity. *Applied and Environmental Microbiology* 59 (3) (March): 828–836. http://www.pubmedcentral.nih. gov/articlerender.fcgi?artid=202196&tool=pmcentrez&rendertype=abstract.

Vats, P, and U C Banerjee. 2002. Studies on the production of phytase by a newly isolated strain of *Aspergillus niger* var *Teigham* obtained from rotten wood-logs. *Process Biochemistry* 38: 211–217.

Washizu, K, K Ando, S Koikeda, S Hirose, A Matsuura, H Takagi, M Motoki, and K Takeuchi. 1994. Molecular cloning of the gene for microbial transglutaminase from *Streptoverticillium* and its expression in *Streptomyces lividans*. *Bioscience, Biotechnology, and Biochemistry* 58 (1) (January): 82–87. http://www.ncbi.nlm.nih.gov/ pubmed/7765334.

Wichmann, R. 2005. Cofactor regeneration at the lab scale. *Advances in Biochemistry of Engineering/Biotechnology* 92: 225–260. doi:10.1007/b98911.

Wilson, J J, and W M Ingledew. 1982. Isolation and characterization of *Schwanniomyces alluvius* isolation and characterization of *Schwanniomyces alluvius* amylolytic enzymes. *Environmental Microbiology* 22: 301.

Zhu, Y, and J Tramper. 2008. Novel applications for microbial transglutaminase beyond food processing. *Trends in Food Science and Technology* 26 (10): 559–565. doi:10.1016/ j.tibtech.2008.06.006.

2 Development of Probiotics and Prebiotics

Ernesto Hernandez and Severino S. Pandiella

CONTENTS

2.1 INTRODUCTION: FUNCTIONAL PRODUCTS FOR GUT HEALTH (PROBIOTICS, PREBIOTICS, AND SYMBIOTICS)

Consumers around the globe are now more aware of the relation between nutrition and good health. This has led to a number of scientific studies identifying food and food components that have specific health benefits. Functional foods are defined as foods that, in addition to nutrients, supply the organism with components that contribute to the cure of diseases, or to reduce the risk of developing them (Prado et al., 2008). One of the first countries in the world to recognize the concept of functional foods was Japan. The Japanese government defines FOSHU (Foods for Specific Health Use) as Foods which are expected to have certain health benefits, and have been licensed to bear a label in the product claiming the health effect (Sanders, 1998).

The establishment of a microflora rich in bifidobacteria poses an advantage with respect to maintaining and improving health. In the past two to three decades, one of the main targets of the functional food industry has been the development of products aimed at improving gut health. The idea of manipulating the intestinal microflora in order to prolong life can be tracked down to the early part of the twentieth century. In 1907, the Nobel Prize winner Eli Metchnikoff published his book *The Prolongation of Life*, where he postulated that the long life span of Bulgarian people was due to the consumption of fermented milks that contained the microorganism *Bacillus bulgaricus*, currently known as *Lactobacillus delbrueckii*. According to Metchnikoff, the consumption of these bacteria would have the effect of eliminating putrefactive bacteria in the human gastrointestinal tract (GIT). In the same period, the French pediatrician Henry Tissier discovered the predominance of bifidobacteria in the feces of healthy infants, and postulated their effectiveness in preventing infections (Said and Mohamed, 2006). Experimental work on the effect of ingestion of "beneficial microorganisms" on human health was carried out in the 1930s, but the results were not clearly positive, and this line of research was abandoned for 40 years.

The term probiotic, from Greek "for life," was used for the first time by Lilly and Stillwell (1965) to define substances produced by microorganisms which would enlarge the growth phase of other microorganisms. Later, Parker (1974, pp. 366) changed this definition to "organisms and substances that contribute to intestinal balance." Fuller (1989, pp. 366) eliminated the concept of substances from the definition of probiotics, and substituted it for "live microbial feed supplements which beneficially affect the host animal by improving its intestinal microbial balance." This definition was later extended to include a viable mono or mixed culture of bacteria that will beneficially affect the host when applied to human or animals by improving the properties of the indigenous microflora. Contrarily, Salminen et al. (1999, pp. 109) stated that there are documented health effects of nonviable probiotics and even the cell wall components on some probiotic microbes. This led to the establishment of the following definition: "Probiotics are microbial cell preparations or components of microbial cells that have a beneficial effect on the health and well-being of the host." This definition introduces a second novelty with respect to the one proposed by Fuller, and it is the beneficial effect of probiotics on human health in general rather than specifically on intestinal health. The most accepted definition nowadays is that given by the World Health Organization as "live microorganisms that, when administered in adequate amounts confer a health benefit to the host" (Araya et al., 2002, pp. 8).

Prebiotics have been defined as nondigestible food ingredients beneficially affecting the host by selectively stimulating the growth and/or activity of a limited number of bacteria in the colon and thus improving the host health (Gibson et al., 2004b). Probiotics and prebiotics can be combined in symbiotic preparations (symbiotics). These mixtures beneficially affect the host in different ways. For instance, by improving the survival and implantation of live microbial dietary supplements in the GIT; by selectively stimulating the growth and/or by activating the metabolism of one or a limited number of health-promoting bacteria, and thus improving the host well-being (Hamilton-Miller, 2003).

2.2 PROBIOTICS

2.2.1 INTRODUCTION

Food marketed as "functional" contains added ingredients and/or ingredients technologically developed to confer a specific health benefit (Siró et al., 2008). The largest segment of this market comprises foods designed to improve gut health such as probiotics, prebiotics, and symbiotics (Stanton et al., 2005, Agrawal, 2005).

Probiotic products are gaining widespread popularity and acceptance throughout the world and are already well received in Japan, the United States of America, Europe, and Australia (Siró et al., 2008). Among these, dairy-based products are the key segment of this sector as it is estimated that they account for about 74% of probiotic product market shares (Frost-Sullivan, 2007). Interests in applying nondairy substrates for the creation of novel probiotic formulations has increased in the past year, aiming at treating lactose intolerance and controlling cholesterol levels (Stanton et al., 2005).

It has been proven that cereal substrates can support the growth of *Lactobacillus* and *Bifidobacterium* (main probiotic microorganisms) to probiotic levels (Charalampopoulos et al., 2002a,b, 2003, Kedia et al., 2007, 2008a,b,c, Patel et al., 2004b, Rozada-Sánchez et al., 2008). Thus, cereals currently have a great potential for the creation of new probiotic beverages.

There is limited information about the effective dose for particular probiotic strains, so large numbers of viable bacteria are recommended in probiotic foods. Scientific evidence recommends that probiotic bacteria consumed at a range of 10^9–10^{11} CFU/day can lower the severity of some intestinal disorders (Fooks et al., 1999, Tannock, 1998, Zubillaga et al., 2001). Also probiotic products must be able to have a satisfactory shelf life containing cell counts grater that 10^6 CFU/mL till the time of consumption (Agrawal, 2005). In Japan, a standard has been created by the Fermented Milks and Lactic Acid Bacteria Beverages Association stipulating that a product containing $\geq 1 \times 10^7$ viable bacteria/g or mL could be considered a probiotic food, while a therapeutic minimum dose of 1×10^5 viable cells/g or mL of product has been proposed by other scientists. However, the minimum dose required for a product to achieve a probiotic effect may be determined by the food selected as carrier and the strain used (Stanton et al., 2001).

The application of probiotics in animal health has also been exploited as they could replace feed antibiotics which are commonly used as supplements in the diets of farm animals. This comes into place as possible antibiotic residues in the animal products and cross-resistance with human pathogens are significant health threats encouraging the development and use of nonantibiotic products (Nousiainen and Setälä, 1998).

2.2.2 BENEFICIAL EFFECTS OF PROBIOTIC MICROORGANISMS AND MECHANISMS OF ACTION

Several beneficial effects of probiotic microorganisms have been described and reviewed frequently. Some of these effects are described in the following sections.

2.2.2.1 Treatment of Lactose Intolerance

Lactose intolerance is due to a deficient performance of the enzyme lactase in the human gut, causing abdominal distension, excessive flatulence, and diarrhea. The consumption of yogurts reduces lactose intolerance, which is attributed to two mechanisms: on the one hand, the presence of bacterial lactase (β-galactosidase) in yogurt helps the cleavage of lactose; on the other, the gastrointestinal (GI) transit of yogurt is slower than that of milk, and therefore, the contact time between lactase on enterocytes and lactose is prolonged, giving more time for the lactose to be cleaved.

2.2.2.2 Prevention and Treatment of *Helicobacter pylori* Infection

H. pylori are Gram-negative, spiral rod-shaped bacteria present in the human gastric mucosa. If the pH of the stomach increases, the main barrier to colonization by these bacteria disappears, and its metabolites cause the appearance of ulcers and eventually cancer. Currently, the treatment for *H. pylori* infections consists of the administration of a mixture of antibiotics together with proton pump inhibitors. The use of probiotics as an alternative therapy is being extensively investigated. Several strains have been found to be effective, such as *Lactobacillus acidophilus* CRL 639 or *Lactobacillus johnsonii* La1 (Felley and Michetti, 2003, Hamilton-Miller, 2003).

2.2.2.3 Treatment of Food Allergies

Th1 cells are responsible for cell-mediated immunity, while Th2 cells are ultimately responsible for antibody production. An imbalance in the Th1/Th2 responses to exogenous antigens can direct the defense mechanism for parasitic infection toward the excess production of antibodies which might bind food or respiratory antigens, provoking a reaction which includes vasodilatation, tissue fluid exudation, smooth muscle contraction and mucous secretion. *Bifidobacterium lactis* Bb-12, *Lactobacillus rhamnosus* GG, and *Lactobacillus paracasei*-33 have been used successfully in different clinical studies for the treatment of atopic diseases. It is thought that probiotics reduce the increased intestinal permeability found in atopic patients to normal values, improving processing of antigens ingested in the diet and increase the uptake of antigens by Peyer's patches by reducing intestinal inflammation and antibody production (Heyman and Ménard, 2002).

2.2.2.4 Prevention of Colon Cancer

Although not the only factor, the ingestion of a diet rich in red meat is thought to play a major role in the development of colon cancer. This is due to the formation of genotoxic carcinogens such as heterocyclic aromatic amines, during the cooking of meat, and to the transformation of protein residues from meat by bacterial enzymes such as NADPH dehydrogenase, nitroreductase, β-glucosidase, β-glucuronidase, or 7-α-dehydroxylase into toxic and genotoxic compounds. These enzymes are more heavily expressed by enterobacteria, *Bacteroides*, and clostridia than by *Lactobacilli* and bifidobacteria (Rafter, 2003). Several mechanisms as to how probiotics can be effective against colon cancer have been suggested. Different microorganisms will however, have different modes of action (Klaver and van der Meer, 1993). The evidence to sustain the prevention or elimination of colon cancer by probiotics is only indirect, and more research is required in this area.

2.2.2.5 Prevention of Coronary Heart Disease

High levels of cholesterol in the blood are associated with a high risk of coronary heart disease. Not many *in vivo* studies are available regarding the possible effect of probiotics in the prevention of coronary heart disease. However, *in vitro* studies indicate that probiotics have the potential to aid in the lowering of cholesterol blood levels by deconjugation of bile acids and subsequent coprecipitation of cholesterol and bile salts (Noh et al., 1997) or by the absorption of cholesterol by the probiotic strain (as is the case for *Lactobacillus acidophilus*) (Said and Mohamed, 2006). This effect is however, controversial and further work is needed in this field (Reid, 2006).

2.2.2.6 Treatment of Urogenital Infections

The vaginal flora is rich in *Lactobacilli*. Disruption of the normal vaginal flora, caused by a change in the vaginal pH, provides a favorable environment for the development of amine-producing microbiota. This may cause problems such as pre-term labor and increase the risk of acquiring sexually transmitted diseases. Several Lactobacilli strains have shown to be able to colonize the vagina, but only four fulfill the probiotic requirement: *Lactobacillus rhamnosus* GR1, *Lactobacillus reuteri* RC-14 and B-54, and *Lactobacillus crispatus* CTV05 (Sartor, 2004).

2.2.2.7 Inflammatory Bowel Diseases

Inflammatory bowel diseases (IBD), including Crohns disease, pouchitis, and ulcerative colitis are thought to occur due to an exacerbated cell-mediated immune response from the host towards commensal luminal bacteria in genetically predisposed individuals (Heyman and Ménard, 2002). This results in a lower content of lactobacilli and bifidobacteria in the normal flora, which suggests that the use of probiotics could be used in the treatment of IBD (Heyman and Ménard, 2002, Sartor, 2004). However, a clear beneficial effect of probiotics in the treatment of Crohns disease is less evident (Sartor, 2004).

2.2.2.8 Treatment and Prevention of Diarrhea

Diarrhea commonly appears in patients subjected to antibiotic treatment within a period of up to 2 months after the treatment has ceased, and it is due to the depletion of the normal microflora by antibiotics and overgrowth of pathogens, normally *Clostridium difficile*. A recent meta-analysis has found evidence of efficacy from *Saccharomyces boulardii*, *Lactobacillus rhamnosus* GG, and a combination of *Bifidobacterium lactis* and *Streptococcus thermophilus* in the prevention of antibiotic-associated diarrhea in children (Said and Mohamed, 2006). Another recent meta-analysis study found evidence of efficacy in the prevention of other types of acute diarrhea, nontraveler's and nonantibiotic-associated diarrhea (McFarland, 2007). Another study found probiotics to be effective in the prevention of traveller's diarrhea (Salminen et al., 1998b). Some clinical studies on the effect of probiotics on human health are summarized in Table 2.1.

2.2.3 DESIRABLE PROBIOTIC CHARACTERISTICS

A probiotic microorganism should meet a number of predefined criteria and it must hold the GRAS (generally recognized as safe) status before being considered for the development of a probiotic product (Farnworth, 2003, Nout, 1992). The theoretical basis for the selection of probiotic microorganisms are shown in Figure 2.1. A probiotic microorganism, despite promoting human health, should fulfill several characteristics such as demonstrating good technological properties in order to be manufactured and incorporated into food products without losing viability and functionality or creating unpleasant flavors and textures. Probiotic bacteria should survive the passage through the harsh conditions of GIT arriving in sufficient numbers to its site of action and must be capable of withstanding the gut environment. Reliable techniques are also required to establish the effect of the probiotic strain in

TABLE 2.1
Clinical Studies on the Effects of Probiotics in Humans

Disorders	Strains	Clinical Outcomes	References
Diarrhea due to pathogenic bacteria and viruses	L. rhamnosus GG, L. casei, B. lactis BB-12, B. bifidum, S. thermophilus, L. reuteri	Prevention and treatment of acute diarrhea caused by rotaviruses in children. Enhanced immune response, and reduced rotavirus shedding	Gupta et al. (2000); Isolauri et al. (1991); Majamaa et al. (1995); Pedone et al. (1999); Saavedra et al. (1994); Shornikova et al. (1997); Shornikova et al. (1997); Szajewska et al. (2001)
Traveler's diarrhea	L. acidophilus La-5, B. lactis Bb-12, L. GG	Reduced incidence of diarrhea	Black et al. (1989); Hilton et al. (1997); Oksanen et al. (1990)
Incorrect lactose digestion	Yogurt, L. acidophilus, B. spp., L. bulgaricus	Improved lactose digestion, reduced/absence of symptoms	Johansson et al. (1993); Kolars et al. (1984); Vesa et al. (1996)
H. pylori infection and complications	L. johnsonii La1, L. salivarius, L. acidophilus LB	Inhibition of the pathogen growth and decreased urease enzyme activity necessary for the pathogen to remain in the acidic environment of the stomach	Aiba et al. (1998); Coconnier et al. (1998); Michetti et al. (1999); Midolo et al. (1995)
Inflammatory diseases and bowel syndromes	L. rhamnosus GG	Remediation in inflammatory conditions through modulation of the gastrointestinal microflora	Gionchetti et al. (2000); Shanahan (2000)
Cancer in gastrointestinal tract	L. rhamnosus GG, L.rhamnosus LC-705, L.casei Shirota, L.acidophilus LA2, B. spp. Propionibacterium sp.	Prevention or delay of the onset of certain cancers	Aso et al. (1995); El-Nezami et al. (2000); Hosoda et al. (1996); Oatley et al. (2000)
	L. casei Shirota	Reduced recurrence	Aso et al. (1995); Aso and Akazan (1992)

continued

TABLE 2.1 (continued)

Clinical Studies on the Effects of Probiotics in Humans

Disorders	Strains	Clinical Outcomes	References
Superficial bladder cancer, cervical cancer	*L. casei* Shirota LC9018	Prolonged survival and relapse-free interval	Okawa et al. (1993)
Mucosal immunity	*L. casei Shirota, L.rhamnosus* HN001, *L.acidophilus* HN017, *B. lactis* HN019	Enhancement of immune parameters	Arunachalam et al. (2000); Donnet-Hughes et al. (1999); Gill et al. (2001); Matsuzaki and Chin (2000); Perdigon et al. (1999); Sheih et al. (2001)
Immune modulation	*L. GG, L. johnsonii* Lj-1, *B. lactis* Bb-12	Higher level of specific IgA secreting cells to rotavirus, adjuvant activity, enhanced phagocytosis	Kaila et al. (1995); Pelto et al. (1998); Schiffrin et al. (1995)
Allergic symptoms	*L. rhamnosus* GG, *B. lactis* BB-12	Prevention of allergic disease onset	Isolauri et al. (2000); Kalliomaki et al. (2001); Majamaa et al. (1995); Majamaa and Isolauri (1997)
Cardiovascular disease	*L.* spp.	Prevention and therapy of ischemic heart syndromes	De Roos and Katan (2000); Oxman et al. (2001)
Bacterial and yeast vaginitis	*L. acidophilus, L.rhamnosus* GG	Eradication of vaginitis through restoration of dominated vaginal flora	Hilton et al. (1992); Hilton et al. (1995); Reid et al. (2001); Sieber and Dietz (1998)
Urinary tract infections	*L. GR-1, L.* B-54, *L.* RC-14	Lower risk of urinary tract infections through restoration of dominated vaginal flora	Reid et al. (2001)
Fecal enzyme activity	*L. GG, Yogurt, L. gasseri* (ADH) *L. casei* Shirota	Reduction of urease, glycocholic acid hydrolase, b-glucuronidase, nitroreductase, and azoreductase activity	Link-Amster et al. (1994); Spanhaak et al. (1998)
Modulation of intestinal flora	*L. plantarum* DSM9843 +(299v)	Increase in fecal short-chain fatty acid content	Johansson et al. (1993); Johansson et al. (1998)

Source: Adapted from Ouwehand, A. C. et al. 1999. *International Dairy Journal*, 9, 43–52; Prado, F. C. et al. 2008. *Food Research International*, 41, 111–123; Saarela, M. et al. 2000. *Journal of Biotechnology*, 197–215.

FIGURE 2.1 Theoretical basis for the selection of probiotic microorganisms. (Adapted from Farnworth, E. R. 2003. *Handbook of Fermented Functional Foods,* Boca Raton, Florida, CRC Press; Nout, M. J. R. 1992. *International Journal of Food Microbiology,* 16, 313–322.)

other members of the intestinal microbiota and most importantly, the host (Saarela et al., 2000).

2.2.4 MECHANISMS OF ACTION

With the understanding of the probiotic mechanism it could be demonstrated that many gastrointestinal (GI) disorders are based on malfunctions of intestinal microflora (Gismondo et al., 1999). The human GIT is the habitat of a complex microbiota consisting of either facultative or obligate anaerobic microorganisms including streptococci, lactobacilli, and yeasts. These are settled in different quantities throughout the GIT and possess a wide spectrum of differences (Figure 2.2). The human GIT could be considered as a traveling bioreactor, as the number of bacteria that colonize the human body is so large that researchers have estimated that the human body contains 10^{14} cells, of which only 10% are not bacteria (Puupponen-Pimiä et al., 2002, Teitelbaum and Walker, 2002). Some of the probiotic mechanisms of action for the prevention of GI disorders are the suppression of harmful bacteria and viruses through the production of bacterocins, stimulation of local and systemic immunity, and the modification of gut microbial metabolic activity (Gismondo et al., 1999).

A positive feature of probiotic bacteria is their ability to produce organic acids at levels that have bactericidal and bacteriostatic effects that inhibit the growth of pathogenic organisms. Probiotic bacteria also compete against harmful pathogens for a

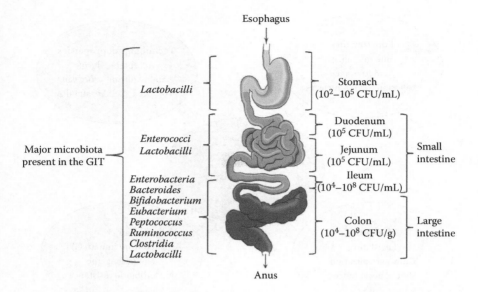

Esophagus

Lactobacilli	Stomach $(10^2$–10^5 CFU/mL)
Enterococci *Lactobacilli*	Duodenum $(10^5$ CFU/mL)
	Jejunum $(10^5$ CFU/mL)
Enterobacteria *Bacteroides* *Bifidobacterium* *Eubacterium* *Peptococcus* *Ruminococcus* *Clostridia* *Lactobacilli*	Ileum $(10^4$–10^8 CFU/mL)
	Colon $(10^4$–10^8 CFU/g)

Major microbiota present in the GIT

Small intestine

Large intestine

Anus

FIGURE 2.2 Microbial colonization of the human gastrointestinal tract (GIT). (Adapted from Holzapfel, W. H. et al. 1998. *International Journal of Food Microbiology,* 41, 85–101.)

position in the intestinal mucosa during the colonization of the GIT which prompts the state of wellbeing in the host (Fooks et al., 1999, Tuohy et al., 2003). Further metabolic activities of the GIT microbiota that beneficially affect the host include continued degradation of food components, vitamin production, and production of short-chain fatty acids (SCFAs) (Salminen et al., 1998a). The degradation of food components is mainly by fermenting dietary carbohydrates that have escaped digestion in the upper GIT such as resistant starch, cellulose, hemicellulose, pectins, and gums.

2.2.5 Probiotic Microorganisms

The microbes used as probiotics belong to a broad genera although the principal strains used belong to the heterogeneous group of lactic acid bacteria (LAB), which comprises lactobacilli, enterococci, and bifidobacteria from which lactobacilli are most commonly used during the formulation of probiotic products (Table 2.2). LAB have a long and safe history of application in the formulation of fermented foods and beverages (Leroy and DeVuyst, 2004).

Despite there being several microbial organisms that have claimed to possess probiotic properties, only a reduced number of these have been industrialized and are used in the formulation of food and beverages (Table 2.3). Amongst these strains, the most frequently used are *Lactobacillus acidophilus, Lactobacillus casei, Lactobacillus plantarum, Lactobacillus reuteri, Lactobacillus plantarum, Bifidobacterium bifidum,* and *Bifidobacterium lactis* which are all of human origin and are recognized by distinct brand names (Frost-Sullivan, 2007).

TABLE 2.2
Commonly Used Probiotic Microorganisms

Lactobacillus sp.	*Bifidobacterium* sp.	Others
L. acidophilus	*B. adolescentis*	*Bacillus cereus*
L. amylovorus	*B. animalis*	*Clostridium butyricum*
L. brevis	*B. breve*	*Enterococcus faecalis*
L. bulgaricus		
L. casei	*B. bifidum*	*Enterococcus faecium*
L. casei sp. *Rhamnosus*	*B. infantis*	*Escherichia coli*
L. crispatus	*B. lactis*	*Lactococcus lactis* sp. *Cremoriss*
L. delbrueckii sp. *Bulgaricus*	*B. longum*	*Lactococcus lactis* sp. *Lactis*
L. fermentum		*Pediococcus acidilactici*
L. gasseri		*Propionibacterium freudenreichii*
L. helveticus		*Saccharomyces boulardii (yeast)*
L. johnsonii		*Saccharomices cerevisiae* (yeast)
L. lactis		VSL#3 (four strains of lactobacilli, three
		strains of bifidobacteria, one strain
		Streptococcus salivarius sp. *Thermophilus*)

Source: Adapted from Ouwehand, A., Salminen, S. and Isolauri, E. 2002. *Antonie van Leeuwenhoek 82,* 82, 279–289; Heyman, M. and Ménard, S. 2002. *Cellular and Molecular Life Sciences,* 59, 1151–1165.

2.2.6 COMMERCIAL PROBIOTICS

2.2.6.1 Dairy-Based Probiotics

In Europe, the probiotic food and beverage market has developed with outstanding success within the functional food industry. This has happened since the consolidation of this market around the mid-1980s. Although the application of bacterial strains claiming to improve consumers' health dates back to the 1920s, the first serious attempts to commercialize probiotics were performed in Japan in the 1950s with the foundation of Yakult Honsha Co. Ltd. At present, the application of new technologies such as microencapsulation have encouraged numerous food producers to integrate probiotic strains in their products (Frost-Sullivan, 2007). Some of these commercial probiotics are shown in Table 2.4. It should be noted that there are about 52 major companies involved in the European food and beverage probiotics market.

Commercially, yogurt drinks are the most popular products within the probiotic market as they have a great acceptance among consumers. The sector also holds the greatest market shares among the bacteria-friendly products as shown in Figure 2.3 (Frost-Sullivan, 2007).

2.2.6.2 Nondairy and Cereal-Based Probiotic Products

Lactose intolerance and cholesterol control are two major downsides encountered by dairy-based probiotic products, and with an increase in the number of vegetarian consumers there is also a demand for vegetarian probiotic products (Heenan et al.,

TABLE 2.3

Popular Commercial Strains Used in the Probiotics Market

Genera	Species	Strain	Company (Owned by)
Lactobacillus	acidophilus	La5	Chr Hansen A/S
	acidophilus	NCFM (HOWARU™ Dophilus)	Danisco A/S
	acidophilus	LAFTI L10	DSM
	casei	Shirota	Yakult Honsha Co. Ltd
	coryne	Hereditum	Puleva Biotech SA
	delbrueckii	subsp. Bulgaricus	Centro Sperimentale del Latte S.p.A.
	fortis		Nestle SA
	paracasei	F-19	Medipharm AB
	plantarum	299v	Probi AB (& Institut Rosell)
	reuteri	ATCC 55730	BioGaia AB
	reuteri	RC-14	Chr Hansen A/S/Urex Biotech
Lactobacillus	rhamnosus	HN001 (HOWARU™ Rhamnosus)	Danisco A/S
	rhamnosus	ATCC 53103 (Gefilus)	Valio
Bifidobacterium	bifidum	Actif essensis	Danone (proprietary)
	bifidum	BB12	Chr Hansen A/S
	lactis	HN019 (HOWARU™ Bifido)	Danisco A/S
	lactis	B94	DSM
	regularis	(Activia-proprietary)	Danone
Saccharomyces	boulardii		Institut Rosell
	boulardii	(ULTRA-LEVURE/Florastor)	Biocodex
Streptococcus	salivarius	K12	BLIS Technologies

Source: Adapted from Frost-Sullivan 2007. *Strategic Analysis of the European Food and Beverage Probiotics Markets (#B956–88).* London: Frost & Sullivan Ltd.

2004, Yoon et al., 2006). The fortification of fruit juices with probiotic bacteria has been recently tried and these products already have a presence in the probiotic sector (Tuorila and Cardello, 2002). However, sensorial studies performed in these products have shown that they can develop off-flavors such as medicinal and savoury, which could negatively affect their performance within the market of probiotic foods (Luckow and Delahunty, 2004a, Luckow et al., 2005, 2006).

Cereals present a further alternative for the creation of nondairy probiotic foods. Besides, their recognized health-promoting effects can be exploited thus leading to the development of cereal-based probiotic products. Cereals can be applied as sources of nondigestible carbohydrates that besides promoting several beneficial physiological effects can also stimulate the growth of lactobacilli and bifidobacteria (Charalampopoulos et al., 2002b, Kedia et al., 2008c, Patel et al., 2004b).

Probiotic bacteria such as lactobacilli and bifidobacteria require several nutrients for their metabolism such as carbohydrates, amino acids, vitamins, and metal ions.

TABLE 2.4
Commercial Examples of Probiotic Products

Brand	Description	Producer
Actimel	Probiotic drinking yogurt with *L. casei* Imunitass® cultures	Danone, France
Activia	Creamy yogurt containing Bifidus ActiRegularis®	Danone, France
Gefilus	A wide range of LGG products	Valio, Finland
Hellus	Dairy products containing *L. fermentum* ME-3	Tallinna Piimatoööstuse AS, Estonia
Jovita Probiotisch	Blend of cereals, fruit, and probiotic yogurt	H&J Bruggen, Germany
Pohadka	Yogurt milk with probiotic cultures	Valašské Meziříčí Dairy, Czech Republic
ProViva	Refreshing natural fruit drink and yogurt in many different flavors containing *L. plantarum*	Skåne mejerier, Sweden
Rela	Yogurts, cultured milks and juices with *L. reuteri*	Ingman Foods, Finland
Revital Active	Yogurt and drink yogurt with probiotics	Olma, Czech Republic
Snack Fibra	Snacks and bars with natural fibers and extra minerals and vitamins	Celigüeta, Spain
SOYosa	Range of products based on soy and oats and includes a refreshing drink and a probiotic yogurt-like soy–oat product	Bioferme, Finland
Soytreat	Kefir-type product with six probiotics	Lifeway, USA
Yakult	Milk drink containing *L. casei* Shirota	Yakult, Japan
Yosa	Yogurt-like oat product flavored with natural fruits and berries containing probiotic bacteria (*L. acidophilus, B. lactis*)	Bioferme, Finland
Vitality	Yogurt with pre- and probiotics and omega-3	Müller, Germany
Vifit	Drink yogurts with LGG, vitamins, and minerals	Campina, the Netherlands

Source: Adapted from Siró, I. 2008. *Appetite,* 51, 456–467.

Cereals have within their structure simple sugars, proteins, complex oligosaccharides, and polysaccharides that could efficiently support the growth of probiotic bacteria (Henry, 2001). Malt, wheat, barley, and oat extracts have been studied to evaluate their performance as substrates for the growth of human-derived LAB. The results indicate that cereals are able to support bacterial growth and increase their acid and bile tolerance (Charalampopoulos et al., 2002b, 2003, Patel et al., 2004a,b, Rozada-Sánchez et al., 2008). In addition, further potential benefits of cereal extracts and cereal fiber for the development of new functional products has been studied showing that these substrates can improve the tolerance of a *Lactobacillus plantarum* strain to gastric and bile juices (Michida et al., 2006).

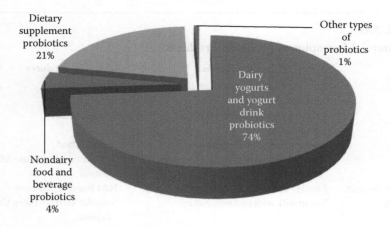

Dietary
supplement
probiotics
21%

Other types
of
probiotics
1%

Dairy
yogurts
and yogurt
drink
probiotics
74%

Nondairy
food and
beverage
probiotics
4%

FIGURE 2.3 European probiotic market share. (Adapted from Frost-Sullivan 2007. *Strategic Analysis of the European Food and Beverage Probiotics Markets (#B956–88).* London: Frost & Sullivan Ltd.)

The potential use of mixed cultures for the fermentation of cereal substrates and the evaluation of the fermentability of cereal fractions by lactobacilli strains has also been studied. Results also showed that by using different cereal fractions, higher probiotic levels can be achieved when compared to using the whole grain (Kedia et al., 2007, 2008a,b,c). Therefore, cereals could be used to design cereal-based fermented beverages with probiotic characteristics if these formulations fulfill the probiotic requirements and have acceptable physicochemical characteristics and organoleptic properties.

2.2.7　Flavor Formation by LAB

An important aspect that should not be underestimated during the development of novel probiotic products is flavor. Consumers purchasing health-promoting products are likely to choose those that have acceptable sensory qualities. Flavor evaluation of some probiotic dairy products and fruit juices has been previously reported (Luckow and Delahunty, 2004b, Lin et al., 2004, Macedo et al., 1999, Martin-Diana et al., 2003, Modzelewska-Kapituła et al., 2008, Pelletier et al., 2007). However, the flavor characterization of cereal-based probiotic beverages has never been attempted.

The quality of fermented products is largely determined by the sensory perception. Sensory perception is a multifaceted process, which is influenced by many factors, such as the content of flavor components, texture, and appearance (Smit et al., 2005). The formation of flavors in these products is a complex, and in some cases a slow process involving various chemicals and biochemical conversions of the raw material components. Three main pathways can be identified: the conversions of carbohydrates (glycolysis and/or pentose phosphate), proteins (proteolysis), and fats (lipolysis). The enzymes involved in these pathways are predominantly derived from the starter cultures used in these fermentations (Dieguez et al., 2002). LAB has a long and safe history of application in the production of fermented foods and beverages (Adams and Marteau, 1995). Today, the major LAB starter cultures used by the food industry are composed of aero-tolerant and acid-tolerant cocci (*Streptococcus, Pediococeus, Leuconostoc*) and rods (*Lactobacillus, Bifidobacterium*), which exhibit different fermentation pathways (Lacerda et al., 2005, Plessas et al., 2005).

The flavor compounds produced during lactic acid fermentation are classified into two categories. Nonvolatile compounds including organic acids produced by homo- and heterofermentative bacteria which decrease pH inhibiting the growth of food spoilage bacteria. The second category is the volatile compounds which include alcohols, aldehydes, ketones, esters, and sulfur components (Salimur et al., 2006). In this respect, lactic acid is among the main flavoring substances that give fermented food their particular sensorial characteristics (Breslin, 2001, Fernandez and McGrgor, 1999). However, this bacterial metabolism leads to the generation of many other flavor metabolites such as ethanol, diacetyl, acetic acid, and acetaldehyde. These secondary metabolites merge with the organic acids to produce a unique flavor distinctive of a fermented product (Leroy and DeVuyst, 2004).

2.2.7.1 Carbohydrate Catabolism by LAB

The conversion of carbohydrates into lactate by LAB may well be considered as the most important fermentation process employed in food technology (Kandler, 1983). However, it is clear that LAB adapts to various conditions and changes its metabolism accordingly leading to significantly different end-product patterns. As mentioned before there are two major pathways for hexose (e.g., glucose) fermentation occurring within LAB. The Embden–Meyerhof pathway (glycolysis) generates lactic acid practically as the only end-product. The second major pathway is known as the pentose phosphate pathway characterized by the creation of 6-phosphogluconate which is further divided into glyceraldehyde-3-phosphate (GAP) and acetyl phosphate. GAP is metabolized to lactic acid while acetyl phosphate is reduced to ethanol and the important flavor compound acetaldehyde (Axelsson, 1998). Furthermore, through this heterofermentative process, equimolar amounts of CO_2, lactate, and acetate or ethanol are formed from hexose fermentation. The ratio of acetate/ethanol depends on the oxidation reduction potential of the system (Kandler, 1983). During the fermentation of complex substrates such as fruit juices and vegetables which contain compounds other than hexoses, for example, pentoses or organic acids, the production of lactate, acetate, and CO_2 will be of different ratios. Pyruvate may, moreover, not only be reduced to lactate but also converted into several other products by alternative mechanisms depending on the growth conditions and properties of the particular microorganisms (Kandler, 1983).

2.2.7.2 Pyruvate and Lactate Catabolism

Pyruvate is mainly reduced to lactate, catalyzed by lactate dehydrogenase (LDH) in LAB. Therefore, pyruvate can alternatively be converted by LAB to other flavoring compounds such as diacetyl, acetoin, acetaldehyde, acetic acid, formate, ethanol, and 2,3-butanediol (Liu, 2003, Smit et al., 2005). The pathways of formation of these minor products from pyruvate are shown in Figure 2.4.

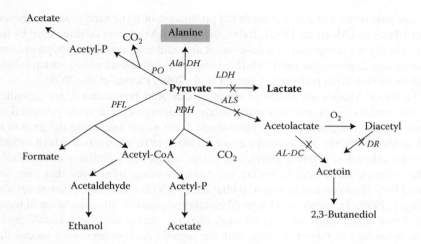

FIGURE 2.4 Pyruvate metabolism and metabolic engineering in lactic acid bacteria. LDH, lactate dehydrogenase; ALS, α-acetolactate synthetase; Al-DC, α-acetolactate decarboxylase; DR, diacetyl reductase; Ala-DH, alanine dehydrogenase; PO, pyruvate oxidase; PFL, pyruvate–formate lyase; PDH, pyruvate dehydrogenase. X indicates potential sites for metabolic engineering. (Adapted from Liu, S.-Q. 2003. *International Journal of Food Microbiology,* 115–131.)

Although lactate is the end-product of lactic acid fermentation, it can be catabolized under aerobic conditions by lactate oxidase or NAD$^+$-independent LDH in some LAB to produce pyruvate, which is further catabolized (Kandler, 1983):

$$\text{lactate} \xrightarrow[\text{lactate oxidase}]{O_2} \text{pyruvate} \xrightarrow[\text{pyruvate oxidase}]{} \text{acetate} + CO_2$$

$$\text{lactate} \xrightarrow[\text{LDH}]{O_2} \text{pyruvate} \xrightarrow[\text{pyruvate oxidase}]{} \text{acetate} + CO_2$$

Under anaerobic conditions, lactate can also be catabolized via NAD$^+$-independent LDH by some lactobacilli. Such lactate degrading lactobacilli use other electron acceptors (e.g., shikimate or oxaloacetate derived from citrate and 3-hydroxypropionaldehye derived from glycerol) for anaerobic catabolism of lactate as follows (Liu, 2003):

$$\text{lactate} \rightarrow \text{pyruvate} \rightarrow \text{acetate} + CO_2$$

or

$$\text{lactate} \rightarrow \text{pyruvate} \rightarrow \text{acetate} + \text{formate}$$

The products formed from pyruvate degradation are dependent upon the presence of a particular enzyme(s) in a particular LAB, pyruvate oxidase or pyruvate-formate lyase, or both (Liu, 2003).

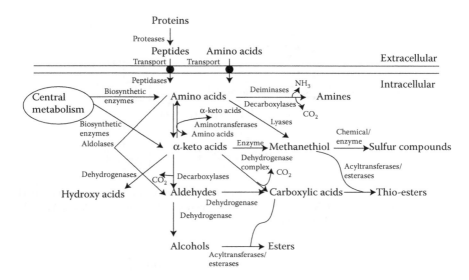

FIGURE 2.5 Overview of pathways leading to the formation of flavor compounds from protein degradation and amino acid conversions. (Adapted from Smit, G. et al. 2002. *Australian Journal of Dairy Technology,* 57, 61–68.)

2.2.7.3 Amino Acid Catabolism

The typical flavor of fermented foods as cheeses, wines, and sausages encompasses flavor compounds produced by the microbial catabolism of amino acids. Commonly, the same microorganisms also produce enzymes to release the amino acids from proteins, and of special interest are the LAB that need amino acids to survive. From the catabolism of amino acids both pleasant and unpleasant aroma compounds are produced (Urbach, 1993). Amino acids can be converted into various volatile flavor compounds in many different ways by enzymes such as deaminases, decarboxylases, transaminases (aminotransferases), and lyases (Figure 2.5). Transamination of amino acids results in the formation of α-keto acids that can be converted into aldehydes by decarboxylation and, subsequently, into alcohols or carboxylic acids by dehydrogenation. Furthermore, other reactions may occur, for example, by hydrogenase activity toward α-keto acids resulting in the formation of hydroxyl acids, which hardly contribute to the flavor (Dieguez et al., 2002).

2.2.7.4 Formation of α-Keto Acids

Catabolism of amino acids commonly starts with removal of the amino group, which is performed by aminotransferase. LAB are equipped with aminotransferases that are specific to different groups of amino acids, such as branched chain (Leu, Ile, Val), aromatic (Phe, Tyr, Trp), sulfuric (Cys, Met), or acidic (Asp). Aminotransferases of *Lactococcus* and *Lactobacillus* strains have been shown to be PLP (pyridoxal phosphate)-dependent enzymes that use α-ketoglutarate as an amino group acceptor and thereby produce Glu. The α-ketoacids of the branched-chain amino acids have been recognized to have cheesy flavors (Figures 2.6 and 2.7) (Ardö, 2006).

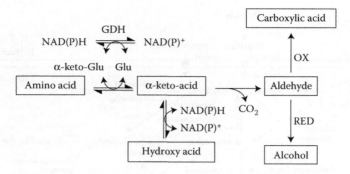

FIGURE 2.6 Catabolism of amino acids starting with aminotransferase activity. GDH, glutamate dehydrogenase; OX, oxidation; RED, reduction. (Adapted from Ardö, Y. 2006. *Biotechnology Advances,* 24, 238–242.)

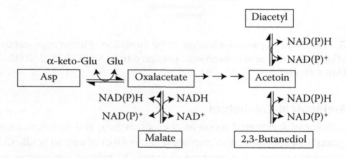

FIGURE 2.7 Catabolism of the amino acid aspartic acid. (Adapted from Ardö, Y. 2006. *Biotechnology Advances,* 24, 238–242.)

2.2.7.5 Transamination and Deamination

Glu, Asp, and Ala are involved in aminotransferase activities using α-ketoglutarate, oxaloacetate, and pyruvate as amino group acceptors, respectively, and if pyruvate or oxaloacetate is used, the Ala or Asp produced commonly is transaminated afterwards with the use of α-ketoglutarate and the production of Glu. A metabolic beneficial consequence is that only Glu has to be removed from the cell, or regenerated into α-ketoglutarate by a deamination process using the enzyme glutamate dehydrogenase and oxidized NAD(P)$^+$. Regeneration of the thereby produced NAD(P)H in turn needs oxygen or oxidized compounds, which sets limits for the processes in anaerobic environments such as the interior of cheeses or sausages. The metabolic ways leading from aldehydes to acids and alcohol, may be performed by dehydrogenases that commonly use the cofactor NAD(P)H or NAD(P)$^+$ for their redox reactions (Figure 2.6). Alcohol may be produced as long as NAD(P)$^+$ is needed, while production of carboxylic acids generates NAD(P)H. Flavor formation starting with deamination mainly takes place in an aerobic environment (Ardö, 2006).

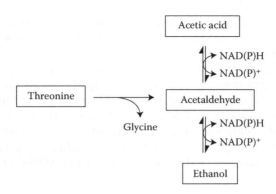

FIGURE 2.8 Catabolism of the amino acid threonine. (Adapted from Ardö, Y. 2006. *Biotechnology Advances*, 24, 238–242.)

2.2.7.6 Systems to Regenerate Reduced Cosubstrates

Aminotransferase activity on Asp produces the α-keto acid oxaloacetate, and this may be catabolized into diacetyl, acetoin, and 1,3-butanediol by, for example, *Lactobacillus paracasei* strains. Diacetyl contributes highly to a buttery flavor, but it has been claimed to also have another role in flavor formation in cheese by influencing the redox potential and thereby the chemistry of some highly volatile sulfuric compounds. Reactions between acetoin, diacetyl, and 2,3-butane diol are used to oxidize or reduce NAD(P)/NAD(P)H, whatever is needed (Figure 2.7) (Ardö, 2006).

Asn may be transformed into Asp by asperginases that have been identified in some LAB. A similar system for regenerating NAD(P)/NAD(P)H in both directions uses acetaldehyde that is produced by threonine aldolase activity directly on Thr (Figure 2.8). This activity also produces Gly, which should be beneficial for yogurt bacteria that produce large amounts of acetaldehyde contributing to the characteristic flavor of yogurt (Ardö, 2006).

2.2.7.7 Sulfur and Thioester Production

Sulfur compounds, which are potent odorants in many fermented and ripened foods, contribute to flavors in boiled cabbage and potatoes, meat, garlic, and egg, and they are produced from the sulfur-containing amino acids (Met, Cys). LAB catabolism of Met may follow the same pattern as branched chain and aromatic amino acids but methane thiol may also be released directly from the amino acid side group of Met. Methane thiol is also further converted into other sulfur compounds, and it may react with carboxy acids to produce thioesters. These in turn have boiled cabbage- and cauliflower-like flavors (Ardö, 2006).

2.3 PREBIOTICS

2.3.1 Introduction

Many food components can be evaluated for prebiotic properties by considering the criteria required. Prebiotic is defined as a nondigestible food ingredient that beneficially affects the host by selectively stimulating the growth and/or activity of one or a limited number of bacteria in the colon, and thus improves host health (Gibson and Roberfroid, 1995, Liong, 2008). The potential prebiotic substrates can be evaluated by following the three main criteria: (1) nondigestibility, (2) fermentation by intestinal microflora, and (3) selective stimulation of growth and/or activity of intestinal bacteria.

Although long-chained carbohydrates, peptides, proteins, and lipids are candidates to prebiotics, so far only the first group has been examined in depth. At the time of publication of the definition of prebiotics, the only recognized prebiotic belonged to the family of fructooligosaccharides. Currently, inulin, transgalacooligosaccharides (TOS), and lactulose are well-established prebiotics (Gibson et al., 2004b).

2.3.2 CRITERIA AND METHODOLOGIES FOR THE EVALUATION OF PREBIOTIC INGREDIENTS

2.3.2.1 Nondigestibility

In vitro methods: *In vitro* demonstration includes determining resistance to acidic conditions in the stomach and enzymatic hydrolysis of saliva and pancreatic enzymes. After an appropriate incubation, hydrolysis products are assayed using standard methods (Gibson et al., 2004a).

In vivo methods: Resistance to any endogenous digestive process can be shown by measuring the recovery in feces of an oral dose given to germ-free rats or after antibiotic pretreatment to suppress the intestinal flora (Nilsson and Bjorck, 1988).

2.3.2.2 Fermentation by Intestinal Microflora

In vitro methods: The most commonly used *in vitro* models to study anaerobic fermentation of carbohydrates by mixed bacterial populations, particularly fecal bacteria, are batch and continuous culture fermentation systems. Batch culture fermenters are inoculated with either pure cultures of selected species of bacteria or, preferably, with a fecal slurry and the carbohydrate to be studied. Multichamber continuous culture systems have been developed to reproduce the physical, anatomical, and nutritional characteristics of GI regions. These models are useful for predicting both the extent and site of prebiotic fermentation (Gibson et al., 2004a).

In vivo methods: The *in vivo* fermentation of nondigestible carbohydrates can be studied in livestock animals and in human subjects. In laboratory animals, often rats, the prebiotic under investigation is added to the food or drinking water. Animals are then anesthetized and killed at predetermined time points. Fecal samples, and the contents of the GI segments, are collected for analysis. In human subjects, the fermentation of dietary carbohydrates can be tested directly by oral feeding. The fecal samples are collected and the recovery of the tested carbohydrate is measured (Gibson et al., 2004a).

2.3.2.3 Selective Stimulation of Growth and/or Activity of Intestinal Bacteria

This criterion could be evaluated through fermentation studies. The most commonly used *in vitro* models to study anaerobic fermentation of carbohydrates by mixed bacterial populations, particularly fecal bacteria, are batch and continuous culture fermentation systems (Gibson et al., 2004a). These models are valuable for studying the gut micro-biota as fermentation substrates can be added under controlled conditions, which may not be feasible in a human trial. Moreover, these studies provide measurements that can be used to assess the prebiotic effect by quantifying the microbial population changes. The prebiotic index (PI), based upon changes in beneficial bacteria (bifidobacteria and lactobacilli) and harmful bacteria such as *Clostridia* and *Bacteroides* (Palframan et al., 2003), was the first and simplest example of quantitative analysis. It is assumed that an increase in the populations of bifidobacteria and/or lactobacilli is a positive effect while an increase in clostridia (*histolyticum* subgroup) and bacteroides is a negative effect. However, the beneficial effect of prebiotics is not only seen through their ability to increase or influence numbers of bacteria but also through activities of these bacteria, particularly, SCFA production (Cardarellia et al., 2007).

2.3.3 CARBOHYDRATES AND HEALTH

Through digestion, starch is hydrolyzed into simple carbohydrates. Carbohydrates also add taste, texture, and variety to our food. The various sources of carbohydrates are cereals, sugars, fruits, vegetables, and legumes. A landmark report recommends that at least 55% of daily energy intake should come from these carbohydrates. Carbohydrates in the form of sugars, starches, oligo- and polysaccharides, and fibers form one of the three major macronutrients that supply the human body with energy (Rolfe, 2000).

Nutrition educators agree that consumers should increase their intake of complex carbohydrates and dietary fiber, both for beneficial health effects of these dietary ingredients and as a means to reduce dietary fat intake. It is important to maintain an appropriate balance between energy intake and expenditure; research suggests that people who eat a high carbohydrate diet are less likely to accumulate body fat compared to those on a low-carbohydrate/high-fat diet. The reasons for these obser-vations include:

1. The lower energy density of high-carbohydrate diets, as carbohydrates have less calories weight for weight than fat. Fiber-rich foods also tend to be bulky and therefore, are physically filling.
2. Studies have found that carbohydrates work quickly to aid satiety, therefore those consuming high carbohydrate diets are less likely to overeat.
3. It has also been suggested that very little dietary carbohydrate is converted to body fat mainly because it is a very inefficient process for the body. Instead, carbohydrate tends to be preferentially used by the body for energy.

Carbohydrates can be classified on the basis of their degree of polymerization (DP) as shown in Table 2.5. Such an approach is a compromise between a chemical classification and a physiological classification (Cummings et al., 2001). If carbohy-drates were reported on the basis of their chemical nature, according to the scheme proposed, much of the terminology currently used could be eliminated since such a terminology is based on physiological and/or methodological distinctions. It is thus desirable to move away from such terminology because its meanings vary for dif-ferent regions of the world, and this leads to confusion. For example, the expression

TABLE 2.5
Classifications of Carbohydrates

Class (DP)	Subgroup	Component	Foods
Sugars (1–2)	Monosaccharides	Glucose, galactose, fructose	Honey, fruit
	Disaccharides	Sucrose, lactose, trehalose	Table sugar, milk
Oligosaccharides (3–9)	Maltooligosaccharides	Maltodextrins	Soya, artichokes, onions
	Other oligosaccharides	Raffinose, stachyose fructooligosaccharides	
Polysaccharides (≥9)	Starch	Amylose, amylopectin	Rice, bread, potato
	Nonstarch polysaccharides	Cellulose, hemicelluloses, pectins	All vegetables and fruits

Source: Cummings, J. H., Macfarlane, G. T. and Englyst, H. N. 2001. *American Journal of Clinical Nutrition*, 73, 415S–420S.

"complex carbohydrate" has come to describe starch in North America but, in Europe, it includes all polysaccharides (Cummings et al., 2001).

2.3.3.1 Oligosaccharides

As classified in the previous section an oligosaccharide is a "Molecule containing a small number (2 to about 10) of monosaccharide residues connected by glycosidic linkages." During the past two decades, many types of oligosaccharides, such as fructo-, galacto-, and xylooligosaccharides, were actively developed. Research on their characteristics found that these oligosaccharides had excellent physiological properties that are both scientifically interesting and beneficial to human health (Bailey and Olis, 1986).

Oligosaccharides can be divided into digestible and nondigestible oligosaccharides depending upon their digestion in the human body. Figure 2.9 shows the difference in the metabolic pathway between digestible and nondigestible oligosaccharides. This difference is an essential property to understand the physiological functions of oligosaccharides (Hirayama, 2002). The left column in Figure 2.9 shows the digestive tract, and the right rectangle shows the inside of the body. Food moves from the mouth to the stomach, then to the small intestine, and then to the large intestine. In the small intestine, digestible oligosaccharides, such as sucrose and starch, undergo the digestion and absorption process. Digestible enzymes convert these saccharides to monosaccharides, which are eventually metabolized and exhaled as breath carbon dioxide or excreted in urine. On the other hand, nondigestible oligosaccharides pass through the small intestine tract. In the large intestine, they enter a fermentation and absorption process (Cummings et al., 2001). In this process, intestinal microbes transform the oligosaccharides into short-chain fatty acids (SCFAs). Acetate, propionate, and butyrate are the most common components. Subsequently, the SCFAs are absorbed and metabolized into carbon dioxide (Hirayama, 2002).

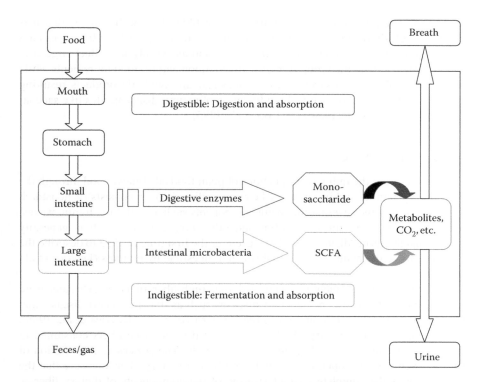

FIGURE 2.9 Difference in the metabolic pathway between digestible and nondigestible oligosaccharides. (From Hirayama, M. 2002. *Pure and Applied Chemistry,* 74, 1271–1279.)

2.3.3.2 Nondigestible Oligosaccharides

Nondigestible oligosaccharides (NDOs, also referred as "resistant oligosaccharides") are complex carbohydrates, which are resistant to hydrolysis by acids and enzymes in the human digestive tract due to the configuration of their osidic bonds. Most dietary oligosaccharides are quantitatively hydrolyzed in the upper part of the GIT. The resulting monosaccharides are transported via the portal blood to the liver and subsequently to the circulatory system. Such carbohydrates are essential for health as they serve both as substrates and regulators of major metabolic pathways. They also trigger hormone secretion. But some of the dietary oligosaccharides do resist, more or less quantitatively, the digestive process. Such carbohydrates reach the colon as they have been eaten and so do not provide the body with monosaccharides (Lee and Prosky, 1995).

Once they have reached the colon most of the nondigestible oligosaccharides are hydrolyzed to small oligomers and monomers that are further metabolized by anaerobic bacteria. Such a metabolic process not only serves the bacteria by providing energy for proliferation but also produces gases (H_2, CO_2, CH_4), which are metabolically useless for the host. SCFAs such as butyrate, propionate, acetate, L-lactate are also produced. About 90–95% of SCFA produced are absorbed through the intestinal wall, in the caecum and the ascending part of the colon (Ruppin et al., 1980).

The most important class of dietary nondigestible oligosaccharides is that it is composed of short-(up to 10 monomers) and medium-(up to 50–60 units) chain length homopolymers. Some NDOs, like inulin and its hydrolysis product oligofructose, which belong to the group fructans, are common natural food ingredients while others like galacto-oligosaccharides or Neosugar are synthetic products resulting from the enzymatic and chemical modification of natural disaccharides like saccharose and lactose, to cite a few (van Loo et al., 1995).

2.3.4 DIETARY FIBER

Dietary fibres are the indigestible portions of plant foods that move food through the digestive system, absorbing water. Chemically, dietary fiber consists of nonstarch polysaccharides and several other plant components such as cellulose, lignin, waxes, chitins, pectins, beta-glucans, inulin, and other oligosaccharides. It is generally agreed that the term "dietary fiber" was used for the first time in 1953 to describe the components of the plant cell wall that was not digestible by humans (Devries et al., 1999).

During the 1970s scientists theorized that high-fiber diets of Africans were responsible for their low incidence of diseases such as diabetes, heart disease, and colon cancer which were common in the Western countries (Devries et al., 1999). Since then the term "dietary fiber" has been used to describe fibers consumed in foods that are important to human health. Recently, The American Association of Cereal Chemists developed an updated definition of dietary fiber to ensure that the term contains the complete characterization of the components of dietary fiber as well as their functions.

"Dietary fiber is the edible part of plants or analogous carbohydrates that are resistant to digestion and absorption in human small intestine with complete or partial fermentation in large intestine. Dietary fiber includes polysaccharides, oligosaccharides, lignin, and associated plant substances. Dietary fibers promote beneficial physiological effects including laxation, blood cholesterol attenuation, and blood glucose attenuation."

Dietary fibers include many components, which are often categorized by their solubility characteristics in the human digestive system. Soluble and insoluble fibers have different characteristics and different physiological effect on the human body. Total dietary fiber (TDF) refers to the total amount of dietary fiber, both soluble and insoluble.

2.3.4.1 Insoluble Dietary Fiber

Insoluble dietary fibers are insoluble in aqueous solutions of enzymes that are designed to stimulate the human digestive system. They are not digested in the human small intestine but may be fermented by bacteria in the large intestine. In general, these fibers increase the bulk in the GIT and aid in waste elimination (McDougall et al., 1996). Insoluble fibers include cellulose, hemi cellulose, lignin, cutin, suberin, chitin and chitosan, and resistant starch.

2.3.4.2   Soluble Dietary Fiber

Soluble dietary fibers are soluble in aqueous solution of enzymes that are typical of the human digestive system. Many soluble dietary fibers can be precipitated in a solution of one part aqueous enzyme solution and four parts ethanol. A number of soluble fibers have been shown to help lower the blood cholesterol levels and regulate the body's use of glucose. Soluble fibers include pectins, β-glucan, gums, and inulin (McDougall et al., 1996).

From the physicochemical standpoint, dietary fibers, especially the soluble ones, can form gels when combined with water. In the small intestine, this action helps to prevent the attachment of pathogens, by trapping them in its gel-like structure (Topping and Clifton, 2001). In the colon, these products are fermented by colonic microbiota, leading to the expansion of commensal bacterial species, such as lactobacillus and bifidobacteria, thus classifying these products as prebiotic.

2.3.4.3   Colonic Fermentation of Nondigestible Oligosaccharides and Dietary Fiber

As discussed in the previous section, nondigestible oligosaccharides and dietary fiber escape the human digestion and reach large intestine where they are fermented by colonic bacteria. These bacteria hydrolyze indigestible carbohydrate products to hexose and SCFAs. In humans, anaerobic fermentation occurs in the colon. The proximal and middle colons have the highest concentration of SCFAs (Macfarlane et al., 1992). Nonstarch polysaccharides (cellulose, hemicellulose, pectins) and resistant starches were found to reach the colon in the same amounts that they are consumed. In Western diets, the average consumption of these components is around 10 g per day. Current estimates suggest that 10 g of fermented carbohydrates yield around 100 mm SCFA and 3 g biomass.

Gibson et al. (1995) used the model of mixed fecal culture for the utilization of fructo-oligosaccharides. A marked proliferation of bifidobacteria and lactobacillus was observed, while populations of bacteroides, clostridia, coliforms, and Gram-positive cocci were maintained at relatively low or even reduced levels. The accepted mechanism by which bifidobacteria and lactobacillus inhibit the growth of other bacteria involves a decrease in pH as a consequence of the intense production of SCFA. It can thus be hypothesized that feeding fructo-oligosaccharides has a major impact on the gut micro flora, by favoring the proliferation of beneficial bacteria like lactobacillus and bifidobacteria and reducing the number of potentially harmful anaerobes. Therefore, they are likely to be beneficial for the host.

2.3.5   SHORT-CHAIN FATTY ACIDS

During the dietary fiber fermentation process, 1–5 carbon SCFA are produced. The beneficial effects of dietary fiber and NDOs have been attributed to SCFA major end product of fermentation (Sakata, 1997). Along with SCFA, carbon dioxide, methane, and hydrogen, are other products of bacterial metabolism of carbohydrates in the distal GIT. Acetate, propionate, and butyrate comprise 83% of the SCFA produced in the colon and are distributed in a molar ratio of 60:25:15, respectively (Cummings,

1984). SCFA are readily absorbed by the intestinal mucosa, metabolized by the intestinal epithelium and the liver, contribute energy (Roediger, 1982), stimulate sodium and water absorption in the colon (Roediger and Moore, 1981, Ruppin et al., 1980), and are trophic to the intestinal mucosa (Kripke et al., 1989, Sakata, 1997). The mechanism by which unionized SCFA cross the mucosa is thought to be by passive diffusion down a concentration gradient (Cummings, 1984).

Sakata (1997) has shown that physiological doses of acetate and butyrate had trophic effects on both the colonic and jejunal epithelium; whereas, propionate was only effective at super-physiological doses. Therefore, it is not known whether the effects are due to the independent action of each SCFA or dependent on the presence of other SCFA. However, butyrate alone has been shown to stimulate epithelial cell proliferation (Sakata, 1997); therefore, it is believed to be the SCFA responsible for the adaptive responses noted with previous research involving the three SCFA. In addition, the exact mechanism for the trophic effects of SCFA has not been elucidated. It has been suggested that the effects are associated with the expression and secretion of specific gut hormones (Reimer and McBurney, 1996, Southon et al., 1987, Tappenden et al., 1996, 1998, Gee et al., 1996).

2.3.5.1 Production of Short-Chain Fatty Acids in the Colon

In colonic fermentation energy-rich compounds yield metabolizable energy for microbial growth and maintenance. SCFA (acetate, propionate, butyrate) and gases including carbon dioxide, methane, and hydrogen are produced (Cummings and Macfarlane, 1991). The general (unbalanced) reaction for SCFA production is as follows:

$$C_6H_{12}O_6 \rightarrow CH_3COOH + CH_3CH_2COOH + CH_3CH_2CH_2COOH$$
$$+ CO_2 + CH_4 + H_2 + \text{heat} + \text{additional microbes}$$

All fibers are not equally fermented; thus, production of SCFA is dependent on the amount and type of fiber (Topping and Clifton, 2001). In addition, SCFA production depends on the species of colonic bacteria, which vary greatly depending on the colonization of the GIT. Thus, the amount of SCFA produced is dependent on the diet consumed and the microflora present.

2.3.5.2 Role of Colonic Substrates in Short-Chain Fatty Acid Production

Substrate availability in the colon is probably the most important determinant of SCFA production (Cummings, 1991). The amount of nonstarch polysaccharides that reaches the colon for most Western diets is between 10 and 20 g/day (Bingham, 1990). A maximal yield of 0.6 g of SCFA is estimated to be produced per gram of carbohydrate fermented (Livesey et al., 1995). However, the amount of SCFA produced per gram of substrate varies from 34% to 59% (Englyst et al., 1987). In addition, the molar ratios of acetate, propionate, and butyrate differ markedly among the carbohydrates. Table 2.6 lists the amounts of SCFA produced from a variety of carbohydrates incubated in batch culture for either 24 or 48 h using human fecal inocula (Englyst et al., 1987, Titgemeyer et al., 1991). Pectin produces almost all acetate (84%) but very little butyrate (2%), compared with starch that produces 50% acetate

TABLE 2.6
Production of SCFAs from Various Substrates

Substrate	Acetate (μmol/g)	Propionate (μmol/g)	Butyrate (μmol/g)	Total (μmol/g)
Starch[a]	4163	1755	2383	8301
Arabinogalactan[a]	3164	2700	454	6318
Xylan[a]	6994	1350	227	8571
Pectin[a]	4496	810	114	5420
Soy Fiber[a]	1385	400	158	1943
Sugarbeet Fiber[b]	1074	77	9	1160
Gum arabic[b]	6015	2234	1135	9384
Arabic—Guar mixture[b]	5217	2371	985	8573
Apple pectin[b]	5446	763	569	6778
Oat Fiber[b]	147	64	36	247
Corn bran[b]	513	218	57	788

[a] Englyst et al. (1987)—Short-chain fatty acids produced, μmol/g substrate, incubated in batch culture for 24 h using human fecal inocula.

[b] Titgemeyer et al. (1991)—Short-chain fatty acids produced, μmol/g substrate, incubated in batch culture for 48 h using human fecal inocula.

and 29% butyrate. Fleming and coworkers (1985) reported that the average person consuming 33 g of total dietary fiber produces 809 mmol of SCFA per kg of faeces, constituting 57%, 20%, and 23%, acetate, propionate, and butyrate, respectively. However, infants have a very different distribution of SCFA production (Table 2.7). In formula-fed infants, acetate (74%) is the major acid in feces, with considerable

TABLE 2.7
Fiber Intakes and Fecal Short-Chain Fatty Acids in Adult and Infant Populations

Population and Diet	Fiber	Acetate (mmol/kg)	Propionate (mmol/kg)	Butyrate (mmol/kg)	Total (mmol/kg)
Adult					
Self-selected	33 g/day	458	165	186	809
Infants					
Breast-fed[b]	NA	46.8	1.20	0.00	48.0
Formula-fed[b]	NA	43.6	13.4	2.20	59.2
Breast-fed[c]	NA	53.9	2.90	0.40	57.2
Formula-fed[c]	NA	52.6	16.2	2.20	71.0

[a] Fleming et al. (1985)—Short-chain fatty acids produced, mmol/kg wet feces.

[b] Edwards et al. (1994)—Short-chain fatty acids produced, mmol/kg wet feces, 4-week-old infants.

[c] Siigur et al. (1993)—Short-chain fatty acids produced, mmol/kg wet feces, 2-month-old infants.

amounts of propionate (23%) and very little butyrate (3%) detected (Edwards et al., 1994, Siigur et al., 1993). In contrast, breast-fed infants produce virtually no butyrate, with acetate comprising ~96% of the total SCFA production (Edwards et al., 1994, Siigur et al., 1993). The type of substrates available contributes to the differences between formula and breast-fed infants. In addition, it appears that fermentation is slower in infants than in adults, but by 2 years of age, an adult SCFA profile has emerged reflecting changes in substrate in the infant (Midtvedt and Midtvedt, 1992).

2.3.5.3 Role of the Microbiota in Short-Chain Fatty Acid Production

At birth, an infant's GIT is sterile but rapidly becomes colonized with organisms acquired from the mother or the local environment. Colonization is time-dependent, with enterobacteria and streptococci predominating during the first 1–3 days after birth (Hoy et al., 2000, Rotimi and Duerden, 1981, Yoshioka et al., 1983). Bifidobacteria then become predominant in most breast-fed infants, constituting 47.6% of microbes (Rubaltelli et al., 1998), while enterococci predominate in formula-fed infants (Balmer and Wharton, 1989). On weaning, bifidobacteria decrease and a more adult profile develops reflecting dietary changes (Mitsuoka and Kaneuchi, 1977). The bacterial population of the adult human cecum and colon is estimated at 10^{13} CFU comprising 40–50% of solid matter (Stephen and Cummings, 1980). More than 50 genera and over 400 species of bacteria are present in the colon (Savage, 1986). The dominant organisms in terms of numbers are anaerobes including bacteroides, bifidobacteria, eubacteria, streptococci, and lactobacilli. Generally, bacteroides are the most numerous and can comprise more than 30% of the total (Topping and Clifton, 2001). When excess substrate is available, clostridia produce more butyrate than acetate (Rogers, 1986); in contrast, *Bacteroides ovatus* produces more acetate and limits the production of propionate during periods of substrate excess (Macfarlane and Gibson, 1991). Thus, microbes endow the colon with a potential for considerable metabolic diversity (Macfarlane, 1991). Bacterial numbers, fermentation, and proliferation are greatest in the proximal colon where substrates are highest, evidenced by the decline in SCFA production during transit from the right to the left colon (Macfarlane et al., 1992). Microbiological factors, such as the rate of depolymerization of carbohydrates and the rate at which the substrate becomes available for assimilation by the bacteria, affect the generation of fermentation products (Macfarlane, 1991). Moreover, because different bacteria produce different fermentation products, the individual substrate specificities and affinities of different intestinal species and their ability to compete for these substrates ultimately determine the types of SCFA that can be produced (Macfarlane, 1991).

2.3.5.4 Absorption and Metabolism of Short-Chain Fatty Acids

Less than 5% of bacterially derived SCFA appear in the feces (McNeil et al., 1978, Ruppin et al., 1980); therefore, the other 95% are readily absorbed by the colonocyte (Cummings, 1981, Ruppin et al., 1980). Once the SCFA are absorbed by the colonocyte, they are either locally used as fuel for the colonic mucosa or enter the portal bloodstream (Cummings, 1984). Their oxidation supplies about 60–70% of the colonocytes' energy needs and suppresses glucose oxidation by the inhibition of pyruvate dehydrogenase (Roediger, 1982). Butyrate is used preferentially over propionate and acetate in a ratio of 90:30:50, with 70–90% of the butyrate produced being metabolized in the epithelium (Cummings, 1984). Acetate and propionate are less avidly metabolized than butyrate and are transported to the liver. Once in the liver, the majority of acetate is used for the synthesis of fatty acids. Propionate is either oxidized in the citric acid cycle or it is used as a precursor for gluconeogenesis. Very little butyrate is transported to the liver; however, what is transported, is thought to be used for fatty acid synthesis.

2.3.5.5 Trophic Effects of Short-Chain Fatty Acids

SCFA have been shown to increase epithelial proliferation and mucosal growth in both the large and small intestines when administered orally, intracolonically, or intravenously (Frankel et al., 1994, Koruda et al., 1988, Kripke et al., 1989, 1991, Sakata, 1989). Supplementation of enteral nutrition with short-chain triglycerides, which are rapidly hydrolyzed by pancreatic lipase to their constituent SCFA, significantly increased the mucosal wet weight (629 ± 40 versus 337 ± 37 g/100 g body weight) and segment weight (1062 ± 83 vs. 884 ± 77 mg) along with mucosal protein (48.8 ± 8.0 versus 37.5 ± 3.1 mg/100 g body weight) of rats after a 60% distal resection compared to unresected controls (Kripke et al., 1991). This indicates that SCFA may be acting as a fuel source to the intestinal mucosa, with butyrate providing energy directly to the colon and acetate and propionate providing energy indirectly via hepatic metabolism.

Kripke et al. (1988) found that a daily instillation of a mixture of acetate, propionate, and butyrate not only increased colon tissue mass by 18%, but also increased the weight of the mucosa and submucosa by 22% and 14%, respectively, along with increasing the mucosal surface area by 31% (Kissmeyer-Nielsen et al., 1995). A 3- to 4-fold increase in epithelial cell proliferation is stimulated in the small and large intestines by daily or continuous administration of SCFA into the colon (Sakata, 1997). The strength of the effect is in the order of butyrate > propionate > acetate with the effect appearing within 1–2 days and lasting for at least 2 weeks (Sakata, 1987).

ACKNOWLEDGMENTS

The authors acknowledge the support of the European Science Foundation (ESF), in the framework of the Research Networking Programme, The European Network for Gastrointestinal Health Research. Severino S. Pandiella is a participant in the FA1005 COST Action INFOGEST on food digestion.

REFERENCES

Adams, M. R. and Marteau, P. 1995. On the safety of lactic acid bacteria from food. *International Journal of Food Microbiology,* 27, 263–264.

Agrawal, R. 2005. Probiotics: An emerging food supplement with health benefits. *Food Biotechnology,* 19, 227–246.

Aiba, Y., Suzuki, N., Kabir, A. M. A., Takagi, A. and Koga, Y. 1998. Lactic acid-mediated suppression of *Helicobacter pylori* by the oral administration of *Lactobacillus salivarius* as a probiotic in a gnotobiotic murine model. *American Journal of Gastroenterology,* 93, 2097–2101.

Araya, M., Morelli, L., Reid, G., Sanders, M. E., Stanton, C., Pineiro, M. and Ben Embarkek, P. 2002. *Guidelines for the Evaluation of Probiotics in Food.* London, Ontario: FAO/WHO.

Ardö, Y. 2006. Flavour formation by amino acid catabolism. *Biotechnology Advances,* 24, 238–242.

Arunachalam, K., Gill, H. S. and Chandra, R. K. 2000. Enhancement of natural immune function by dietary consumption of *Bifidobacterium lactis* (HN019). *European Journal of Clinical Nutrition,* 54, 1–5.

Aso, Y., Akaza, H., Kotake, T., Tsukamoto, T., Imai, K. and Naito, S. 1995. Preventive effect of a *Lactobacillus casei* preparation on the recurrence of superficial bladder cancer in a double-blind trial. *European Urology,* 27, 104–109.

Aso, Y. and Akazan, H. 1992. Prophylactic effect of a *Lactobacillus casei* preparation on the recurrence of superficial bladder cancer. *Urologia Internationalis,* 49, 125–129.

Axelsson, L. 1998. Lactic acid bacteria: Classification and physiology. In: Salmínen, S. and von Wright, A. (eds.) *Lactic Acid Bacteria: Microbiology and Functional Aspects.* 2nd ed. New York: Marcel Dekker, Inc.

Bailey, J. E. and Olis, D. F. 1986. *Biochemical Engineering Fundamentals.* New York: McGraw-Hill, Inc.

Balmer, S. E. and Wharton, B. A. 1989. Diet and faecal flora in the newborn: Breast milk and infant formula. *Archives of Disease in Childhood,* 64, 1672–1677.

Bingham, S. A. 1990. Mechanisms and experimental and epidemiological evidence relating dietary fibre (non-starch polysaccharides) and starch to protection against large bowel cancer. *Proceedings of the Nutrition Society,* 49, 153–171.

Black, F. T., Andersen, P. L., Orskov, J., Orskov, F., Gaarslev, K. and Laulund, S. 1989. Prophylactic efficacy of lactobacilli on traveller's diarrhoea. *Travel Med.* 333–335.

Breslin, P. A. S. 2001. Human gustation and flavour. *Flavour and Fragrance Journal,* 16, 439–456.

Cardarellia, H. R., Saadb, S. M. I., Gibson, G. R. and Vulevic, J. 2007. Functional petit-suisse cheese: Measure of the prebiotic effect. *Anaerobe,* 13, 200–207.

Charalampopoulos, D., Pandiella, S. S., Wang, R. and Webb, C. 2002a. Application of cereals and cereal components in functional foods: A review. *International Journal of Food Microbiology,* 79, 131–141.

Charalampopoulos, D., Pandiella, S. S. and Webb, C. 2002b. Growth studies of potentially probiotic lactic acid bacteria in cereal-based substrates. *Journal of Applied Microbiology,* 92, 851–859.

Charalampopoulos, D., Pandiella, S. S. and Webb, C. 2003. Evaluation of the effect of malt, wheat and barley extracts on the viability of potentially probiotic lactic acid bacteria under acidic conditions. *International Journal of Food Microbiology,* 82, 133.

Coconnier, M. H., Lievin, V., Hemery, E. and Servin, A. L. 1998. Antagonistic activity against *Helicobacter* infection *in vitro* and *in vivo* by the human *Lactobacillus acidophilus* strain LB. *Applied and Environmental Microbiology,* 64, 4573–4580.

Cummings, J. H. 1981. Short chain fatty acids in the human colon. *Gut,* 22, 763–779.

Cummings, J. H. 1984. Colonic absorption: The importance of short chain fatty acids in man. *Scandinavian Journal of Gastroenterology Supplement,* 93, 89–99.

Cummings, J. H. 1991. Production and metabolism of short-chain fatty acids in humans. In: Roche, A. F. (ed.) *Short-Chain Fatty Acids: Metabolism and Clinical Importance.* pp. 11–17, Columbus OH: Ross Laboratories.

Cummings, J. H. and Macfarlane, G. T. 1991. The control and consequences of bacterial fermentation in the human colon. *Journal of Applied Bacteriology,* 70, 443–459.

Cummings, J. H., Macfarlane, G. T. and Englyst, H. N. 2001. Prebiotic digestion and fermentation. *American Journal of Clinical Nutrition,* 73, 415S–420S.

De Roos, N. M. and Katan, M. B. 2000. Effects of probiotic bacteria on diarrhea, lipid metabolism, and carcinogenesis: A review of papers published between 1988 and 1998. *American Journal of Clinical Nutrition,* 71, 405–411.

Devries, J. W., Prosky, L., Li, B. and Cho, S. 1999. A historical perspective on defining dietary fiber. *Cereal Foods World,* 44, 367–369.

Dieguez, S. C., Diaz, L. D., de la Pena, M. L. G., Gomez, E. F. and 2002. Variation of volatile organic acids in spirits during storage at low and room temperatures. *Lebensmittel-Wissenschaft und-Technologie,* 35, 452–457.

Donnet-Hughes, A., Rochat, F., Serrant, P., Aeschlimann, J. M. and Schiffrin, E. J. 1999. Modulation of nonspecific mechanisms of defense by lactic acid bacteria: Effective dose. *Journal of Dairy Science,* 82, 863–869.

Edwards, C. A., Parrett, A. M., Balmer, S. E. and Wharton, B. A. 1994. Faecal short chain fatty acids in breast-fed and formula-fed babies. *Acta Paediatrica,* 83, 459–462.

El-Nezami, H., Mykkänen, H., Kankaanpää, P., Salminen, S. and Ahokas, J. 2000. Ability of Lactobacillus and Propionibacterium strains to remove aflatoxin B1 from the chicken duodenum. *Journal of Food Protection,* 63, 549–552.

Englyst, H. N., Hay, S. and Macfarlane, G. T. 1987. Polysaccharide breakdown by mixed populations of human faecal bacteria. *FEMS Microbiol Ecol,* 95, 163.

Farnworth, E. R. 2003. *Handbook of Fermented Functional Foods,* Boca Raton, Florida: CRC Press.

Felley, C. and Michetti, P. 2003. Probiotics and *Helicobacter pylori. Best Practice and Research Clinical Gastroenterology,* 17, 785.

Fernandez, E. and Mcgrgor, J. 1999. Determination of organic acids during the fermentation and cold storage of yogurt. *Journal of Dairy Science,* 11, 2934–2939.

Fleming, S. E., O'donnell, A. U. and Perman, J. A. 1985. Influence of frequent and long-term bean consumption on colonic function and fermentation. *The American Journal of Clinical Nutrition,* 41, 909–918.

Fooks, L., Fuller, R. and Gibson, G. R. 1999. Prebiotics, probiotics and human gut microbiology. *International Dairy Journal,* 9, 53–61.

Frankel, W. L., Zhang, W., Singh, A., Klurfeld, D. M., Don, S., Sakata, T., Modlin, I. and Rombeau, J. L. 1994. Mediation of the trophic effects of short-chain fatty acids on the rat jejunum and colon. *Gastroenterology,* 106, 375–380.

Frost-Sullivan 2007. *Strategic Analysis of the European Food and Beverage Probiotics Markets (#B956–88).* London: Frost & Sullivan Ltd.

Fuller, R. 1989. Probiotics in man and animal. *Journal of Applied Bacteriology,* 66, 365–368.

Gee, J. M., Lee-Finglas, W., Wortley, G. W. and Johnson, I. T. 1996. Fermentable carbohydrates elevate plasma enteroglucagon but high viscosity is also necessary to stimulate small bowel mucosal cell proliferation in rats. *Journal of Nutrition,* 126, 373–379.

Gibson, G. R., Beatty, E. R., Wang, X. and Cummings, J. H. 1995. Selective stimulation of bifidobacteria in the human colon by oligofructose and inulin. *Gastroenterology,* 108, 975–982.

Gibson, G. R., Probert, H. M., Loo, J. V., Rastall, R. A. and Roberfroid, M. B. 2004a. Dietary modulation of the human colonic microbiota: Updating the concept of prebiotics. *Nutrition Research Reviews,* 17, 259–275.

Gibson, G. R., Probert, H. M., van Loo, J., Rastall, R. A. and Roberfroid, M. G. 2004b. Dietary modulation of the human colonic microbiota: Updating the concept of prebiotics. *Nutrition Research Reviews,* 17, 259–257.

Gibson, G. R. and Roberfroid, M. B. 1995. Dietary modulation of the human colonic microbiota: Introducing the concept of prebiotics. *Journal of Nutrition,* 125, 1401–1412.

Gill, H. S., Cross, M. L., Rutherfurd, K. J. and Gopal, P. K. 2001. Dietary probiotic supplementation to enhance cellular immunity in the elderly. *British Journal of Biomedical Science*, 58, 94–96.

Gionchetti, P., Rizzello, F., Venturi, A., Brigidi, P., Matteuzzi, D., Bazzocchi, G., Poggioli, G., Miglioli, M. and Campieri, M. 2000. Oral bacteriotherapy as maintenance treatment in patients with chronic pouchitis: A double-blind, placebo-controlled trial. *Gastroenterology*, 119, 305–309.

Gismondo, M. R., Drago, L. and Lombardi, A. 1999. Review of probiotics available to modify gastrointestinal flora. *International Journal of Antimicrobial Agents*, 12, 287–292.

Gupta, P., Andrew, H., Kirschner, B. S. and Guandalini, S. 2000. Is Lactobacillus GG helpful in children with Crohn's disease? Results of a preliminary, open-label study. *Journal of Pediatric Gastroenterology and Nutrition*, 31, 453–457.

Hamilton-Miller, J. M. T. 2003. The role of probiotics in the treatment and prevention of *Helicobacter pylori* infection. *International Journal of Antimicrobial Agents*, 22, 360.

Heenan, C. N., Adams, M. C., Hosken, R. W. and Fleet, G. H. 2004. Survival and sensory acceptability of probiotic microorganisms in a nonfermented frozen vegetarian dessert. *Lebensmittel-Wissenschaft und-Technologie*, 37, 461–466.

Henry, R. J. 2001. Biotechnology, cereal and cereal products quality. *In:* Owens, G. (ed.) *Cereals processing technology*. Cambridge: Woodhead Publishing Limited.

Heyman, M. and Ménard, S. 2002. Probiotic microorganisms: How they affect intestinal pathophysiology. *Cellular and Molecular Life Sciences*, 59, 1151–1165.

Hilton, E., Isenberg, H. D., Alperstein, P., France, K. and Borenstein, M. T. 1992. Ingestion of yogurt containing Lactobacillus acidophilus as prophylaxis for candidal vaginitis. *Annals of Internal Medicine*, 116, 353–357.

Hilton, E., Kolakowski, P., Singer, C. and Smith, M. 1997. Efficacy of *Lactobacillus* GG as a diarrheal preventive in travelers. *Journal of Travel Medicine*, 4, 41–43.

Hilton, E., Rindos, P. and Isenberg, H. D. 1995. *Lactobacillus* GG vaginal suppositories and vaginitis [2]. *Journal of Clinical Microbiology*, 33, 1433.

Hirayama, M. 2002. Novel physiological functions of oligosaccharides. *Pure and Applied Chemistry*, 74, 1271–1279.

Holzapfel, W. H., Haberer, P., Snel, J., Schillinger, U. and Huis IN'T Veld, J. H. J. 1998. Overview of gut flora and probiotics. *International Journal of Food Microbiology*, 41, 85–101.

Hosoda, M., Hashimoto, H., He, F., Morita, H. and Hosono, A. 1996. Effect of administration of milk fermented with *Lactobacillus acidophilus* LA-2 on fecal mutagenicity and microflora in the human intestine. *Journal of Dairy Science*, 79, 745–749.

Hoy, C. M., Wood, C. M., Hawkey, P. M. and Puntis, J. W. 2000. Duodenal microflora in very-low-birth-weight neonates and relation to necrotizing enterocolitis. *Journal of Clinical Microbiology*, 38, 4539–4547.

Isolauri, E., Arvola, T., Sutas, Y., Moilanen, E. and Salminen, S. 2000. Probiotics in the management of atopic eczema. *Clinical and Experimental Allergy*, 30, 1604–1610.

Isolauri, E., Juntunen, M., Rautanen, T., Sillanaukee, P. and Koivula, T. 1991. A human *Lactobacillus* strain (*Lactobacillus casei* sp strain GG) promotes recovery from acute diarrhea in children. *Pediatrics*, 88, 90–97.

Johansson, M. L., Molin, G., Jeppsson, B., Nobaek, S., Ahrne, S. and Bengmark, S. 1993. Administration of different *Lactobacillus* strains in fermented oatmeal soup: *In vivo* colonization of human intestinal mucosa and effect on the indigenous flora. *Applied and Environmental Microbiology*, 59, 15–20.

Johansson, M. L., Nobaek, S., Berggren, A., Nyman, M., Bengmark, I., Ahrne, S., Jeppsson, B. and Molin, G. 1998. Survival of *Lactobacillus* plantarum DSM 9843 (299v), and effect on the short-chain fatty acid content of faeces after ingestion of a rose-hip drink with fermented oats. *International Journal of Food Microbiology*, 42, 29–38.

Kaila, M., Isolauri, E., Saxelin, M., Arvilommi, H. and Vesikari, T. 1995. Viable versus inactivated *Lactobacillus* strain GG in acute rotavirus diarrhoea. *Archives of Disease in Childhood,* 72, 51–53.

Kalliomaki, M., Salminen, S., Arvilommi, H., Kero, P., Koskinen, P. and Isolauri, E. 2001. Probiotics in primary prevention of atopic disease: A randomised placebo-controlled trial. *Lancet,* 357, 1076–1079.

Kandler, O. 1983. Carbohydrate metabolism in lactic acid bacteria. *Antonie van Leeuwenhoek,* 49, 209–224.

Kedia, G., Vázquez, J. A. and Pandiella, S. S. 2008a. Enzymatic digestion and *in vitro* fermentation of oat fractions by human lactobacillus strains. *Enzyme and Microbial Technology,* 43, 355–361.

Kedia, G., Vázquez, J. A. and Pandiella, S. S. 2008b. Evaluation of the fermentability of oat fractions obtained by debranning using lactic acid bacteria. *Journal of Applied Microbiology,* 105, 1227–1237.

Kedia, G., Vázquez, J. A. and Pandiella, S. S. 2008c. Fermentability of whole oat flour, PeriTec flour and bran by *Lactobacillus plantarum*. *Journal of Food Engineering,* 89, 246.

Kedia, G., Wang, R., Patel, H. and Pandiella, S. S. 2007. Use of mixed cultures for the fermentation of cereal-based substrates with potential probiotic properties. *Process Biochemistry,* 42, 65–70.

Kissmeyer-Nielsen, P., Mortensen, F. V., Laurberg, S. and Hessov, I. 1995. Transmural trophic effect of short chain fatty acid infusions on atrophic, defunctioned rat colon. *Diseases of the Colon and Rectum,* 38, 946–951.

Klaver, F. A. and van der Meer, R. 1993. The assumed assimilation of cholesterol by *Lactobacilli* and *Bifidobacterium bifidum* is due to their bile salt-deconjugating activity. *Applied and Environmental Microbiology,* 59, 1120–1124.

Kolars, J. C., Levitt, M. D., Aouji, M. and Savaiano, D. A. 1984. Yogurt—An autodigesting source of lactose. *New England Journal of Medicine,* 310, 1–3.

Koruda, M. J., Rolandelli, R. H., Settle, R. G., Zimmaro, D. M. and Rombeau, J. L. 1988. Effect of parenteral nutrition supplemented with short-chain fatty acids on adaptation to massive small bowel resection. *Gastroenterology,* 95, 715–720.

Kripke, S. A., De Paula, J. A., Berman, J. M., Fox, A. D., Rombeau, J. L. and Settle, R. G. 1991. Experimental short-bowel syndrome: Effect of an elemental diet supplemented with short-chain triglycerides. *The American Journal of Clinical Nutrition,* 53, 954–962.

Kripke, S. A., Fox, A. D., Berman, J. M., Depaula, J., Birkhahn, R. H., Rombeau, J. L. and Settle, R. G. 1988. Inhibition of TPN-associated intestinal mucosal atrophy with mono-acetoacetin. *Journal of Surgical Research,* 44, 436–444.

Kripke, S. A., Fox, A. D., Berman, J. M., Settle, R. G. and Rombeau, J. L. 1989. Stimulation of intestinal mucosal growth with intracolonic infusion of short-chain fatty acids. *JPEN Journal of Parenteral and Enteral Nutrition,* 13, 109–116.

Lacerda, I. C. A., Miranda, R. L., Borelli, B. M., Nunes, A. C., Nardi, M. R. D., Lachance, M. A. L. and Rosa, C. A. 2005. Lactic acid bacteria and yeasts associated with spontaneous fermentations during the production of sour cassava starch in Brazil. *International Journal of Food Microbiology,* 105, 213–219.

Lee, S. C. and Prosky, L. 1995. International survey on dietary fiber: Definition, analysis, and reference materials. *Journal of AOAC International,* 78, 22–36.

Leroy, F. and De Vuyst, L. 2004. Lactic acid bacteria as functional starter cultures for the food fermentation industry. *Trends in Food Science and Technology,* 15, 67–78.

Lilly, D. M. and Stillwell, R. H. 1965. Probiotics: Growth-promoting factors produced by microorganisms. *Science,* 147, 747–748.

Lin, F.-M., Chiu, C.-H. and Tzu-Ming, P. 2004. Fermentation of a milk–soymilk and Lycium chinense Miller mixture using a new isolate of *Lactobacillus paracasei* subsp. paracasei

NTU101 and *Bifidobacterium longum*. *Journal of Industrial Microbiology and Biotechnology*, 31, 559–564.

Link-Amster, H., Rochat, F., Saudan, K. Y., Mignot, O. and Aeschlimann, J. M. 1994. Modulation of a specific humoral immune response and changes in intestinal flora mediated through fermented milk intake. *FEMS Immunology and Medical Microbiology*, 10, 55–63.

Liong, M. T. 2008. Roles of probiotics and prebiotics in colon cancer prevention: Postulated mechanisms and in-vivo evidence. *International Journal of Molecular Sciences*, 9, 854–863.

Liu, S.-Q. 2003 Practical implications of lactate and pyruvate metabolism by lactic acid bacteria in food and beverage fermentations. *International Journal of Food Microbiology*, 83, 115–131.

Livesey, G., Wilkinson, J. A., Roe, M., Faulks, R., Clark, S., Brown, J. C., Kennedy, H. and Elia, M. 1995. Influence of the physical form of barley grain on the digestion of its starch in the human small intestine and implications for health. *The American Journal of Clinical Nutrition*, 61, 75–81.

Luckow, T. and Delahunty, C. 2004a. Consumer acceptance of orange juice containing functional ingredients. *Food Research International*, 37, 805–814.

Luckow, T. and Delahunty, C. 2004b. Which juice is "healthier"? A consumer study of probiotic non-dairy juice drinks. *Food Quality and Preference*, 15, 751–759.

Luckow, T., Sheehan, V., Delahunty, C. and Fitzgerald, G. 2005. Determining the odor and flavor characteristics of probiotic, health-promoting ingredients and the effects of repeated exposure on consumer acceptance. *Journal of Food Science*, 70, S53–S59.

Luckow, T., Sheehan, V., Fitzgerald, G. and Delahunty, C. 2006. Exposure, health information and flavour-masking strategies for improving the sensory quality of probiotic juice. *Appetite*, 47, 315–323.

Macedo, R. F., Freitas, R. J. S., Pandey, A. and Soccol, C. R. 1999. Production and shelf-life studies of low cost beverage with soymilk, buffalo cheese whey and cow milk fermented by mixed cultures of *Lactobacillus casei* ssp. *shirota* and *Bifidobacterium adolescentis*. *Journal of Basic Microbiology*, 39, 243–251.

Macfarlane, G. T. 1991. Fermentation reactions in the large intestine. *Tenth Ross Conference on Medical Research*.

Macfarlane, G. T. and Gibson, G. R. 1991. Co-utilization of polymerized carbon sources by *Bacteroides ovatus* grown in a two-stage continuous culture system. *Applied and Environmental Microbiology*, 57, 1–6.

Macfarlane, G. T., Gibson, G. R. and Cummings, J. H. 1992. Comparison of fermentation reactions in different regions of the human colon. *Journal of Applied Bacteriology*, 72, 57–64.

Majamaa, H. and Isolauri, E. 1997. Probiotics: A novel approach in the management of food allergy. *Journal of Allergy and Clinical Immunology*, 99, 179–185.

Majamaa, H., Isolauri, E., Saxelin, M. and Vesikari, T. 1995. Lactic acid bacteria in the treatment of acute rotavirus gastroenteritis. *Journal of Pediatric Gastroenterology and Nutrition*, 20, 333–338.

Marteau, P., Flourie, B., Pochart, P., Chastang, C., Desjeux, J. F. and Rambaud, J. C. 1990. Effect of the microbial lactase (EC 3.2.1.23) activity in yoghurt on the intestinal absorption of lactose: An *in vivo* study in lactase-deficient humans. *British Journal of Nutrition*, 64, 71–79.

Martin-Diana, A. B., Janer, C., Pellaez, C. and Requena, T. 2003. Development of a fermented goat's milk containing probiotic bacteria. *International Dairy Journal*, 13, 827–833.

Matsuzaki, T. and Chin, J. 2000. Modulating immune responses with probiotic bacteria. *Immunology and Cell Biology*, 78, 67–73.

Mcdougall, G. J., Morrison, I. M., Stewart, D. and Hillman, J. R. 1996. Plant cell walls as dietary fibre: Range, structure, processing and function. *Journal of the Science of Food and Agriculture*, 70, 133–150.

Mcfarland, L. V. 2007. Meta-analysis of probiotics for the prevention of traveler's diarrhea. *Travel Medicine and Infectious Disease*, 5, 97.

Mcneil, N. I., Cummings, J. H. and James, W. P. 1978. Short chain fatty acid absorption by the human large intestine. *Gut*, 19, 819–822.

Michetti, P., Dorta, G., Wiesel, P. H., Brassart, D., Verdu, E., Herranz, M., Felley, C., Porta, N., Rouvet, M., Blum, A. L. and Corthésy-Theulaz, I. 1999. Effect of whey-based culture supernatant of *Lactobacillus acidophilus* (*johnsonii*) La1 on *Helicobacter pylori* infection in humans. *Digestion*, 60, 203–209.

Michida, H., Tamalampudi, S., Pandiella, S. S., Webb, C., Fukuda, H. and Kondo, A. 2006. Effect of cereal extracts and cereal fiber on viability of *Lactobacillus plantarum* under gastrointestinal tract conditions. *Biochemical Engineering Journal*, 28, 73.

Midolo, P. D., Lambert, J. R., Hull, R., Luo, F. and Grayson, M. L. 1995. *In vitro* inhibition of *Helicobacter pylori* NCTC 11637 by organic acids and lactic acid bacteria. *Journal of Applied Bacteriology*, 79, 475–479.

Midtvedt, A. C. and Midtvedt, T. 1992. Production of short chain fatty acids by the intestinal microflora during the first 2 years of human life. *Journal of Pediatric Gastroenterology and Nutrition*, 15, 395–403.

Mitsuoka, T. and Kaneuchi, C. 1977. Ecology of the bifidobacteria. *The American Journal of Clinical Nutrition*, 30, 1799–1810.

Modzelewska-Kapituła, M., Kłębukowska, L. and Kornacki, K. 2008. Evaluation of the possible use of potentially probiotic *Lactobacillus* strains in dairy products. *International Journal of Dairy Technology*, 61, 165–169.

Nilsson, U. and Bjorck, I. 1988. Availability of cereal fructans and inulin in the rat intestinal tract. *Journal of Nutrition*, 118, 1482–1486.

Noh, D. O., Kim, S. H. and Gilliland, S. E. 1997. Incorporation of cholesterol into the cellular membrane of *Lactobacillus acidophilus* ATCC 43121. *Journal of Dairy Science*, 80, 3107–3113.

Nousiainen, J. and Setälä, J. 1998. Lactic acid bacteria as animal probiotics. In: Salminen, S. and von Wright, A. (eds.) *Lactic Acid Bacteria: Microbiology and Functional Aspects*. New York, USA: Marcel Dekker Inc.

Nout, M. J. R. 1992. Accelerated natural lactic fermentation of cereal-based formulas at reduced water activity. *International Journal of Food Microbiology*, 16, 313–322.

Oatley, J. T., Rarick, M. D., Ji, G. E. and Linz, J. E. 2000. Binding of aflatoxin B1 to bifidobacteria in vitro. *Journal of Food Protection*, 63, 1133–1136.

Okawa, T., Niibe, H., Arai, T., Sekiba, K., Noda, K., Takeuchi, S., Hashimoto, S. and Ogawa, N. 1993. Effect of LC9018 combined with radiation therapy on carcinoma of the uterine cervix: A phase III, multicenter, randomized, controlled study. *Cancer*, 72, 1949–1954.

Oksanen, P. J., Salminen, S., Saxelin, M., Hamalainen, P., Ihantola-Vormisto, A., Muurasniemi-Isoviita, L., Nikkari, S., Oksanen, T., Porsti, I., Salminen, E., Siitonen, S., Stuckey, H., Toppila, A. and Vapaatalo, H. 1990. Prevention of travellers' diarrhoea by *Lactobacillus* GG. *Annals of Medicine*, 22, 53–56.

Ouwehand, A. C., Kirjavainen, P. V., Shortt, C. and Salminen, S. 1999. Probiotics: Mechanisms and established effects. *International Dairy Journal*, 9, 43–52.

Ouwehand, A., Salminen, S. and Isolauri, E. 2002. Probiotics: An overview of beneficial effects. *Antonie van Leeuwenhoek*, 82, 82, 279–289.

Oxman, T., Shapira, M., Klein, R., Avazov, N. and Rabinowitz, B. 2001. Oral administration of *Lactobacillus* induces cardioprotection. *Journal of Alternative and Complementary Medicine*, 7, 345–354.

Palframan, R., Gibson, G. R. and Rastall, R. A. 2003. Development of a quantitative tool for the comparison of the prebiotic effect of dietary oligosaccharides. *Letters in Applied Microbiology,* 37, 281–284.

Parker, R. B. 1974. Probiotics: The other half of the antibiotic story. *Animal and Nutrition Health,* 29, 4–8.

Patel, H. M., Pandiella, S. S., Wang, R. H. and Webb, C. 2004a. Influence of malt, wheat, and barley extracts on the bile tolerance of selected strains of lactobacilli. *Food Microbiology,* 21, 83.

Patel, H. M., Wang, R., Chandrashekar, O., Pandiella, S. S. and Webb, C. 2004b. Proliferation of *Lactobacillus* plantarum in solid-state fermentation of oats. *Biotechnology Progress,* 20, 110–116.

Pedone, C. A., Bernabeu, A. O., Postaire, E. R., Bouley, C. F. and Reinert, P. 1999. The effect of supplementation with milk fermented by *Lactobacillus casei* (strain DN-114 001) on acute diarrhoea in children attending day care centres. *International Journal of Clinical Practice,* 53, 179–184.

Pelletier, J., Faurie, J., François, A. and Teissier, P. 2007. Lait fermenté: la technologie au service du goût. [Fermented milk: Technology with the taste service.] *Cahiers de Nutrition et de Diététique,* 42, 15–20.

Pelto, L., Isolauri, E., Lillus, E. M., Nuutila, J. and Salminen, S. 1998. Probiotic bacteria down-regulate the milk-induced inflammatory response in milk-hypersensitive subjects but have an immunostimulatory effect in healthy subjects. *Clinical and Experimental Allergy,* 28, 1474–1479.

Perdigon, G., Vintini, E., Alvarez, S., Medina, M. and Medici, M. 1999. Study of the possible mechanisms involved in the mucosal immune system activation by lactic acid bacteria. *Journal of Dairy Science,* 82, 1108–1114.

Plessas, S., Bekatorou, A., Kanellaki, M., Psarianos, C. and Koutinas, A. 2005. Cells immobilized in a starch–gluten–milk matrix usable for food production. *Food Chemistry,* 89, 175–179.

Prado, F. C., Parada, J. L., Pandey, A. and Soccol, C. R. 2008. Trends in non-dairy probiotic beverages. *Food Research International,* 41, 111–123.

Puupponen-Pimiä, R., Aura, A.-M., Oksman-Caldentey, K.-M., Myllärinen, P., Saarela, M., Mattila-Sandholm, T. and Poutanen, K. 2002. Development of functional ingredients for gut health. *Trends in Food Science and Technology,* 13, 3–11.

Rafter, J. 2003. Probiotics and colon cancer. *Best Practice and Research Clinical Gastroenterology,* 17, 849.

Reid, G. 2006. Prevention and treatment of urogenital infections and complications: Lactobacilli's multi-pronged effects. *Microbial Ecology in Health and Disease,* 18, 181–186.

Reid, G., Beuerman, D., Heinemann, C. and Bruce, A. W. 2001. Probiotic *Lactobacillus* dose required to restore and maintain a normal vaginal flora. *FEMS Immunology and Medical Microbiology,* 32, 37–41.

Reimer, R. A. and Mcburney, M. I. 1996. Dietary fiber modulates intestinal proglucagon messenger ribonucleic acid and postprandial secretion of glucagon-like peptide-1 and insulin in rats. *Endocrinology,* 137, 3948–3956.

Roediger, W. E. 1982. Utilization of nutrients by isolated epithelial cells of the rat colon. *Gastroenterology,* 83, 424–429.

Roediger, W. E. W. and Moore, A. 1981. Effect of short-chain fatty acid on sodium absorption in isolated human colon perfused through the vascular bed. *Digestive Diseases and Sciences,* 26, 100–106.

Rogers, P. 1986. Genetics and biochemistry of *Clostridium* relevant to development of fermentation processes. *Advances in Applied Microbiology,* 21, 1.

Rolfe, R. D. 2000. The role of probiotic cultures in the control of gastrointestinal health. *Journal of Nutrition,* 130, 396S–402S.

Rotimi, V. O. and Duerden, B. I. 1981. The development of the bacterial flora in normal neonates. *Journal of Medical Microbiology,* 14, 51–62.

Rozada-Sánchez, R., Sattur, A. P., Thomas, K. and Pandiella, S. S. 2008. Evaluation of *Bifidobacterium* spp. for the production of a potentially probiotic malt-based beverage. *Process Biochemistry,* 43, 848–854.

Rubaltelli, F. F., Biadaioli, R., Pecile, P. and Nicoletti, P. 1998. Intestinal flora in breast- and bottle-fed infants. *Journal of Perinatal Medicine,* 26, 186–191.

Ruppin, H., Bar-Meir, S. and Soergel, K. H. 1980. Absorption of short-chain fatty acids by the colon. *Gastroenterology,* 78, 1500–1507.

Saarela, M., Mogensen, G., Fodén, R., Mättö, J. and Mattila-Sandholm, T. 2000. Probiotic bacteria: Safety, functional and technological properties. *Journal of Biotechnology,* 84, 197–215.

Saavedra, J. M., Bauman, N. A., Oung, I., Perman, J. A. and Yolken, R. H. 1994. Feeding of *Bifidobacterium bifidum* and *Streptococcus thermophilus* to infants in hospital for prevention of diarrhoea and shedding of rotavirus. *Lancet,* 344, 1046–1049.

Said, H. M. and Mohamed, Z. M. 2006. Intestinal absorption of water soluble vitamins: An update. *Current Opinion in Gastroenterology,* 26, 140–146.

Sakata, T. 1987. Stimulatory effect of short-chain fatty acids on epithelial cell proliferation in the rat intestine: A possible explanation for trophic effects of fermentable fibre, gut microbes and luminal trophic factors. *British Journal of Nutrition,* 58, 95–103.

Sakata, T. 1989. Stimulatory effect of short-chain fatty acids on epithelial cell proliferation of isolated and denervated jejunal segment of the rat. *Scandinavian Journal of Gastroenterology,* 24, 886–890.

Sakata, T. 1997. Influence of short chain fatty acids on intestinal growth and functions. *Advances in Experimental Medicine and Biology,* 427, 191–199.

Salim UR, R., Paterson, A. and Piggott, J. R. 2006. Flavour in sourdough breads: A review. *Trends in Food Science and Technology,* 17, 557–566.

Salminen, S., Bouley, C., Boutron-Ruault, M. C., Cummings, J. H., Franck, A., Gibson, G. R., Isolauri, E., Moreau, M. C., Roberfroid, M. and Rowland, I. 1998a. Functional food science and gastrointestinal physiology and function. *British Journal of Nutrition,* Suppl. I, S147–S171.

Salminen, S., Ouwehand, A., Benno, Y. and Lee, Y. K. 1999. Probiotics: How should they be defined? *Trends in Food Science and Technology,* 10, 107.

Salminen, S., von Wright, A., Morelli, L., Marteau, P., Brassart, D., de Vos, W. M., Fonden, R., Saxelin, M., Collins, K., Mogensen, G., Birkeland, S.-E. and Mattila-Sandholm, T. 1998b. Demonstration of safety of probiotics—A review. *International Journal of Food Microbiology,* 44, 93.

Sanders, M. E. 1998. Overview of functional foods: Emphasis on probiotic bacteria. *International Dairy Journal,* 8, 341–347.

Sartor, R. B. 2004. Therapeutic manipulation of the enteric microflora in inflammatory bowel diseases: Antibiotics, probiotics, and prebiotics. *Gastroenterology,* 126, 1620.

Savage, D. C. 1986. Gastrointestinal microflora in mammalian nutrition. *Annual Review of Nutrition,* 6, 155–178.

Schiffrin, E. J., Rochat, F., Link-Amster, H., Aeschlimann, J. M. and Donnet-Hughes, A. 1995. Immunomodulation of human blood cells following the ingestion of lactic acid bacteria. *Journal of Dairy Science,* 78, 491–497.

Shanahan, F. 2000. Probiotics and inflammatory bowel disease: Is there a scientific rationale? *Inflammatory Bowel Diseases,* 6, 107–115.

Sheih, Y. H., Chiang, B. L., Wang, L. H., Liao, C. K. and Gill, H. S. 2001. Systemic immunity-enhancing effects in healthy subjects following dietary consumption of the lactic acid bacterium *Lactobacillus rhamnosus* HN001. *Journal of the American College of Nutrition,* 20, 149–156.

Shornikova, A. V., Casas, I. A., Isolauri, E., Mykkanen, H. and Vesikari, T. 1997. *Lactobacillus reuteri* as a therapeutic agent in acute diarrhea in young children. *Journal of Pediatric Gastroenterology and Nutrition,* 24, 399–404.

Sieber, R. and Dietz, U. T. 1998. *Lactobacillus acidophilus* and yogurt in the prevention and therapy of bacterial vaginosis. *International Dairy Journal,* 8, 599–607.

Siigur, U., Ormisson, A. and Tamm, A. 1993. Faecal short-chain fatty acids in breast-fed and bottle-fed infants. *Acta Paediatrica,* 82, 536–538.

Siró, I., Kápolna, E., Kápolna, B. and Lugasi, A. 2008. Functional food. Product development, marketing and consumer acceptance—A review. *Appetite,* 51, 456–467.

Smit, G., Smit, B. A. and Engels, W. J. M. 2005. Flavour formation by lactic acid bacteria and biochemical flavour profiling of cheese products. *Federation of European Microbiological Societies (FEMS) Microbiology Reviews,* 29, 591–610.

Smit, G., van Hylckama Vlieg, J. E. T., Smit, B. A., Ayad, E. H. E. and Engels, W. J. M. 2002. Fermentative formation of flavour compounds by lactic acid bacteria. *Australian Journal of Dairy Technology,* 57, 61–68.

Southon, S., Gee, J. M. and Johnson, I. T. 1987. The effect of dietary protein source and guar gum on gastrointestinal growth and enteroglucagon secretion in the rat. *Britrish Journal of Nutrition,* 58, 65–72.

Spanhaak, S., Havenaar, R. and Schaafsma, G. 1998. The effect of consumption of milk fermented by *Lactobacillus casei* strain *Shirota* on the intestinal microflora and immune parameters in humans. *European Journal of Clinical Nutrition,* 52, 899–907.

Stanton, C., Gardiner, G., Meehan, H., Collins, K., Fitzgerald, G., Lynch, P. B. and Ross, R. P. 2001. Market potential for probiotics. *American Journal of Clinical Nutrition,* 73, 476S–483S.

Stanton, C., Ross, R. P., Fitzgerald, G. F. and van Sinderen, D. 2005. Fermented functional foods based on probiotics and their biogenic metabolites. *Current Opinion in Biotechnology,* 16, 198–203.

Stephen, A. M. and Cummings, J. H. 1980. The microbial contribution to human faecal mass. *Journal of Medical Microbiology,* 13, 45–56.

Szajewska, H., Kotowska, M., Mrukowicz, J. Z., Armaìnska, M. and Mikolajczyk, W. 2001. Efficacy of *Lactobacillus* GG in prevention of nosocomial diarrhea in infants. *Journal of Pediatrics,* 138, 361–365.

Tannock, G. W. 1998. Studies of the intestinal microflora: A prerequisite for the development of probiotics. *International Dairy Journal,* 8, 527–533.

Tappenden, K. A., Drozdowski, L. A., Thomson, A. B. and Mcburney, M. I. 1998. Short-chain fatty acid-supplemented total parenteral nutrition alters intestinal structure, glucose transporter 2 (GLUT2) mRNA and protein, and proglucagon mRNA abundance in normal rats. *The American Journal of Clinical Nutrition,* 68, 118–125.

Tappenden, K. A., Thomson, A. B., Wild, G. E. and Mcburney, M. I. 1996. Short-chain fatty acids increase proglucagon and ornithine decarboxylase messenger RNAs after intestinal resection in rats. *JPEN Journal of Parenteral and Enteral Nutrition,* 20, 357–362.

Teitelbaum, J. E. and Walker, A. W. 2002. Nutritional impact of pre- and probiotics as protective gastrointestinal organisms. *Annual Review of Nutrition,* 22, 107–138.

Titgemeyer, E. C., Bourquin, L. D., Fahey, G. C., JR. and Garleb, K. A. 1991. Fermentability of various fiber sources by human fecal bacteria in vitro. *The American Journal of Clinical Nutrition,* 53, 1418–1424.

Topping, D. L. and Clifton, P. M. 2001. Short-chain fatty acids and human colonic function: Roles of resistant starch and nonstarch polysaccharides. *Physiological Reviews,* 81, 1031–1064.

Tuohy, K. M., Probert, H. M., Smejkal, C. W. and Gibson, G. R. 2003. Using probiotics and prebiotics to improve gut health. *Therapeutic Focus,* 15, 692–700.

Tuorila, H. and Cardello, A. V. 2002. Consumer response to an off-flavour in juice in the presence of specific health claims. *Food Quality and Preference,* 13, 561–569.

Urbach, G. 1993. Relations between cheese flavour and chemical composition. *International Dairy Journal,* 3, 389–422.

van Loo, J., Coussement, P., Leenheer, L., Hoebregs, H. and Smits, G. 1995. On the presence of inulin and oligofructose as natural ingredients in the western diet. *Critical Reviews in Food Science and Nutrition,* 35, 525–552.

Vesa, T. H., Marteau, P., Zidi, S., Briet, F., Pochart, P. and Rambaud, J. C. 1996. Digestion and tolerance of lactose from yoghurt and different semi-solid fermented dairy products containing *Lactobacillus acidophilus* and bifidobacteria in lactose maldigesters—Is bacterial lactase important? *European Journal of Clinical Nutrition,* 50, 730–733.

Yoon, K. Y., Woodams, E. E., Hang, Y. D. and Ziemer, C. J. 2006. Production of probiotic cabbage juice by lactic acid bacteria. *Bioresource Technology,* 97, 1427–1430.

Yoshioka, H., Iseki, K. and Fujita, K. 1983. Development and differences of intestinal flora in the neonatal period in breast-fed and bottle-fed infants. *Pediatrics,* 72, 317–321.

Zubillaga, M., Weill, R., Postaire, E., Goldman, C., Caro, R. and Boccio, J. 2001. Effect of probiotics and functional foods and their use in different diseases. *Nutrition Research,* 21, 569–579.

3 Production and Food Applications of Microbial Biopolymers

Filomena Freitas, Vitor D. Alves, Isabel Coelhoso, and Maria A.M. Reis

CONTENTS

3.1 INTRODUCTION

Microbial biopolymers are polymeric substances synthesized by microorganisms. They include polysaccharides, polyamides, polyesters, and polyanhydrides (Rehm, 2010). Depending on their composition and molecular weight, microbial polymers have properties that range from rheology modifiers of aqueous systems to bioplastics, which makes them useful in many industrial applications (e.g., agro-food, cosmetics, pharmaceuticals, textile, paper, and oil recovery) (Moreno et al., 1998).

Many microbial biopolymers have been described in the last few decades having promising functional properties potentially suitable for different applications. However, only a few have found widespread industrial use. An even more limited number was granted permission for food applications, mostly water-soluble microbial polysaccharides (e.g., xanthan gum, gellan gum, and pullulan) that are used as thickening, stabilizing, or gelling agents in food systems or processes (Sworm, 2007). Microbial biopolymers (e.g., pullulan, gellan) are also emerging as materials for natural edible coatings and as encapsulating agents for the delivery of flavors and bioactive ingredients in food systems (Azarakhsh et al., 2012; Nedovic et al., 2011). Another promising area of application is the use of some microbial biopolymers, such as polyhydroxyalkanoates (PHAs), as biodegradable packaging materials for food products (Siracusa et al., 2008).

Hence, microbial biopolymers can be advantageously used as alternatives to other natural biopolymers, such as plant- or algae-derived products, and petrochemical-based products, due to their novel or improved properties. Moreover, these polymers offer the advantage of being produced from renewable resources, under controlled environmental conditions that assure both the quantity and the quality of the final products, and can be produced with properties tailored for specific applications. However, their wide spread use in food applications has been hindered by several factors, mainly their production costs that are higher than for most of the traditional polymers, the restrictions related to their approval for use in food products or processes, and finally, public acceptance that is not always easily achieved. Unless the biopolymer is a generally recognized as safe (GRAS status) material, it must be evaluated by the certified authorities (e.g., U.S. Food and Drug Administration, EFSA—European Food Safety Authority) to determine its safety under the conditions of the intended uses as well as specifications, including purity and physical properties, and limitations in the conditions of use. The components of food-contact materials, including food packaging and processing equipment, must also be evaluated to guarantee that they comply with the regulatory specifications and limitations.

The natural biopolymers' market is dominated by plant- and animal-derived products, mostly starch and gelatin, with annual productions of 1.7–1.8 million metric tons in 2007–2008 (Rehm, 2010). In terms of market value, xanthan gum is the only significant microbial biopolymer, representing around 6% of the total hydrocolloids' market, with an annual production of 100,000 tons in 2008 (Seisun, 2010).

This chapter outlines the main aspects of the biotechnological production of microbial biopolymers, focusing on the industrial bioprocesses used for the production of food-grade biopolymers and their main current applications in food products or processes. The limitations faced by such bioprocesses and their prospects in the rise of innovative products and/or processes are also addressed.

3.2 MICROBIAL BIOPOLYMERS FOR FOOD APPLICATIONS

The microbial polymers that have been proposed for food applications as food additives, processing aids, or packaging/coating materials for food products, include polysaccharides and polyesters. These biopolymers are characterized by different molecular structures (Figures 3.1 and 3.2) that result in a wide range of functional properties valuable in many areas of application, including food products and processes.

3.2.1 POLYSACCHARIDES

The basic structure of microbial polysaccharides is formed by carbohydrates. The most common monosaccharide constituents of microbial polysaccharides are D-glucose, D-galactose, and D-mannose. Glucuronic acid is also found in many

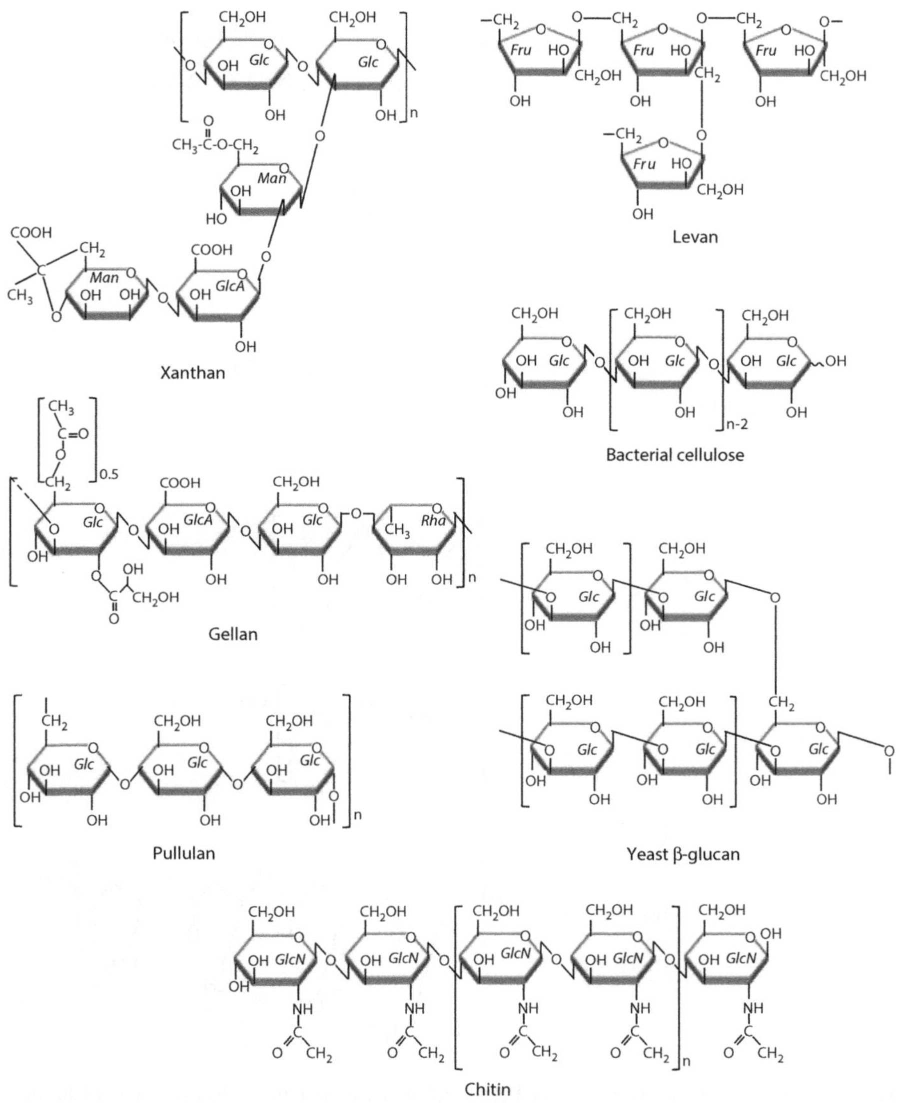

FIGURE 3.1 Chemical structures of some microbial polysaccharides.

polysaccharides (e.g., xanthan, gellan), conferring on them a polyanionic character. Although less common, other uronic acids, such as galacturonic, mannuronic and guluronic acids, as well as some *N*-acetylamino sugars, such as *N*-acetylglucosamine and *N*-acetylgalactosamine, may be found in some polymers (e.g., bacterial alginate, chitin–glucan complex). All these possible sugar components may be linked through different glycosyl linkages and present different configurations, resulting in a wide range of possible molecular structures (Figure 3.1).

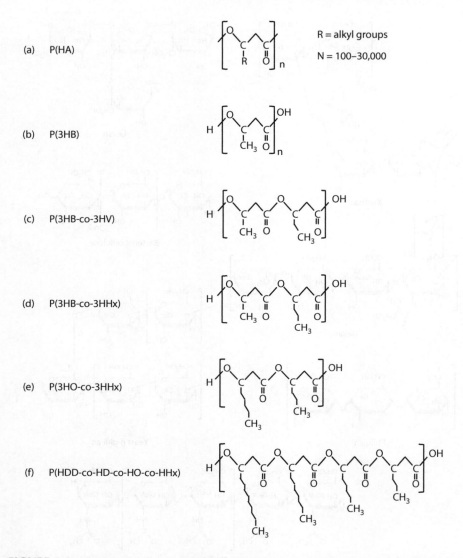

FIGURE 3.2 Chemical structures of PHAs: (a) general structure, (b) examples of homo-polymers, (c, d, e) copolymers, and (f) terpolymers, with different monomer chain lengths.

Microbial polysaccharides may also contain several organic and inorganic sub-stituents, ester-linked substituents (acetyl, glyceryl, succinyl, amino acids), and pyruvate ketals. The presence of some of these acyl groups confers the polymers a polyelectrolyte character, increases their lipophilicity, and affects the capacity to interact with other polymers and with cationic/anionic materials.

Most microbial polysaccharides are composed of repeating units that vary in size from disaccharides to octasaccharides. Polysaccharides of commercial inter-est include both homopolymers (e.g., pullulan) and heteropolymers (e.g., xanthan, gellan). Homopolysaccharides may be linear or branched molecules, whereas heteropolysaccharides frequently contain short side-chains, varying from one to four sugars in length (Figure 3.1). Some bacterial alginates constitute an exception, as they have irregular structures.

The functional properties of microbial polysaccharides are determined not only by their chemical composition, but also by their molecular structure, average molecular weight, and polydispersity. This molecular structure variability is translated into a wide range of different properties (Table 3.1) that offer interesting opportunities for the production of tailor-made polymers with desired properties.

Xanthan gum is the most widely accepted commercial microbial polysaccharide for both food and nonfood applications (Rottava et al., 2009). Other microbial polysaccharides that have been granted food-grade status are currently being commercially exploited in food applications, including gellan, pullulan, and chitin–glucan complex (CGC) (Table 3.1).

Xanthan is a heteropolysaccharide composed of a glucose backbone linked through β-1,4 glycosidic linkages, with trisaccharide side chains on every alternate glucose residue (Figure 3.1). The side chains contain a glucuronic acid residue between two mannose residues. About one-half of the terminal mannose residues are linked to a pyruvil residue, whereas some of the nonterminal mannose is linked to acetyl residues (Becker and Vorholter, 2009; García-Ochoa et al., 2000; Sworm, 2010). The level of acyl group substituents in the molecule are highly dependent on the production conditions and have an impact on the polymer's rheological properties (García-Ochoa et al., 2000). Gellan is a heteropolysaccharide composed of a tetrasaccharide backbone with L-rhamnose, D-glucose, D-glucuronic acid, and D-glucose monomers and side chains of acetyl and glyceryl substituents (Figure 3.1) (Coleman et al., 2008). Both substituents are located on the same glucose residue, and on an average, there is one glycerate per repeat and one acetate per every two repeats. Pullulan is a linear glucan consisting of repeating units of maltotriose joined by α-D-(1 → 6) linkages (Figure 3.1) (Chaen, 2010). CGC is a natural material composed of two types of biopolymers, namely, chitin, a polymer of N-acetylglucosamine units, covalently linked to β-1,3-glucans, a polymer of glucose units (Figure 3.1), resulting in a water- and alkaline-insoluble complex (Roca et al., 2012). Xanthan, gellan, and pullulan are high-molecular-weight polysaccharides ($>1.0 \times 10^6$) (Table 3.1), which greatly contribute by their improved rheological properties (Freitas et al., 2011a; Kim et al., 2000), whereas CGC is characterized by lower average molecular weights ($10^4–10^5$) (Machova et al., 1999).

Other microbial polysaccharides that have been extensively studied but still have not reached full industrial development include (Figure 3.1): bacterial cellulose, a glucose homopolysaccharide, with β-(1,4) linkages (Chawla et al., 2009), bacterial alginates, linear acetylated polysaccharides composed of two uronic acids, namely mannuronic and guluronic acids that form "block structures" of poly-D-mannuronic acid sequences, poly-L-guluronic acid sequences, and mixed sequences (Ruffing and Chen, 2006); levan, a highly branched fructose homopolysaccharide (Kang et al., 2009); succinoglycan, a branched polysaccharide with a backbone composed of glucose and galactose; and tetrasaccharide side chains composed of glucose residues (Simsek et al., 2009).

TABLE 3.1

Physical–Chemical Properties and Food Applications of the Most Widely Used Food-Grade Commercial Microbial Polysaccharides

Polysaccharide	Composition	Molecular Weight	Main Properties	Main Food Applications	References
Xanthan gum	Glucose Mannose Glucuronic acid Acetate Pyruvate	$(2.0–50) \times 10^6$	Water soluble High-viscosity yield at low shear rates even at low concentrations Stability over wide temperature, pH, and salt-concentration ranges Gel formation when blended with locust bean gum and guar gum	Good suspending agent Stabilizing agent for oil/water emulsions Thickening agent in bakery and confectionery products, syrups, ice creams, frozen foods, and dairy products	Freitas et al. (2011b), Sworn (2007), Khan et al. (2007), Becker and Vorholter (2009)
Gellan gum	Glucose Glucuronic acid Rhamnose Acetate Glycerate	5.0×10^5	Water soluble Stability over wide pH range Gelling capacity at low concentration The acylated form produces thermoreversible elastic gels, whereas the deacylated form produces heat-stable brittle gels Fluid gels Transparent gels with good flavor release	Suspending agent and texture modifier Improves physical stability, flavor release, and water-holding capacity of the products Water-based gels (dessert gels, Asiatic gelled products, and aspics) Enhancing mouthfeel in beverages, dairy products, sauces, and confectionery Replacement of gelatin in marshmallows and sweets	Bajaj et al. (2007), Freitas et al. (2011b), Khan et al. (2007), Morris et al. (2012)
Pullulan	Glucose	$10^5–10^6$	Water soluble Nonionic Film former Low-viscosity solutions Stability to pH and sodium chloride Bioadhesive Dietary fiber	Printable edible films with low oxygen permeability for confectionery decoration Carrier system for flavors, colors, and active ingredients Coating of food products Thickener in beverages, creams, icings, and sauces Emulsion stabilizer in dressings and seasonings Low-caloric food ingredient	Chaen (2010), Khan et al. (2007), Kim et al. (2000)
Chitin–glucan	Glucose Glucosamine	$10^4–10^5$	Water insoluble Insoluble in most organic solvents	Functional fiber Wine clarification	Machova et al. (1999)

Additionally, over the last decades, a large number of new microbial polysaccharides have been reported having interesting properties that might render them suitable for many different areas of application, including food products and processes. GalactoPol and FucoPol are examples of two recently reported bacterial extracellular polysaccharides with interesting rheological and film-forming properties (Freitas et al., 2009a, 2011a). Both polymers are high-molecular-weight $(1.0–5.0 \times 10^7)$ heteropolysaccharides with distinct sugar monomer and acyl groups composition.

GalactoPol is mainly composed of galactose (60–90% mol), with lower amounts of mannose (14–23% mol), glucose (7–17% mol), and rhamnose (1–4% mol) (Freitas et al., 2009a; Reis et al., 2008). FucoPol is composed of fucose, galactose, and glucose in a nearly equimolar proportion, with lower amounts of glucuronic acid (~10% mol) (Reis et al., 2011a; Torres et al., 2012). Both polymers have acyl group substituents, namely, acetyl, pyruvil, and succinyl, but their contents are higher in FucoPol (10–20 wt% of the polymer's dry mass) (Alves et al., 2010; Torres et al., 2011) than in GalactoPol (3–5 wt%) (Freitas et al., 2009a,b). The presence of higher pyruvate and succinate contents, together with the presence of glucuronic acid, confer FucoPol an increased anionic character. Furthermore, this biopolymer is rich in fucose, a rare sugar, present, for example, in human-milk oligosaccharides, which renders it an enormous additional potential for use in different areas of application (Vanhooren and Vandamme, 1999).

3.2.2 Polyesters

PHAs are polyoxoesters of hydroxyalkanoic acids (HAs) (Koller et al., 2010). More than 150 PHA-building blocks have been identified, including short-chain-length monomers (scl), containing 3–5 carbon atoms (e.g., hydroxybutyrate [HB], hydroxyvalerate [HV]) and medium-chain length (mcl), containing 6–14 carbon atoms (e.g., HHX, HO, and HD) (Chan et al., 2006; Koller et al., 2010) (Figure 3.2). These monomers can have aliphatic, saturated, unsaturated, straight, or branched side chains. Some microorganisms can synthesize PHAs with aromatic, halogenic, pseudohalogenic, or alkoxy groups (Koller et al., 2010). All these wide diversities of possible HA monomers, arranged into different molecular structures (homopolyesters, copolyesters, and terpolyesters) result in biopolymers presenting thermoplastic and/or elastomeric properties. Additionally, the polymers' molecular weight will also modulate their properties.

3.3 PRODUCTION OF MICROBIAL BIOPOLYMERS

Biopolymers are naturally synthesized by microorganisms with different functions in the microbial cell, including intracellular carbon or energy storage reserves (e.g., glycogen, polyesters, cyanophycin, and polyphosphate), structural cell-wall components (e.g., chitin, β-glucans), and extracellular biopolymers (e.g., exopolysaccharides, ε-poly-L-lysine, and poly-γ-glutamate) (Rehm, 2010), often secreted as protective mechanisms in response to environmental conditions.

3.3.1 Production Bioprocesses

Industrially, most microbial biopolymers are produced in single-strain systems. The bioprocesses for their production usually occur under aerobic conditions. However, some microorganisms require maximal aeration for optimal production (e.g., xanthan), whereas others maximize polymer synthesis under microaerophilic conditions (e.g., bacterial alginate) (Freitas et al., 2011b). Batch or fed-batch modes are the most commonly used. Although continuous production could allow for high and constant productivities with little variation on product yield and quality, it is seldom employed because of the difficulty in maintaining strain stability and the risk of contamination (Rehm, 2010).

Xanthan, the most widely used food-grade microbial polymer, is synthesized by the bacterium *Xanthomonas campestris* using sucrose or glucose as feedstocks in the industrial process. *X. campestris* is a very efficient polysaccharide producer because it is able to reach high productivities and substrate conversion into biopolymers (Table 3.2). Nowadays, xanthan's major producers include CP Kelco, Merck, Pfizer, Rhone Poulenc, Sanofi-Elf, and Jungbunzlauer and the polymer is marketed with different purity grades for applications ranging from food to personal-care products and oil recovery.

The bacterial strain used for industrial production of gellan gum is *Sphingomonas paucimobilis* ATCC 31461 (Bajaj et al., 2007). Simple sugars are the most common feedstocks. Compared to xanthan, lower productivities and polymer yields on the substrate are obtained (Table 3.2), which contribute for higher production costs. Nevertheless, gellan gum has received approval for use in the food industry and is marketed by CP Kelco under the trade name KELCOGEL®, being used as a superior gelling agent (Bajaj et al., 2007).

Pullulan is an extracellular glucan produced by the fungus *Aureobasidium pullulans* from starch (Table 3.2). Its commercial production began in 1976, by the Hayashibara Company Ltd., in Japan, where it was used for hard capsules, dental care, and, later, as a food additive. Recently, pullulan was approved as a food additive in the European Union for use in food capsules, tablets, and films and its GRAS status was recognized (Chaen, 2010).

Bacterial cellulose is an extracellular polysaccharide synthesized by *Acetobacter xylinum*, mostly under static conditions, using sugars as the feedstocks (Table 3.2). However, static cultures are not feasible for large-scale production because they require long culture periods and are labor intensive. On the other hand, stirred bioreactor cultures may be used, but cellulose-negative mutants are generated due to rapid growth (Kim et al., 2007). Hence, despite its valuable properties and GRAS status (Khan et al., 2007), bacterial cellulose application is currently limited to high-value products and specialty chemicals, where this polymer's superior performance justifies its high production costs (Chawla et al., 2009).

Microbial levan is synthesized from sucrose by the action of the enzyme levansucrase (Shih and Yu, 2005), an extracellular enzyme synthesized by several bacteria, including species of the genera *Erwinia, Streptococcus, Pseudomonas,* and *Zymomonas* (Table 3.2) (Senthilkumar and Gunasekaran, 2005). It is produced by Montana Polysaccharides Corp. in the United States. In a few countries, levan has

TABLE 3.2

Production-Kinetic Parameters of Some Microbial Polysaccharides

Polysaccharide	Microorganism	Carbon Sources	Polymer $(g\,L^{-1})$	Productivity $(g\,L^{-1}\,d^{-1})$	$Y_{p/s}$ $(g\,g^{-1})$	Industrial Production	Food Grade	References
Xanthan gum	X. campestris	Glucose Sucrose	10.5–30.0	3.1–12.2	0.34–0.81	Yes	Yes (E415)	Garcia-Ochoa et al. (2000), Becker and Vorholter (2009)
Gellan gum	S. paucimobilis	Glucose Sucrose Soluble starch	4.1–23.5	1.27–11.8	0.49–0.57	Yes	Yes (E418)	Wang et al. (2006)
Pullulan	A. pullulans	Starch Glucose Sucrose	<15	<7.5	0.27	Yes	Yes (E1204)	Chaen (2010), Singh et al. (2008)
GalactoPol	P. oleovorans	Glycerol Glycerol by-product	8.1–12.2	2.0–4.6	0.19–0.36	No	No	Freitas et al. (2009a, 2010)
FucoPol	Enterobacter A47 DSM 23139	Glycerol Glycerol by-product	7.4–13.3	2.5–4.0	0.26–0.047	No	No	Alves et al. (2010b), Torres et al. (2011)

been approved as a food additive to support functional soluble fibers. However, mainly due to the limited information about its biological safety and polymeric properties, it has still not reached full industrial development.

Succinoglycan is produced by several soil bacteria, such as *Rhizobium, Alcaligenes, Pseudomonas,* and *Agrobacterium.* Only *Agrobacterium* strains have been considered for the industrial production of succinoglycan due to the good yield and quality of the product (Stredansky et al., 1999). It is not permitted in food, except in Japan.

GalactoPol and FucoPol were synthesized at laboratory scale using glycerol by-product from the biodiesel industry as the sole carbon source (Alves et al., 2010; Freitas et al., 2009a, 2010; Torres et al., 2011). The producing strains are *Pseudomonas oleovorans* NRRL B-14682 and *Enterobacter* A47 DSM 23139, respectively. Typically, both bioprocesses are based on bioreactor cultivations with an initial batch phase (10–24 h), followed by a fed-batch phase, wherein the bioreactor is fed with fresh culture medium (Freitas et al., 2010; Torres et al., 2011). The objective is to ensure that, after sufficient cell growth has occurred, the carbon source concentration is kept nearly constant, concomitantly with a growth-limiting nitrogen concentration (<0.1 g L^{-1}). During the fed-batch phase, the culture is also subjected to a limitation of the available oxygen (dissolved oxygen concentration was controlled below 10%). These conditions have shown to enhance exopolysaccharide production by several microorganisms (Kumar and Mody, 2009).

The lab-scale production of GalactoPol and FucoPol, under nonoptimized conditions, resulted in productivities and yields comparable to some of the commercial microbial bacterial polysaccharides, such as xanthan and gellan (Table 3.2). The optimization of these biopolymers' production bioprocesses can further improve their productivity and can drive them toward industrial production.

CGC is a copolymer synthesized by yeast and fungi as a structural material that confers rigidity and stability to the cell wall. It combines the properties of β-glucans and chitin polymers into a biomaterial that has recently started to emerge as a valuable biopolymer for several applications, including food products and processes. It can also be regarded as a source of non-animal chitin, avoiding the allergen potential of crustacean products. Currently, chitin is mainly obtained from crustacean, but the seasonal character of those raw materials and the variability of the composition of the organisms, make the processes rather expensive with a lower reproducibility. Microbial production allows for the production of such biopolymers with no availability restrictions and for the continuous optimization of the process with high cell density. Moreover, both the composition and the properties of the polymers are more stable than the polymers obtained by the traditional extraction method from crustacean (Roca et al., 2012).

Recently, CGC isolated from the cell wall of the fungus *Aspergillus niger* was granted GRAS status and it has been approved as a food ingredient by the EFSA. KitoZyme markets CGC obtained from the postfermentation of *A. niger* biomass, arising as a waste of food-grade citric acid production. It is used as a food supplement. It has also been recently approved by the Organization of Vine and Wine (OIV) for wine clarification and fining.

Another CGC source recently reported is the yeast *Pichia pastoris* (Roca et al., 2012). This methylotrophic yeast presents the advantage of reaching high cellular densities (>100 g L^{-1}) during fermentation and, consequently, a relatively high amount of CGC can be extracted after biomass production. Moreover, this biopolymer is produced using glycerol, a low-cost by-product arising from the biodiesel industry, under controlled fermentation conditions (Reis et al., 2010; Roca et al., 2012). The physical–chemical characteristics of *P. pastoris* CGC were similar to the product derived from *A. niger* (Roca et al., 2012), thus making it potentially suitable for food uses. Moreover, *P. pastoris* is GRAS status, which may contribute to the recognition of its products as safe for food applications.

PHAs are produced by a great number of prokaryotes (*Cupriavidus necator, Pseudomonas putida, P. oleovorans, Burkholderia cepacia*, etc.) as intracellular reserve materials (Gorenflo et al., 2001; Keenan et al., 2006; Lee et al., 2008; Rai et al., 2011). By choosing the appropriate microorganism, carbon source, cosubstrate, and culture conditions, a variety of homopolymers, copolymers, ter-polyesters, and so on, can be obtained with different physical and chemical properties (Gorenflo et al., 2001; Keenan et al., 2006) that offer a wide range of applications. Polyhydroxybutyrate (PHB) and poly(3HB-*co*-3HV) have been commercially available for several years, their major manufacturers being Monsanto and Metabolix, in the United States and TianAn, in China.

Although PHAs industrial production is mostly based on the use of single strains (e.g., *C. necator, Alcaligenes latus*) due to the high polymer contents (>80%) that can be achieved in the biomass, nowadays, mixed microbial consortia (MMC) are emerging as potential producing systems (Reis et al., 2011b). MMC have the advantage of being able to use a wider variety of complex substrates, such as industrial feedstocks containing compounds of undefined composition and have no sterility requirements. Although it was not yet implemented at the industrial scale, laboratory and pilot tests show that the process is stable and the polymer contents can reach up to 89% of PHA (Johnson et al., 2009).

3.3.2 BIOPROCESS LIMITATIONS AND IMPROVEMENT STRATEGIES

The high cost of the most commonly used carbon sources for the cultivation of microorganisms for biopolymer production has a direct impact on production costs (Kumar et al., 2007). To decrease production costs and to make microbial biopolymers production bioprocesses economically viable, it is critical to search for less-expensive carbon sources, such as wastes and by-products, some of which have already demonstrated the ability to support microbial growth and to be adequate for the production of several biopolymers. A wide range of agro-food and industrial wastes/by-products have been proposed as alternative substrates for the production of microbial biopolymers, including molasses, cheese, whey, palm date syrup, olive mill wastewater (OMW), glycerol by-product from the biodiesel industry, corn-steep liquor, spent malt grains, apple and grape pomaces, citrus peels, peach pulp, used oils, and several acid-hydrolysate wastes (e.g., melon, watermelon, cucumber, tomato, and rice), among others (Freitas et al., 2011b; Verlinden et al., 2011).

On the other hand, the use of low-cost agro-food and industrial wastes/by-products must be carefully considered when food applications are envisaged, since non-reacted components or impurities may accumulate in the broth, increasing the risk of their carryover to the final product. For that reason, if wastes/by-products are used, higher investments may be necessary in downstream procedures to guarantee that the biopolymers produced are food grade.

The use of low-value substrates and MMC for PHAs production are expected to contribute to the decrease of biopolymer production costs, by reducing the cost of the substrate, saving energy (since no sterilization is required), and reducing fermentation equipment costs (less-expensive materials and control are needed) (Reis et al., 2011b).

During the production of most extracellular microbial polysaccharides, the rheology of the fermentation broth drastically changes from an initial Newtonian fluid behavior, with a viscosity near that of water, to a highly viscous fluid with shear-thinning behavior (Freitas et al., 2009a). This viscosity increase causes a loss of bulk homogeneity, making it very difficult to maintain the appropriate mixing, aeration, or control of bioreactor parameters (Freitas et al., 2011b), frequently leading to premature termination of the bioreaction, especially at large-scale production systems. During the production of intracellular (e.g., PHAs) or cell-wall microbial biopolymers (e.g., chitin–glucan), high cell densities are usually achieved, which also pose homogeneity and mass-transfer problems (Ozturk, 1996).

The optimization of mechanical mixing by using different paddle configurations or increasing the stirring rate is commonly used to improve the hydrodynamics of fermentation broths. However, the use of these strategies can result in cell rupture as a result of the increased mechanical stress (Ntwampe et al., 2010) or can alter the polymer properties (Cheng et al., 2009; Pena et al., 2008). Lately, several emerging technologies (e.g., water-in-oil cultivation technology, use of surfactants, and magnetic nanoparticles) have been reported for some microbial production biosystems (Arockiasamy and Banik, 2008; Kuttuva et al., 2004). Nevertheless, the impact of these promising approaches upon polymer yield and quality must be evaluated.

3.4 DOWNSTREAM PROCESSING

The recovery of extracellular microbial polysaccharides from the culture broth are commonly achieved by procedures involving cell removal, usually achieved by centrifugation or filtration, followed by polymer precipitation from the cell-free supernatant by the addition of a precipitating agent (e.g., methanol, ethanol, isopropanol, and acetone). Finally, the precipitated polymer is dried by freeze drying (laboratory scale) or drum drying (industrial scale) (Freitas et al., 2011b).

Owing to the high broth viscosity usually achieved in such processes, cell removal is facilitated by the dilution of the culture broth by the addition of deionized water prior to the centrifugation/filtration or by subjecting it to heat treatment (up to 90–95°C) at the end of the fermentation process, prior to cell removal (Bajaj et al., 2007). However, this approach increases operating costs, as considerably higher volumes of the cell-free supernatant are generated and, consequently, higher volumes of the precipitating agent are required (Freitas et al., 2011b).

To obtain pure polymers, additional procedures are commonly performed, including reprecipitation of the polymer from diluted aqueous solution (<1.0 g L^{-1}), deproteinization by chemical (e.g., salting out or protein precipitation with trichloroacetic acid) or enzymatic methods (e.g., proteases) (Ayala-Hernández et al., 2008; Wang et al., 2007), and membrane processes, such as dialysis, ultrafiltration, and diafiltration (Bahl et al., 2010; Freitas et al., 2011a; Kumar et al., 2007), wherein low-molecular-weight compounds, coproduced or added during the production processes are removed (Freitas et al., 2011b). The choice of the most appropriate procedure must be made carefully because some of these purification procedures may decrease product recovery or may have a negative impact on the polymer's properties. As such, research effort is still needed, either for the improvement of the existing extraction and purification processes, or for the development of new approaches, focused on the specifications required for the final product (Freitas et al., 2011b).

GalactoPol and FucoPol have been purified by treating the cell-free supernatant, either by dialysis against deionized water (laboratory scale) or by diafiltration using micro/ultrafiltration membranes (pilot scale plate and frame membrane modules), with a complete removal of salts and a rather efficient removal of proteins (<5% of final products by dry weight) (Freitas et al., 2011a). The product was then processed by freeze drying without any intermediate precipitation step. This approach proved to be more efficient for salts and proteins removal than the previous methodology used comprising solvent extraction with ethanol or acetone in which protein and ash contents were up to 30% of the polymers' dry weight (Freitas et al., 2009b).

The extraction of microbial cell-wall polysaccharides, such as CGC, also begins with cell separation from the cultivation broth. Afterwards, the cells are subjected to treatment with an alkaline solution (pH above 10), at temperatures between 65°C and 90°C, to remove lipids, proteins, and nucleic acids. Lipids removal is enhanced by including organic solvents (e.g., ethanol, methanol, and acetone) or detergents in the alkaline reaction mixture. During this procedure, alkaline-soluble polysaccharides, namely highly branched α- and β-glucans and galactomannans are discarded with the supernatant. The obtained insoluble material, mostly composed of CGC, is washed with water and organic solvents for the removal of adsorbed organic materials and also inorganic salts reminiscent from the broth. Further purification may be achieved by additional enzymatic or chemical treatments.

The recovery and purification of PHAs significantly contribute to the overall economics of the process, with the associated costs representing up to 50% of the cost of the final product (Grage et al., 2009). Most current extraction processes use organic solvents, such as chloroform and dichloromethane that, being volatile and toxic, have a high environmental impact. On the other hand, these methods have shown to be economically not viable because they use large volumes of the solvent (Fiorese et al., 2009).

PHA-extraction techniques that minimize the amount of solvent used or replace them by less toxic solvents and with less environmental impact, would make PHA production process less environmental hazardous and more sustainable. Some reports suggest alternative solvent-free means for polymer recovery (de Koning and Witholt, 1997; Marchessault et al., 1995) that, however, have not generated to date much practical innovation in PHA production.

Nonsolvent methods based on the use of hydrolytic enzymes, such as lysozyme, phospholipase, lecithinase, and proteinase, have been tested. These processes consist of thermal treatment of biomass containing PHA, enzymatic digestion, and washing with an anionic surfactant to solubilize the cellular material other than PHA. Supercritical fluids (SFC) have also started to emerge as potential extraction techniques for PHAs recovery from microbial biomass (Kunasundari and Sudesh, 2011). Supercritical–carbon dioxide (SC–CO_2) can advantageously be used due to its low toxicity and reactivity, moderate critical temperature and pressure, low cost, and availability. Although this technique is still under development, promising results have already been obtained for the isolation of PHB from *C. necator* cells (Hejazi et al., 2003).

3.5 BIOPOLYMERS PROPERTIES AND APPLICATIONS

Microbial biopolymers, mainly polysaccharides (Table 3.1), are used in the food industry due to their ability to confer the products good sensory properties, extended shelf life, and easier processing, as they may function as thickening, stabilizing, texturizing, or gelling agents (Sworm, 2007). Many microbial polysaccharides (e.g., xanthan, gellan, and pullulan) are hydrocolloids, which are hydrophilic biopolymers with high affinity to water, some of them presenting a polyelectrolyte behavior. Other applications for some microbial polysaccharides are based on their film-forming capacity for the preparation of natural edible coatings or as encapsulating agents for the delivery of flavors and bioactive ingredients in food systems (McMaster et al., 2005; Rokka and Rantamaki, 2010; Vidhyalakshmi et al., 2009; Walle et al., 2001, Whitehouse et al., 2002). Another promising area of application is the use of some microbial biopolymers, such as PHAs, as biodegradable packaging materials for food products (Yalpani, 1993a,b).

3.5.1 RHEOLOGY MODIFIERS IN FOOD SYSTEMS

The preparation of homogeneous dispersions and particle hydration are the initial steps in all industrial applications of hydrocolloid thickeners. The main factors that affect the thickener hydration are the degree of particle dispersion, mixing speed, particle size, and the composition of the solvent (e.g., presence of salts and sugars). Poor dispersion leads to lumping of particles as a result of entanglements between partly hydrated macromolecules, resulting in a gel-like layer at the surface preventing particles separation and hindering the entrance of water. These facts require high-energy input on mixing and a viscosity buildup is observed over time as a consequence of the slow dissolution. Dispersion may be enhanced by blending the powder with ingredients, such as sugar and starch.

In the case of xanthan, the particulate material produced by processing the biopolymer by twin-screw extrusion cooking under mild conditions, has shown a substantial dispersibility improvement. The extruded xanthan gum swells allowing an apparently homogeneous dispersion to be achieved after few seconds, even under low shear (Sereno et al., 2007). Xanthan gum is soluble in cold or hot water and has the ability to develop an extremely high viscosity even at low concentration. Its aqueous solutions present a high low-shear viscosity, with a marked shear-thinning behavior. When compared to other hydrocolloids (e.g., guar gum, locust bean gum, and alginate), it presents a higher low-shear viscosity and is more shear thinning at

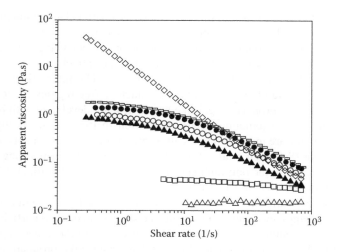

FIGURE 3.3 Apparent viscosity as a function of shear rate for different polysaccharide solutions (1% wt in 0.1 M NaCl), measured at 25°C in a controlled stress rheometer (HAAKE RS75, Germany) using a cone-plate geometry (diameter 35 mm, angle 2°): (◇) xanthan gum, (▢) Fucogel, (●) GalactoPol, (○) guar gum, (▲) FucoPol, (□) carboxymethylcellulose (CMC), and (△) alginate.

the shear rates typical of food processes (Sworm, 2010), which makes it the most widely used gum in the food industry (Figure 3.3). The shear thinning enhances sensory qualities (e.g., flavor release and mouthfeel) and guarantees an easy pouring and mixing. In addition, the entangled network formed by the associated chains makes xanthan gum an efficient stabilizer for suspensions and emulsions.

Several factors may influence the rheology of xanthan gum solutions, and consequently, the range of applications. The temperature of dissolution has a great impact on the viscosity. At dissolution temperatures below 40°C, xanthan gum shows an ordered structure and a low viscosity is observed. However, at temperatures between 40°C and 60°C, a change from an ordered to a disordered molecular conformation takes place, producing a solution with higher thickening behavior (García-Ochoa et al., 2000). As such, the temperature affects either the processing of products or their final properties. Xanthan solutions are quite stable during heating, even in the presence of salts and/or acid media. As the temperature increases, the viscosity of the xanthan solution decreases, but it is recovered upon cooling. A good stability is also observed over a wide range of pH values (between 4 and 10) (Sworn, 2010).

Regarding the effect of salts, the viscosity slightly decreases when a small amount of salt is added to xanthan solution at low polymer concentration (below ~0.3% wt). On the other hand, there is a viscosity increase at higher xanthan concentration with the presence of salts. This effect is normally observed for low sodium chloride content, but the viscosity is maintained regardless of the salt concentration for sodium chloride content above 0.1% wt (Kang and Pettit, 1993). These features enable the application of xanthan as a thickener in a range of products presenting quite dissimilar ionic-strength values.

Xanthan is able to establish synergetic interactions with galactomannans, namely with guar gum and locust bean gum. These gums are neutral polysaccharides consisting of a linear mannan backbone, the mannose unit being linked together by β-(1,4) glycosidic bonds. Single galactose units are attached as side chains to the mannan backbone by α-(1,6) glycosidic linkages. The interactions take place between the ordered xanthan helices and the galactose-free regions of the mannose backbone. Synergistic properties (gel-like properties) are developed with both gums when the xanthan/galactomannan mass proportion in the mixture is higher than 10/90, although they are much stronger with xanthan/locust bean gum at comparable molecular weight (Scborsch et al., 1997). Xanthan is able to form strong gels with locust bean gum, being the optimum gel strength found at xanthan/locust bean gum mass ratio around 60/40 (Sworn, 2010).

The largest application area of xanthan in the food industry is as a thickener and a stabilizer in dressings and sauces. It prevents the separation of solid particles and stabilizes oil-in-water emulsions by increasing the viscosity of the aqueous phase. The shear-thinning flow behavior provided by xanthan enables the production of dressings and sauces that are easy to pour, maintaining at the same time a good adhesion to meat and salads. The stability of xanthan to salt, temperature, and pH also helps to maintain product viscosity and texture during processes such as pasteurization, ultra-high temperature (UHT), microwave cooking, as well as during the product shelf life (Sworn, 2010).

In beverages, xanthan is used for suspending fruit pulp, delivering an improved mouthfeel to the drink. In addition, it plays an important role in the compensation of viscosity reduction in low-calorie drinks, caused by the use of artificial sweeteners instead of sugars. Xanthan is also applied in low-fat food products (e.g., mayonnaise, bakery fillings, spreads, dairy products, and processed cheese), taking advantage of the emulsion-stabilizing and water-binding abilities of xanthan, to maintain products' sensory properties. In bakery, xanthan contributes to smoothness, air incorporation, and water retention of baked products. It helps to obtain cakes with higher volume and reduced crumbliness.

Gellan gum molecules are composed of a straight chain based on repeating glucose, rhamnose, and glucuronic acid units. In its original form (high acyl gellan—HA), two acyl substituents—acetate and glycerate—are present. Low acyl gellan gum (LA) is obtained with the removal of acyl groups by a brief exposure to alkali (KOH) at high temperature (around 95°C) (Morris et al., 2012).

It is a good gelling agent, forming gels upon cooling, which properties are quite dependent on acyl groups content. Generally, the HA form produces soft, elastic, nonbrittle, and thermoreversible gels, whereas the LA form produces firm, nonelastic, brittle, and thermo-stable gels. The gelation mechanism of gellan involves the conversion of the polymer from a disordered coil state into a double-helix form, followed by the association of double helices into stable aggregates (Figure 3.4). The helices aggregation may be promoted by pH reduction or may be mediated by cations, either by site binding between pairs of carboxylate groups on neighboring helices or by suppressing electrostatic repulsion by binding to individual helices (Morris et al., 2012).

The properties of the gels prepared with LA gellan are quite dependent on the pH, dissolved sugar content, and the presence of cations (e.g., Ca^{2+}, Na^+, K^+, and Mg^{2+}).

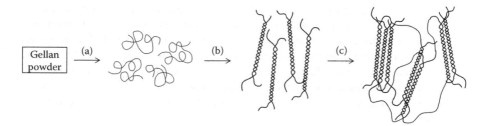

FIGURE 3.4 General model for gelation of gellan: (a) gum hydration by mixing with water and heating, (b) transition from random coil to double-helix conformation upon cooling, and (c) gel formation by association of double helices in the presence of cations, under cooling.

At neutral pH and low sugar concentration, the carboxyl groups on the gellan gum molecule interact with cations resulting in a high degree of molecular association and the formation of a long-range network. LA gellan gum is an effective gel agent in these conditions, creating true and thermo-irreversible gels. Increasing sugar composition at neutral pH, the gellan gum molecules become dispersed in a high-viscosity environment in which sugars have a plasticizing effect. The establishment of associations between gellan gum chains is slower and less extensive. The gels are still formed upon cooling, but the maturation may take weeks to be completed (Valli and Clark, 2010). At low pH (e.g., below 3.7) and low sugar content, most of the carboxyl groups on the gellan gum molecule are protonated, which disable their interaction with cations. In these conditions, the gel formation is acid induced and occurs rapidly upon cooling, without the requirement of cations. The resulting gels are less thermo-stable and more brittle when compared to those obtained at neutral pH. Increasing sugar concentration, at low pH, results in the formation of less-brittle gels due to the plasticizing effect of dissolved sugar. The sugar content significantly influences the amount of gellan needed for gel formation, either at low or neutral pH. While at low sugar content, gels are created at gum concentrations as low as 0.05% wt; for high sugar amounts, a minimum gellan concentration of 0.3% wt is required (Valli and Clark, 2010). Regarding the HA form of gellan, it behaves quite differently, as in this case, gel formation is not as sensitive as LA to changes in ionic environment and in sugar content.

There is a wide range of commercial uses of gellan gum in the food industry, namely in water-based gels (dessert gels, Asiatic gelled products, and aspics), jams and jellies (reduced-calorie jams, imitation jams, and bakery fillings), and bakery (bakery dry mixes, bakery fillings). Furthermore, soft-fluid gellan gels may be prepared by shearing while cooling and may be used in products requiring a low-viscosity suspension, to enhance mouthfeel and flavor, such as beverages (chocolate milk, pulp-containing fruit drinks, or suspended insoluble minerals), dairy foods (ice cream, stirred yogurts, and low-fat spreads), sauces, and toppings (Valli and Clark, 2010; KELCOGEL®).

Pullulan has valuable properties (adhesive, film-forming ability) that make it useful in several food applications, such as the preparation of edible films and coatings in healthy foods and dietary supplements (Table 3.1). It is commercialized in the form of a white nonhygroscopic powder. It dissolves easily in cold or hot water, producing viscous aqueous solutions without forming gel networks. Its thickening capacity is similar to that of Arabic gum. The viscosity of pullulan aqueous solutions is proportional to the molecular weight of the pullulan molecules and it is not affected by heating, changes in pH, and most metal ions, including sodium chloride. Its thickening ability enables its application as low-viscosity filler in beverages and sauces, as well as a texture-stabilizing agent in mayonnaise (Singh et al., 2008).

It can be used in low-caloric solid and liquid food products (e.g., pasta and baked goods) to replace starch, due to its capacity to introduce the characteristics derived from starch, such as consistency, dispersibility, and moisture content. Pullulan enables the enhancement of products shelf life since it is not easily assimilated as carbon source by microorganisms, which retards microbial spoilage and possesses a higher water-binding capacity than starch, restraining the loss of moisture.

Pullulan is considered as a dietary fiber, as it is resistant to mammalian amylases and it is referred to function as a prebiotic promoting the growth of beneficial *Bifidobacteria*. Studies indicate that this biopolymer is slowly digested by human enzymes, releasing glucose, which is gradually absorbed into the blood. As so, pullulan may be incorporated into snack foods designed for diabetics or patients who have weakened glucose tolerance (Singh et al., 2008).

The exopolysaccharides FucoPol and GalactoPol demonstrated to have a shear-thinning behavior, immediately recovering the viscosity when the shear rate was decreased (Freitas et al., 2011b) (Figure 3.3). The mechanical spectra indicate a viscous behavior constituted by entangled polymer chains. The steady-state and dynamic shear behaviors are similar to the ones demonstrated by guar gum solutions. These novel polysaccharides produce aqueous solutions that are thermorheologically simple and stable, suggesting their application in processes in which thermal variations may occur (Freitas et al., 2011b). Additionally, aqueous solutions (0.5% wt) of Fucopol have shown a good oil-in-water emulsion-stabilizing capacity for some food-grade oils, namely sunflower oil, corn oil, and rice bran oil (Freitas et al., 2011a).

β-Glucans from different natural sources (e.g., bacteria, yeast, fungi, algae, and plants) can be used as food thickeners or fat replacers, dietary fiber, gelling or stabilizing agents. Baker's or brewer's yeast (*Saccharomyces cerevisiae*) that has been used for centuries for the manufacture of food products (e.g., bread, beer) is a good source of β-glucan due to its GRAS status. Hence, *S. cerevisiae* β-glucan is used in salad toppings, frozen desserts, sauces, yogurts and other milk products, meat products, and beverages (Zechner-Krpan et al., 2010). Chitin–glucan derived from *A. niger* has also recently started to be marketed as a functional food additive.

PHAs macrocolloids comprising water-dispersible nonaggregated particles prepared from PHB and P(HB-*co*-HV) have also been proposed as fat replacers for use in low-fat food products, being claimed that they impart a creamy mouthfeel (Yalpani, 1993b). PHAs have also been proposed as environmentally friendly chewing gum base for the replacement of the typically used nondegradable elastomers (e.g., polybutadiene, butadiene–styrene copolymers, etc.) (Li et al., 1998). However, to date, these applications have not had commercial development.

3.5.2 FILMS AND COATINGS

To extend their shelf life, food products are usually protected by packages or coatings that control the moisture transfer with the environment, reduce the rate of metabolism, structurally reinforce the product, and protect against temperature fluctuations. The common packaging materials for food products (e.g., high-density polyethylene) have several disadvantages, including water condensation due to storage, temperature fluctuations, and fermentation due to the depletion of oxygen (Shahidi et al., 1999). Hence, the demand to increase product shelf life and enhance microbial safety of food products has prompted the development of edible films and coatings. Edible coatings are thin layers of material applied to the surface of a food that can be eaten as part of the whole product, providing a barrier against water vapor, gases, aroma compounds, and can also serve as carriers of functional ingredients, including antimicrobial agents and antioxidants. In addition, they may improve mechanical integrity, handling, and appearance of the food product. As edible coatings form an integral part of the food product, they should not have an impact on the sensory characteristics of the food and their components should be GRAS.

Edible coatings can be applied by different methods such as panning, fluidized bed, dipping, and spraying, which are the most used. All these techniques exhibit several advantages and disadvantages and their performance principally depends on the characteristics of the foods to be coated and the physical properties of the coating (viscosity, density, and surface tension) (Andrade et al., 2012).

Coatings can be opaque (milky) or transparent (clear), but usually, the latter are preferred by consumers. Coatings used for food products must have appropriate moisture–barrier properties, water or lipid solubility, color, appearance, mechanical, and rheological characteristics. The main components of our everyday foods (e.g., proteins, carbohydrates, and lipids) can fulfill the requirements for the preparation of edible films and coatings. As a general rule, fats are used to reduce water transmission; polysaccharides are used to control oxygen and other gases transfer, whereas protein films provide mechanical stability (Pavlath and Orts, 2009).

Many materials have been used for forming edible films and coatings, namely, polysaccharides (starch, cellulose, pectin, alginate, carrageenan, xanthan, gellan, pullulan, and chitosan), proteins (gelatin, casein, wheat gluten, zein, and soy protein), and lipids (shellac, beeswax, acetylated monoglycerides, fatty alcohols, and fatty acids). Minor components are usually included, such as polyols acting as plasticizers (glycerol or polyethylene glycol) or acid/base compounds used to regulate pH (acetic or lactic acid) (Falguera et al., 2011; Nieto, 2009).

Films and coatings of different materials are used in the food industry. For example, collagen films are used for sausage casings; some hydroxymethyl cellulose films, such as soluble pouches, are used for dried food ingredients. Shellac and wax coatings are most commonly used on fruits and vegetables, zein coatings are used on candies, and sugar coatings are used on nuts (Krochta and De Mulder-Johnston, 1997).

Polysaccharides and proteins are good barriers of O_2 and CO_2, which is an advantage when dealing with post-harvested fruits and vegetables. However, they possess a limited resistance to water. In contrast, lipids are hydrophobic compounds with better moisture–barrier properties. Nowadays, the novel edible coatings using nanotechnology processes are under development, for instance, nanocomposite edible films, which include nanoparticles to improve mechanical and barrier properties and nanolaminates to create multilayered systems that could be used to coat highly hydrophilic food systems (Bilbao-Sainz et al., 2011; Rojas-Grau et al., 2009; Tunç and Duman, 2011).

Edible films and coatings can also incorporate active ingredients such as antioxidants, antimicrobial agents, colorants, flavors, and/or spices. There are several categories of antimicrobials that can be potentially incorporated into edible films and coatings: organic acids (acetic, benzoic, lactic, propionic, and sorbic); fatty acid esters (glyceryl monolaurate); polypeptides (lysozyme, peroxidase, lactoferrin, and nisin); plant-essential oils (cinnamon, oregano, and lemongrass); and nitrites and sulfites. Within these categories, plant-essential oils are good alternatives and their use in foods meets consumer demands for minimally processed natural products (Martin-Belloso et al., 2009).

A composite film containing these active compounds can improve the shelf life and maintain the quality of a food product after the packaging is opened, by protecting against moisture change, oxygen uptake, and aroma loss. Thus, coated foods may not require high-barrier packaging materials, allowing the packaging structure to be reduced and/or simplified.

The use of carboxymethylcellulose and methylcellulose as matrices for coatings is predominant and some commercial edible coatings are available to be applied on a wide range of foods: Tal Pro-long®, Semprefresh F®, Nu-Coat, FloC®, BrilloshineC®, Snow-White®, and White-Wash® (Andrade et al., 2012).

Despite significant benefits from using edible coatings for extending product shelf life and enhancing the quality and microbial safety of foods, commercial applications are still very limited (Lin and Zhao, 2007). One of the main obstacles is cost, restricting its application to high-value products. Thus, the food industry is looking for edible films and coatings that can be used on a wide variety of foods, add value to their products, increase product shelf life, and/or reduce packaging.

Microbial polymers represent attractive alternatives, since the production parameters are easily controlled, high growth rates can be obtained, and the products can be easily modulated. The most used in the food industry for the preparation of films and/or coatings are pullulan, gellan, and xanthan (Nieto, 2009).

Pullulan films are transparent and have good mechanical properties. Owing to their excellent oxygen-barrier properties, they can be used to entrap flavors and colors and to stabilize other active ingredients. Several pullulan-based films are commercially available in a variety of colors (marketed by Hayashibara).

Gellan film is transparent and is insoluble in cold water. If gellan is de-esterified (partial or complete removal of its acyl groups), the linear gellan polymer can become completely insoluble in cold water and will require higher temperature or boiling to dissolve.

Alginate- and gellan-based coatings with antioxidant agents, including cysteine, glutathione, ascorbic and citric acids, and *Bifidobacteria* were applied to fresh-cut fruits (Rojas-Graü et al., 2009).

Xanthan gum is soluble in cold water, even in the presence of high levels of sugar, salt, or alcohol. It can have a soft, gel-like consistency, depending on the concentration and shear applied, typical of a pseudoplastic fluid. Xanthan gum coatings, containing high concentrations of calcium and vitamin E, were developed for enhancing nutritional and sensory qualities of fresh vegetables (Mei et al., 2002).

Many other microbial polysaccharides with filmogenic capacity have been reported, including, for example, levan and GalactoPol (Alves et al., 2011; Freitas et al., 2009a; Lacroix and Le Tien, 2005; Roberts and Garegg, 1998) that may find applications in the food industry as packaging and/or coating materials. Levan films are clear to white, depending on the thickness, and are good oxygen barriers. This biopolymer can also be used for coating of food products (www.polysaccharides.us).

The films obtained with GalactoPol were transparent and were quite flexible and tough when handled. Their permeability to carbon dioxide was quite low and the barrier properties to water vapor were typical of hydrophilic polysaccharide films (Alves et al., 2011). The results obtained are rather promising regarding the ability of the novel microbial polysaccharide to produce biodegradable films, which have an economic advantage over other microbial polysaccharides, as they can be produced by bacteria that use an abundant and low-cost carbon source.

The most common type of PHA is PHB, which is the most popular PHA used in food packaging. PHB has properties similar to polypropylene (PP) in relation to melting temperature (175–180°C) and mechanical behavior. However, it is stiffer and more brittle. PHA copolymers provide the best balance of properties, with poly(3-hydroxybutyrate-*co*-3-hydroxyhexanoate), now being the most important of such biodegradable thermoplastics (Nodax™). Transparent films derived from PHBHx that adhere to polyolefins without tie layers, are heat sealable and printable without surface treatment.

The most well-known application of PHB and poly(3HB-*co*-3HV) is as substitutes for conventional, nonbiodegradable plastics used for packaging purposes and derived products. Owing to its water resistance, it is also possible to use PHB and poly(3HB-*co*-3HV) for coating paper or cardboard, protecting it from damages caused by moisture of the packaged food, or by the environment.

The recent applications based on mcl PHAs include biodegradable cheese coatings. The conventional cheese coatings, typically a copolymer of polyvinyl acetate and dibutyl maleic acid, have a fixed permeability, necessitating rigorous storage and ripening conditions of the cheeses. A further advantage of the new PHA-based cheese coatings is that the water permeability can be tailored to optimize ripening conditions and to protect against mold growth (Walle et al., 2001; Whitehouse et al., 2002).

3.5.3 ENCAPSULATION OF BIOACTIVE FOOD INGREDIENTS

Encapsulation is a process for the entrapment of target substances within a carrier material. In food technology, it is a very useful tool to protect and deliver bioactive molecules (e.g., flavors, antioxidants, and vitamins) and living cells (e.g., probiotics) in food systems. Their stability in the final products and during processing is preserved, by preventing the reaction with other components (e.g., oxygen, water, and enzymes) and protecting against unfavorable pH, temperature, and mechanical stresses. Furthermore, encapsulates enable bioactives protection throughout the gut until being assimilated in the organs where their functions are needed. Encapsulation also enables masking unpleasant tastes and odors.

For food applications, the encapsulation material, commonly known as the shell, coating, wall, or membrane, must be food grade, biodegradable, and able to form a barrier between the internal phase (core) and the surroundings. Several factors contribute to the selection of the most appropriate material, including the nature of the bioactive molecule to be encapsulated and the physical/chemical properties of the encapsulating material (solubility, molecular weight, glass melting temperatures, crystallinity, barrier to diffusion, film-forming and emulsifying capacities), as well as stability requirements and cost constraints (Gharsallaoui et al., 2007). Although many materials possess these properties, only some of them have been approved for food applications.

The most widely used materials used for the encapsulation of sensitive bioactive compounds in the food industry are polysaccharides (e.g., starch, cellulose derivatives, and plant exudates), but proteins (e.g., casein, gluten, and gelatin), lipids (e.g., waxes, fatty acids, and phospholipids), and other materials (e.g., paraffin, inorganic materials) are also used (Beirão da Costa et al., 2012; Nedovic et al., 2011). Depending on the type of material, different structures may be produced, such as microparticles and liposomes.

Although starches, maltodextrins, and corn syrup solids are normally used for the encapsulation of food ingredients due to their low viscosities at high solid contents and good solubility, these materials suffer from poor interfacial properties (Gharsallaoui et al., 2007). In contrast, polysaccharide gums have better interfacial properties, film-forming, and emulsifying capacities. Among all gums, Arabic gum is the most widely used, despite its high cost, limited supply, and quality variations that have hindered its use for encapsulation.

Microbial exopolysaccharides, such as xanthan, gellan, and pullulan have also been exploited as materials for the encapsulation of food ingredients (Zuidam and Nedovic, 2010). Xanthan, gellan, and mixtures of both gums have been reported to be adequate for the encapsulation of probiotic bacteria, greatly improving their survival when exposed to acidic conditions and bile salts (McMaster et al., 2005; Rokka and Rantamaki, 2010; Vidhyalakshmi et al., 2009).

Although less common, PHAs have also been proposed as encapsulating materials for the delivery of bioactive ingredients in food systems. For example, Yalpani (1993b) has disclosed a composition based on PHA microparticles for the delivery of fat-soluble flavors (e.g., vanillin) or colors in low-fat and no-fat food products (e.g., ice cream, yogurt, mayonnaise, cheese, sauces, etc.).

3.6   FINAL REMARKS

Microorganisms are a versatile source of biopolymers since they are able to synthesize many different polymeric structures from a wide range of feedstocks, under controlled conditions that guarantee the final products quantity and quality. Owing to their new or improved properties, microbial biopolymers can replace plant, algae, and animal products, which currently dominate the market, either in their traditional or in new applications. Moreover, microbial processes can be easily manipulated to drive production toward specific tailored polymer composition that will fit the intended product applications.

Although many microbial biopolymers possess valuable properties that make them useful in food products or processes, only a very limited number has found widespread use in such applications. The main limitations are the high production costs, which can be lowered by optimizing the bioprocesses and/or using low-cost feedstocks, and the difficulties to obtain permission from the competent authorities for commercialization, as well as to gain public acceptance. While keeping their thoroughness, the evaluation process for new products should be more flexible and less time-consuming to allow new polymers to reach the market and to prove themselves as valuable materials for food applications.

REFERENCES

Alves V.D., Ferreira A.R., Costa N., Freitas F., Reis M.A.M., Coelhoso I.M. 2011. Characterization of biodegradable films from the extracellular polysaccharide produced by *Pseudomonas oleovorans* grown on glycerol byproduct. *Carbohydr. Pol.* 83, 1582–1590.

Alves V.D., Freitas F., Torres C.A.V., Cruz M., Marques R., Grandfils C., Gonçalves M.P., Oliveira R., Reis M.A.M. 2010. Rheological and morphological characterization of the culture broth during exopolysaccharide production by *Enterobacter* sp. *Carbohydr. Pol.* 81, 758–764.

Andrade R.D., Skurtys O., Osorio F.A. 2012. Atomizing spray systems for application of edible coatings. *Compr. Rev. Food Sci. Food Saf.* 11, 323–337.

Arockiasamy S., Banik R.M. 2008. Optimization of gellan gum production by *Sphingomonas paucimobilis* ATCC 31461 with nonionic surfactants using central composite design. *J. Biosci. Bioeng.* 105, 204–210.

Ayala-Hernández I., Hassan A., Goff H.D., Mira de Orduña R., Corredig M. 2008. Production, isolation and characterization of exopolysaccharides produced by *Lactococcus lactis* subsp. *cremoris* JFR1 and their interaction with milk proteins: Effect of pH and media composition. *Int. Dairy J.* 18, 1109–1118.

Azarakhsh N., Osman A., Ghazali H.M., Tan C.P., Adzahan M.N. 2012. Optimization of alginate and gellan-based edible coating formulations for fresh-cut pineapples. *Int. Food Res. J.* 19(1), 279–285.

Bahl M.A., Schultheis E., Hempel D.C., Nörtemann B., Franco-Lara E. 2010. Recovery and purification of the exopolysaccharide PS-EDIV from *Sphingomonas pituitosa* DSM 13101. *Carbohydr. Pol.* 80, 1037–1041.

Bajaj I.B., Survase S.A., Saudagar P.S., Singhal R.S. 2007. Gellan gum: Fermentative production, downstream processing and applications. *Food Technol. Biotechnol.* 45, 341–354.

Becker A., Vorholter F.-J. 2009. Xanthan biosynthesis by *Xanthomonas* bacteria: An overview of the current biochemical and genomic data. In: *Microbial Production of Biopolymers and Polymer Precursors: Applications and Perspectives*, edited by B.H.M. Rehm, Caister Academic Press, United Kingdom, pp. 1–12.

Beirão-da-Costa S., Duarte C., Pinheiro A.C., Bourbon A.I., Serra A.T., Moldão-Martins M., Vicente A., Delgadillo I., Duarte C.M.M., Beirão da Costa M.L. 2012. The effect of the matrix system in the delivery and *in-vitro* bioactivity of microencapsulated oregano essential oil. *J. Food Eng.* 110, 190–199.

Bilbao-Sainz C., Bras J., Williams T., Sénechal T., Orts W. 2011. HPMC reinforced with different cellulose nanoparticles. *Carbohydr. Pol.* 86(4), 1549–1557.

Chaen H. 2010. Pullulan. In: *Food Stabilisers, Thickeners and Gelling Agents*, edited by A. Imeson, Wiley-Blackwell, United Kingdom, Chapter 14.

Chan P., Yu V., Wai L., Yu H. 2006. Production of medium-chain-length polyhydroxyalkano-ates by *Pseudomonas aeruginosa* with fatty acids and alternative carbon sources. *Appl. Biochem. Biotechnol.* 132(1–3), 933–941.

Chawla, P.R., Bajaj I.B., Survase S.A., Singhal R.S. 2009. Microbial cellulose: Fermentative production and applications. *Food Technol. Biotechnol.* 47, 107–124.

Cheng K.-C., Catchmark J.M., Demirci A. 2009. Enhanced production of bacterial cellulose by using a biofilm reactor and its material property analysis. *J. Biol. Eng.* 3, 12.

Coleman R.J., Patel Y.N., Harding N.E. 2008. Identification and organization of genes for diutan polysaccharide synthesis from *Sphingomonas* sp. ATCC 53159. *J. Ind. Microbiol. Biotechnol.* 35, 263–274.

de Koning G.J.M., Witholt B. 1997. A process for the recovery of poly(hydroxyalkanoates) from *Pseudomonads*. Part 1: Solubilization. *Bioproc. Biosys. Eng.* 17(1), 7–13.

Falguera V., Quintero J.P., Jiménez A., Muñoz J.A., Ibarz A. 2011. Edible films and coatings: Structures, active functions and trends in their use. *Trends Food Sci. Technol.* 22(6), 292–303.

Fiorese M.L., Freitas F., Pais J., Ramos A.M., Reis M.A.M., Aragão G.M.F. 2009. Recovery of P(3HB) produced by *Ralstonia eutropha* by solvent extraction with propylene carbon-ate. *Eng. Life Sci.* 9(6), 454–461.

Freitas F., Alves V.D., Carvalheira M., Costa N., Oliveira R., Reis M.A.M. 2009b. Emulsifying behaviour and rheological properties of the extracellular polysaccharide produced by *Pseudomonas oleovorans* grown on glycerol byproduct. *Carbohyd. Pol.* 78, 549–556.

Freitas F., Alves V.D., Pais J., Carvalheira M., Costa N., Oliveira R., Reis M.A.M. 2010. Production of a new exopolysaccharide (EPS) by *Pseudomonas oleovorans* NRRL B-14682 grown on glycerol. *Proc. Biochem.* 45, 297–305.

Freitas F., Alves V.D., Pais J., Costa N., Oliveira C., Mafra L., Hilliou L., Oliveira R., Reis M.A.M. 2009a. Characterization of an extracellular polysaccharide produced by a *Pseudomonas* strain grown on glycerol. *Biores. Technol.* 100, 859–865.

Freitas F., Alves V.D., Reis M.A.M. 2011b. Advances in bacterial exopolysaccharides: From production to biotechnological applications. *Trends Biotechnol.* 29(8), 388–398.

Freitas F., Alves V., Torres C.A.V., Cruz M., Sousa I., Melo M.J., Ramos A.M., Reis M.A.M. 2011a. Fucose-containing exopolysaccharide produced by the newly isolated *Enterobacter* strain A47 DSM 23139. *Carbohyd. Pol.* 83, 159–165.

García-Ochoa F., Santos V.E., Casas J.A., Gómez E. 2000. Xanthan gum: Production, recov-ery, and properties. *Biotechnol. Adv.* 18, 549–579.

Gharsallaoui A., Roudaut G., Chambin O., Voilley A., Saurel R. 2007. Applications of spray-drying in microencapsulation of food ingredients: An overview. *Food Res. Int.* 40, 1107–1121.

Gorenflo V., Schmack G., Vogel R., Steinbuchel A. 2001. Development of a process for the biotechnological large-scale production of 4-hydroxyvalerate-containing polyesters and characterization of their physical and mechanical properties. *Biomacromolecules* 2(1), 45–57.

Grage K., Peters V., Palanisamy R., Rehm B.H.A. 2009. Polyhydroxyalkanoates: From bacte-rial storage compound via renewable plastic to bio-bead. In: *Microbial Production of Biopolymers and Polymer Precursors: Applications and Perspectives*, edited by B.H.M. Rehm, Caister Academic Press, United Kingdom, pp. 255–288.

Hejazi P., Vasheghani-Farahani E., Yamini Y. 2003. Supercritical fluid disruption of *Ralstonia eutropha* for poly(β-hydroxybutyrate) recovery. *Biotechnol. Prog.* 19, 1519–1523.

Johnson K., Jiang Y., Kleerebezem R., Muyzer G., van Loosdrecht M.C.M. 2009. Enrichment of a mixed bacterial culture with a high polyhydroxyalkanoate storage capacity. *Biomacromolecules* 10, 670–676.

Kang S.A., Jang K.-H., Seo J.-W., Kim K.H., Kim Y.H., Rairakhwada D., Seo M.Y. et al. 2009. Levan: Applications and perspectives. In: *Microbial Production of Biopolymers and*

Polymer Precursors: Applications and Perspectives, edited by B.H.M. Rehm, Caister Academic Press, United Kingdom, pp. 145–162.

Kang K.S., Pettit D.J. 1993. Xanthan, gellan, wellan, and rhamsan. In: *Industrial Gums*, edited by R.L Whistler, J.N. BeMiller, Academic Press, New York, NY, pp. 341–398.

Keenan T.M., Nakas J.P., Tanenbaum S.W. 2006. Polyhydroxyalkanoate copolymers from forest biomass. *J. Ind. Microbiol. Biotechnol.* 33(7), 616–626.

Khan T., Park J.K., Kwon J.-H. 2007. Functional biopolymers produced by biochemical technology considering applications in food engineering. *Korean J. Chem. Eng.* 24(5), 816–826.

Kim J.-H., Kim M.-R., Lee J.-H., Lee J.-W., Kim S.-K. 2000. Production of high molecular weight pullulan by *Aureobasidium pullulans* using glucosamine. *Biotechnol. Lett.* 22(12), 987–990.

Kim J.Y., Kim J.N., Wee Y.J., Park D.H., Ryu H.W. 2007. Bacterial cellulose production by *Gluconacetobacter* sp. RKY5 in a rotary biofilm contactor. *Appl. Biochem. Biotechnol.* 137, 529–537.

Koller M., Salerno A., Dias M., Reiterer A., Braunegg G. 2010. Modern biotechnological polymer synthesis: A review. *Food Technol. Biotechnol.* 48(3), 255–269.

Krochta J.M., De Mulder-Johnston C. 1997. Edible and biodegradable polymer films: Challenges and opportunities. *Food Technol.* 51(2), 61–74.

Kumar A.S., Mody K. 2009. Microbial exopolysaccharides: Variety and potential applications. In: *Microbial Production of Biopolymers and Polymer Precursors: Applications and Perspectives*, edited by B.H.M. Rehm, Caister Academic Press, United Kingdom, pp. 229–255.

Kumar A.S., Mody K., Jha B. 2007. Bacterial exopolysaccharides—A perception. *J. Basic Microbiol.* 47, 103–117.

Kunasundari B., Sudesh K. 2011. Isolation and recovery of microbial polyhydroxyalkanoates. *Express Pol. Lett.* 5(7), 620–634.

Kuttuva, S.G., Restrepo A.S., Ju L.K. 2004. Evaluation of different organic phases for water-in-oil xanthan fermentation. *Appl. Microbiol. Biotechnol.* 64, 340–345.

Lacroix M., Le Tien C. 2005. Edible films and coatings from non-starch polysaccharides. *Innov. Food Packaging*, 338–361.

Lee W.H., Loo C.Y., Nomura, C.T., Sudesh K. 2008. Biosynthesis of polyhydroxyalkannoate copolymers from mixtures of plant oils and hydroxyvalerate precursors. *Biores. Technol.* 99(15), 6844–6851.

Li W., Orfan C.P., Liu J., Foster J.W. 1998. Environmentally friendly chewing gum bases including polyhydroxyalkanoates. US patent 9802511.

Lin D., Zhao Y. 2007. Innovations in the development and application of edible coatings for fresh and minimally processed fruits and vegetables. *Compr. Rev. Food Sci. Food Saf.* 6(3), 60–75.

Machova E., Kogan G., Soltes L., Kvapilova K., Sandula J. 1999. Ultrasonic depolymerization of the chitin–glucan isolated from *Aspergillus niger*. *Reactive Funct. Pol.* 42(3), 265–271.

Marchessault R.H., LePoutre P.F., Wrist P.E. 1995. Latex of poly-β-hydroxyalkanoates for treating fiber constructs and coating paper. US Patent 5451456.

Martín-Belloso O., Rojas-Graü M.A., Soliva-Fortuny R. 2009. Delivery of flavor and active ingredients using edible films and coatings. In: *Edible Films and Coatings for Food Applications*, edited by M.E. Embuscado and K.C. Huber, Springer, New York, NY, pp. 295–314.

McMaster L.D., Kokkot S.A., Slatter P. 2005. Micro-encapsulation of *Bifidobacterium lactis* for incorporation into soft foods. *W. J. Microbiol. Biotechnol.* 21, 723–728.

Mei Y., Zhao Y., Yang J., Furr H.C. 2002. Using edible coating to enhance nutritional and sensory qualities of baby carrots. *J. Food Sci.* 67(5), 1964–1968.

Moreno J., Vargas M.A., Olivares H., Rivas J., Guerrero M.G. 1998. Exopolysaccharide production by the cyanobacterium *Anabaena* sp. ATCC 33047 in batch and continuous culture. *J. Biotechnol.* 60, 175–182.

Morris E.R., Nishinari K., Rinaudo R. 2012. Gelation of gellan—A review. *Food Hydrocoll.* 28, 373–411.

Nedovic V., Kalusevic A., Manojlovic V., Levic S., Bugarski B. 2011. An overview of encapsulation technologies for food applications. *Proc. Food Sci.* 1, 1806–1815.

Nieto M.B. 2009. Structure and function of polysaccharide gum-based edible films and coatings. In: *Edible Films and Coatings for Food Applications*, edited by M.E. Embuscado and K.C. Huber, Springer, New York, NY, pp. 57–112.

Ntwampe S.K.O., Williams C.C., Sheldom M.S. 2010. Water-immiscible dissolved oxygen carriers in combination with pluronic F 68 in bioreactors. *Afr. J. Biotechnol.* 9, 1106–1114.

Ozturk S.S. 1996. Engineering challenges in high density cell culture systems. *Cytotechnology* 22(1–3), 3–16.

Pavlath A.E., Orts W. 2009. Edible films and coatings: Why, what, and how?, In: *Edible Films and Coatings for Food Applications*, edited by M.E. Embuscado and K.C. Huber, Springer, New York, NY, pp. 1–24.

Pena C., Millán M., Galindo E. 2008. Production of alginate by *Azotobacter vinelandii* in a stirred fermentor simulating the evolution of power input observed in shake flasks. *Process Biochem.* 43, 775–778.

Rai R., Yunos D., Boccaccini A.R., Knowles J.C., Barker A., Howdle S.M., Tredwell G.D., Keshavarz T., Roy I. 2011. Poly-3-hydroxyoctanoate P(3HO), a medium chain length polyhydroxyalkanoate homopolymer from *Pseudomonas mendocina*. *Biomacromolecules* 12(6), 2126–2136.

Rehm B.H.A. 2010. Bacterial polymers: Biosynthesis, modifications and applications. *Nat. Rev. Microbiol.* 8, 578–592.

Reis M., Albuquerque M., Villano M., Majone M. 2011a. Mixed culture processes for polyhydroxyalkanote production from agro-industrial surplus/organic wastes as feedstocks. In: *Encyclopaedia Comprehensive Biotechnology*, 2nd edition, Murray Moo-Young (editor-in-chief), vol. 6, Spiros Agathos, Elsevier.

Reis M.A.M., Oliveira R., Freitas F., Alves V.D., Pais J., Oliveira C. 2008. Galactose-rich polymer, process for the production of the polymer and its applications. International Publication under the Patent Cooperation Treaty WO 2008127134.

Reis M.A.M., Oliveira R., Freitas F., Alves V.D. 2011b. Fucose-containing bacterial biopolymer. International Publication under the Patent Cooperation Treaty WO 2011073873.

Reis M.A.M., Oliveira R., Freitas F., Chagas B., Cruz L. 2010. Process for the co-production of chitin, its derivatives and polymers containing glucose, mannose and/or galactose, by the fermentation of the yeast *Pichia pastoris*. International Publication under the Patent Cooperation Treaty WO 2010/013174.

Roberts E.J., Garegg P.J. 1998. Levan derivatives, their preparation, composition and applications including medical and food applications. International Publication under the Patent Cooperation Treaty WO 9803184.

Roca C., Chagas B., Farinha I., Freitas F., Mafra L., Aguiar F., Oliveira R., Reis M.A.M. 2012. Production of yeast chitin–glucan complex from biodiesel industry byproduct. *Proc. Biochem.* 47, 1670–1675.

Rojas-Grau M.A., Soliva-Fortuny R., Martín-Belloso O. 2009. Edible coatings to incorporate active ingredients to fresh-cut fruits: A review. *Trends Food Sci. Technol.* 20(10), 438–447.

Rokka S., Rantamaki P. 2010. Protecting probiotic bacteria by microencapsulation: Challenges for industrial applications. *Eur. Food Res. Technol.* 231, 1–12.

Rottava I., Batesini G., Silva M.F., Lerin L., de Oliveira D., Padilha F.F., Toniazzo G. et al. 2009. Xanthan gum production and rheological behavior using different strains of *Xanthomonas* sp. *Carbohydr. Pol.* 77(1), 65–71.

Ruffing A., Chen R.R. 2006. Metabolic engineering of microbes for oligosaccharide and polysaccharide synthesis. *Microb. Cell Factories* 5:25.

Scborsch C., Gamier C., Doublier J.-L. 1997. Viscoelastic properties of xanthan/galactomannan mixtures: Comparison of guar gum with locust bean gum. *Carbohydr. Pol.* 34, 165–175.

Seisun D. 2010. Introduction. In: *Food Stabilisers, Thickeners and Gelling Agents*. Edited by A. Imeson, Wiley-Blackwell, United Kingdom, Chapter 1.

Senthilkumar V., Gunasekaran P. 2005. Influence of fermentation conditions on levan production by *Zymomonas mobilis* CT2. *Indian J. Biotechnol.* 4, 491–496.

Sereno N.M., Hill S.E., Mitchell J.R. 2007. Impact of the extrusion process on xanthan gum behavior. *Carbohydr. Res.* 342, 1333–1342.

Shahidi F., Arachchi J.K.V., Jeon Y.-J. 1999. Food applications of chitin and chitosans. *Trends Food Sci. Technol.* 10, 37–51.

Shih I.-L., Yu Y.-T. 2005. Simultaneous and selective production of levan and poly(g-glutamic acid) by *Bacillus subtilis*. *Biotechnol. Lett.* 27, 103–106.

Simsek S., Mert B., Campanella O.H., Reuhs B. 2009. Chemical and rheological properties of bacterial succinoglycan with distinct structural characteristics. *Carbohydr. Pol.* 76, 320–324.

Singh R.S., Saini G.K., Kennedy J.F. 2008. Pullulan: Microbial sources, production and applications. *Carbohydr. Pol.* 73, 515–531.

Siracusa V., Rocculi P., Romani S., Rosa M.D. 2008. Biodegradable polymers for food packaging: A review. *Trends Food Sci. Technol.* 19, 634–643.

Stredansky M., Conti E., Bertocchi C., Navarini L., Matulova M., Zanetti F. 1999. Fed-batch production and simple isolation of succinoglycan from *Agrobacterium tumefaciens*. *Biotechnol. Techn.* 13(1), 7–10.

Sworm G. 2007. Natural thickeners. In: *Handbook of Industrial Water Soluble Polymers*, edited by P.A. Williams, Blackwell Publishing Ltd., Oxford, United Kingdom, Chapter 2.

Sworm G. 2010. Xanthan gum. In: *Food Stabilisers, Thickeners and Gelling Agents*. Edited by A. Imeson, Wiley-Blackwell, Oxford, United Kingdom, Chapter 17.

Torres C.A.V., Antunes S., Ricardo A.R., Grandfils C., Alves V.D., Freitas F., Reis M.A.M. 2012. Study of the interactive effect of temperature and pH on exopolysaccharide production by *Enterobacter* A47 using multivariate statistical analysis. *Biores. Technol.* 119, 148–156.

Torres C.A.V., Marques R., Antunes S., Alves V.D., Sousa I., Ramos A.M., Oliveira R., Freitas F., Reis M.A.M. 2011. Kinetics of production and characterization of the fucose-containing exopolysaccharide from *Enterobacter* A47. *J. Biotechnol.* 156, 261–267.

Tunç S., Duman O. 2011. Preparation of active antimicrobial methyl cellulose/carvacrol/montmorillonite nanocomposite films and investigation of carvacrol release. *Food Sci. Technol.* 44(2), 465–472.

Valli R., Clark R. 2010. Gellan gum. In: *Food Stabilisers, Thickeners and Gelling Agents*. Edited by A. Imeson, Wiley-Blackwell, United Kingdom, Chapter 8.

Vanhooren P.T., Vandamme E.J. 1999. L-Fucose: Occurrence, physiological role, chemical, enzymatic and microbial synthesis. *J. Chem. Technol. Biotechnol.* 74, 479–497.

Verlinden R.A.J., Hill D.J., Kenward M.A., Williams C.D., Piotrowska-Seget Z., Radecka I.K. 2011. Production of polyhydroxyalkanoates from waste frying oil by *Cupriavidus necator*. *AMB Express* 2011, 1:11

Vidhyalakshmi R., Bhakyaraj R., Subhasree R.S. 2009. Encapsulation "The future of probiotics"—A review. *Adv. Biol. Res.* 3(3–4), 96–103.

Walle G.A.M., de Koning G.J.M., Weusthuis R.A., Eggink G. 2001. Properties, modifications and applications of biopolyesters. *Adv. Biochem. Eng./Biotechnol.* 71, 264–291.

Wang X., Yuan Y., Wang K., Zhang D., Yang Z., Xu P. 2007. Deproteinization of gellan gum produced by *Sphingomonas paucimobilis* ATCC 31461. *J. Biotechnol.* 128, 403–407.

Whitehouse R.S., Zhong L., Daughtry S. 2002. Compositions comprising low molecular weight polyhydroxyalkanoates and methods employing same. International Publication under the Patent Cooperation Treaty WO 0234857.

Yalpani M. 1993a. Polyhydroxyalkanoate cream substitute. US Patent 9108720.

Yalpani M. 1993b. Polyhydroxyalkanoate flavor delivery system. US Patent 5225227.

Zechner-Krpan V., Petravic-Tominac V., Galovic P., Galovic V., Filipovic-Grcic J., Srecec S. 2010. Application of different drying methods on β-glucan isolated from spent brewer's yeast using alkaline procedure. *Agr. Conspectus Scientificus* 75(1), 45–50.

Zuidam N.J., Nedovic V.A. 2010. *Encapsulation Technologies for Active Food Ingredients and Food Processing.* Springer, New York.

4 Advanced Fermentation Processes

Leona Paulová, Petra Patáková, and Tomáš Brányik

CONTENTS

4.1 INTRODUCTION

The most common meaning of fermentation is the conversion of a sugar into an organic acid or an alcohol. Fermentation occurs naturally in many foods and humans have intentionally used it since ancient times to improve both the preservation and organoleptic properties of food. However, the term "fermentation" is also used in a broader sense for the intentional use of microorganisms such as bacteria, yeast, and fungi to make products useful to humans (biomass, enzymes, primary and secondary metabolites, recombinant products, and products of biotransformation) on an industrial scale.

Modern industrial fermentation processes used in the food and beverage industry can be described according to different perspectives. In the center of these processes are usually bioreactors, which can be classified with respect to the feeding of the bioreactor (batch, fed-batch, and continuous mode of operation), immobilization of the biocatalyst (free or immobilized cells/enzymes), the characteristic state of matter in the system (submerged or solid substrate fermentations), single strain/mixed culture processes, mixing of the bioreactor (mechanical, pneumatic, and hydraulic agitation), or the availability of oxygen (aerobic, microaerobic, and anaerobic processes). The decision as to which bioreactor or fermentation process should be implemented in any particular application involves considering the advantages and disadvantages of each setup. This includes examining the properties and availability of the primary raw materials, any necessary investment and operating costs, sustainability, availability of a competent workforce, as well as the desired productivity and return on investment (Inui et al., 2010). Since in large-scale applications, each fermentation system needs to operate efficiently and reliably, the major criterion for the selection of a bioreactor/fermentation process remains the minimum for capital costs per unit of product recovered. Simultaneously, with efficient design and operation, in large-scale processes, the issues concerning by-product and wastewater management are inevitable.

4.2 TYPES OF FERMENTATION PROCESSES

4.2.1 Submerged Cultivation

Submerged cultivation of microbial cells in bioreactors guarantees a controlled environment for the efficient production of high-quality end products and to achieve optimum productivity and yield. Industrial bioreactors operated in batch, fed-batch, or continuous mode are utilized to culture different types of microorganisms producing a wide range of products. In the following sections, different approaches to submerged cultivation of microorganisms in bioreactors are discussed briefly and the typical features, benefits, and drawbacks of each cultivation mode are highlighted. Finally, the relevant applications for batch, fed-batch, and continuous cultivation of microorganisms in liquid media used in the production of different types of food industry products are demonstrated.

4.2.1.1 Batch Cultivation

Batch culture represents a closed system in which the medium, nutrients, and inoculum are added to the bioreactor, mostly under aseptic conditions, at the beginning of cultivation (Figure 4.1a), that is, the volume of the culture broth in the bioreactor is theoretically constant during cultivation (practically, small deviations in culture volume are caused by a low feed rate of acid/base solutions to keep the pH at a desired level and by sampling or introducing air/gas into the culture; on balance, such changes are usually ignored due to their small value relative to the total working volume of the bioreactor).

Typically, at the beginning of batch cultivation, a known number of viable cells are inoculated into the bioreactor that is already filled with sterilized medium

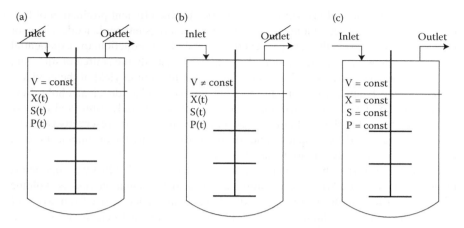

FIGURE 4.1 Simplified scheme of (a) batch, (b) fed-batch, and (c) continuous cultivation.

containing all nutrients. After inoculation, the cell culture follows the classical growth curve described by Monod (1949), which is divided into four main phases. As the lag phase is an "inefficient" stage of culture (even though the cells are metabolically active—they are adapting their enzymatic apparatus to a new environment; no significant increases in biomass concentration, substrate consumption, or product synthesis are observed), it is desirable to shorten it as much as possible. The length of the lag phase is influenced mainly by the concentration of cells in the inoculum and their physiological state, the composition of the inoculation and cultivation medium (mainly the source of carbon and energy, pH, and temperature), and the size of the inoculum. The exponential (logarithmic growth) phase is characterized by rapid cell proliferation (biomass concentration is an exponential function of time), constant specific growth rate, which is equal to the maximum specific growth rate of the culture under conditions of absence of growth limitation (growth rate is not limited because all nutrients are present in excess, while also not attaining growth-inhibiting concentrations), fast consumption of the source of carbon and energy, and a high rate of primary metabolite production. The depletion of nutrients by the end of the exponential phase (in the case of aerobically grown cultures, these are signaled by a rapid increase in dissolved oxygen concentration) causes a progressive reduction in the specific growth rate and a transition to the stationary phase, characterized by the stagnation of growth and utilization of endogenous reserves of carbon and energy; this phase is important for the synthesis of secondary metabolites. Most industrial bioreactors are operated in batch mode due to the relative simplicity of this process. The whole batch operation consists of several steps, including medium formulation, filling the bioreactor, sterilization in place (SIP systems), inoculation, cultivation, product harvesting, and bioreactor cleaning in place (CIP systems). For efficient performance of batch operation, it is important to minimize all nonproductive steps (all steps listed above except cultivation), achieve a high rate of product synthesis, optimize productivity, and maximize the yield of the end product. The performance of any particular batch operation is thus influenced by the type of end product—an

extension of exponential growth is advantageous for the efficient production of biomass (baker's yeasts, feed biomass) or primary metabolites (ethanol, acetic, citric, or
lactic acids), whereas in the case of secondary metabolite production, the exponential
phase is shortened (by the limitation of one nutrient, usually the source of nitrogen)
and the stationary phase is prolonged to achieve the maximum yield of the product.

Submerged batch cultivation can be used for the production of alcoholic beverages (beer, wine, and distilled spirits such as whisky, brandy, rum, and others),
organic acids used in the food industry either as acidifiers or as preservatives (citric,
acetic (vinegar), and lactic acids), and amino acids used as flavor enhancers (e.g.,
monosodium glutamate) or sweeteners (e.g., aspartate).

For distilled spirits, the fermentation of wort during Scotch whisky production is
taken as an example. Washbacks, simple cylindrical fermentation vessels (volume
250–500 m^3) for the production of distilled spirits are made either from wood or
from stainless steel. Although wood washbacks are difficult to clean and sanitize,
they are still used, especially in malt whisky distilleries. Wort to be fermented is
pumped to the washback, cooled to 20°C, and inoculated with either fresh or dried
yeast cells (Campbell, 2003).

The global production of citric acid reached 1.8×10^6 tonnes in 2010 (F.O. Licht
data); 90% of this was produced by microbial (*Aspergillus niger*) synthesis from
sugar- or starch-containing materials (sugar beet, sugarcane molasses, and corn)
and about 60% of this amount was consumed in the food industry. Although citric
acid can be produced at an industrial scale using surface liquid cultivation, solid-
state cultivation, or submerged liquid cultivation, nowadays, the latter predominates.
Submerged cultivation is carried out in stirred bioreactors (capacity 150–200 m^3)
or bubble columns (capacity up to 1000 m^3), usually operating aerobically for 4–10
days until the citric acid concentration reaches 10–15% w/v (Moresi and Parente,
2000; Soccol et al., 2006).

4.2.1.2 Fed-Batch Cultivation

Fed-batch culture represents a semi-open system in which one or more nutrients are
aseptically and gradually added to the bioreactor while the product is retained inside
(Figure 4.1b); that is, the volume of the culture broth in the bioreactor increases
within this time. The main advantages of fed-batch over batch cultures are: (a) the
possibility to prolong product synthesis, (b) the ability to achieve higher cell densities and thus increase the amount of the product, which is usually proportional to
the concentration of the biomass, (c) the capacity to enhance yield or productivity by
controlled sequential addition of nutrients, and (d) the feature of prolonged productive cultivation over the "unprofitable periods" when the bioreactor would normally
be prepared for a new batch.

Fed-batch is advantageously used in processes (a) where substrate inhibition or
catabolic repression is expected; this problem can be overcome by using a "safe"
concentration of the substrate in batch mode followed by feeding the remaining substrate within fed-batch operation, (b) where a Crabtree effect (repression of yeast
respiratory enzymes by high concentrations of glucose) is expected (de Deken,
1966); by gradual feeding of the substrate, the production of ethanol by yeasts can be
eliminated under aerobic conditions, (c) where a high cell density is required; a high

and constant specific growth rate can be maintained by exponential feeding of the substrate, (d) where a high production rate should be achieved; cell metabolism can be regulated by precise sequential feeding of nutrients, and (e) where a high viscosity of culture broth is expected (e.g., production of dextran or xanthan); a gradual dilution of the medium can overcome the problems of mixing and oxygen transfer.

There are many methods of adding a substrate to the bioreactor (either as a concentrated solution of a sole carbon and energy source or as a medium containing carbon plus other nutrients); the proper choice of the nutrient feeding rate can enhance the culture performance considerably since it influences cellular growth rate, cell physiology, and the rate of product formation. The common feeding strategies are: (a) discontinuous feeding, achieved by regular or irregular pulses of substrates and (b) regular continuous feeding of nutrients designed according to a precalculated profile (Figure 4.2) or based on the feedback control of online measured variables associated with cell growth and metabolism, for example, dissolved oxygen concentration, pH, CO_2, evolution rate, and biomass concentration.

The typical food fed-batch fermentations are large-scale production of baker's yeast, pure ethanol, which is further utilized for alcoholic beverages produced by mixing ingredients such as liquors or cordials, and submerged acetification for vinegar production.

Baker's yeast (*Saccharomyces cerevisiae*), which is distributed as compressed, dried, or instant biomass, is valued for its dough-leavening ability.

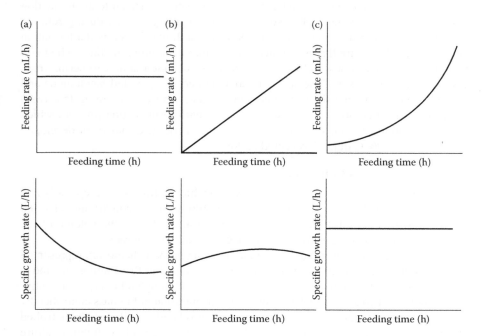

FIGURE 4.2 Relationship between precalculated profiles of substrate feeding and specific growth rates of cell culture. (a) Constant feeding rate, (b) linearly increasing feeding rate, and (c) exponential feeding rate.

Yeast production is the only technology in which the respiratory metabolism of *S. cerevisiae*, leading to high biomass yield, is stressed. Nevertheless, due to the Crabtree effect, that is, the formation of ethanol under aerobic conditions in the presence of excess substrate, the only alternative for producing baker's yeast is fed-batch cultivation. The main bottleneck of large-scale baker's yeast production is the control of nutrient medium inflow, which was traditionally based on empirical data. Currently, many feeding methods utilizing different approaches have been developed, for example, a logistic feeding profile (Borowiak et al., 2012), evolutionary optimization of genetic feeding algorithms (Yüzgeç et al., 2009), fuzzy control (Karakuzu et al., 2006), pulsed feeding optimization (Kasperski and Miśkiewitz, 2008), and others.

In the European Union (EU), the annual production of vinegar, comprising 10% (w/v) acetic acid, is estimated to be about 5×10^6 hL (García-García et al., 2009). Human knowledge of vinegar production dates back to ancient history and several distinct production methods have been used, for example, surface oxidation (Orleans process, slow vinegar production), the quick vinegar process using a trickle-bed bioreactor, or submerged acetification. The most modern process is submerged acetification, in which acetic acid (vinegar) is produced by *Acetobacter*-mediated oxidation of ethanol, and takes place in special "acetator" bioreactors. The most common acetator, a Frings Acetator®, differs from the usual bioreactors by using a special rotor/stator turbine aerator. The aerator with a double function (aeration and mixing), consists of a rotor placed under the bioreactor and connected to an air-suction pipe surrounded by a stator. The aerator is self respiring, that is, during rotation (speed about 1500 rpm), it sucks air and pumps liquid, which causes the formation of an air–liquid mixture that is radially injected into the culture medium. The foam is broken by a mechanical defoamer and because oxidation is an exothermal process, cooling is necessary. The acetators are operated in repeat fed-batch mode and one cultivation cycle in a single acetator can produce vinegar containing 15% (w/v) of acetic acid. In addition, a dual-stage high-strength fermentation process, which allows the culture to generate up to 20.5% (w/v) acetic acid, has been developed (Ebner et al., 1996; García-García et al., 2009).

4.2.1.3 Continuous Cultivation

Continuous culture represents an open system in which nutrients are aseptically and continuously added to the bioreactor, and the culture broth (containing cells and metabolites) is removed at the same time (Figure 4.1c), that is, the volume of the culture broth is constant due to a constant feed-in and feed-out rate.

Frequently, continuous culture is used as a synonym for a chemostat, represented by a constant specific growth rate of cells, which is equal to the dilution rate and is controlled by the availability of the limiting nutrient, although other types of continuous operation such as turbidistat (a constant concentration of biomass controlled by the dilution rate) or nutristat (a constant parameter related to cell growth controlled by the dilution rate) can be employed. The main advantages of continuous culture (chemostat) over the batch mode are (a) the possibility to set up optimum conditions for maximum and long-term product synthesis, (b) the ability to achieve stable product quality (the steady state is characterized by a homogeneous cell culture represented by a constant concentration of biomass and metabolites), and (c) a distinct reduction in "unprofitable" periods of the bioreactor operation.

In spite of these advantages, there are also several problems that hamper the extensive utilization of continuous operation on a large scale. These include (a) increased risk of contamination due to the pumping of the medium in and out of the bioreactor, (b) the danger of genetic mutations in the production strain in a long-term operation, and (c) additional investments may be required for technical facilities.

Despite the producer's proclamations, only a few food or feed production systems employing microorganisms are operated in a genuine continuous mode, where continuous fermentation is defined as a process running in one or more bioreactors at a stable dilution rate. One of the rare examples of this is fodder yeast (*Kluyveromyces fragilis*) production using spent sulfite liquor as the substrate, which is operated in the Czech Republic (http://www.biocel.cz/e_html/index.htm). The main reason why this process can be performed continuously is the low sensitivity of the production medium to possible contamination. At the same time, this process can be taken as a rare example of industrial-scale production using a highly valued substrate, lignocellulosic hydrolysate, which is obtained as a waste product from pulp production.

In other food applications such as in modern distilleries, a semicontinuous fermentation operated in a series of fermentors is usually used instead of genuine continuous fermentation. The current estimates are that only about 16% of ethanol in North America is produced in a continuous mode due to problems with contamination (Ingledew and Lin, 2011). Nevertheless, both batch and semicontinuous mode fermentations permit continuous distillation, which is the reason why the whole distillery production is often considered as continuous.

Continuous fermentation systems based on immobilized cell technology have also been studied in beer production. Although continuous beer fermentation has been tested as a promising technology for several decades, the number of industrial applications is still limited. The reasons include engineering problems (excess biomass and problems with CO_2 removal, optimization of operating conditions, clogging, and channeling of the reactor) and unrealized cost benefits (carrier price, complex, and unstable operation). The major obstacle hindering the extensive industrial exploitation of this technology is the difficulty in achieving the correct balance of sensory compounds in a short time typical for continuous systems. However, recent developments in reactor design and in our understanding of immobilized cell physiology, together with applications of novel carrier materials, could provide a new stimulus to research and potential applications of this promising technology (Brányik et al., 2008).

4.2.2 Solid Substrate Fermentation

The term "solid substrate fermentation" (SSF) or "solid substrate cultivation" (SSC) is used for systems where microorganisms are cultured on the surface of a concentrated water-insoluble substrate (usually containing polysaccharides as a carbon and energy source) with a low level of free water. This technique was developed in the Eastern countries, where it has been used for centuries for the production of traditional foods such as soy sauce, koji, miso, or sake, using different substrates and microorganisms. In the Western countries, it has not been widely exploited and its application is limited mainly to the production of industrial enzymes, certain food products, or feed supplements. In the following section, typical features, advantages, and problems of SSF together with relevant applications in both the food and feed industries are discussed.

4.2.2.1 Typical Features of Solid Substrate Fermentation

Solid substrate fermentation is characterized by very low water activity (the relative humidity of the gaseous phase in equilibrium with the moist solid is significantly below 1) (Hölker and Lenz, 2005); thus, the main features of this system are substantially different compared with classical submerged cultivation as shown in Table 4.1.

There are several advantages (Ali et al., 2011; Barrios-Gonzáles, 2012) of SSF over the conventional submerged technology such as (a) the use of a concentrated medium, resulting in a smaller reactor volume and lower capital investment costs, (b) the lower risk of contamination with yeasts and bacteria due to low moisture levels and substrate complexity, (c) the simplicity of the technology and low production of effluent water from the process, (d) the higher product yield and easier product recovery, and (e) the use of agricultural wastes as substrates for certain applications (e.g., feed supplements and cellulolytic enzymes).

4.2.2.2 Microorganisms and Substrates Used in SSF Processes

Filamentous fungi are preferable for SSF processes, mainly due to their abilities to (a) grow on substrates with reduced water activity, (b) penetrate their hyphae into

---

TABLE 4.1

Main Differences between Solid Substrate Fermentation and Submerged Cultivation

Solid Substrate Fermentation	Submerged Cultivation
Low water content of the cultivation medium (40–80%)	Liquid cultivation medium (~95% water content)
Three-phase system: gas–liquid–solid	Two-phase system: gas–liquid
Complex substrate insoluble in water, high local concentration of nutrients	Nutrients are dissolved in water, concentration of nutrients is lower
Nonhomogeneous system, gradient of nutrients	Homogeneous system
Microorganisms are grown on the surface of the solid substrate	Microorganisms are grown in the liquid medium
Gas–liquid and liquid–solid oxygen transfer	Gas–liquid oxygen transfer
Limitations in heat, oxygen, and nutrient transfer	Transport processes are usually not limited (exception can be oxygen transfer)
Heat is removed by using a stream of air or by placing the bioreactor into the temperature-controlled chamber	Cooling is achieved by bioreactor jacket cooling system
Process monitoring and control are difficult	Online monitoring and control of the process are common
High concentration of the product	Product is dissolved in the liquid phase

the solid substrate, and (c) produce exoenzymes (e.g., amylolytic and cellulolytic enzymes), which decompose the polysaccharides (the main carbon source often present in solid substrates). When using yeast, it is necessary to integrate a material pretreatment step (such as steam explosion, acid or alkali treatment, followed by enzymatic digestion or a combination of these) into the process (Jacob, 1991) or to use a mixed culture, wherein the complex cellulosic or starchy material is first degraded by other organisms (usually molds) that are able to produce extracellular enzymes and the released glucose is then consumed by yeasts (or less frequently by bacteria) yielding the desired product.

The efficiency of SSF is highly influenced by the selection of the solid substrate. The substrates suitable for SSF should ideally meet the following requirements: (a) have a porous solid matrix with a large surface area per unit volume (10^3–10^6 m²/cm³), (b) should sustain gentle compression and mixing, (c) should contain biodegradable carbohydrates, (d) its matrix should absorb water in the proportion of 1 to several times its dry weight, (e) should have relatively high water activity on the solid/gas interface to support microbial growth, and (f) should absorb the additionally added nutrients such as nitrogen sources (ammonia, urea, and peptides) and mineral salts (Orzua et al., 2009; Raimbault, 1998; Singhania et al., 2009). On the basis of these characteristics, the ideal material consists of small granular or fibrous particles that do not break or stick together. The commonly used solid substrates include wheat, wheat bran, soybean, rice barley, oats, and other cereals (Chisti, 1999).

4.2.2.3 Limitations and Challenges of Solid Substrate Fermentation

Although SSF has many advantages over liquid cultivation, the main challenges are poor heat and mass transfers within the substrate, and limited potential to monitor, online, key cultivation parameters (temperature, pH, dissolved oxygen, nutrient concentrations, or water content), and thus an inability to precisely control the microbial environment (Mitchel et al., 1999). The heat generated by microbial metabolism is a major bottleneck for process scale-up; the released heat can reach up to 3000 kcal from 1 kg of assimilated substrate, causing a radial gradient of 5°C/cm at the center of the reactor (Bellon-Maurel et al., 2003). Low water content in the system and poor heat conductivity of SSF substrates promote the heat gradient, which is accentuated by limited agitation (the molds are very sensitive to shear stress) and is reduced mostly by evaporative cooling, which can conversely exacerbate further water loss. In addition, the pH gradient caused mainly by the production of organic acids and the utilization of proteins poses another challenge; pH monitoring and control is difficult because no existing pH electrodes can operate in the absence of free water (Bellon-Maurel et al., 2003). Therefore, the variations in pH should be prevented by increasing substrate buffering capacity (e.g., by the addition of $CaCO_3$) or by the addition of urea, which can counteract acidification. Oxygen gradients can be reduced by aeration with moist air, which also plays a role in the desorption of carbon dioxide and the regulation of temperature and moisture levels.

4.2.2.4 Industrial Applications of SSF

In Europe, SSF in the food and feed contexts is usually used for the production of mold and ripening cheeses, production of fermented vegetables (e.g., sauerkraut or pickled cucumbers) or silage (preserved cattle feed), and in Asia for the production of fermented products from soya, rice, or corn.

An example of Oriental SSF is red yeast rice (red rice, ang-kak, anka, ankak, beni-koji, and other names), which is a product obtained after SSF of rice with different *Monascus* species, in the most usual case with *Monascus purpureus*. This product has been known in various Asian countries (China, Japan, Thailand, Indonesia, Taiwan, and Philippines) for centuries and can be used for food coloring, such as red koji (for the production of Kaoliang liqueur) or as a food supplement with an anticholesterolemic effect. The cultivation conditions of *Monascus* SSC of rice differ depending on the intended use. For food coloring, high pigment content (especially oligoketide, red-colored compounds, monascorubramine and rubro-punctamine, and their complexes with the amino group-containing compounds) is required; for koji production, the creation of hydrolytic enzymes is most important and in red yeast rice food supplement, a high concentration of monacolin K (lovastatin) is a necessary prerequisite (Martinkova and Patakova, 2000; Patakova, 2005). The main drawback of red yeast rice is its possible contamination with citrinin, which can be overcome by the selection of an appropriate producer strain or by its modification (Jia et al., 2010).

Traditionally, red yeast rice used for coloring of fish meals, soya products, vinegar, Beijing duck, and more recently, also frankfurters or sausages, is produced by successive washing, soaking, draining, and steaming of nonglutinous rice, which is then followed by inoculation with *Monascus*, 7-days incubation at 30°C, and product drying at 45°C or 60°C. Originally, the production took place in wobbled bamboo trays covered with banana leaves in which rice kernels were manually mixed or moistened if necessary (Lin et al., 2008). Nowadays, different types of trays, roller drums, or fluid-bed bioreactors are being used (Chiu et al., 2006).

Mushroom cultivation, which is well accepted by the public, can be considered as a product of a special single-cell protein (SCP) system, which utilizes various types of agricultural and forestry wastes and is relatively simple to perform. The special types of SSF are used for the production of edible mushrooms of *Agaricus, Pleurotus, Lentinula, Flammulina*, and other genera. These comprise the following steps: maintenance of mushroom culture, seed or spawn production, substrate pretreatment, growing, and cropping. The primary rot fungi such as oyster mushroom (*Pleurotus* spp.), shiitake (*Lentinula edodes*), winter mushroom (*Flammulina velutipes*), or paddy straw mushroom (*Volvariella* spp.) degrade the moistened lignocellulosic material such as straw, corn stover, saw dust, wood logs, or stumps as substrates (Rai, 2003). The button mushroom (*Agaricus bisporus*), unlike other species, is a secondary rot fungus that requires compost preparation prior to culture (Rai, 2003). To get high mushroom yields, and because many mushroom species need changes in the environmental conditions to form fruiting bodies, the original methods of outdoor or extensive cultivation have been replaced by intensive mushroom farming that requires the construction of specialized facilities that allow the precise control of environmental factors (temperature, humidity, light, and atmospheric gases). After thermal pretreatment, the substrate is placed into different immobile (shelves) or mobile (plastic bags or containers) systems. Substrate handling, spawn spreading (inoculation), substrate mixing, and moistening, along with cropping, can be mechanized, for example, special combined harvesters are used for button mushrooms (Chakravarty, 2011; Sánchez, 2004).

Among a wide range of SSF applications, the processes yielding protein-enriched agro-industrial materials that can be used as animal feed play an important role (Ugwuanyi et al., 2009). In the literature, there are reports on the production of protein-enriched animal feeds by SSF using starchy materials as substrates. Besides their characteristics described above, all organisms intended for animal consumption must comply with certain nutritional requirements such as amino acid composition and digestibility, and the absence of toxins, antibiotics, and mainly mycotoxins (Ghorai et al., 2009).

SSF has also been described as a protein enhancement factor for cereal grain and potato residues (Gélinas and Barrete, 2007). The protein content of wheat bran was increased fourfold by SSF of *Aspergillus terreus* (Sabry, 1993).

4.3 INTENSIFICATION OF FERMENTATION PROCESSES

There are several strategies available for the intensification of bioprocesses. Some of them focus on engineering aspects, whereas others exploit the tools of physiological modulation (selection or adaptation of microorganisms), mutagenesis, or genetic manipulation to improve the production strains. Examples of areas where a significant improvement of the fermentation processes can be achieved by engineering approaches are improved mass and heat transfer, reduction of power consumption, high-density cell cultures, and low-shear mixing (Chisti and Moo-Young, 1996). The performance of the bioprocess is both individually and synergistically influenced by all components of the production unit and related know-how (strain, bioreactor, media composition, feeding strategy, etc.). In addition, the biological elements of the process (microorganisms, animal and plant cells, and enzymes) are subject to many processing constraints (fragility, temperature, pH range, and hygienic design of the equipment). These facts place important practical limitations on the bioreactor and bioreaction engineering.

In the last few decades, there has been a significant progress in the area of process control and instrumentation for bioreactors. This has an economic importance because the optimum operation of a fermentation process is associated with improved productivity (high product concentration, high production rate) and savings in product separation. The ability to operate a process at high productivity requires a sound understanding of the biological requirements, process kinetics (limitation, inhibition), and transport phenomena (Erickson, 2011). The following sections provide the principles and examples of some bioprocess intensification methods.

4.3.1 IMMOBILIZED CELL TECHNOLOGY

The increased productivity of bioprocesses can be achieved through controlled contact of substrates with a high concentration of the active biocatalyst, enzyme, or microbial cells. These high-cell-density cultures can be created by feeding strategies, cell retention/recycle, or cell immobilization (Bumbak et al., 2011; Schiraldi et al., 2003; Verbelen et al., 2006). Among the strategies to create high cell-density cultures, cell immobilization is the most widely studied and applied in the food and

TABLE 4.2

Examples of Immobilized Cell Applications in Food and Beverage Production

Application	References
Wine	Yokotsuka et al. (1997); Divies and Cachon (2005)
Vinegar	de Ory et al. (2004)
Malolactic fermentation	Kosseva and Kennedy (2004)
Cider	Nedovic et al. (2000)
Beer	Willaert and Nedovic (2006); Brányik et al. (2012)
Meat processing	McLoughlin and Champagne (1994)
Aroma compounds	Rozenbaum et al. (2006)
Pigments	Fenice et al. (2000)
Sweetener	Kawaguti and Sato (2007); Krastanov et al. (2007)
Dairy products	Dimitrellou et al. (2009)
Baking	Plessas et al. (2007)
Nutraceuticals	Tsen et al. (2004)

beverage industries (Karel et al., 1985; Kosseva, 2011). Table 4.2 lists some references of research papers on food applications of immobilized cells.

The maximum immobilized biomass concentration achieved in continuous beer fermentation was up to 10 times greater than the free cell concentration at the end of the conventional batch fermentation (Nakanishi et al., 1993). However, the immobilization of microorganisms provokes different physiological responses when compared to low-cell-density cultures of free cells (Junter et al., 2002) and therefore, their application has to be carefully considered in terms of product quality. For example, the application of immobilized microorganisms in fermentation processes induces modifications in cell physiology due to mass transfer limitations, concentration gradients created by an immobilization matrix, and by aging of the immobilized biomass.

4.3.1.1 Immobilized Cell Physiology

An important factor influencing the growth and metabolic activity of immobilized cells is the microenvironment of the solid immobilization matrix, represented by parameters such as water activity, pH, oxygen, substrate and product concentration gradients, and mechanical stress. The interplay between the appropriate production strains and immobilization methods is very important in immobilized cell reactors and their suitable combination can improve both system performance and product quality. The importance of careful matching of the chosen yeast strain with the immobilization method and the suitable reactor arrangement was demonstrated in beer production (Mota et al., 2011). Although there are a variety of methods for investigating the metabolic state of immobilized cells (monitoring of cellular activity, microscopic, noninvasive, and destructive methods), acquiring the reliable data is still the limiting factor for process optimization (Kosseva, 2011; Pilkington et al., 1998). In addition, the data concerning the physiological conditions of immobilized microbial cells are rather complex, due to different matrices and variable system configurations, and therefore their interpretation is difficult (Junter and Jouenne, 2004).

Immobilization has been reported to activate some metabolic functions (substrate uptake, product formation, enzyme expression, and activity) of microbial cells (Lohmeier-Vogel et al., 1996; Norton and D'Amore, 1994 ; Van Iersel et al., 2000). According to some authors, the enhanced metabolic activity can also be attributed to surface-sensing responses in immobilized microbial cells (Prakasham et al., 1999) but the reasons are still a matter of controversy. Overall, conclusions should be very carefully drawn from the results, since sampling and sample treatment may also influence the measurements of immobilized cell physiology.

It has been shown that immobilized cells exhibit increased levels of deoxyribonucleic acid (DNA), structural carbohydrates (Doran and Bailey, 1986), glycogen (Galazzo and Bailey, 1990), and fatty acids (Hilge-Rotmann and Rehm, 1991), as well as modifications of cell proteome, cell wall, and cell membrane composition (Jirků, 1995; Parascandola et al., 1997). Not surprisingly, alterations in plasma membrane composition have a profound impact on several enzymes, sensor proteins, transporters, and membrane fluidity. Many reports also underline increased stress resistance of immobilized cells (Junter and Jouenne, 2004; Reimann et al., 2011). The increased resistance to inhibit substances (ethanol, pollutants, antimicrobial agents, etc.) can be ascribed to changes in the composition and organization at the level of the cell wall and plasma membrane (Jirků, 1999) and/or to the protective effect of the immobilization support (Norton et al., 1995).

4.3.1.2 Mass Transfer in Immobilized Cell Systems

The diffusional resistance to substrate transport from the bulk solution to the biocatalyst and the hindered diffusion of products in the opposite direction may represent the most significant mass transfer limitations arising from the use of immobilized cell technology. These mass transfer limitations constitute the most evident hypothesis to explain the often-observed decrease in immobilized cell growth rate and specific productivities as compared to free cell cultures (Abdel-Naby et al., 2000; Taipa et al., 1993). The typical immobilization materials exhibiting internal mass transfer limitations are polymeric matrices (Willaert and Baron, 1996). In these materials used for cell entrapment, the internal mass transfer limitations of cells by nutrients can be further influenced by the position of the cells, bead size, and structure of the polymer. Mass transfer limitations are crucial in immobilized cell systems when oxygen supply to cells and the removal of carbon dioxide are required. Oxygen transfer from the gas phase to the immobilized biocatalyst has long been recognized as the major rate-limiting step in aerobic immobilized cell processes. The most common option to improve mass transfer in these systems is to reduce the bead diameter (Groboillot et al., 1994).

Unlike polymeric matrices, the preformed porous (sintered glass) and nonporous (DEAE-cellulose, wood chips, and spent grains) carriers do not have the additional gel-diffusion barrier. However, depending on the porosity of the carrier and on the amount of biomass adsorbed in the pores, internal mass transfer limitations may also occur (Norton and D'Amore, 1994). In the case of nonporous carriers, internal mass transfer problems vary with the thickness of the cell layer (biofilm). The yeast adhered in a single layer of DEAE-cellulose showed similar metabolic activities (Šmogrovičová and Dömény, 1999) whereas multilayers of yeast attached to spent grains had a significantly lower specific sugar consumption rate as compared to free cells (Brányik et al., 2004).

4.3.2 Engineering Aspects of Process Intensification

Fermentation processes can be divided into two main categories based on the characteristic state of matter of the medium: solid substrate fermentation (Section 4.2.2) and submerged fermentation. Among the latter, the most common bioreactor configurations used in food and beverage applications are stationary particle bioreactors, such as packed-bed/fibrous-bed, trickle-bed reactors, and mixed (particle) bioreactors, such as fluidized bed, gas lift, bubble column, and stirred tank (Kosseva, 2011; Raspor and Goranovic, 2008; Willaert and Nedovic, 2006). The stationary particle bioreactors are either operated with immobilized cells (enzymes) or a mixture of free and immobilized cells (enzymes), and their typical internal mass transfer issues are discussed in Section 4.3.1.2. Mixed bioreactors may contain solely free or immobilized cells as well as their mixture. There are also bioreactor configurations that do not fit into the two previous categories. For example, rotating biological contactors (RBCs), also classified among moving surface reactors, where a biofilm grows on rotating disks partially or completely immersed in a liquid medium. The use of RBCs in food applications is rare, and is limited to the production of citric acid (Wang, 2000).

The selection of a suitable reactor design from numerous available types and configurations (Zhong, 2011) is a complex task and depends on various factors (Table 4.3). The importance of individual factors may change depending on the process requirements and the product characteristics. However, there is a need to have a fundamental understanding of the kinetics and transport limitations when a bioreactor is selected or when a new bioreactor is designed and constructed. Two-phase bioreactors are generally limited to anaerobic processes or to processes where

---

TABLE 4.3
Factors Influencing the Selection of Bioreactor Type

Free Cell Bioreactors

Nature of substrate and product	Biological requirements
Kinetics of product formation	Process conditions (T, pH, and Δp)
Mass transfer considerations	Hygienic considerations
Heat transfer considerations	Scale-up considerations
Hydrodynamic considerations	Ease of fabrication and reactor costs
Process control	Running costs

Immobilized Cell Bioreactors (Additional Factors)

Method of immobilization	Internal mass transfer in the biocatalyst
Carrier costs	Biocatalyst replacement/regeneration

gas–liquid mass transfer plays a marginal role. Conversely, in three-phase bioreactors, efficient mass transfer usually requires an intimate mixing of all three phases. Three-phase bioreactor design is an area in which significant process intensification can be achieved through the enhancement of gas–liquid mass transfer (Chisti and Moo-Young, 1996; Suresh et al., 2009).

4.3.2.1 Gas–Liquid Mass Transfer Considerations

In aerobic bioprocesses, oxygen is the key substrate due to its low solubility in aqueous media. Consequently, a continuous supply of oxygen into aerobic bioreactors is often needed. Therefore, the oxygen transfer rate (OTR) should be predicted prior to the choice/design and scale-up of a bioreactor. Many studies have been conducted to estimate the efficiency of oxygen transfer in different bioreactors and these have been reviewed in various works (Clarke and Correia, 2008; Garcia-Ochoa and Gomez, 2009; Kantarci et al., 2005; Suresh et al., 2009). Another area where gas–liquid mass transfer rate is crucially important (CO_2 supply and O_2 removal) is the construction and operation of photobioreactors used for the cultivation of photoautotrophic microorganisms (microalgae) with nutritional potential (Carvalho et al., 2006).

The dissolved oxygen concentration in aerobic cultures depends on the rate of oxygen transfer from the gas phase (usually air bubbles) to the liquid, on the rate at which oxygen is transported into the cells, and on rate of oxygen uptake by the microorganism. The transport of oxygen from air bubbles to the site of oxygen consumption can be described in a number of steps, among which oxygen diffusion through the liquid film surrounding the bubble shows the greatest resistance. The gas–liquid mass transfer rate is usually modeled according to the two-film theory and is characterized by the volumetric (gas–liquid) mass (oxygen) transfer coefficient ($k_L a$), while the driving force of the process is the difference between the concentration of oxygen at the interface (C^*) and that in the bulk liquid (C_L). In the case of large microbial pellets, immobilized cells, or fungal hyphae, the resistance in the liquid film surrounding the solid can also be significant (Blanch and Clark, 1997).

Oxygen transfer in aerobic bioprocesses is strongly influenced by the hydrodynamic conditions in the bioreactors. These conditions are known to be affected by the operational conditions (stirrer speed, superficial gas velocity, liquid circulation velocity, etc.), physicochemical properties of the culture (viscosity, density, and surface tension), bioreactor geometry, and also by the presence of oxygen-consuming cells (Garcia-Ochoa and Gomez, 2009).

Stirred tank bioreactors (STBRs) are widely used in a large variety of bioprocesses taking advantage of free cell (enzyme) suspensions. An industrial-scale STBR usually consists of a stainless-steel vessel, motor-driven impeller, and gas sparger positioned below the impeller. Aerated STBRs generally have high mass and heat transfer coefficients, good homogenization, and the capability of handling a wide range of superficial gas velocities. Mass transfer and mixing in STBRs are most significantly affected by stirrer speed, type and number of stirrers, and the gas flow rate (Garcia-Ochoa et al., 2011).

In bioreactors with a height-to-diameter ratio (H/T) above two, standard single-impeller systems were often found to have unsuitable operating parameters. Oxygen transfer in these geometries can be improved by using multiple-impeller configurations (approximately one impeller per each H/T = 1) that exhibit efficient gas distribution, increased gas holdup, superior liquid-flow characteristics, and lower power consumption per impeller as compared to single-impeller systems (Gogate et al., 2000).

Pneumatic bioreactors consist of a cylindrical vessel, into the bottom (usually) of which air (gas) is introduced to ensure aeration, mixing, and liquid circulation, without any moving mechanical parts. In pneumatically agitated reactors such as bubble columns (random liquid circulation) and airlift reactors (streamlined liquid circulation), the homogeneous shear environment compared to the local shear extremes in STBRs has enabled the successful cultivation of shear-sensitive cells such as mammalian and plant cells or mycelial fungi (Guieysse et al., 2011). In contrast, the lack of mechanical agitation can cause poor mixing in a highly viscous medium and serious foaming under high aeration. Airlift and stirred tank reactors exhibit comparable mass transfer capacities; however, airlift reactors can be superior to STBRs in terms of operating costs because of lower power consumption (Chisti, 1998). A further increase in the overall volumetric gas–liquid oxygen transfer coefficient ($k_L a$) in bubble column and airlift reactors was achieved by the installation of static mixers (Thakur et al., 2003) into the draft tubes and riser sections, respectively (Chisti et al., 1990; Goto and Gaspillo, 1992). The improvement of OTR achieved by static mixers is a result of air bubble breakup increasing the specific gas–liquid interfacial area (a). Industrial applications include the cultivation of a filamentous mold (Gavrilescu and Roman, 1995) and ethanol production in an airlift reactor (Vicente et al., 1999). However, in the second example, increased ethanol productivity was also achieved as a consequence of size reduction of yeast flocs, and thus improved liquid–solid mass transfer, provoked by the new riser design.

The predictions of OTR determined by a dynamic method in sterile culture medium in the absence of biomass, often underestimates the $k_L a$ value for the real bioprocess (Djelal et al., 2006). In fermentation processes, an enhancement of OTR was found to be due to oxygen consumed by the microorganisms, leading to a lower dissolved oxygen concentration in the bulk liquid (C_L). Simultaneously, mass transfer enhancement was also attributed to the presence of a dispersed phase (microorganisms) adsorbed onto the gas–liquid interface, influencing the oxygen adsorption rate and gas–liquid interfacial area. The extent of this enhancement was expressed as the biological enhancement factor (E). According to some studies, the E value of $k_L a$ can be up to 1.3 times that of the mass transfer coefficient determined for the system without microbial cells (Garcia-Ochoa and Gomez, 2005).

4.4 FUTURE PERSPECTIVES

The development of fermentation processes for the food and beverage industries aims at improving the productivity and product quality by means of process design, strain selection/construction, and process monitoring. In all these areas, there have emerged some very innovative ideas that could lead to economically attractive solutions.

With regard to (online) process monitoring, significant progress is required, particularly in the area of advanced instrumentation and sensor development, for solid substrate fermentations. The innovative techniques described so far include different sensor technologies, respirometry, x-rays, image analysis, infrared spectrometry, magnetic resonance imaging, and so on. However, for some of them, the main drawback is high cost, which makes these techniques unsuitable for large-scale applications (Bellon-Maurel et al., 2003).

One of the significant challenges in the bioreactor design is the improvement of large-scale photobioreactors and phycocultures (seaweed farms) for the production of micro- and macroalgae and algae-derived food products (Carvalho et al., 2006; Luening and Pang, 2003). Another prospective strategy to increase the metabolic productivity in bioprocesses is the use of suitably controlled ultrasonication. The beneficial effects of ultrasound can be exploited at the level of biocatalysts (cells and enzymes) and their function (e.g., cross-membrane ion fluxes, stimulated sterol synthesis, altered cell morphology, and increased enzyme activity) and sonobioreactor performance (mass transfer enhancement) (Chisti, 2003; Kwiatkowska et al., 2011).

The potential for genetic engineering in the field of food fermentation is indisputable and has been reviewed (Leisegang et al., 2006). However, the nutritional status of fermented foods can also be improved by the rational choice of food-fermenting microbes based on the understanding of their interaction with diet and human gastrointestinal microbiota. In this respect, fermented foods can be regarded as an extension of the food digestion and fermentation processes and can be steered toward beneficial health attributes (Vlieg et al., 2011).

ACKNOWLEDGMENT

The authors thank the Ministry of Education, Youth, and Sports for financial support in frame of projects MSM 6046137305 and Kontakt ME10146.

REFERENCES

Abdel-Naby, M.A., Reyad, R.M., and Abdel-Fattah, A.F. 2000. Biosynthesis of cyclodextrin glucosyltransferase by immobilized *Bacillus amyloliquefaciens* in batch and continuous cultures. *Biochem. Eng. J.* 5:1–9.

Ali, H.K.H. and Zulkali, M.M.D. 2011. Design aspects of bioreactors for solid-state fermentation: A review. *Chem. Biochem. Eng. Q.* 25:255–266.

Barrios-Gonzáles, J. 2012. Solid-state fermentation: Physiology of solid medium, its molecular basis and applications. *Process Biochem.* 47:175–185.

Bellon-Maurel, V., Orliac, O., and Christen, P. 2003. Sensors and measurements in solid state fermentation: A review. *Process Biochem.* 38: 881–896.

Blanch, H.W. and Clark, D.S. 1997. *Biochemical Engineering.* Boca Raton: Taylor and Francis.

Borowiak, D., Miśkiewitz, T., Miszczak, W., Cibis, E., and Krzywonos, M. 2012. A straightforward logistic method for feeding a fed-batch baker's yeast culture. *Biochem. Eng. J.* 60:36–43.

Brányik, T., Silva, D.P., Baszczyński, M., Lehnert, R., and Almeida e Silva, J.B. 2012. A review of methods of low alcohol and alcohol-free beer production. *J. Food Eng.* 108: 493–506.

Brányik, T., Vicente, A., Dostálek, P., and Teixeira, J. 2008. A review of flavour formation in continuous beer fermentations. *J. Inst. Brew.* 114:3–13.

Brányik, T., Vicente, A.A., Machado Cruz, J.M., and Teixeira, J.A. 2004. Continuous primary fermentation of beer with yeast immobilized on spent grains—The effect of operational conditions. *J. Am. Soc. Brew. Chem.* 62:29–34.

Bumbak, F., Cook, S., Zachleder, V., Hauser, S., and Kovar, K. 2011. Best practices in heterotrophic high-cell-density microalgal processes: Achievements, potential and possible limitations. *Appl. Microbiol. Biotechnol.* 91:31–46.

Campbell, I. 2003. Yeast and fermentation. In *Whisky, Technology, Production and Marketing*, ed. I. Russell, 115–150. London: Academic Press.

Carvalho, A.P., Meireles, L.A., and Malcata, F.X. 2006. Microalgal reactors: A review of enclosed system designs and performances. *Biotechnol. Prog.* 22:1490–1506.

Chakravarty, B. 2011. Trends in mushroom cultivation and breeding. *Austr. J. Agric. Eng.* 2:102–109.

Chisti, Y. 1998. Pneumatically agitated bioreactors in industrial and environmental bioprocessing: Hydrodynamics, hydraulics, and transport phenomena. *Appl. Mech. Rev.* 51:33–112.

Chisti, Y. 1999. Solid substrate fermentations, enzyme production, food enrichment. In *Encyclopedia of Bioprocess Technology —Fermentation, Biocatalysis and Bioseparation,* eds., M.C. Flickinger, and S.W. Drew, 2446–2460. Wiley & Sons. Online version available at: http://www.knovel.com/web/portal/browse/display?_EXT_KNOVEL_DISPLAY_bookid=678&VerticalID=0

Chisti, Y. 2003. Sonobioreactors: Using ultrasound for enhanced microbial productivity. *Trends Biotechnol.* 21:89–93.

Chisti, Y., Kasper, M., and Moo-Young, M., 1990. Mass transfer in external-loop airlift bioreactors using static mixers. *Can. J. Chem. Eng.* 68:45–50.

Chisti, Y. and Moo-Young, M. 1996. Bioprocess intensification through bioreactor engineering. *Chem. Eng. Res. Des.* 74(A5):575–583.

Chiu, C.-H., Ni, K.-H., Guu, Y.-K., and Pan, T.-M. 2006. Production of red mould rice using a modified Nagata type koji maker. *Appl. Microbiol. Biotechnol.* 73:297–304.

Clarke, K.G. and Correia, L.D.C. 2008. Oxygen transfer in hydrocarbon–aqueous dispersions and its applicability to alkane bioprocesses: A review. *Biochem. Eng. J.* 39:405–429.

de Deken, R.H. 1966. The Crabtree effect: A regulatory system in yeast. *J. Gen. Microbiol.* 44:149–156.

de Ory, I., Romero, L.E., and Cantero, D. 2004. Optimization of immobilization conditions for vinegar production. Siran, wood chips and polyurethane foam as carriers for *Acetobacter aceti. Process Biochem.* 39:547–555.

Dimitrellou, D., Kourkoutas, Y., Koutinas, A.A., and Kanellaki, M. 2009. Thermally-dried immobilized kefir on casein as starter culture in dried whey cheese production. *Food Microbiol.* 26:809–820.

Divies, C. and Cachon, R. 2005. Wine production by immobilized cell systems. In *Applications of Cell Immobilization Biotechnology*, eds., V. Nedovic, and R. Willaert, 285–293. Heidelberg: Springer Verlag.

Djelal, H., Larher, F., Martin, G., and Amrane, A. 2006. Effect of the dissolved oxygen on the bioproduction of glycerol and ethanol by *Hansenula anomala* growing under salt stress conditions. *J. Biotechnol.* 125:95–103.

Doran, P. and Bailey, J.E. 1986. Effects of immobilization on growth, fermentation properties, and macromolecular properties of *Saccharomyces cerevisiae* attached to gelatin. *Biotechnol. Bioeng.* 28:73–87.

Ebner, H., Sellmer, S., and Follmann, H. 1996. Acetic acid. In *Biotechnology: Products of Primary Metabolism,* eds. H.J. Rehm, and G. Reed, 382–401. Weinheim: Wiley-VCH Verlag GmbH.

Erickson, L.E. 2011. Bioreactors for commodity products. In *Comprehensive Biotechnology (2nd ed.),* eds. M. Moo-Young, M. Butler, and C. Webb et al., 179–197. Amsterdam: Elsevier B.V.

Fenice, M., Federici, F., Selbmann, L., and Petruccioli, M. 2000. Repeated batch production of pigments by immobilized *Monascus purpureus. J. Biotechnol.* 80:271–276.

Galazzo, J.L. and Bailey, J.E. 1990. Growing *Saccharomyces cerevisiae* in calcium–alginate beads induces cell alterations which accelerate glucose conversion to ethanol. *Biotechnol. Bioeng.* 36:417–426.

García-García, I., Santos-Dueñas, I.M., Jiménez-Ot, C., Jiménez-Hornero, J.E., and Bonilla-Venceslada, J.L. 2009. Vinegar engineering. In *Vinegars of the World*, eds. L. Solieri,, and P. Giudici, 97–120. Milan, Italy: Springer-Verlag.

Garcia-Ochoa, F. and Gomez, E. 2005. Prediction of gas–liquid mass transfer in sparged stirred tank bioreactors. *Biotechnol. Bioeng.* 92:761–772.

Garcia-Ochoa, F. and Gomez, E. 2009. Bioreactor scale-up and oxygen transfer rate in microbial processes: An overview. *Biotechnol. Adv.* 27:153–176.

Garcia-Ochoa, F., Santos, V.E., and Gomez, E. 2011. Stirred tank bioreactors. In *Comprehensive Biotechnology (2nd ed.)*, eds., M. Moo-Young, M. Butler, and C. Webb et al., 179–197. Amsterdam: Elsevier B.V.

Gavrilescu, M. and Roman, R.V. 1995. Cultivation of a filamentous mould in an airlift reactor. *Acta. Biotechnol.* 15:323–335.

Gélinas, P. and Barrete, J. 2007. Protein enrichment of potato processing waste through yeast fermentation. *Bioresour. Technol.* 98:1138–1143.

Ghorai, S. Banik, S.P., Verma, D., Chowdhury, S., Mukherjee, S., and Khowala, S. 2009. Fungal biotechnology in food and feed processing. *Food Res. Int.* 42:577–587.

Gogate, P.R., Beenackers, A.A.C.M., and Pandit, A.B. 2000. Multiple-impeller systems with a special emphasis on bioreactors: A critical review. *Biochem. Eng. J.* 6:109–144.

Goto, S., and Gaspillo, P.D. 1992. The effect of static mixer on mass transfer in draft tube bubble column and in external loop column. *Chem. Eng. Sci.* 47:3533–3539.

Groboillot, A., Boadi, D.K., Poncelet, D., and Neufeld, R.J. 1994. Immobilization of cells for application in the food industry. *Crit. Rev. Biotechnol.* 14:75–107.

Guieysse, B., Quijano, G., and Munoz, R. 2011. Airlift bioreactors. In *Comprehensive Biotechnology (2nd ed.)*, eds. M. Moo-Young, M. Butler, and C. Webb et al., 199–212. Amsterdam: Elsevier B.V.

Hilge-Rotmann, B. and Rehm, H.J. 1991. Relationship between fermentation capability and fatty acid composition of free and immobilized *Saccharomyces cerevisiae*. *Appl. Microbiol. Biotechnol.* 34:502–508.

Hölker, U. and Lenz, J. 2005. Solid-state fermentation—Are there any biotechnological advantages? *Curr. Opin. Microbiol.* 8:301–306.

Ingledew, W.M.M. and Lin, Y.-H. 2011. Ethanol from starch-based feedstocks. In *Comprehensive Biotechnology (2nd ed.)*, eds. M. Moo-Young, M. Butler, and C. Webb et al., 37–49. Amsterdam: Elsevier B.V.

Inui, M., Vertes, A.A., and Yukawa, H. 2010. Advanced fermentation technologies. In *Biomass to Biofuels*, eds. A.A. Vertes, N. Qureshi, H.P. Blashek, and H. Yukawa, 311–330. Oxford, UK: Blackwell Publishing, Ltd.

Jacob, Z. 1991. Enrichment of wheat bran by *Rhodotorula gracilis* through solid-state fermentation, *Folia Microbiol.*, 36(1):86–91.

Jia, X.Q., Xu, Z.N., Zhou, L.P., and Sung, C.K. 2010. Elimination of the mycotoxin citrinin in the industrial important strain *Monascus purpureus* SM001. *Metabolic Eng.* 12:1–7.

Jirků, V. 1995. Covalent immobilization as a stimulus of cell wall composition changes. *Experientia* 51:569–571.

Jirků, V. 1999. Whole cell immobilization as a means of enhancing ethanol tolerance. *J. Ind. Microbiol. Biotechnol.* 22:147–151.

Junter, G.A., Coquet, L., Vilain, S., and Jouenne, T. 2002. Immobilized-cell physiology: Current data and the potentialities of proteomics. *Enzyme Microb. Technol.* 31:201–212.

Junter, G.A. and Jouenne, T. 2004. Immobilized viable microbial cells: From the process to the proteome... or the cart before the horse. *Biotechnol. Adv.* 22:633–658.

Kantarci, N., Borak, F., and Ulgen, K.O. 2005. Bubble column reactors. *Process. Biochem.* 40:2263–2283.

Karakuzu, C., Türker, M., and Öztürk, S. 2006. Modelling, on-line state estimation and fuzzy control of production scale fed-batch baker's yeast fermentation. *Control Eng. Pract.* 14:959–974.

Karel, S., Libicki, S., and Robertson, C. 1985. The immobilization of whole cells: Engineering principles. *Chem. Eng. Sci.* 40:1321–1354.

Kasperski, A. and Miśkiewitz, T. 2008. Optimization of pulsed feeding in a baker's yeast process with dissolved oxygen concentration as a control parameter. *Biochem. Eng. J.* 40:321–327.

Kawaguti, H.Y. and Sato, H.H. 2007. Palatinose production by free and Ca–alginate gel immobilized cells of *Erwinia* sp. *Biochem. Eng. J.* 36:202–208.

Kosseva, M.R. and Kennedy, J.F. 2004. Encapsulated lactic acid bacteria for control of malolactic fermentation in wine. *Artif. Cell. Blood Sub.* 32:55–65.

Kosseva, M.R. 2011. Immobilization of microbial cells in food fermentation processes. *Food Bioprocess. Technol.* 4:1089–1118.

Krastanov, A., Blazheva, D., and Stanchev, V. 2007. Sucrose conversion into palatinose with immobilized *Serratia plymuthica* cells in a hollow-fibre bioreactor. *Process Biochem.* 42:1655–1659.

Kwiatkowska, B., Bennett, J., Akunna, J., Walker, G.M., and Bremner, D.H. 2011. Stimulation of bioprocesses by ultrasound. *Biotechnol. Adv.* 29:768–780.

Leisegang, R., Nevoigt, E., Spielvogel, A., Kristan, G., Niederhaus, A., and Stahl, U. 2006. Fermentation of food by means of genetically modified yeast and filamentous fungi. In *Genetically Engineered Food (2nd ed.)*, eds. K.J. Heller, 64–94. Weinheim: Wiley-VCH Verlag GmbH & Co.

Lin, Y.-L., Wang, T.-H., Lee, M.-H., and Su, N.-W. 2008. Biologically active components and nutraceuticals in the *Monascus*-fermented rice: A review. *Appl. Microbiol. Biotechnol.* 77:965–973.

Lohmeier-Vogel, E.M., McIntyre, D.D., and Vogel, H.J. 1996. Phosphorus-31 and carbon-13 nuclear magnetic resonance studies of glucose and xylose metabolism in cell suspensions and agarose-immobilized cultures of *Pichia stipitis* and *Saccharomyces cerevisiae*. *Appl. Environ. Microbiol.* 62:2832–2838.

Luening, K. and Pang, S. 2003. Mass cultivation of seaweeds: Current aspects and approaches. *J. Appl. Phycol.* 15:115–119.

Martinkova, L. and Patakova, P. 2000. Monascus. In *Encyclopedia of Food Microbiology*, eds., R.K. Robinson, C.A. Batt, and P.D. Patel, 1481–1487. San Diego: Academic Press.

McLoughlin, A. and Champagne, C.P. 1994. Immobilized cells in meat fermentation. *Crit. Rev. Biotechnol.* 14:179–192.

Mitchell, D.A., Stuart, D.M., and Tanner, R.D. 1999. Solid state fermentation, microbial growth kinetics. In *Encyclopedia of Bioprocess Technology—Fermentation, Biocatalysis and Bioseparation*, eds. M.C. Flickinger and S.W. Drew, 2408–2428. USA: Wiley & Sons.

Monod, J. 1949. The growth of bacterial cultures. *Annu. Rev. Microbiol.* 3: 371–394.

Moresi, M. and Parente, E. 2000. Fermentation (industrial)/production of organic acids. In *Encyclopedia of Food Microbiology*, eds., R.K. Robinson, C.A. Batt, and P.D. Patel, 705–717. San Diego: Academic Press.

Mota, A., Novák, P., Macieira, F., Vicente, A.A., Teixeira, J.A., Šmogrovićová, D., and Brányik, T. 2011. Formation of flavor active compounds during continuous alcohol-free beer production: The influence of yeast strain, reactor configuration and carrier type. *J. Am. Soc. Brew. Chem.* 69:1–7.

Nakanishi, K., Murayama, H., Nagara, A., and Mitsui, S. 1993. Beer brewing using an immobilized yeast bioreactor system. *Bioprocess Technol.* 16:275–289.

Norton S. and D'Amore, T. 1994. Physiological effects of yeast cell immobilization: Applications for brewing. *Enzyme Microb. Technol.* 16:365–375.

Norton, S., Watson, K., and D'Amore, T. 1995. Ethanol tolerance of immobilized brewers' yeast cells. *Appl. Microbiol. Biotechnol.* 43:18–24.

Orzua, M.C., Mussato, S.I., Contreras-Esquivel, J.C., Rodriguez, R., de la Garza, H., Teixeira, J.A., and Aguilar, C.N. 2009. Exploitation of agro industrial wastes as immobilization carrier for solid-state fermentation. *Ind. Crop. Prod.* 30:24–27.

Parascandola, P., De Alteriis, E., Sentandreu, R., and Zueco, J. 1997. Immobilization and ethanol stress induce the same molecular response at the level of the cell wall in growing yeast. *FEMS Microbiol. Lett.* 150:121–126.

Patakova, P. 2005. Red yeast rice. In *McGraw-Hill Yearbook of Science and Technology*, 286–288, New York: McGraw-Hill.

Pilkington, P.H., Margaritis, A., Mensour, N.A., and Russel, I. 1998. Fundamentals of immobilized yeast cells for continuous beer fermentation: A review. *J. Inst. Brew.* 104:19–31.

Plessas, S., Bekatorou, A., and Kanellaki, M. et al. 2007. Use of immobilized cell biocatalysts in baking. *Process Biochem.* 42:1244–1249.

Prakasham, R.S., Kuriakose, B., and Ramakrishna, S.V. 1999. The influence of inert solids on ethanol production by *Saccharomyces cerevisiae*. *Appl. Biochem. Biotechnol.* 82:127–134.

Rai, R.D. 2003. Production of edible fungi. In *Fungal Biotechnology in Agricultural, Food and Environmental Applications,* eds. D.K. Arora, P.D. Bridge, and D. Bhatnagar, 382–404. New York: CRC Press.

Raimbault, M. 1998. General and microbiological aspects of solid substrate fermentation. *Electron. J. Biotechnol.* 1:174–188.

Raspor, P. and Goranovic, D. 2008. Biotechnological applications of acetic acid bacteria. *Crit. Rev. Biotechnol.* 28:101–124.

Reimann, S., Grattepanche, F., and Benz, R. et al. 2011. Improved tolerance to bile salts of aggregated *Bifidobacterium longum* produced during continuous culture with immobilized cells. *Bioresour. Technol.* 102:4559–4567.

Rozenbaum, H.F., Patitucci, M.L., Antunes, O.A.C., and Pereira, N. 2006. Production of aromas and fragrances through microbial oxidation of monoterpenes. *Braz. J. Chem. Eng.* 23:273–279.

Sabry, S.A. 1993. Protein-enrichment of wheat bran using *Aspergillus terreus*. *Microbiologia* 9:125–133.

Sánchez, C. 2004. Modern aspects of mushroom culture technology. *Appl. Microbiol. Biotechnol.* 64:756–762.

Schiraldi, Ch., Adduci, V., Valli, V., Maresca, C., Giuliano, M., Lamberti, M., Carteni, M., and De Rosa, M. 2003. High cell density cultivation of probiotics and lactic acid production. *Biotechnol. Bioeng.* 82:213–222.

Singhania, R.R., Patel, A.K., Soccol, C.R., and Pandey, A. 2009. Recent advances in solid-state fermentation. *Biochem. Eng. J.* 44:13–18.

Šmogrovičová, D. and Dömény, Z. 1999. Beer volatile by-product formation at different fermentation temperature using immobilized yeast. *Process Biochem.* 34:785–794.

Soccol, C.R., Vandenberghe, L.P.S., Rodrigues, C., and Pandey, A. 2006. New perspectives for citric acid production and application. *Food Technol. Biotechnol.* 44:141–149.

Suresh, S., Srivastava, V.C., and Mishra, I.M. 2009. Critical analysis of engineering aspects of shaken flask bioreactors. *Crit. Rev. Biotechnol.* 29:255–278.

Taipa, M.A., Cabral, J.M.S., and Santos, H. 1993. Comparison of glucose fermentation by suspended and gel-entrapped yeast cells: An *in vivo* nuclear magnetic resonance study. *Biotechnol. Bioeng.* 41:647–653.

Thakur, R.K., Vial, C.H., Nigam, K.D.P., Nauman, E.B., and Djelveh, G. 2003. Static mixers in the process industries—A review. *Chem. Eng. Res. Des.* 81(A7):787–826.

Tsen, J.-H., Lin, Y.-P., and King, V.A.-E. 2004. Fermentation of banana media by using κ-carrageenan immobilized *Lactobacillus acidophilus*. *Int. J. Food Microbiol.* 91:215–220.

Ugwuanyi, J.O., McNeil, B., and Harvey, L.M. 2009. Production of protein-enriched feed using agro-industrial residues as substrates. In *Biotechnology for Agro-Industrial Residues Utilization*, eds., P. Singh Nigam and A. Pandey, 77–104. Springer Science +Business Media B.V.

Van Iersel, M.F.M., Brouwer-Post, E., Rombouts, F.M., and Abee, T. 2000. Influence of yeast immobilization on fermentation and aldehyde reduction during the production of alcohol-free beer. *Enzyme Microb. Technol.* 26:602–607.

Verbelen, P.J., De Schutter, D.P., Delvaux, F., Verstrepen, K.J., and Delvaux, F.R. 2006. Immobilized yeast cell systems for continuous fermentation applications. *Biotechnol. Lett.* 28:1515–1525.

Vicente, A.A., Dluhy, M., and Teixera, J.A. 1999. Increase of ethanol productivity in an airlift reactor with a modified draught tube. *Can. J. Chem. Eng.* 77:497–502.

Vlieg, J.E.T.H., Veiga, P., Zhang, C.H., Derrien, M., and Zhao, L. 2011. Impact of microbial transformation of food on health—From fermented foods to fermentation in the gastrointestinal tract. *Curr. Opin. Biotech.* 22:211–219.

Wang, J. 2000. Production of citric acid by immobilized *Aspergillus niger* using rotating biological contactor (RBC). *Bioresour. Technol.* 75:245–247.

Willaert, R.G. and Baron, G.V. 1996. Gel entrapment and micro-encapsulation: Methods, applications and engineering principles. *Rev. Chem. Eng.* 12:5–205.

Willaert, R. and Nedovic, V. 2006. Primary beer fermentation by immobilized yeast—A review on flavor formation and control strategies. *J. Chem. Technol. Biotechnol.* 81:1353–1367.

Yokotsuka, K., Yajima, M., and Matsudo, T. 1997. Production of bottle-fermented sparkling wine using yeast immobilized in double-layer gel beads or strands. *Am. J. Enol. Viticult.* 48:471–481.

Yüzgeç, U., Türker, M., and Hocalar, A. 2009. On-line evolutionary optimization of an industrial fed-batch yeast fermentation process. *ISA Trans.* 48:79–92.

Zhong, J-J. 2011. Bioreactor engineering. In *Comprehensive Biotechnology (2nd ed.)*, eds. M. Moo-Young, M. Butler, and C. Webb et al., 165–177. Amsterdam: Elsevier B.V.

5 Meet the Stem Cells
Production of Cultured Meat from a Stem Cell Biology Perspective

Bas Brinkhof, Bernard A.J. Roelen,
and Henk P. Haagsman

CONTENTS

5.1   CONCEPT OF *IN VITRO* MEAT

5.1.1   Meat Consumption

According to a latest estimate, the world's population will have grown to 9 billion people in 2050 and the life expectancy will have increased from 67.6 in 2009 to 75.5 years (United Nations 2009). Approximately, 70% of this population will be urban and it is expected that gross income levels will have increased. To feed the increased, more urban, and richer population, it has been calculated that food production must increase by 70%. As a result of urbanization and income growth, food diets will most likely change. Indeed, meat consumption is expected to grow with an extra 74% in 2050 to 470 million tons (Food and Agriculture Organization of the United Nations 2009).

5.1.2   Environmental Damage Associated with Meat Production

Since the relative use of natural resources is high and because their waste considerably contributes to global pollution, the production of meat has a rather dramatic impact on the environment. Currently, the livestock uses 26% of ice-free land for grazing and the area used for feedcrop production is about 33% of all cropland, totaling to 30% of the planet's land surface. A further increase in land use is bound to come at the cost of natural habitats, including rain forests. The livestock is also responsible for 18% of the global greenhouse gas emissions, in particular carbon dioxide, methane, and nitrous oxide. Indeed, in terms of carbon dioxide equivalents, the impact of gaseous emissions from livestock production is more than that from the total transportation sector. In addition, meat production is responsible for over 8% of human global water use, the latter mainly for feedcrop irrigation (Food and Agriculture Organization of the United Nations 2009).

In a business as-usual perspective: (1) livestock production will continue to grow, (2) the pressure of crop agriculture to expand will remain high and the associated negative environmental impact will grow, (3) livestock contribution to anthropogenic greenhouse gas emissions will increase, and (4) livestock-induced desertification of arid and semiarid areas will continue.

Since the livestock's environmental impact is considerable, it is important to find alternatives for the conventional meat production (Steinfeld et al. 2006). The possible alternatives are: (1) make the current meat production more sustainable, (2) increase product quality (and reduce meat consumption), (3) replace animal protein by plant or fungal proteins, and (4) make animal protein without animals. All these solutions require innovative research. Here, an approach is presented to produce meat without the use of animals, by using stem cell and tissue regeneration technology instead.

5.1.3   Cultured Meat

Many problems associated with animal production could be circumvented by making edible products from skeletal muscle cells, cultured from stem cells, and outside the animal (*in vitro*) in a bioreactor. The idea of culturing the animal tissue *in vitro* for human consumption is not new. Already in the early twentieth century, Winston Churchill proposed the idea to grow the animal tissue instead of whole animals (Churchill 1932, 24–27). Unfortunately, this idea was at that time never experimentally explored. Supported by National Aeronautics and Space Administration (NASA), in 2002, the possibilities were explored to increase goldfish (*Carassius auratus*) muscle explants by *in vitro* culture and a limited increase in muscle mass was established. After preparation by culinary chefs, a test panel judged the products acceptable as food, although according to Food and Drug Administration rules, the tissue could not be consumed (Benjaminson et al. 2002, 879–889). Although promising, this study was not continued.

A prerequisite for *cultured meat* is a group of proliferating cells. The skeletal muscle tissue consists of several different cell types, but in adult organisms, most of these cells are differentiated and have lost proliferative capacity. In contrast, stem cells exhibit two important characteristics: (1) the capacity to self-renew and (2) the ability to give rise to at least one differentiated cell type. It has been hypothesized that stem cells can be used to replace damaged or diseased tissue and thereby can have an enormous potential in regenerative medicine. Alternatively, stem cells from farm animals may be used to produce large masses of undifferentiated cells that upon differentiation produce muscle fibers that comprise animal protein: cultured meat.

Different types of stem cells exist; perhaps the simplest subdivision in stem cells that can be made is that between embryonic stem (ES) and nonembryonic (adult) stem cells. Apart from these, a type of stem cell can be generated by the introduction of genes or their products in differentiated cells. These so-called induced pluripotent stem (iPS) cells have been demonstrated to be very similar to ES cells. ES cells are derived from preimplantation embryos and have the capacity to differentiate to all cell types of the animal: this is referred to as *pluripotency*. Adult stem cells are derived from tissues of a fetal or postnatal animal and are considered to have reduced developmental potential. In contrast to ES cells, the proliferating pace of adult stem cells is slow and the differentiation capacity is restricted (Katsumoto et al. 2010, 115–129). As in most tissues, adult stem cells in the skeletal muscle are a rare cell type, which makes identification, harvesting, and culturing rather complex and laborious. Nevertheless, adult stem cells have been considered as the most promising cell type for use in the production of cultured meat, so far (Bhat and Bhat 2011, 441–459).

In this chapter, we try to focus on the aspects of stem cell biology aimed at the *in vitro* generation of the skeletal muscle tissue that can be used for the production of cultured meat.

5.2 STEM CELLS

Currently, what we consider meat for consumption is made of the muscle tissue, composed of skeletal muscle, fiber, blood, connective tissue, and fat. Most of the cells that form these tissues are differentiated and have lost the capacity to divide. On the other hand, stem cells can divide, differentiate, and, importantly, self-renew. This self-renewing capacity means that as a result of cell division, either both daughter cells (symmetric division) or one daughter cell (asymmetric division) retain the stem cell characteristics and have the same developmental potential (Shenghui et al. 2009, 377–406).

The developmental potency of stem cells can vary from pluripotent to unipotent (Figure 5.1) (Stocum 2001, 429–442).

1. Pluripotent: These cells can differentiate into all cells of the fetal and adult organism. ES cells are an example of pluripotent stem cells (Seydoux and Braun 2006, 891–904). Apart from the capacity to differentiate to germ cells, the progeny of these cells is committed to either one of the three germ layers; (1) endoderm; eventually forming, for instance, the stomach, the liver, the colon, and the lungs; (2) ectoderm; differentiating into, for example, the central nervous system (CNS), hair, skin, and mammary glands; and (3) mesoderm; from what, among others, blood cells, bone, heart tissue, and skeletal muscle are formed (Gilbert 1994, 323–368).

2. Multipotent: These stem cells have the potential to give rise to multiple, though limited cell lineages. As an example, hematopoietic stem cells can differentiate into several types of blood cells but in principle, they cannot differentiate into nerve cells or the brain tissue (Zandstra and Nagy 2001, 275–305).

3. Unipotent: A cell of this type can only differentiate into one cell type. Unipotent stem cells can be found, for example, in the skin (Blanpain et al. 2007, 445–458).

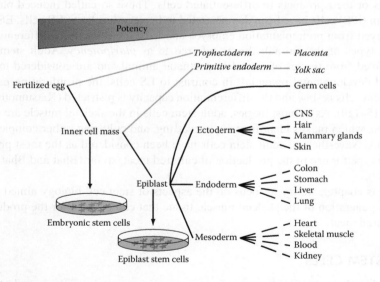

FIGURE 5.1 Developmental potential of cell types during embryogenesis. The cleavage divisions of the fertilized egg lead to the formation of extraembryonic (gray) and embryonic (black) structures. The ICM is composed of pluripotent cells capable of the formation of all cells and tissues of an organism. The position of the cell type is relative and not absolute. The dashed line represents a pathway with intermediate cell types.

5.3 EMBRYONIC STEM CELLS

During mammalian embryonic development, the fertilized egg (zygote) develops into a blastocyst-stage embryo by a series of cleavage divisions (Figure 5.2). The blastocyst is composed of morphologically distinct tissue types: (1) trophectoderm (TE) and (2) inner cell mass (ICM) from which a selection of cells will further differentiate into (3) the primitive endoderm (PE). The rest of the ICM becomes the epiblast and will eventually form the entire embryo. During the initial development of the blastocyst stage, the cell fate is demarcated by the expression of several proteins. The expression of the transcription factor CDX2 marks TE differentiation whereas the expression of the transcription factor OCT4 represents ICM fate. These two proteins mutually inhibit each other's expression, thereby actively promoting lineage segregation. During a second-fate decision, GATA6 expression in cells antagonizes NANOG expression and these GATA6-expressing cells segregate from the ICM to form the primitive endoderm, whereas the opposite mechanism results in the formation of the epiblast (Zernicka-Goetz et al. 2009, 467–477). *In vivo*, the TE and PE develop into extraembryonic tissues that constitute a part of the placenta and yolk sacs whereas the ICM develops into the fetus and finally into an organism.

ES cells are derived from ICM cells *in vitro* and bear almost indefinite self-renewal potential as well as the ability to generate all cell types of the body. These characteristics distinguish ES cells from the tissue-specific adult stem cells, which have more limited self-renewal and developmental potential (Shenghui et al. 2009,

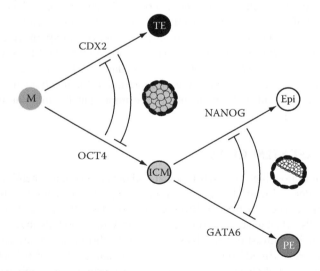

FIGURE 5.2 Fate decision during blastocysts formation. During cleavage divisions, a zygote develops into a morula. During two fate-determining stages, first TE and ICM cells are formed depending primarily on Cdx2 and Oct4 expressions, respectively. During the second stage from ICM cells, PE cells emerge predominantly expressing Gata6 whereas the epiblast cells express Nanog. Epi, epiblast; ICM, inner cell mass; M, morula; PE, primitive endoderm; and TE, trophectoderm.

377–406). ES cells were generated for the first time in 1981 from mouse blastocyst-stage embryos (Evans and Kaufman 1981, 154–156; Martin 1981, 7634–7638). It only seemed possible to generate ES cells from few mouse strains. The efforts to use other strains and species remained unsuccessful. Finally, in 1995, researchers had been able to generate ES cells from monkey embryos and later in 1998, they were able to generate ES cells from human embryos (Thomson et al. 1995, 7844–7848; 1998, 1145–1147). Again, it took a decade to obtain ES cells from another species' embryos; the rat (Buehr et al. 2008, 1287–1298). In almost 30 years, from only two species of rodents and two species of primates, ES cell lines have been generated, illustrating the complexity. The conditions to generate and maintain the ES cells from these species are not the same as will be discussed in Section 5.5. The differences demonstrate the difficulties in generating ES cells and therefore demonstrate the difficulty in generating livestock species derived ES cells.

To demonstrate the establishment of a *bona fide* ES cell line, the cells should fulfill a number of criteria. For instance, the cells should express several genes and their corresponding proteins that have been associated with pluripotency, such as *Nanog*, *Oct4*, and *Sox2*. To further demonstrate the pluripotent character of ES cells, these cells can be injected into an immune deficient mouse. If the cells are indeed pluripotent, the mouse will develop tumors that originated from the injected cells and these tumors will consist of tissues from the three germ layers, that is, endoderm, ectoderm, and mesoderm (Figure 5.1). To further demonstrate the pluripotent character of the ES cells, they can be introduced into a blastocysts-stage embryo that after transfer to a pseudopregnant female develops further and is carried to term. If indeed the ES cells are pluripotent, the newborn animal will be a *chimera* with tissues composed of cells originating from both the host blastocyst and the ES cells. If the chimerism is also present in the germ line (i.e., oocytes or sperm cells), breeding with these animals will lead to live offspring completely derived from the ES cells (Kuijk et al. 2011, 254–271). Unfortunately, to date, no ES cells exist from farm animals. Therefore, it is important to learn more about the mechanisms involved in embryonic development and which factors are important for maintaining pluripotency in these species.

5.4   EPIBLAST STEM CELLS

The cells with almost the same self-renewal and developmental potential as ES cells are stem cells derived from the epiblast of a mouse postimplantation embryo, the so-called epiblast stem cells (EpiSCs) (Brons et al. 2007, 191–195; Tesar et al. 2007, 196–199). These cells, when generated from female embryos, exhibit the inactivation of one X-chromosome in contrast to ES cells where both X-chromosomes are active leaving EpiSCs in a more "primed" state compared to the "naïve" state of ES cells (Nichols and Smith 2009, 487–492; Tilo 2011, 241–242). Furthermore, although expressing the core pluripotency factor genes *Oct4*, *Sox2,* and *Nanog*, EpiSCs differ from ES cells in their expression of several other transcripts (Nichols and Smith 2009, 487–492). EpiSCs are also different from ES cells in their developmental and functional potential since they are less competent to contribute to chimeras (Brons et al. 2007, 191–195; Tesar et al. 2007, 196–199; Guo et al. 2009, 1063–1069). It has been postulated that in certain aspects, human ES cells are more similar to mouse EpiSCs than to mouse ES cells (Tesar et al. 2007, 196–199; Nichols and Smith 2009, 487–492).

5.5   ES CELL SELF-RENEWAL AND SIGNALING PATHWAYS

The core of the regulatory circuit that promotes the expression of genes maintaining pluripotency while repressing genes inducing differentiation is formed by three transcription factors: OCT4, SOX2, and NANOG (Figure 5.3; white). The expression of these proteins needs to be carefully regulated as increased OCT4 expression in human ES cells causes differentiation into PE and mesoderm whereas repression leads to the differentiation of TE and a loss of pluripotency (Niwa et al. 2000, 372–376). Furthermore, the downregulation of SOX2 decreases OCT4 expression eventually leading to TE differentiation (Masui et al. 2007, 625–635). The enhanced expression of the protein NANOG inhibits differentiation to PE both in ES cells and in ICM where it promotes self-renewal and epiblast formation, respectively (Mitsui et al. 2003, 631–642). Taken together, these results suggest that maintaining the self-renewal potential of ES cells is accomplished by the inhibition of differentiation (Katsumoto et al. 2010, 115–129).

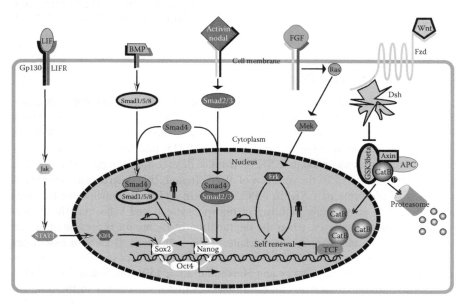

FIGURE 5.3   Pathways involved in ES cell self-renewal and pluripotency. In ES cells derived from human and rodent preimplantation embryos, the Wnt pathway (double headed arrows) has comparable effects on the self-renewing machinery. The factors important in maintaining pluripotency and self-renewal in human ES cells (wide arrow heads) have no or opposing effects in mouse ES cell maintenance, where mouse ES cell-specific pathways (black and white arrow heads) may result in differentiation or may have no effect in human ES cells. All signals regulate the expression of the core pluripotency genes; Sox2, Nanog, and Oct4, which are also self-regulatory (white). The narrow headed, hooked arrows represent transcription. (Further explanation provided in the text.)

5.5.1 LIF Signaling Pathway

The cellular signaling pathways (Figure 5.3) modulating proliferation and differentiation capable of changing the balance from self-renewal to differentiation have been partially established (Avery et al. 2006, 729–740; Ohtsuka and Dalton 2007, 74–81; Pera and Tam 2010, 713–720). At the time when the first mouse ES cells were established, these cells were cultured on a layer of nondividing feeder cells: mitotically inactivated fibroblasts producing the factors required for maintaining self-renewal capacity and pluripotency. Later, one factor produced by the feeder cells responsible for maintaining the pluripotent character of the ES cells was discovered to be leukemia inhibitory factor (LIF) (Williams et al. 1988, 684–687). LIF is a member of the interleukin-6 (IL6) family of cytokines and binds to the common transmembrane receptor for this family of cytokines (Figure 5.3; black and white arrow heads). This receptor is called gp130 and forms a heterodimeric complex with the specific LIF receptor. The subsequent activation of intracellular janus–activated kinases (JAKs) results in homodimerization and nuclear translocation of signal transducers and activators of transcription 3 (STAT3) (Avery et al. 2006, 729–740). STAT3 functions as a transcription factor for pluripotency genes such as Krüppel-like transcription factor 4 (Klf4) of which the protein products serve as a regulator of one of the pluripotency genes *Sox2* (Kim et al. 2008, 1049–1061; Hall et al. 2009, 597–609; Niwa et al. 2009, 118–122).

5.5.2 WNT Signaling Pathway

A cellular signaling pathway that likewise appears to be involved in the maintenance of pluripotency in various mammalian species is the Wnt signaling pathway (Singh and Brivanlou 2011, 87–94). Through this pathway, the secreted protein WNT is bound by the transmembrane receptor Frizzled (FZD) that leads to the sequestering of the cytoplasmic protein GSK3β from a destruction complex (Figure 5.3; double headed arrows). In the absence of WNT ligand, this destruction complex is composed of the proteins GSK3β, APC, and Axin and is responsible for the degradation via the proteasome pathway of phosphorylated and subsequently ubiquitinated β-catenin. However, when the degradation of β-catenin is prevented, it can accumulate and translocate to the nucleus of the cell. Here, β-catenin associates with the pluripotency factor OCT4 to activate OCT4-dependent transcription that is important for pluripotency (Wray and Hartmann 2011, 159–168; Sokol 2011, 4341–4350). Alternatively, β-catenin can bind to TCF3 that results either in the removal of TCF3 from the target promoters or in the removal of corepressors from TCF3 and transcriptional derepression. Another model proposed is the switch model in which TCF3 is replaced by TCF1 resulting in the activation of pluripotency (Wray and Hartmann 2011, 159–168; Sokol 2011, 4341–4350; Niwa 2011, 1024–1026). The inhibition of Wnt signaling, for instance, by the exposure to only the soluble domain of the Wnt receptor Fzd (Fz8CRD) or by the molecule IWP2, results in the differentiation of mouse ES cells to EpiSCs (Ten Berge et al. 2011, 1070–1075). The inhibition of GSK3β also results in elevated nuclear β-catenin levels in ES cells (Bennett et al. 2002, 30998–31004). Together with a MAP/ERK kinase (MEK) inhibitor, it has been demonstrated that this so-called 2i technique permits de novo derivation of ES cells from mouse blastocysts as well as the generation of ES cells from rat embryos (Buehr et al. 2008, 1287–1298; Ying et al. 2008, 519–523; Blair et al. 2011, e1002019). Surprisingly, although human ES cells resemble EpiSCs, it has been reported that the activation of the Wnt pathway in human ES cells also inhibits differentiation (Sato et al. 2004, 55–63).

5.5.3 FGF Signaling Pathway

Another signaling pathway important for ES cells is mediated through a secreted protein growth factor termed fibroblast growth factor (FGF) (Bellot et al. 1991, 2849–2854; Schlessinger 2000, 211–225) (Figure 5.3; wide arrow heads). The activation of FGF receptors can lead to multiple signaling events; one of these is mediated via the intracellular MAP kinase (MAPK—mitogen-activated protein kinases) signaling pathway in which the G-protein Ras stimulates MEK and ERK (extracellular signal-regulated kinases) (Ong et al. 2000, 979–989; Browaeys-Poly et al. 2001, 363–368). In mouse ES cells, the activation of this cascade reduces the self-renewal capacity whereas the inhibition of MAPK signaling enhances self-renewal (Burdon et al. 1999, 30–43; Kunath et al. 2007, 2895–2902) (Figure 5.3; FGF path & mouse pictogram). Although MAPK inhibition by itself is not sufficient to maintain mouse ES cells in a pluripotent state, MAPK inhibition in combination with GSK3β inhibition (2i) and LIF has resulted in the successful generation of ES cells from the previously recalcitrant mouse and rat strains (Chen et al. 2006, 17266–17271; Ying et al. 2008, 519–523; Blair et al. 2011, e1002019).

In mouse preimplantation embryos, FGF stimulation causes all cells of the ICM to express GATA6, representing PE fate, with no cells expressing NANOG. On the other hand, the inhibition of MEK signaling results in all ICM cells expressing NANOG defining a total epiblast fate (Yamanaka et al. 2010, 715–724). That early embryonic development is not driven by exactly the same molecular pathways in different mammalian species has been demonstrated by Kuijk et al. (2012, 871–882). When they stimulated FGF signaling in bovine preimplantation embryos, all cells of the ICM expressed GATA6 similar to the results obtained with mouse embryos (Yamanaka et al. 2010, 715–724). However, when MEK signaling was inhibited, a considerable amount of ICM cells retained GATA6 expression, although a larger percentage of the ICM cells expressed NANOG compared to the control embryos (Kuijk et al. 2012, 871–882). This suggests that other pathways than those active in mouse embryos are involved and can be important for the derivation of ES cells from livestock. Interestingly, in human ES cells, it seems that FGF signaling is indispensable (Figure 5.3; FGF path & man pictogram) but the inhibition of MEK in human embryos did not alter the ratio of GATA6/NANOG-expressing cells in the ICM (Kuijk et al. 2012, 871–882).

5.5.4 ACTIVIN/NODAL Signaling Pathway

Although the presence of FGF alone is not sufficient to maintain pluripotency in human ES cells, in combination with either ACTIVIN or NODAL, it can maintain pluripotency and can support human ES cell culture (Figure 5.3; wide arrow heads) for prolonged periods of time even in the absence of feeder cells (Vallier et al. 2005, 4495–4509). Both ACTIVIN and NODAL function through receptor activation by which SMAD2/3 is translocated to the nucleus resulting in NANOG transcription (Xu et al. 2008, 196–206; Vallier et al. 2009, 1339–1349). In mouse ES cells, ACTIVIN/NODAL signaling is essential for proper commitment of several lineages, but dispensable for maintaining self-renewal capacity (Fei et al. 2010, 1306–1318).

5.5.5 BMP SIGNALING PATHWAY

It has been suggested that bone morphogenetic proteins (BMPs) are involved in the maintenance of pluripotency in mouse ES cells by inhibiting specific transcription factors that regulate fate determination (Ying et al. 2003, 281–292). BMPs function by binding to two types of transmembrane receptors (Figure 5.3; black and white arrow heads). Upon ligand binding, these receptors form a heterodimeric complex that phosphorylates and thereby activates different types of signaling molecules termed SMADs. The activated SMADs form a complex together with a co-SMAD (SMAD4) that translocates to the nucleus where it can regulate transcription (Kretzschmar et al. 1997, 618–622; Shi and Massagué 2003, 685–700). In contrast to mouse ES cells, the activation of BMP signaling downregulates NANOG expression in human ES cells (Xu et al. 2008, 196–206) (Figure 5.3; mouse pictogram vs. man pictogram).

5.6 ES-LIKE CELLS

Social resistance to cultured meat obtained from existing ES cells (i.e., from rodent and primate embryos) will probably be considerable and will not result in a marketable product (Bhat and Bhat 2011, 441–459). However, until so far, attempts to generate well-characterized ES cell lines from domestic ungulates such as cattle and pigs have been rather unsuccessful (Keefer et al. 2007, 147–168). Instead, the generation of cell lines with ES-like characteristics from livestock animals has been reported. In contrast to *bona fide* ES cells, all these ES-like cell lines showed limited self-renewal potential and undefined developmental potency. The generation of germ-line chimeras from these cells, for instance, has not been demonstrated. Moreover, the expression patterns of key genes and proteins that represent pluripotency are different from the expression patterns of the existing ES cell lines. These ES-like cells cannot be cultured *in vitro* for a long period, which means that these are not suitable to produce a continuous supply of sufficient cells. Furthermore, the differentiation potential of these cells to the skeletal muscle cells is unknown, but is predicted to be variable. For these reasons, currently, there are no ES or ES-like cells that can be used in the production of cultured meat.

5.6.1 BOVINE ES-LIKE CELLS

The generation of various ES-like cell lines from different embryonic developmental stages in cattle has been described (Van Stekelenburg-Hamers et al. 1995, 444–454; Saito et al. 2003, 104–113; Roach et al. 2006, 21–37; Muñoz et al. 2008, 1159–1164;

Cao et al. 2009, 368–376; Gong et al. 2010, 151–160). Some of these cell lines could be expanded and could be maintained in an undifferentiated state for several passages (Muñoz et al. 2008, 1159–1164; Cao et al. 2009, 368–376). Indeed, when some of the ES-like cell lines were cultured in the absence of feeder cells or other differentiation-inhibiting factors, they formed embryo-like structures known as embryoid bodies made up of a wide variety of differentiated cell types, which suggest that the cells were pluripotent (Stice et al. 1996, 100–110; Roach et al. 2006, 21–37; Gong et al. 2010, 151–160). For several ES-like cells, the formation of chimeric animals has been reported but this chimerism was only found in one or a few tissues indicating that the developmental potential of these cells was limited (Cibelli et al. 1998, 642–646; Iwasaki et al. 2000, 470–475). Long-term culture of bovine ES-like cells has not been demonstrated; therefore, the currently available cells are not suitable for the production of cultured meat.

5.6.2 Porcine ES-Like Cells

Pigs have many physiological as well as anatomical similarities with humans, for instance, the size of their organs. Therefore, apart from being an important livestock species, the pig is a much appreciated model in translational research, tissue engineering, and even xenotransplantation (Matsunari and Nagashima 2009, 225–230). Indeed, efforts have been made to establish porcine ES cell lines particularly for tissue regeneration purposes but with limited success. Culture and maintenance for prolonged periods of undifferentiated porcine ICM-derived cells have been reported using human-recombinant LIF and other heterologous cytokines (Piedrahita et al. 1990a, 865–877; Hochereau-de Reviers and Perreau 1993, 475–483; Moore and Piedrahita 1997, 62–71; Puy et al. 2010, 61–70; du Puy et al. 2011, 513–526). As in mouse ES cells, most of these cell lines could differentiate *in vitro* and several cell lines were capable of the formation of carcinomas consisting of cell types from all germ layers upon transplantation to immunocompromised mice (Hochereau-de Reviers and Perreau 1993, 475–483; Gerfen and Wheeler 1995, 1–14). Although some genetic markers were expressed similar to the expression in mouse ES cells, a putative porcine ES cell line was documented to have an epithelial appearance (phenotype) as indicated by the expression of cytokeratin (Piedrahita et al. 1990b, 879–901).

For farm animals, most embryos that have been used in attempts to generate ES cell lines were the so-called *in vitro*-derived embryos. Usually, oocytes are collected from ovaries obtained as the left-over tissue from slaughterhouses. Oocytes within ovarian follicles at final stages before ovulation, the antral follicles, can be aspirated from the ovaries and cultured *in vitro*. These oocytes are not yet ready to be fertilized as they still need to undergo a substantial part of meiosis. After maturation *in vitro*, the oocytes can be fertilized *in vitro* and the developing embryos can be cultured *in vitro* until the blastocyst stage. The *in vitro* culture steps are suboptimal when compared to the steps as they occur *in vivo* and indeed, the quality in terms of numbers of viable cells of *in vitro* blastocysts is reduced in comparison with the quality of *in vivo* blastocysts (Rubio Pomar et al. 2005, 2254–2268). Therefore, possibly, *in vivo*-obtained embryos are a better source for the generation of porcine

ES cells but as a disadvantage, these embryos can be obtained only from timely inseminated and slaughtered animals. When ICMs from *in vivo*-obtained blastocysts were cultured in the presence of FGF, colonies appeared that could be cultured for ~24 passages but after that, the cells differentiated into neural and other lineages (Puy et al. 2010, 61–70). Interestingly, ICM cells of the pig early blastocysts do not express NANOG yet suggesting that later-stage pig embryos would be more suitable for the derivation of pluripotent cell lines (Kuijk et al. 2008, 918–927). Therefore, pig embryos from blastocyst and epiblast embryo stages have been cultured under conditions that are needed to maintain mouse EpiSCs in culture, that is, in the presence of FGF and ACTIVIN. Although colonies with an ES-like morphology that expressed NANOG appeared, none of these colonies could be maintained in their undifferentiated state (du Puy et al. 2011, 513–526). In addition, although pig chimeras have been described when porcine ES-cell-like cells had been introduced into blastocyst-stage embryos, the level of chimerism and therefore, the developmental potency of these cells have not been documented (Wheeler 1994, 563–568). Therefore, none of the pig ES-like cell lines generated until now are thought to be suitable for the generation of cultured meat.

5.7   GENERATING PLURIPOTENT CELL LINES FROM LIVESTOCK SPECIES

Mouse ES cells differ from human ES cells in several aspects such as the signaling pathways important for the maintenance of pluripotency and the expression of marker proteins. Therefore, it is not unlikely that in different mammalian species, different signaling pathways need to be activated or repressed to generate ES cell lines from embryos. In addition, there are important differences in pre- and peri-implantation developments of the different mammalian species, particularly in the timing of developmental processes. Most preimplantation embryos of livestock species are generated by *in vitro* fertilization and it has been suggested that it would be difficult to establish cell lines from pig and cattle embryos because of the low number of cells present in the ICM of *in vitro*-produced embryos (Anderson et al. 1994, 204–212; Van Soom et al. 1996, 171–182). Unfortunately, a well-characterized cell line from an embryo of livestock that fulfills the criterion of a *bona fide* ES cell line has not been established yet (Telugu et al. 2010, 31–41). Theoretically, one such ES cell line obtained from a farm-animal species would be sufficient to generate an almost unlimited amount of edible protein since its regenerative potential would eliminate the need to harvest more preimplantation embryos.

The identification of pluripotency-related candidate genes/gene products and their pathways by comparative studies (including microarray studies) might aid in the establishment of embryo-derived livestock of stem cell lines (Keefer et al. 2007, 147–168). Cytokines and growth factors involved in the inhibition of differentiation in mouse and primate ES cell lines such as LIF and FGF apparently have different functions in ungulate ICM or epiblast cultures indicating that other pathways need to be activated or repressed to induce the self-renewing capacity in these species (Moore and Piedrahita 1997, 62–71; Talbot et al. 1995, 35–52). Although bovine ICM outgrowths express the LIF receptor and the transmembrane signal transducer gp130, the stimulation of this pathway has not succeeded in the establishment of bovine ES cells (Pant and Keefer 2006, 110; Keefer et al. 2007, 147–168). Therefore, the pathways demonstrated to be involved in ES cell-line establishment and maintenance can have different effects in livestock embryo-derived cells or require different concentrations of the proteins involved in these pathways. The difficulty to formulate culture conditions (most importantly, media components) is presumably one of the most important reasons for the absence of ES cell lines from the livestock.

5.8 TISSUE-SPECIFIC STEM CELLS

It is generally thought that stem cells reside in all tissue types. The function of these tissue-specific stem cells is to help regenerate the tissue that is damaged by, for instance, physical trauma or disease. Although these cells have self-renewal capacity, it does not mean that they actually self-renew extensively under physiological conditions.

Hematopoietic stem cells were the first adult cells identified that can self-renew and can also give rise to differentiated cells (Becker et al. 1963, 452–454). These hematopoietic stem cells are exemplary as adult stem cells; they reside in the bone marrow and can give rise to all types of blood cells *in vivo*. This makes them extremely suitable in regenerative medicine such as replenishing blood cells after the treatment of diseases including leukemia and autoimmune disorders (Bryder et al. 2006, 338–346). A pool of cells residing in the bone marrow as well is known as mesenchymal stem cells and these cells were initially identified to be capable of bone formation *in vivo* (Friedenstein et al. 1966, 381–390). Apart from the observation that these cells can form additional cell types *in vivo,* it became apparent that mesenchymal stem cells can also be derived from sources other than the bone marrow (Caplan 1991, 641–650; Erices et al. 2000, 235–242; Zuk et al. 2002, 4279–4295; In't Anker et al. 2004, 1338–1345; Lee et al. 2011, 689–699). These mesenchymal stem cells can proliferate extensively *in vitro* undergoing extensive self-renewal and differentiation into a wide range of tissues such as the skeletal muscle (Caplan 1991, 641–650; Pittenger et al. 1999, 143–147). Thus, these cells might be useful in the generation of cultured meat (Langelaan et al. 2010, 59–66).

Stem cells derived from adipose tissue share characteristics of mesenchymal stem cells, but are named adipose-derived stem cells (Kim et al. 2006, 386–392; Gimble et al. 2007, 1249–1260). As adult stem cells have reduced proliferative capacity for the efficient production of cultured meat, it would be necessary to harvest new cells regularly. The derivation of cells from the adipose tissue would be easier and less invasive for an animal than harvesting from the muscle tissue. Still, in comparison to ES cells, the proliferative capacity of adult stem cells is limited and adult stem cells have the tendency to differentiate spontaneously *in vitro* (Langelaan et al. 2010, 59–66).

To no surprise, the muscle contains stem cells as well and as it seems, it contains more than one type (Asakura 2003, 123–128). Numerous reports have been published that describe satellite cells and their function and this will be addressed later in this chapter. Most of the research on satellite cells has been performed with rodent cells. To investigate the satellite cell machinery, satellite cell lines have been established for the mice and the rat (Numann et al. 1994, 4226–4236; Düsterhöft et al. 1999, 203–208). Reports of porcine or bovine muscle stem cells are scarce (Dodson et al. 1987, 159–166; Wilschut et al. 2008, 1228–1239).

Indeed, stem cells have been isolated from skeletal muscle of neonatal pigs and these cells have been cultured *in vitro* for several months, which represented over 120 population doublings, while maintaining a capacity to self-renew (Wilschut et al. 2008, 1228–1239). The proliferation and differentiation capacity of pig muscle-derived stem cells was discovered to be the most optimal when the cells were cultured under conditions that mimic the natural niche for muscle stem cells such as the presence of extracellular matrix proteins (Wilschut et al. 2010, 341–352). Pig muscle stem cells that expressed receptors, such as integrins, on their cell surfaces that can bind extracellular matrix components were discovered to be more myogenic. More specifically, pig muscle stem cells that lacked the expression of the integrin alpha-6 had a reduced capacity to fuse, and therefore, a reduced myogenic capacity. In addition, when alpha-6 intergrin function was blocked in pig muscle stem cells, the cells were no longer able to form myotubes (Wilschut et al. 2011, 112–123).

5.9   INDUCED PLURIPOTENT STEM CELLS

For long, it was considered that differentiation is a one-way process, which once mammalian cells have differentiated, they cannot dedifferentiate. However, the birth of the cloned sheep Dolly in 1996 unequivocally demonstrated that differentiated somatic cells can be reprogrammed by the introduction of the somatic nucleus into an oocyte from which the genetic material was removed. The underlying technique is known as cloning by somatic cell nuclear transfer and since then, a variety of mammalian species have been cloned using this technique (Wilmut et al. 1997, 810–813; Campbell et al. 2005, 256–268). In 2006, it was further demonstrated that differentiated somatic cells can be reprogrammed by the introduction of only four factors (Takahashi and Yamanaka 2006, 663–676). After this reprogramming, the cells demonstrated self-renewal capacity and pluripotency very similar to ES cells. These cells are known as iPS cells and in addition to mouse and human cells, iPS cells have also been generated from the pig (Takahashi and Yamanaka 2006, 663–676; Takahashi et al. 2007, 861–872; Wu et al. 2009, 46–54; Esteban et al. 2009, 17634–17640; Ezashi et al. 2009, 10993–10998).

The initial factors discovered to be required for reprogramming somatic cells are the transcription factors OCT4, SOX2, KLF4, and c-MYC (Takahashi and Yamanaka 2006, 663–676). However, since these seminal studies, other combinations of similar transcription factors have also been demonstrated as successful in the nuclear reprogramming of differentiated cells (Robinton and Daley 2012, 295–305). Most iPS cells to date have been generated by using integrating retroviral vectors to deliver the genes coding for transcription factors to the cells. One disadvantage of such an approach is that these genes are integrated into the genome of the targeted cells resulting in an alteration of the cellular genome. If the cells have been efficiently reprogrammed, the virus-mediated genes will be inactivated but the genome remains altered. Since by the retroviral method the position of the integration is a stochastic process, the possibility exists that the genes are integrated into, and lead to the inactivation of, important genes. Also, depending on the cell and species type, the viral gene silencing can be incomplete (Robinton and Daley 2012, 295–305). Since c–MYC is an oncogene, the continuous expression increases the risk of cancerous behavior of the iPS cells (Telugu et al. 2010, 31–41).

For iPS cells to be applied safely in future regenerative therapies, the inserted transgenes need to be deleted after reprogramming, for instance, by making use of the *piggy*Bac transposon system (Woltjen et al. 2009, 766–770). Alternatively, the generation of iPS cells can be facilitated by means of nonintegrating vectors (Okita et al. 2008, 949–953; Junying et al. 2009, 797–801), by the introduction of messenger ribonucleic acid (mRNA) coding for pluripotency proteins or the proteins themselves rather than the genes (Kim et al. 2009, 472–476; Zhou et al. 2009, 381–384), or small molecules (Shi et al. 2008, 568–574; Huangfu et al. 2008, 795–797). The aforementioned drawbacks of iPS cells in regenerative medicine also apply to their use in the production of cultured meat, making them less suitable.

5.10 DIFFERENTIATION TO MUSCLE CELLS

Upon the establishment of a stem cell line, these cells need to differentiate into skeletal muscle to obtain an edible protein product such as *in vitro* meat. This requires a detailed knowledge of the processes and factors that steer muscle differentiation. To understand the differentiation of muscle cells, studies have been conducted in developing embryos and also in adult organisms investigating proliferation and differentiation of muscle stem cells after injury (Grefte et al. 2007, 857–868; Tedesco et al. 2010, 11–19; Ten Broek et al. 2010, 7–16; Bentzinger et al. a008342). Different from most mammalian tissues, skeletal muscle is generated by the fusion of cells, more specifically, myoblasts, instead of only cell division (Mintz and Baker 1967, 592–598). Myoblasts can fuse with other myoblasts, resulting in myotubes and, subsequently, myofibers or they can fuse with preexisting myofibers. Therefore, these myofibers contain multiple cell nuclei.

5.10.1 EMBRYONIC MYOGENESIS

In a developing embryo, the first muscle tissues are formed during a process known as myogenesis. Although this process, as all differentiation processes, involves a reduction in developmental potency, with every next step during myogenesis, some cells are set aside and can be reactivated in adulthood in a way rather comparable with embryonic myogenesis. During embryogenesis, the mesoderm is separated into paraxial, intermediate, and lateral mesoderm (Aulehla and Pourquié 2006, S3–S8). From the paraxial mesoderm, marked by *Pax3* expression (Williams and Ordahl 1994, 785–796), somites are formed (Figure 5.4). The somites are located adjacent to the notochord, a primitive axis in the embryo of vertebrates. Dorsal to the notochord lies the neural tube that later in embryonic development will form the CNS. Together with the surrounding surface ectoderm, these structures secrete signaling molecules responsible for the further development of the somites. Sonic hedgehog (Shh), a member of the hedgehog family of proteins excreted by the notochord, instructs the ventral part of the somite to form the sclerotome (Borello et al.

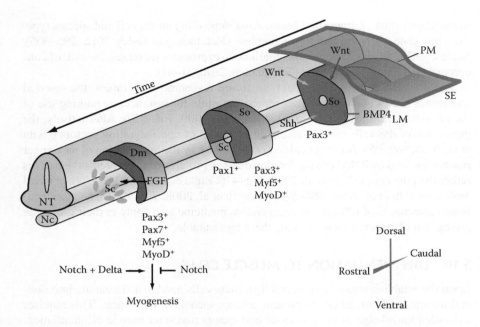

FIGURE 5.4 Embryonic myogenesis. Several extrinsic factors determine the differentiation of the somite into dermamyotome and subsequent commitment to myogenesis. Each stage and structure is identified by their expression of stage-(and location-)specific genes (details in the text). Dm, dermamyotome; LM, lateral mesoderm; Nc, notochord; NT, neural tube; PM, paraxial mesoderm; Sc, sclerotome; SE, surface ectoderm; and So, somite.

2006, 3723–3732; Bentzinger et al. 2012, a008342). The sclerotome, marked by *Pax1* expression, is committed to form the axial skeleton later in development (Brent et al. 2005, 515–528). Secreted members of the WNT family of proteins function to upregulate the expression of the transcription factors MYF5 and MYOD, both considered as markers of terminal differentiation into the muscle lineage (Pownall et al. 2002, 747–783). Shortly after the formation of the sclerotome, *Myf5* expression is synergistically activated by Shh. On the other hand, BMP4, secreted from the lateral mesoderm, delays the induction of *Myf5* and *MyoD* and maintains *Pax3* expression retaining the undifferentiated state of the somite (Pourquie et al. 1995, 3219–3223). Therefore, it is suggested that *Bmp4* expression favors the expansion of the pool of myogenic progenitor cells before further commitment is initiated (Bentzinger et al. 2012, a008342). The development into the *Pax3*- and *Pax7*-expressing dermomyotome depends on the balanced levels of the signaling molecules WNT, BMP, and Shh (Wang and Rudnicki 2012, 127–133). The dermomyotome is the source of dermal and endothelial precursor cells, as well as all skeletal muscles in the body (Kassar-Duchossoy et al. 2005, 1426–1431). During subsequent stages of development, the dermomyotome disappears thereby forming muscle progenitor cells in the myotome and a small portion of cells that resides in postnatal skeletal muscle (Gros et al. 2005, 954–958; Relaix et al. 2005, 948–953). Dermomyotomal cells expressing NOTCH, a fate-determining signal, will undergo myogenesis only when they make contact with other surrounding cells that express proteins of the Delta family (Hirsinger et al. 2001, 107–116; Schuster-Gossler et al. 2007, 537–542; Rios et al. 2011, 532–535). NOTCH expression alone results in the suppression of *MyoD* and promotes the expansion of dermomyotomal cells while terminal differentiation is being prevented (Jarriault et al. 1995, 355–358; Kuroda et al. 1999, 7238–7244; Vasyutina et al. 2007, 1451–1454; Bentzinger et al. 2012, a008342).

When the expression of *Notch* is downregulated, the induced expression of myogenic regulatory factors (MRFs) such as MYOD and MYF5 results in the downregulation of *Pax3* and *Pax7* expression (Wang and Rudnicki 2012, 127–133) and subsequently stimulates the formation of muscle precursors known as myoblasts (Tajbakhsh 2003, 413–422). Further muscle differentiation into myofibers is mediated by myogenin, MyoD, and Mrf4 (Tajbakhsh and Buckingham 2000, 225–268).

5.10.2 POSTNATAL MYOGENESIS

Relatively few cells that originate from the dermomyotome localize between the formed myofibers and the surrounding basal lamina (Schultz et al. 1978, 451–456) (Figure 5.5) and become a distinct cell type known as the satellite cell (Gros et al.

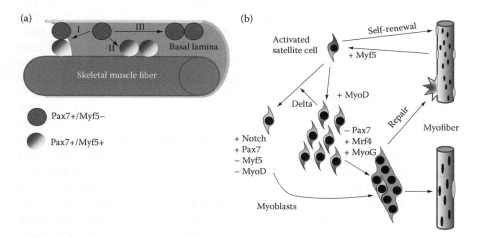

FIGURE 5.5 (a) Satellite cell commitment and (b) postnatal myogenesis. Satellite cells expressing *Pax7* are quiescent and reside between the myofiber and the basal lamina of the muscle. (a) Cell division can result in muscle lineage commitment of the daughter cell when coexpressing *Pax7* and *Myf5*. This coincides with an apical location of the cell. The cells retaining their basal location remain uncommitted and do not express *Myf5*. A satellite cell division can result in (1) one committed and one uncommitted satellite cell (asymmetric division) or via symmetric division in (2) two committed satellite cells or (3) two uncommitted satellite cells. (b) In de novo myogenesis or upon injury, *Pax7*-expressing satellite cells are activated and express *Myf5*. *MyoD*-positive myoblasts can upregulate *Notch* expression in a sister cell by the Delta family protein expression resulting in regaining *Pax7* expression and losing *Myf5* and *MyoD* expressions. These cells and the myoblasts fuse resulting either in repair or formation of new myofibers.

2005, 954–958; Ciemerych et al. 2011, 473–527). Satellite cells, already identified in 1961 by Katz and Mauro (Katz 1961, 221–240; Mauro 1961, 493–495), are located within the muscle tissue and provide new myonuclei to growing muscle fibers during postnatal development (Moss and Leblond 1971, 421–435) similar to what myoblasts do during embryonic development. Satellite cells are mitotically quiescent in normal adult muscle (Schultz et al. 1978, 451–456), but retain the ability to proliferate and differentiate in response to muscle injury (Grounds et al. 1992, 99–104; Collins et al. 2005, 289–301; Montarras et al. 2005, 2064–2067).

Other cells, for instance those derived from hematopoietic lineages, can also contribute to adult muscle regeneration (Ferrari et al. 1998, 1528–1530; Gussoni et al. 1999, 390–394; Sampaolesi et al. 2003, 487–492; Dellavalle et al. 2007, 255–267). Nevertheless, it appears that *Pax7*-expressing satellite cells are indispensable in muscle repair (Oustanina et al. 2004, 3430–3439; Relaix et al. 2006, 91–102; Lepper et al. 2011, 3639–3646). Therefore, it has been suggested that PAX7 acts as a key regulator in satellite cell biogenesis (Bryson-Richardson and Currie 2008, 632–646). The commitment to muscle-progenitor fate begins with cellular coexpression of both *Pax7* and *Myf5* (Figure 5.5a). *Pax7*–expressing satellite cells that do not express *Myf5* are not yet committed to muscle cells. These cells can divide either asymmetrically, where only one daughter cell becomes $Pax7^+/Myf5^+$ and the other remains $Pax7^+/Myf5^-$ and therefore uncommitted, or symmetrically giving rise to either two committed $Pax7^+/Myf5^+$ satellite cells or two uncommitted $Pax7^+/Myf5^-$ satellite cells (Kuang et al. 2007, 999–1010; Bryson-Richardson and Currie 2008, 632–646). The direction of commitment of each of these cells is associated with their individual location: the muscle stem cell niche. Satellite cells on the apical base (adjacent to the muscle fiber) express M-cadherin whereas satellite cells on the basal surface express $\alpha7\beta1$ integrin, a protein that can bind laminin (Collo et al. 1993, 19019–19024; Bornemann and Schmalbruch 1994, 119–125; Irintchev et al. 1994, 326–337; Burkin and Kaufman 1999, 183–190). The cells inheriting the Numb protein from the mother cell after cell division can either retain self-renewing capacity or, in a differentiating myoblast, it can repress Notch expression, a stem cell self-renewal promoting gene (Conboy and Rando 2002, 397–409; Shinin et al. 2006, 677–682; Jory et al. 2009, 2769–2780; Ciemerych et al. 2011, 473–527). Another factor that influences satellite cell fate is the Notch ligand Delta that in asymmetrical division upregulates Notch expression in the sister cell (Conboy et al. 2003, 1575–1577). This leads to increased *Pax7* expression and therefore enhanced self-renewal capacity (Ciemerych et al. 2011, 473–527) (Figure 5.5b). It is reported that in contrast, WNT directs satellite cells into differentiation resulting in the formation of fibrous connective tissue rather than myogenic tissue (Brack et al. 2007, 807–810). This "aging" process can be suppressed by Wnt inhibitors such as Klotho that binds to and antagonizes Wnts (Liu et al. 2007, 803–806). Yet another poorly understood factor is hepatocyte growth factor (HGF). It has been suggested that HGF activates quiescent satellite cells both *in vivo* and *in vitro* (Tatsumi et al. 2006, C1487–C1494; Li et al. 2009, 2284–2292; Gill et al. 2010, 1873–1883).

In the mouse, activation induced by muscle injury elevates *MyoD* and myogenin expressions 6 h after injury and *MyoD* expression further increased after 24 h returning to preinjury levels after 8 days (Grounds et al. 1992, 99–104). It has also been reported that *MyoD* expression is upregulated in *in vitro*-cultured mouse muscle progenitor cells and in isolated mouse muscle myofibers correlating with satellite cell activation and induction of differentiation (Yablonka-Reuveni et al. 1999, 440–455; Zammit et al. 2004, 347–357; Ciemerych et al. 2011, 473–527).

Several signaling molecules seem to have similar effects in pre- and postnatal muscle (re-) generation. The information gathered from embryonic development can be used to fill in the gaps in the knowledge of adult muscle regeneration and vice versa. Therefore, information about the extrinsic and intrinsic pathways in muscle development will be helpful in regenerative medicine as well as in the generation of *in vitro* meat.

5.11 CULTURING MEAT

5.11.1 CELL CULTURE

Cells can be isolated from tissue and can be maintained in culture and decades of experiments and improvements have led to different protocols for *in vitro* cell culture. Typically, mammalian cells are cultured at a temperature of around 37°C and a gas mixture of 5% CO_2 in air and these conditions can be varied in a cell incubator. A prerequisite for cell culture is a liquid growth medium and this needs to be adapted to the specific cell requirements or to the specific cell phenotype upon differentiation.

To remain viable, every cell needs water, salts, glucose, growth factors, and amino acids. This is provided by the liquid medium called culture medium. Cell culture usually takes place in commercially available culture media supplemented with animal sera from adult, newborn, or fetal source, with fetal bovine serum (FBS) as the standard (Coecke et al. 2005, 261–287).

5.11.2 SERUM

Serum production for use in the biotechnology industry and research sectors is highly regulated. The collection and movement of all animal-derived products globally are strictly controlled. Veterinary control of animal-derived products largely follows the regulations set by the European Union (DG SANCO) and the United States (USDA). Milk and meat themselves contain serum; thus, if they are well controlled, there is no major health risk in using FBS. On the other hand, serum is an expensive ingredient and because of its *in vivo* source, the composition and therefore, the efficacy varies between batches. More importantly, in generating a sustainable animal-derived additive-free product, serum-free conditions are a prerequisite.

In the past, the role and components of FBS have been identified as reviewed by Brunner et al. (2010, 53–62). In short, FBS or serum in general (1) provides factors to stimulate cell growth and proliferation, (2) facilitates attachment and spreading, and (3) stabilizes pH and osmolarity. These functions are facilitated by hormones, growth factors, cytokines, vitamins, enzymes, and so on in the serum. Because of its biological origin, FBS is subjected to lot-to-lot differences in the biochemical composition (Jayme et al. 1988, 547–548). The identification of serum components and their function and the recombinant production of these factors (Gstraunthaler 2003,

275–281), facilitated the design and production of chemically well-defined, serum-free media as has been accomplished for several cell types (Taub 1990, 213–225; Van der Valk et al. 2010, 1053–1063). Serum-free media will (1) further reduce the risk of infectious diseases, (2) have well-defined constituents minimizing possible variation, and (3) improve cost-effectiveness. By providing an alternative for the growing demand of livestock in the form of *in vitro* meat, the use of (unborn) animals as a supplier of cell culture ingredients is not an option, making it essential to develop serum-free cell culture media.

Several serum-free media have been developed to support myosatellite cell cultures from different species (Dodson et al. 1988, 909–918; McFarland et al. 1991, 163–167; Doumit et al. 1993, 326–332; Gawlitta et al. 2008, 161–171) indicating that there might be species-dependent requirements (Dodson et al. 1996, 107–126). Ultroser G is a serum substitute demonstrated to have various beneficial effects on growth and maturation in mammalian myosatellite culture in comparison to animal serum (Benders et al. 1991, 284–294). Because of its sole use in academic research, its current price might not qualify Ultroser G as a serum-replacement candidate in large-scale production unless bulk prices are reduced dramatically. The growth factors for cell culture also may have their origin from bacterial or fungal species. Indeed, it has been demonstrated that maitake mushroom (*Grifola frondosa*) extracts can be used to culture fish muscle tissue (Benjaminson et al. 2002, 879–889).

5.11.3 SCAFFOLDS

Some cells, such as white blood cells, can be kept in suspension whereas others, adherent cells, require a surface to attach to. The latter results in a single cell layer and is therefore called a monolayer or two-dimensional (2D) culture. In a living animal, the tissue can adapt to a complex three-dimensional (3D) culture because of connective tissue and an intricate system of arteries, blood vessels, and capillaries, combined with a circulating blood flow. The blood provides all cells with oxygen and nutrients while it removes metabolic waste products. Cultured cells lack such a circulating blood system and it is for this reason that the culture of 3D tissue is difficult if not impossible in a bioreactor. In the absence of a circulating blood flow, oxygen and nutrients can only reach the cells by passive diffusion. As a consequence, when tissue cultures reach a thickness of several cell layers, the cells too far away from the culture media die because of a lack of essential factors.

After the cells have reached high enough numbers to generate a substantial form of meat, the cultured undifferentiated muscle stem cells need to differentiate to produce muscle cells. This requires the activation of specific pathways by, for instance, growth factors. Again, the effects of these factors are species dependent (Burton et al. 1999, 51–61). Skeletal muscle fibers result from proliferating, differentiating, and fusion of myoblasts but are attachment dependent. Differentiation and fusion of myoblasts can be initiated by lowering the levels of growth factors indicating that the composition of the culture media must be adaptable depending on the requirements.

A solution to the culture of tissues in three dimensions might be provided by the use of cellular scaffolds that are either biodegradable or edible. Micropatterned surfaces can promote myoblast fusion and alignment resulting in a 2D structure similar to meat (Lam et al. 2006, 4340–4347). Electrospinning, the process of extracting very fine fibers from liquids, may produce a microfibrous meshwork that upon coating with extracellular matrix proteins such as collagen or fibronectin, promotes the attachment of myocytes and myofiber morphology (Riboldi et al. 2005, 4606–4615). This and other attempts to create (3D) structures facilitating expanded cell cultures were found to be less applicable in mass production due to the microfabrication required (Datar and Betti 2010, 13–22). Scaffolds can be made of either edible polymers such as cellulose or inedible polymers that need to be removed. Cell removal from a surface can be done enzymatically, but this may damage the cells and extracellular matrix resulting in the breakdown of the cultured cell sheets (Canavan et al. 2005, 1–13). Thermal liftoff is a technique applied to thermoresponsive coatings that upon lowering temperature change from hydrophilic to hydrophobic, releasing the cultured cell sheet intact. Unfortunately, this technique has led to the aggregation of the cells as a result of the imbalanced contractile forces exerted by the cytoskeleton of the myocyte sheet (da Silva et al. 2007, 577–583). A fibrin hydrogel might be an alternative as cells can migrate, proliferate, and produce their own extracellular matrix within it and at the same time, they can degrade the excess fibrin (Lam et al. 2009, 1150–1155). By stacking these 2D sheets, a 3D product could be generated. Bian and Bursac succeeded to generate muscle tissue at least twice the thickness obtained in regular 2D cultures by generating a 3D scaffold consisting of polydimethylsiloxane, a silicon also used in food as an antifoaming agent, and fibrin (Bian and Bursac 2009, 1401–1412). Even larger myofibers were obtained in $6 \times 3 \times 2.5$ mm generated collagen sponges, seeded with mouse myogenic cells and cultured *in vitro*. The final maturation was obtained only after engraftment in an injured mouse muscle regenerating a fully functional muscle (Kroehne et al. 2008, 1640–1648).

5.12 FUTURE PROSPECTS

It will take many years before cultured meat can be produced on a large scale. In this chapter, we have focused on the stem cells that are required for the production of cultured meat. Our knowledge with respect to the generation of suitable stem cells from farm animals is still rudimentary. Much effort is needed to obtain the right stem cells for large-scale production and to grow huge amounts of these cells. Next, cell biologists should devise methods to efficiently differentiate these stem cells to muscle cells and small muscle fibers. Tissue and food engineers must develop ways to grow these small muscle fibers to larger pieces that form the basis of edible products. At the same time, methods should be worked out to obtain cheap nutrients and factors that are required to culture stem cells and muscle fibers. If these efforts have resulted in a technology to make edible protein products from livestock-derived stem cells, this knowledge must be used to produce cultured meat in bioreactors (Figure 5.6). Currently, animal cells are grown on a relatively large scale for the production of pharmaceutical products. However, making large quantities of animal tissue for consumption will pose a challenge for engineers. Last but not the least, communication with consumers and societal embedding during the early stages of the research and development of cultured

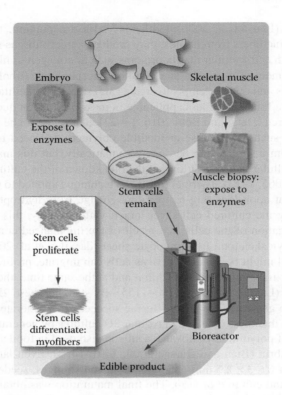

FIGURE 5.6 Schematic cartoon of the commercialized production of meat. From a farm animal, a biopsy is taken from the muscle and muscle stem cells are isolated or ES cells are generated from a preimplantation embryo. The stem cells are induced to proliferate in a bioreactor to obtain quantities ready to differentiate into myofibers and produce large masses of an edible product.

meat are essential. Only then will the technology be viable. The first steps have been taken and the road is long. Major investments are needed to develop cultured meat. However, given the enormous problems associated with the production of conventional meat, the technology to produce cultured meat may be established sooner than we think.

REFERENCES

Anderson, G. B., S. J. Choi, and R. H. BonDurant. 1994. Survival of porcine inner cell masses in culture and after injection into blastocysts. *Theriogenology* 42(1): 204–12.

Asakura, A. 2003. Stem cells in adult skeletal muscle. *Trends in Cardiovascular Medicine* 13(3): 123–8.

Aulehla, A., and O. Pourquié. 2006. On periodicity and directionality of somitogenesis. *Anatomy and Embryology* 211(1): S3–8.

Avery, S., K. Inniss, and H. Moore. 2006. The regulation of self-renewal in human embryonic stem cells. *Stem Cells and Development* 15(5): 729–40.

Becker, A. J., E. A. McCulloch, and J. E. Till. 1963. Cytological demonstration of the clonal nature of spleen colonies derived from transplanted mouse marrow cells. *Nature* 197(4866): 452–4.

Bellot, F., G. Crumley, J. M. Kaplow, J. Schlessinger, M. Jaye, and C. A. Dionne. 1991. Ligand-induced transphosphorylation between different FGF receptors. *EMBO Journal* 10(10): 2849–54.

Benders, A. A. G. M., T. H. M. S. M. Van Kuppevelt, A. Oosterhof, and J. H. Veerkamp. 1991. The biochemical and structural maturation of human skeletal muscle cells in culture: The effect of the serum substitute Ultroser G. *Experimental Cell Research* 195(2): 284–94.

Benjaminson, M. A., J. A. Gilchriest, and M. Lorenz. 2002. *In vitro* edible muscle protein production system (mpps): Stage 1, fish. *Acta Astronautica* 51(12): 879–89.

Bennett, C. N., S. E. Ross, K. A. Longo, L. Bajnok, N. Hemati, K. W. Johnson, S. D. Harrison, and O. A. MacDougald. 2002. Regulation of Wnt signaling during adipogenesis. *Journal of Biological Chemistry* 277(34): 30998–1004.

Bentzinger, F. C., Y. X. Wang, and M. A. Rudnicki. 2012. Building muscle: Molecular regulation of myogenesis. *Cold Spring Harbor Perspectives in Biology* 4(2): a008342.

Bhat, Z. F., and H. Bhat. 2011. Animal-free meat biofabrication. *American Journal of Food Technology* 6: 441–59.

Bian, W., and N. Bursac. 2009. Engineered skeletal muscle tissue networks with controllable architecture. *Biomaterials* 30(7): 1401–12.

Blair, K., J. Wray, and A. Smith. 2011. The liberation of embryonic stem cells. *PLoS Genetics* 7(4): e1002019.

Blanpain, C., V. Horsley, and E. Fuchs. 2007. Epithelial stem cells: Turning over new leaves. *Cell* 128(3): 445–58.

Borello, U., B. Berarducci, P. Murphy, L. Bajard, V. Buffa, S. Piccolo, M. Buckingham, and G. Cossu. 2006. The Wnt/β-catenin pathway regulates gli-mediated Myf5 expression during somitogenesis. *Development* 133(18): 3723–32.

Bornemann, A., and H. Schmalbruch. 1994. Immunocytochemistry of M-cadherin in mature and regenerating rat muscle. *The Anatomical Record* 239(2): 119–25.

Brack, A. S., M. J. Conboy, S. Roy, M. Lee, C. J. Kuo, C. Keller, and T. A. Rando. 2007. Increased Wnt signaling during aging alters muscle stem cell fate and increases fibrosis. *Science* 317(5839): 807–10.

Brent, A. E., T. Braun, and C. J. Tabin. 2005. Genetic analysis of interactions between the somitic muscle, cartilage and tendon cell lineages during mouse development. *Development* 132(3): 515–28.

Brons, I. G. M., L. E. Smithers, M. W. B. Trotter, P. Rugg-Gunn, B. Sun, S. M. Chuva De Sousa Lopes, S. K. Howlett et al. 2007. Derivation of pluripotent epiblast stem cells from mammalian embryos. *Nature* 448(7150): 191–5.

Browaeys-Poly, E., K. Cailliau, and J.-P Vilain. 2001. Transduction cascades initiated by fibroblast growth factor 1 on xenopus oocytes expressing MDA-MB-231 mRNAs: Role of Grb2, phosphatidylinositol 3-kinase, src tyrosine kinase, and phospholipase Cγ. *Cellular Signalling* 13(5): 363–8.

Brunner, D., J. Frank, H. Appl, H. Schöffl, W. Pfaller, and G. Gstraunthaler. 2010. Serum-free cell culture: The serum-free media interactive online database. *ALTEX* 27(1): 53–62.

Bryder, D., D. J. Rossi, and I. L. Weissman. 2006. Hematopoietic stem cells: The paradigmatic tissue-specific stem cell. *American Journal of Pathology* 169(2): 338–46.

Bryson-Richardson, R. J., and P. D. Currie. 2008. The genetics of vertebrate myogenesis. *Nature Reviews Genetics* 9(8): 632–46.

Buehr, M., S. Meek, K. Blair, J. Yang, J. Ure, J. Silva, R. McLay, J. Hall, Q.-L Ying, and A. Smith. 2008. Capture of authentic embryonic stem cells from rat blastocysts. *Cell* 135(7): 1287–98.

Burdon, T., C. Stracey, I. Chambers, J. Nichols, and A. Smith. 1999. Suppression of SHP-2 and ERK signalling promotes self-renewal of mouse embryonic stem cells. *Developmental Biology* 210(1): 30–43.

Burkin, D. J., and S. J. Kaufman. 1999. The a7ß1 integrin in muscle development and disease. *Cell and Tissue Research* 296(1): 183–90.

Burton, N. M., J. L. Vierck, L. Krabbenhoft, K. Byrne, and M. V. Dodson. 1999. Methods for animal satellite cell culture under a variety of conditions. *Methods in Cell Science* 22(1): 51–61.

Campbell, K. H. S., R. Alberio, I. Choi, P. Fisher, R. D. W. Kelly, J. -H Lee, and W. Maalouf. 2005. Cloning: Eight years after Dolly. *Reproduction in Domestic Animals* 40(4): 256–68.

Canavan, H. E., X. Cheng, D. J. Graham, B. D. Ratner, and D. G. Castner. 2005. Cell sheet detachment affects the extracellular matrix: A surface science study comparing thermal liftoff, enzymatic, and mechanical methods. *Journal of Biomedical Materials Research - Part A* 75(1): 1–13.

Cao, S., F. Wang, Z. Chen, Z. Liu, C. Mei, H. Wu, J. Huang, C. Li, L. Zhou, and L. Lin. 2009. Isolation and culture of primary bovine embryonic stem cell colonies by a novel method. *Journal of Experimental Zoology Part A: Ecological Genetics and Physiology* 311(5): 368–76.

Caplan, A. I. 1991. Mesenchymal stem cells. *Journal of Orthopaedic Research* 9(5): 641–50.

Chen, S., J. T. Do, Q. Zhang, S. Yao, F. Yan, E. C. Peters, H. R. Schöler, P. G. Schultz, and S. Ding. 2006. Self-renewal of embryonic stem cells by a small molecule. *Proceedings of the National Academy of Sciences of the United States of America* 103(46): 17266–71.

Churchill, W. S. 1932. Fifty years hence. In *Thoughts and Adventures*, 24–27. London: Thornton Butterworth.

Cibelli, J. B., S. L. Stice, P. J. Golueke, J. J. Kane, J. Jerry, C. Blackwell, F. A. Ponce De León, and J. M. Robl. 1998. Transgenic bovine chimeric offspring produced from somatic cell-derived stem-like cells. *Nature Biotechnology* 16(7): 642–6.

Ciemerych, M., K. Archacka, I. Grabowska, and M. Przewoźniak. 2011. Cell cycle regulation during proliferation and differentiation of mammalian muscle precursor cells. In *Cell Cycle in Development.*, ed. J. Z. Kubiak. Vol. 53, 473–527. Springer: Berlin.

Coecke, S., M. Balls, G. Bowe, J. Davis, G. Gstraunthaler, T. Hartung, R. Hay et al. 2005. Guidance on good cell culture practice: A report of the second ECVAM task force on good cell culture practice. *Alternatives to Laboratory Animals* 33(3): 261–87.

Collins, C. A., I. Olsen, P. S. Zammit, L. Heslop, A. Petrie, T. A. Partridge, and J. E. Morgan. 2005. Stem cell function, self-renewal, and behavioral heterogeneity of cells from the adult muscle satellite cell niche. *Cell* 122(2): 289–301.

Collo, G., L. Starr, and V. Quaranta. 1993. A new isoform of the laminin receptor integrin alpha 7 beta 1 is developmentally regulated in skeletal muscle. *Journal of Biological Chemistry* 268(25): 19019–24.

Conboy, I. M., M. J. Conboy, G. M. Smythe, and T. A. Rando. 2003. Notch-mediated restoration of regenerative potential to aged muscle. *Science* 302(5650): 1575–7.

Conboy, I. M., and T. A. Rando. 2002. The regulation of notch signaling controls satellite cell activation and cell fate determination in postnatal myogenesis. *Developmental Cell* 3(3): 397–409.

da Silva, R. M. P., J. F. Mano, and R. L. Reis. 2007. Smart thermoresponsive coatings and surfaces for tissue engineering: Switching cell-material boundaries. *Trends in Biotechnology* 25(12): 577–83.

Datar, I., and M. Betti. 2010. Possibilities for an *in vitro* meat production system. *Innovative Food Science and Emerging Technologies* 11(1): 13–22.

Dellavalle, A., M. Sampaolesi, R. Tonlorenzi, E. Tagliafico, B. Sacchetti, L. Perani, A. Innocenzi et al. 2007. Pericytes of human skeletal muscle are myogenic precursors distinct from satellite cells. *Nature Cell Biology* 9(3): 255–67.

Dodson, M. V., E. L. Martin, and M. A. Brannon. 1987. Optimization of bovine satellite cell-derived myotube formation *in vitro*. *Tissue and Cell* 19(2): 159–66.

Dodson, M. V., B. A. Mathison, M. A. Brannon, E. L. Martin, B. A. Wheeler, and D. C. McFarland. 1988. Comparison of ovine and rat muscle-derived satellite cells: Response to insulin. *Tissue and Cell* 20(6): 909–18.

Dodson, M. V., D. C. McFarland, A. L. Grant, M. E. Doumit, and S. G. Velleman. 1996. Extrinsic regulation of domestic animal-derived satellite cells. *Domestic Animal Endocrinology* 13(2): 107–26.

Doumit, M. E., D. R. Cook, and R. A. Merkel. 1993. Fibroblast growth factor, epidermal growth factor, insulin-like growth factors, and platelet-derived growth factor-BB stimulate proliferation of clonally derived porcine myogenic satellite cells. *Journal of Cellular Physiology* 157(2): 326–32.

du Puy, L., S. M. Chuva de Sousa Lopes, H. P. Haagsman, and B. A. J. Roelen. 2011. Analysis of co-expression of OCT4, NANOG and SOX2 in pluripotent cells of the porcine embryo, *in vivo* and *in vitro*. *Theriogenology* 75(3): 513–26.

Düsterhöft, S., C. T. Putman, and D. Pette. 1999. Changes in FGF and FGF receptor expression in low-frequency-stimulated rat muscles and rat satellite cell cultures. *Differentiation* 65(4): 203–8.

Erices, A., P. Conget, and J. J. Minguell. 2000. Mesenchymal progenitor cells in human umbilical cord blood. *British Journal of Haematology* 109(1): 235–42.

Esteban, M. A., J. Xu, J. Yang, M. Peng, D. Qin, W. Li, Z. Jiang et al. 2009. Generation of induced pluripotent stem cell lines from Tibetan miniature pig. *Journal of Biological Chemistry* 284(26): 17634–40.

Evans, M. J., and M. H. Kaufman. 1981. Establishment in culture of pluripotential cells from mouse embryos. *Nature* 292(5819): 154–6.

Ezashi, T., B. P. V. L. Telugu, A. P. Alexenko, S. Sachdev, S. Sinha, and R. M. Roberts. 2009. Derivation of induced pluripotent stem cells from pig somatic cells. *Proceedings of the National Academy of Sciences of the United States of America* 106(27): 10993–8.

Fei, T., S. Zhu, K. Xia, J. Zhang, Z. Li, J. -D J. Han, and Y. -G Chen. 2010. Smad2 mediates Activin/Nodal signaling in mesendoderm differentiation of mouse embryonic stem cells. *Cell Research* 20(12): 1306–18.

Ferrari, G., G. Cusella-De Angelis, M. Coletta, E. Paolucci, A. Stornaiuolo, G. Cossu, and F. Mavilio. 1998. Muscle regeneration by bone marrow-derived myogenic progenitors. *Science* 279(5356): 1528–30.

Food and Agriculture Organization of the United Nations. 2009. How to feed the world in 2050. *Paper Presented at Proceedings of the Expert Meeting on How to Feed the World in 2050*, FAO Headquarters, Rome.

Friedenstein, A. J., I. I. Piatetzky-Shapiro, and K. V. Petrakova. 1966. Osteogenesis in transplants of bone marrow cells. *Journal of Embryology and Experimental Morphology* 16(3): 381–90.

Gawlitta, D., K. J. M. Boonen, C. W. J. Oomens, F. P. T. Baaijens, and C. V. C. Bouten. 2008. The influence of serum-free culture conditions on skeletal muscle differentiation in a tissue-engineered model. *Tissue Engineering – Part A* 14(1): 161–71.

Gerfen, R. W., and M. B. Wheeler. 1995. Isolation of embryonic cell-lines from porcine blastocysts. *Animal Biotechnology* 6(1): 1–14.

Gilbert, S. F. 1994. Early vertebrate development. In *Developmental Biology*. 4th ed., 323–368. Sunderland, MA: Sinauer Associates.

Gill, R., L. Hitchins, F. Fletcher, and G. K. Dhoot. 2010. Sulf1A and HGF regulate satellite-cell growth. *Journal of Cell Science* 123(11): 1873–83.

Gimble, J. M., A. J. Katz, and B. A. Bunnell. 2007. Adipose-derived stem cells for regenerative medicine. *Circulation Research* 100(9): 1249–60.

Gong, G., M. L. Roach, L. Jiang, X. Yang, and X. C. Tian. 2010. Culture conditions and enzymatic passaging of bovine ESC-like cells. *Cellular Reprogramming* 12(2): 151–60.

Grefte, S., A. M. Kuijpers-Jagtman, R. Torensma, and J. W. Von Den Hoff. 2007. Skeletal muscle development and regeneration. *Stem Cells and Development* 16(5): 857–68.

Gros, J., M. Manceau, V. Thomé, and C. Marcelle. 2005. A common somitic origin for embryonic muscle progenitors and satellite cells. *Nature* 435(7044): 954–8.

Grounds, M. D., K. L. Garrett, M. C. Lai, W. E. Wright, and M. W. Beilharz. 1992. Identification of skeletal muscle precursor cells *in vivo* by use of MyoD1 and myogenin probes. *Cell and Tissue Research* 267(1): 99–104.

Gstraunthaler, G. 2003. Alternatives to the use of fetal bovine serum: Serum-free cell culture. *Alternativen Zu Tierexperimenten* 20(4): 275–81.

Guo, G., J. Yang, J. Nichols, J. S. Hall, I. Eyres, W. Mansfield, and A. Smith. 2009. Klf4 reverts developmentally programmed restriction of ground state pluripotency. *Development* 136(7): 1063–9.

Gussoni, E., Y. Soneoka, C. D. Strickland, E. A. Buzney, M. K. Khan, A. F. Flint, L. M. Kunkel, and R. C. Mulligan. 1999. Dystrophin expression in the mdx mouse restored by stem cell transplantation. *Nature* 401(6751): 390–4.

Hall, J., G. Guo, J. Wray, I. Eyres, J. Nichols, L. Grotewold, S. Morfopoulou et al. 2009. Oct4 and LIF/Stat3 additively induce Krüppel factors to sustain embryonic stem cell self-renewal. *Cell Stem Cell* 5(6): 597–609.

Hirsinger, E., P. Malapert, J. Dubrulle, M. -C Delfini, D. Duprez, D. Henrique, D. Ish-Horowicz, and O. Pourquié. 2001. Notch signalling acts in postmitotic avian myogenic cells to control MyoD activation. *Development* 128(1): 107–16.

Hochereau-de Reviers, M. T., and C. Perreau. 1993. *In vitro* culture of embryonic disc cells from porcine blastocysts. *Reproduction, Nutrition, Development* 33(5): 475–83.

Huangfu, D., R. Maehr, W. Guo, A. Eijkelenboom, M. Snitow, A. E. Chen, and D. A. Melton. 2008. Induction of pluripotent stem cells by defined factors is greatly improved by small-molecule compounds. *Nature Biotechnology* 26(7): 795–7.

In't Anker, P. S., S. A. Scherjon, C. Kleijburg-Van Der Keur, G. M. J. S. De Groot-Swings, F. H. J. Claas, W. E. Fibbe, and H. H. H. Kanhai. 2004. Isolation of mesenchymal stem cells of fetal or maternal origin from human placenta. *Stem Cells* 22(7): 1338–45.

Irintchev, A., M. Zeschnigk, A. Starzinski-Powitz, and A. Wernig. 1994. Expression pattern of M-cadherin in normal, denervated, and regenerating mouse muscles. *Developmental Dynamics* 199(4): 326–37.

Iwasaki, S., K. H. S. Campbell, C. Galli, and K. Akiyama. 2000. Production of live calves derived from embryonic stem-like cells aggregated with tetraploid embryos. *Biology of Reproduction* 62(2): 470–5.

Jarriault, S., C. Brou, F. Logeat, E. H. Schroeter, R. Kopan, and A. Israel. 1995. Signalling downstream of activated mammalian notch. *Nature* 377(6547): 355–8.

Jayme, D. W., D. A. Epstein, and D. R. Conrad. 1988. Fetal bovine serum alternatives. *Nature* 334(6182): 547–8.

Jory, A., I. Le Roux, B. Gayraud-Morel, P. Rocheteau, M. Cohen-Tannoudji, A. Cumano, and S. Tajbakhsh. 2009. Numb promotes an increase in skeletal muscle progenitor cells in the embryonic somite. *Stem Cells* 27(11): 2769–80.

Junying, Y., H. Kejin, S. -O Kim, T. Shulan, R. Stewart, I. I. Slukvin, and J. A. Thomson. 2009. Human induced pluripotent stem cells free of vector and transgene sequences. *Science* 324(5928): 797–801.

Kassar-Duchossoy, L., E. Giacone, B. Gayraud-Morel, A. Jory, D. Gomès, and S. Tajbakhsh. 2005. Pax3/Pax7 mark a novel population of primitive myogenic cells during development. *Genes and Development* 19(12): 1426–31.

Katsumoto, K., N. Shiraki, R. Miki, and S. Kume. 2010. Embryonic and adult stem cell systems in mammals: Ontology and regulation. *Development Growth and Differentiation* 52(1): 115–29.

Katz, B. 1961. The terminations of the afferent nerve fibre in the muscle spindle of the frog. *Philosophical Transactions of the Royal Society of London, Series B, Biological Sciences* 243(703): 221–40.

Keefer, C. L., D. Pant, L. Blomberg, and N. C. Talbot. 2007. Challenges and prospects for the establishment of embryonic stem cell lines of domesticated ungulates. *Animal Reproduction Science* 98(1–2): 147–68.

Kim, M., Y. S. Choi, S. H. Yang, H. -N Hong, S. -W Cho, S. M. Cha, J. H. Pak, C. W. Kim, S. W. Kwon, and C. J. Park. 2006. Muscle regeneration by adipose tissue-derived adult stem cells attached to injectable PLGA spheres. *Biochemical and Biophysical Research Communications* 348(2): 386–92.

Kim, J., J. Chu, X. Shen, J. Wang, and S. H. Orkin. 2008. An extended transcriptional network for pluripotency of embryonic stem cells. *Cell* 132(6): 1049–61.

Kim, D., C.-H. Kim, J.-I. Moon, Y.-G. Chung, M.-Y. Chang, B.-S. Han, S. Ko et al. 2009. Generation of human induced pluripotent stem cells by direct delivery of reprogramming proteins. *Cell Stem Cell* 4(6): 472–6.

Kretzschmar, M., J. Doody, and J. Massagué. 1997. Opposing BMP and EGF signalling pathways converge on the TGF-β family mediator Smad1. *Nature* 389(6651): 618–22.

Kroehne, V., I. Heschel, F. Schügner, D. Lasrich, J. W. Bartsch, and H. Jockusch. 2008. Use of a novel collagen matrix with oriented pore structure for muscle cell differentiation in cell culture and in grafts. *Journal of Cellular and Molecular Medicine* 12(5A): 1640–8.

Kuang, S., K. Kuroda, F. Le Grand, and M. A. Rudnicki. 2007. Asymmetric self-renewal and commitment of satellite stem cells in muscle. *Cell* 129(5): 999–1010.

Kuijk, E. W., S. M. Chuva de Sousa Lopes, N. Geijsen, N. Macklon, and B. A. J. Roelen. 2011. The different shades of mammalian pluripotent stem cells. *Human Reproduction Update* 17(2): 254–71.

Kuijk, E. W., L. Du Puy, H. T. A. Van Tol, C. H. Y. Oei, H. P. Haagsman, B. Colenbrander, and B. A. J. Roelen. 2008. Differences in early lineage segregation between mammals. *Developmental Dynamics* 237(4): 918–27.

Kuijk, E. W., L. T. A. van Tol, H. Van de Velde, R. Wubbolts, M. Welling, N. Geijsen, and B. A. J. Roelen. 2012. The roles of FGF and MAP kinase signaling in the segregation of the epiblast and hypoblast cell lineages in bovine and human embryos. *Development* 139(5): 871–82.

Kunath, T., M. K. Saba-El-Leil, M. Almousailleakh, J. Wray, S. Meloche, and A. Smith. 2007. FGF stimulation of the Erk1/2 signalling cascade triggers transition of pluripotent embryonic stem cells from self-renewal to lineage commitment. *Development* 134(16): 2895–902.

Kuroda, K., S. Tani, K. Tamura, S. Minoguchi, H. Kurooka, and T. Honjo. 1999. Delta-induced notch signaling mediated by RBP-J inhibits MyoD expression and myogenesis. *Journal of Biological Chemistry* 274(11): 7238–44.

Lam, M. T., Y. -C Huang, R. K. Birla, and S. Takayama. 2009. Microfeature guided skeletal muscle tissue engineering for highly organized 3-dimensional free-standing constructs. *Biomaterials* 30(6): 1150–5.

Lam, M. T., S. Sim, X. Zhu, and S. Takayama. 2006. The effect of continuous wavy micropatterns on silicone substrates on the alignment of skeletal muscle myoblasts and myotubes. *Biomaterials* 27(24): 4340–7.

Langelaan, M. L. P., K. J. M. Boonen, R. B. Polak, F. P. T. Baaijens, M. J. Post, and D. W. J. van der Schaft. 2010. Meet the new meat: Tissue engineered skeletal muscle. *Trends in Food Science and Technology* 21(2): 59–66.

Lee, S., S. An, T. H. Kang, K. H. Kim, N. H. Chang, S. Kang, C. K. Kwak, and H. -S Park. 2011. Comparison of mesenchymal-like stem/progenitor cells derived from supernumerary teeth with stem cells from human exfoliated deciduous teeth. *Regenerative Medicine* 6(6): 689–99.

Lepper, C., T. A. Partridge, and C. -M Fan. 2011. An absolute requirement for Pax7-positive satellite cells in acute injury-induced skeletal muscle regeneration. *Development* 138(17): 3639–46.

Li, J., S. A. Reed, and S. E. Johnson. 2009. Hepatocyte growth factor (HGF) signals through SHP2 to regulate primary mouse myoblast proliferation. *Experimental Cell Research* 315(13): 2284–92.

Liu, H., M. M. Fergusson, R. M. Castilho, J. Liu, L. Cao, J. Chen, D. Malide et al. 2007. Augmented Wnt signaling in a mammalian model of accelerated aging. *Science* 317(5839): 803–6.

Martin, G. R. 1981. Isolation of a pluripotent cell line from early mouse embryos cultured in medium conditioned by teratocarcinoma stem cells. *Proceedings of the National Academy of Sciences of the United States of America* 78(12): 7634–8.

Masui, S., Y. Nakatake, Y. Toyooka, D. Shimosato, R. Yagi, K. Takahashi, H. Okochi et al. 2007. Pluripotency governed by Sox2 via regulation of Oct3/4 expression in mouse embryonic stem cells. *Nature Cell Biology* 9(6): 625–35.

Matsunari, H., and H. Nagashima. 2009. Application of genetically modified and cloned pigs in translational research. *Journal of Reproduction and Development* 55(3): 225–30.

Mauro, A. 1961. Satellite cell of skeletal muscle fibers. *The Journal of Biophysical and Biochemical Cytology* 9: 493–5.

McFarland, D. C., J. E. Pesall, J. M. Norberg, and M. A. Dvoracek. 1991. Proliferation of the turkey myogenic satellite cell in a serum-free medium. *Comparative Biochemistry and Physiology - A Physiology* 99(1–2): 163–7.

Mintz, B., and W. W. Baker. 1967. Normal mammalian muscle differentiation and gene control of isocitrate dehydrogenase synthesis. *Proceedings of the National Academy of Sciences of the United States of America* 58(2): 592–8.

Mitsui, K., Y. Tokuzawa, H. Itoh, K. Segawa, M. Murakami, K. Takahashi, M. Maruyama, M. Maeda, and S. Yamanaka. 2003. The homeoprotein Nanog is required for maintenance of pluripotency in mouse epiblast and ES cells. *Cell* 113(5): 631–42.

Montarras, D., J. Morgan, C. Collins, F. Relaix, S. Zaffran, A. Cumano, T. Partridge, and M. Buckingham. 2005. Direct isolation of satellite cells for skeletal muscle regeneration. *Science* 309(5743): 2064–7.

Moore, K., and J. A. Piedrahita. 1997. The effects of human leukemia inhibitory factor (HLIF) and culture medium on *in vitro* differentiation of cultured porcine inner cell mass (PICM). *In Vitro Cellular and Developmental Biology—Animal* 33(1): 62–71.

Moss, F. P., and C. P. Leblond. 1971. Satellite cells as the source of nuclei in muscles of growing rats. *The Anatomical Record* 170(4): 421–35.

Muñoz, M., A. Rodríguez, C. De Frutos, J. N. Caamaño, C. Díez, N. Facal, and E. Gómez. 2008. Conventional pluripotency markers are unspecific for bovine embryonic-derived cell-lines. *Theriogenology* 69(9): 1159–64.

Nichols, J., and A. Smith. 2009. Naive and primed pluripotent states. *Cell Stem Cell* 4(6): 487–92.

Niwa, H. 2011. Wnt: What's needed to maintain pluripotency? *Nature Cell Biology* 13(9): 1024–6.

Niwa, H., J. -I Miyazaki, and A. G. Smith. 2000. Quantitative expression of Oct-3/4 defines differentiation, dedifferentiation or self-renewal of ES cells. *Nature Genetics* 24(4): 372–6.

Niwa, H., K. Ogawa, D. Shimosato, and K. Adachi. 2009. A parallel circuit of LIF signalling pathways maintains pluripotency of mouse ES cells. *Nature* 460(7251): 118–22.

Numann, R., S. D. Hauschka, W. A. Catterall, and T. Scheuer. 1994. Modulation of skeletal muscle sodium channels in a satellite cell line by protein kinase C. *The Journal of Neuroscience* 14(7): 4226–36.

Ohtsuka, S., and S. Dalton. 2007. Molecular and biological properties of pluripotent embryonic stem cells. *Gene Therapy* 15 : 74–81.

Okita, K., M. Nakagawa, H. Hyenjong, T. Ichisaka, and S. Yamanaka. 2008. Generation of mouse induced pluripotent stem cells without viral vectors. *Science* 322(5903): 949–53.

Ong, S. H., G. R. Guy, Y. R. Hadari, S. Laks, N. Gotoh, J. Schlessinger, and I. Lax. 2000. FRS2 proteins recruit intracellular signaling pathways by binding to diverse targets on fibroblast growth factor and nerve growth factor receptors. *Molecular and Cellular Biology* 20(3): 979–89.

Oustanina, S., G. Hause, and T. Braun. 2004. Pax7 directs postnatal renewal and propagation of myogenic satellite cells but not their specification. *The EMBO Journal* 23(16): 3430–9.

Pant, D., and C. Keefer. 2006. Gene expression in cultures of inner cell masses isolated from *in vitro*-produced and *in vivo*-derived bovine blastocysts. *Reproduction, Fertility and Development* 18(2): 110.

Pera, M. F., and P. P. L. Tam. 2010. Extrinsic regulation of pluripotent stem cells. *Nature* 465(7299): 713–20.

Piedrahita, J. A., G. B. Anderson, and R. H. BonDurant. 1990a. Influence of feeder layer type on the efficiency of isolation of porcine embryo-derived cell lines. *Theriogenology* 34(5): 865–77.

Piedrahita, J. A., G. B. Anderson, and R. H. BonDurant. 1990b. On the isolation of embryonic stem cells: Comparative behavior of murine, porcine and ovine embryos. *Theriogenology* 34(5): 879–901.

Pittenger, M. F., A. M. Mackay, S. C. Beck, R. K. Jaiswal, R. Douglas, J. D. Mosca, M. A. Moorman, D. W. Simonetti, S. Craig, and D. R. Marshak. 1999. Multilineage potential of adult human mesenchymal stem cells. *Science* 284(5411): 143–7.

Pourquie, O., M. Coltey, C. Breant, and N. M. Le Douarin. 1995. Control of somite patterning by signals from the lateral plate. *Proceedings of the National Academy of Sciences of the United States of America* 92(8): 3219–23.

Pownall, M. E., M. K. Gustafsson, and C. P. Emerson Jr. 2002. Myogenic regulatory factors and the specification of muscle progenitors in vertebrate embryos. *Annual Review of Cell and Developmental Biology* 18: 747–83.

Puy, L. D., S. M. Chuva De Sousa Lopes, H. P. Haagsman, and B. A. J. Roelen. 2010. Differentiation of porcine inner cell mass cells into proliferating neural cells. *Stem Cells and Development* 19(1): 61–70.

Relaix, F., D. Montarras, S. Zaffran, B. Gayraud-Morel, D. R. Rocancourt, S. Tajbakhsh, A. ansouri, A. Cumano, and M. Buckingham. 2006. Pax3 and Pax7 have distinct and overlapping functions in adult muscle progenitor cells. *The Journal of Cell Biology* 172(1): 91–102.

Relaix, F., D. Rocancourt, A. Mansouri, and M. Buckingham. 2005. A Pax3/Pax7-dependent population of skeletal muscle progenitor cells. *Nature* 435(7044): 948–53.

Riboldi, S. A., M. Sampaolesi, P. Neuenschwander, G. Cossu, and S. Mantero. 2005. Electrospun degradable polyesterurethane membranes: Potential scaffolds for skeletal muscle tissue engineering. *Biomaterials* 26(22): 4606–15.

Rios, A. C., O. Serralbo, D. Salgado, and C. Marcelle. 2011. Neural crest regulates myogenesis through the transient activation of NOTCH. *Nature* 473(7348): 532–5.

Roach, M., L. Wang, X. Yang, and X. C. Tian. 2006. Bovine embryonic stem cells. *Methods in Enzymology* 418: 21–37.

Robinton, D. A., and G. Q. Daley. 2012. The promise of induced pluripotent stem cells in research and therapy. *Nature* 481(7381): 295–305.

Rubio Pomar, F. J., K. J. Teerds, A. Kidson, B. Colenbrander, T. Tharasanit, B. Aguilar, and B. A. J. Roelen. 2005. Differences in the incidence of apoptosis between *in vivo* and *in vitro* produced blastocysts of farm animal species: A comparative study. *Theriogenology* 63(8): 2254–68.

Saito, S., K. Sawai, H. Ugai, S. Moriyasu, A. Minamihashi, Y. Yamamoto, H. Hirayama et al. 2003. Generation of cloned calves and transgenic chimeric embryos from bovine embryonic stem-like cells. *Biochemical and Biophysical Research Communications* 309(1): 104–13.

Sampaolesi, M., Y. Torrente, A. Innocenzi, R. Tonlorenzi, G. D'Antona, M. A. Pellegrino, R. Barresi et al. 2003. Cell therapy of α-sarcoglycan null dystrophic mice through intra-arterial delivery of mesoangioblasts. *Science* 301(5632): 487–92.

Sato, N., L. Meijer, L. Skaltsounis, P. Greengard, and A. H. Brivanlou. 2004. Maintenance of pluripotency in human and mouse embryonic stem cells through activation of Wnt signaling by a pharmacological GSK-3-specific inhibitor. *Nature Medicine* 10(1): 55–63.

Schlessinger, J. 2000. Cell signaling by receptor tyrosine kinases. *Cell* 103(2): 211–25.

Schultz, E., M. C. Gibson, and T. Champion. 1978. Satellite cells are mitotically quiescent in mature mouse muscle: An EM and radioautographic study. *Journal of Experimental Zoology* 206(3): 451–6.

Schuster-Gossler, K., R. Cordes, and A. Gossler. 2007. Premature myogenic differentiation and depletion of progenitor cells cause severe muscle hypotrophy in Delta1 mutants. *Proceedings of the National Academy of Sciences of the United States of America* 104(2): 537–42.

Seydoux, G., and R. E. Braun. 2006. Pathway to totipotency: Lessons from germ cells. *Cell* 127(5): 891–904.

Shenghui, H., D. Nakada, and S. J. Morrison. 2009. Mechanisms of stem cell self-renewal. *Annual Review of Cell and Developmental Biology* 25(1): 377–406.

Shi, Y., C. Desponts, J. T. Do, H. S. Hahm, H. R. Schöler, and S. Ding. 2008. Induction of pluripotent stem cells from mouse embryonic fibroblasts by Oct4 and Klf4 with small-molecule compounds. *Cell Stem Cell* 3(5): 568–74.

Shi, Y., and J. Massagué. 2003. Mechanisms of TGF-β signaling from cell membrane to the nucleus. *Cell* 113(6): 685–700.

Shinin, V., B. Gayraud-Morel, D. Gomès, and S. Tajbakhsh. 2006. Asymmetric division and cosegregation of template DNA strands in adult muscle satellite cells. *Nature Cell Biology* 8(7): 677–82.

Singh, H., and A. H. Brivanlou. 2011. Chapter 4—The molecular circuitry underlying pluripotency in embryonic stem cells and iPS cells. In *Principles of Regenerative Medicine*, eds. A. Atala, R. Lanza, J. A. Thomson, and R. M. Nerem. 2nd ed., 87–94. San Diego, CA: Academic Press.

Sokol, S. Y. 2011. Maintaining embryonic stem cell pluripotency with Wnt signaling. *Development* 138(20): 4341–50.

Steinfeld, H., P. Gerber, T. Wassenaar, V. Castel, M. Rosales, and C. De Haan. 2006. Livestock's long shadow: Environmental issues and options. In *Environment and Development Initiative Livestock*. Rome: Food and Agriculture Organization of the United Nations.

Stice, S. L., N. S. Strelchenko, C. L. Keefer, and L. Matthews. 1996. Pluripotent bovine embryonic cell lines direct embryonic development following nuclear transfer. *Biology of Reproduction* 54(1): 100–10.

Stocum, D. L. 2001. Stem cells in regenerative biology and medicine. *Wound Repair and Regeneration* 9(6): 429–42.

Tajbakhsh, S. 2003. Stem cells to tissue: Molecular, cellular and anatomical heterogeneity in skeletal muscle. *Current Opinion in Genetics and Development* 13(4): 413–22.

Tajbakhsh, S., and M. Buckingham. 2000. The birth of muscle progenitor cells in the mouse: Spatiotemporal considerations. *Current Topics in Developmental Biology* 48 : 225–68.

Takahashi, K., K. Tanabe, M. Ohnuki, M. Narita, T. Ichisaka, K. Tomoda, and S. Yamanaka. 2007. Induction of pluripotent stem cells from adult human fibroblasts by defined factors. *Cell* 131(5): 861–72.

Takahashi, K., and S. Yamanaka. 2006. Induction of pluripotent stem cells from mouse embryonic and adult fibroblast cultures by defined factors. *Cell* 126(4): 663–76.

Talbot, N. C., A. M. Powell, and C. E. Rexroad Jr. 1995. *In vitro* pluripotency of epiblasts derived from bovine blastocysts. *Molecular Reproduction and Development* 42(1): 35–52.

Tatsumi, R., X. Liu, A. Pulido, M. Morales, T. Sakata, S. Dial, A. Hattori, Y. Ikeuchi, and R. E. Allen. 2006. Satellite cell activation in stretched skeletal muscle and the role of nitric oxide and hepatocyte growth factor. *American Journal of Physiology—Cell Physiology* 290(6): C1487–94.

Taub, M. 1990. The use of defined media in cell and tissue culture. *Toxicology In Vitro* 4(3): 213–25.

Tedesco, F. S., A. Dellavalle, J. Diaz-Manera, G. Messina, and G. Cossu. 2010. Repairing skeletal muscle: Regenerative potential of skeletal muscle stem cells. *Journal of Clinical Investigation* 120(1): 11–9.

Telugu, B. P. V. L., T. Ezashi, and R. M. Roberts. 2010. The promise of stem cell research in pigs and other ungulate species. *Stem Cell Reviews and Reports* 6(1): 31–41.

Ten Berge, D., D. Kurek, T. Blauwkamp, W. Koole, A. Maas, E. Eroglu, R. K. Siu, and R. Nusse. 2011. Embryonic stem cells require Wnt proteins to prevent differentiation to epiblast stem cells. *Nature Cell Biology* 13(9): 1070–5.

Ten Broek, R. W., S. Grefte, and J. W. Von Den Hoff. 2010. Regulatory factors and cell populations involved in skeletal muscle regeneration. *Journal of Cellular Physiology* 224(1): 7–16.

Tesar, P. J., J. G. Chenoweth, F. A. Brook, T. J. Davies, E. P. Evans, D. L. Mack, R. L. Gardner, and R. D. G. McKay. 2007. New cell lines from mouse epiblast share defining features with human embryonic stem cells. *Nature* 448(7150): 196–9.

Thomson, J. A., J. Kalishman, T. G. Golos, M. Durning, C. P. Harris, R. A. Becker, and J. P. Hearn. 1995. Isolation of a primate embryonic stem cell line. *Proceedings of the National Academy of Sciences of the United States of America* 92(17): 7844–8.

Thomson, J. A., J. Itskovitz-Eldor, S. S. Shapiro, M. A. Waknitz, J. J. Swiergiel, V. S. Marshall, and J. M. Jones. 1998. Embryonic stem cell lines derived from human blastocysts. *Science* 282(5391): 1145–7.

Tilo, K. 2011. Primed for pluripotency. *Cell Stem Cell* 8(3): 241–2.

United Nations. 2009. *World Population Prospects: The 2008 Revision, Highlights.* Department of Economic and Social Affairs, Population Division, Working Paper No. ESA/P/WP.210.

Vallier, L., M. Alexander, and R. A. Pedersen. 2005. Activin/Nodal and FGF pathways cooperate to maintain pluripotency of human embryonic stem cells. *Journal of Cell Science* 118(19): 4495–509.

Vallier, L., S. Mendjan, S. Brown, Z. Ching, A. Teo, L. E. Smithers, M. W. B. Trotter et al. 2009. Activin/Nodal signalling maintains pluripotency by controlling Nanog expression. *Development* 136(8): 1339–49.

Van der Valk, J., D. Brunner, K. De Smet, Å. Fex Svenningsen, P. Honegger, L. E. Knudsen, T. Lindl et al. 2010. Optimization of chemically defined cell culture media—Replacing fetal bovine serum in mammalian *in vitro* methods. *Toxicology In Vitro* 24(4): 1053–63.

Van Soom, A., M. Boerjan, M.-T. Ysebaert, and A. De Kruif. 1996. Cell allocation to the inner cell mass and the trophectoderm in bovine embryos cultured in two different media. *Molecular Reproduction and Development* 45(2): 171–82.

Van Stekelenburg-Hamers, A. E. P., T. A. E. Van Achterberg, H. G. Rebel, J. E. Flechon, K. H. S. Campbell, S. M. Weima, and C. L. Mummery. 1995. Isolation and characterization of permanent cell lines from inner cell mass cells of bovine blastocysts. *Molecular Reproduction and Development* 40(4): 444–54.

Vasyutina, E., D. C. Lenhard, and C. Birchmeier. 2007. Notch function in myogenesis. *Cell Cycle* 6(12): 1451–4.

Wang, Y. X., and M. A. Rudnicki. 2012. Satellite cells, the engines of muscle repair. *Nature Reviews Molecular Cell Biology* 13(2): 127–33.

Wheeler, M. B. 1994. Development and validation of swine embryonic stem cells: A review. *Reproduction, Fertility and Development* 6(5): 563–8.

Williams, R. L., D. J. Hilton, S. Pease, T. A. Wilson, C. L. Stewart, D. P. Gearing, E. F. Wagner, D. Metcalf, N. A. Nicola, and N. M. Gough. 1988. Myeloid leukaemia inhibitory factor maintains the developmental potential of embryonic stem cells. *Nature* 336(6200): 684–7.

Williams, B. A., and C. P. Ordahl. 1994. Pax-3 expression in segmental mesoderm marks early stages in myogenic cell specification. *Development* 120(4): 785–96.

Wilmut, I., A. E. Schnieke, J. McWhir, A. J. Kind, and K. H. S. Campbell. 1997. Viable offspring derived from fetal and adult mammalian cells. *Nature* 385(6619): 810–3.

Wilschut, K. J., H. P. Haagsman, and B. A. J. Roelen. 2010. Extracellular matrix components direct porcine muscle stem cell behavior. *Experimental Cell Research* 316(3): 341–52.

Wilschut, K. J., S. Jaksani, J. Van Den Dolder, H. P. Haagsman, and B. A. J. Roelen. 2008. Isolation and characterization of porcine adult muscle-derived progenitor cells. *Journal of Cellular Biochemistry* 105(5): 1228–39.

Wilschut, K. J., H. T. A. van Tol, G. J. A. Arkesteijn, H. P. Haagsman, and B. A. J. Roelen. 2011. Alpha 6 integrin is important for myogenic stem cell differentiation. *Stem Cell Research* 7(2): 112–23.

Woltjen, K., I. P. Michael, P. Mohseni, R. Desai, M. Mileikovsky, R. Hämäläinen, R. Cowling et al. 2009. PiggyBac transposition reprograms fibroblasts to induced pluripotent stem cells. *Nature* 458(7239): 766–70.

Wray, J., and C. Hartmann. 2011. WNTing embryonic stem cells. *Trends in Cell Biology* 22(3): 159–68.

Wu, Z., J. Chen, J. Ren, L. Bao, J. Liao, C. Cui, L. Rao et al. 2009. Generation of pig induced pluripotent stem cells with a drug-inducible system. *Journal of Molecular Cell Biology* 1(1): 46–54.

Xu, R. -H, T. L. Sampsell-Barron, F. Gu, S. Root, R. M. Peck, G. Pan, J. Yu et al. 2008. NANOG is a direct target of TGFβ/Activin-mediated SMAD signaling in human ESCs. *Cell Stem Cell* 3(2): 196–206.

Yablonka-Reuveni, Z., M. A. Rudnicki, A. J. Rivera, M. Primig, J. E. Anderson, and P. Natanson. 1999. The transition from proliferation to differentiation is delayed in satellite cells from mice lacking MyoD. *Developmental Biology* 210(2): 440–55.

Yamanaka, Y., F. Lanner, and J. Rossant. 2010. FGF signal-dependent segregation of primitive endoderm and epiblast in the mouse blastocyst. *Development* 137(5): 715–24.

Ying, Q -L., J. Nichols, I. Chambers, and A. Smith. 2003. BMP induction of id proteins suppresses differentiation and sustains embryonic stem cell self-renewal in collaboration with STAT3. *Cell* 115(3): 281–92.

Ying, Q -L., J. Wray, J. Nichols, L. Batlle-Morera, B. Doble, J. Woodgett, P. Cohen, and A. Smith. 2008. The ground state of embryonic stem cell self-renewal. *Nature* 453(7194): 519–23.

Zammit, P. S., J. P. Golding, Y. Nagata, V. Hudon, T. A. Partridge, and J. R. Beauchamp. 2004. Muscle satellite cells adopt divergent fates. *The Journal of Cell Biology* 166(3): 347–57.

Zandstra, P. W., and A. Nagy. 2001. Stem cell bioengineering. *Annual Review of Biomedical Engineering* 3: 275–305.

Zernicka-Goetz, M., S. A. Morris, and A. W. Bruce. 2009. Making a firm decision: Multifaceted regulation of cell fate in the early mouse embryo. *Nature Reviews Genetics* 10(7): 467–77.

Zhou, H., S. Wu, J. Y. Joo, S. Zhu, D. W. Han, T. Lin, S. Trauger et al. 2009. Generation of induced pluripotent stem cells using recombinant proteins. *Cell Stem Cell* 4(5): 381–4.

Zuk, P. A., M. Zhu, P. Ashjian, D. A. De Ugarte, J. I. Huang, H. Mizuno, Z. C. Alfonso, J. K. Fraser, P. Benhaim, and M. H. Hedrick. 2002. Human adipose tissue is a source of multipotent stem cells. *Molecular Biology of the Cell* 13(12): 4279–95.

Section II

Advanced Unit Operations in Food Biotechnology

6 Membrane Filtration

Štefan Schlosser

CONTENTS

6.1 INTRODUCTION

Membrane processes (MP) find increased applications in the food, beverage, and nutraceutical industries (Daufin et al. 2001; Drioli and Giorno 2009, 2010; Pabby et al. 2009; Echavarria et al. 2011; Mohammad et al. 2012; Akin et al. 2012; Lin et al. 2012; Carstensen et al. 2012a). This chapter aims at membrane filtration applications, covering lower-pressure MP, microfiltration (MF, 20–400 kPa), ultrafiltration (UF, 200–1400 kPa), and nanofiltration (NF, 1–4 MPa). Reverse osmosis (RO) as a high-pressure-driven MP (2–10 MPa) with different transport mechanisms will be discussed later as well because it is finding application in the food industry and in biotechnology.

A membrane can be defined as an incomplete barrier between two fluids, which implicitly expresses that not all components in contact with the membrane are transported through it at the same rate; hence separation can be achieved. Transport through the membrane can occur under the action of various driving forces; in case of membrane filtration, it is transmembrane pressure difference (TMP). The separation effect of a membrane is based on

- Sieving effect
- Physical or chemical interactions of separated components with the membrane

Under the sieving effect, the separation of rejected species (ions, molecules, colloid particles, or microparticles) according to their size is considered. This is the dominating separation effect in MF and UF. Among physical interactions are electrostatic repulsions between charged species as divalent ions, amino acids, or charged colloids, which play an important role in the separation by NF or UF when membranes with fixed charge are used. The differences in sorption or solubility are decisive in the solution diffusion mechanism of transport in RO and other MP. Chemical interactions such as the formation of complexes transported through the membrane or catalytic splitting of solutes are the bases for the separation and transport in other MP. An overview of pressure-driven MP and types of species rejected by membranes in a given process is shown in Figure 6.1. The borders between these processes in terms of particle or molecule size are not sharp and strictly defined and serve only for a rough orientation. In complex mixtures typical of food and biotechnology applications, interactions between solutes may influence the separation effect.

Membrane filtration can be applied in several modes of operation as continuous or batch processes. To effectively remove the permeate components from those rejected by the membrane, fresh solvent may be added to the feed to replace the permeate

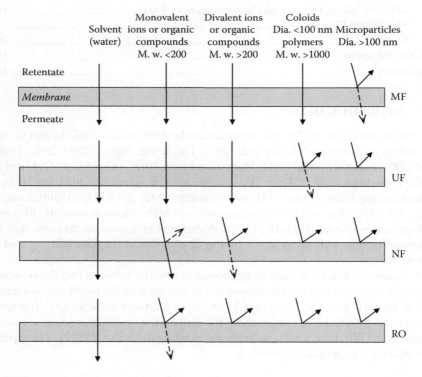

FIGURE 6.1 Pressure-driven membrane processes. Overview of species that pass through the membrane in a given MP or are rejected by the membrane. Dashed lines show species partially transported through the membrane or partially rejected.

volume. This process is referred to as diafiltration (DF) and can be used with MF, UF, or NF. The addition of the solvent can be done continuously at the same rate as the permeate flow rate and the feed volume remains constant. Alternatively, periodical concentration and dilution steps can be used. DF is to some extent analogous to the washing of the filter cake to remove soluble components in classical filtration. DF is widely used in the purification of rejected solutes, for example, in whey proteins for the desalting and removal of lactose (Kelly and Kelly 1995a,b; Daufin et al. 2001; Baldasso et al. 2011). DF can be used for the exchange of the solvent, for example, ethanol for water in human albumin fractionation from plasma (Jaffrin and Charrier 1994). Other applications of DF are shown in Tables 6.1 and 6.3. The optimization of DF is discussed in several papers (Jaffrin and Charrier 1994; Fikar et al. 2010; Paulen et al. 2012; Roman et al. 2012).

Other than pressure-driven MP are also applied in the food and biotechnology industries. Selected reviews and research results on these can be found in references for electrodialysis (Bazinet 2005; Bazinet and Firdaous 2009, 2011), osmosis (also frequently termed forward osmosis) (Cath et al. 2006; Drioli and Cassano 2010; Rastogi and Nayak 2011; Chung et al. 2012; Zhao et al. 2012b), pervaporation (Souchon et al. 2002; Rafia et al. 2011), membrane-based distillation (vacuum or osmotic) (Nagaraj et al. 2006; Kozak et al. 2009; Bagger-Jorgensen et al. 2011; Belafi-Bako and Boor 2011; Alkhudhiri et al. 2012; Cho et al. 2012), and pertraction or membrane-based solvent extraction (Schlosser et al. 2005; Kertész and Schlosser 2005; Viladomat et al. 2006; Schlosser 2009; Grzenia et al. 2012; Mihal' et al. 2012).

There is a wide inventory of membrane materials and structures—from natural polymers and their derivatives (regenerated cellulose, cellulose triacetate), tailored synthetic polymers (polysulfone, polyamides, Teflon, etc.), to ceramic and metallic membranes. Early membranes were homogenous and made of one material; recent membranes are mostly laminated with a very thin active layer of about 1–10 µm thickness, which provides separation, placed on one or more supporting layers that are usually from different materials optimized for their function.

Membranes are fixed in membrane modules with a proper geometry of feed channel and adequate support of membranes, which are in many cases not strong enough mechanically. There are four main types of membrane modules with planar, tubular, tubule (capillary), and hollow fiber membranes. Very important are spiral-wound modules with planar membrane wound together with a support layer and the feed channel spacer on the central collecting tube of the module. More details on membranes, membrane modules, transport mechanisms, and modeling of membrane transport and processes can be found in books, for example, Pabby et al. (2009) and Drioli and Giorno (2010).

An important issue in the application of MP is the build-up of rejected components at the membrane surface, which results in an increase of their concentration by order(s) of magnitude as shown in Figure 6.2. Connected to concentration polarization are two phenomena that can significantly influence the performance of MP: fouling of membrane and decrease of driving force of the process. The control of the concentration polarization and fouling is quite an important task in the food applications of MP and will be discussed in Section 6.5. The pressure difference across a membrane is the driving force not only in pressure-driven MP but also in the cases

TABLE 6.1
Selected Papers on Membrane Filtration in Food and Biotechnology Applications

Application Area/ Feed	MP[a]	Membrane Type[b]	MP Purpose[c]	References
Dairy				
Milk	MF	C	Ster.	Daufin et al. (2001); Brans et al. (2004); Skrzypek and Burger (2010); Tomasula et al. (2011)
Whey	MF	P	Sep.	Rombaut et al. (2007)
	MF	P	Pretr.	Ye et al. (2011)
	UF/DF	P	Conc., Desal.	Baldasso et al. (2011)
	NF/DF	P	Conc., Desal.	Kelly and Kelly (1995a,b); Daufin et al. (2001); Roman et al. (2012)
Proteins	UF	C	Sep., Pur.	Muller et al. (2003a,b)
	EDUF	P	Sep., Conc.	Ndiaye et al. (2010)
Brine in cheese production	MF	C	Sep., Pur.	Piry et al. (2008); Piry et al. (2012)
	MF	C	Sep., Recirc.	Skrzypek and Burger (2010)
CIP solutions	MF	C	Recirc.	Gesan-Guiziou et al. (2007)
Beer				
Tank bottoms	MF	C	Conc., Rec.	Bugan et al. (2000); Stopka et al. (2001); Fillaudeau et al. (2006); Fillaudeau et al. (2008); Yazdanshenas et al. (2010)
Rough beer	MF	C	Clar.	Gan (2001); Fillaudeau et al. (2006); Fillaudeau et al. (2007); Alicieo et al. (2008)
Fruit juices				
Apple	UF	C	Conc.	Bahceci (2012)
Grape	NF	P	Conc.	Santos et al. (2008)
Orange	UF	P	Clar., Conc.	Ruby-Figueroa et al. (2012)
Pineapple	MF	P	Clar.	de Carvalho and da Silva (2010)
Pomegranate	UF	P	Clar.	Cassano et al. (2011a,b); Baklouti et al. (2012)
Wine	MF	P, C	Clar., Ster.	Fillaudeau et al. (2008); El Rayess et al. (2011)
Vinegar	MF	P	Clar.	Lopez et al. (2005)
Sugars				
Beet	RO, NF, UF	P	Conc.	Hinkova et al. (2002)
Sugarcane	UF	P	Conc.	Saha et al. (2007, 2009)
	MF	C	Pur.	Sim et al. (2009)
Enzymes	EUF	P	Sep., Pur.	Enevoldsen et al. (2007a,b)
Algae production	MF	P	Conc.	Bilad et al. (2012b)
Organic acids	NF, NF/DF	P	Sep.	Ecker et al. (2012)
Extracts in organic solvents	NF	P	Sep., Conc.	Shi et al. (2006)
Coffee extract	NF	P	Conc.	Vincze and Vatai (2004)

[a] EDUF, electrodialysis with UF membrane; EUF, electro-ultrafiltration.

[b] P, polymeric; C, ceramic.

[c] Conc., concentration; Clar., clarification; Desal., desalination; Pur., purification; Rec., recovery; Recirc., recirculation; Rem., removal; Sep., separation; Ster., sterilization.

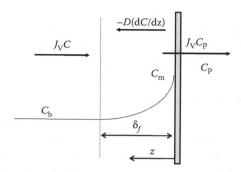

FIGURE 6.2 Concentration polarization of rejected species in a boundary layer at the membrane surface.

of RO and NF and partly also UF (polymers, which are retained in UF, have nonzero osmotic pressure despite their higher molecular mass) it is decreased by the difference of osmotic pressures of solutions in contact with the upstream and downstream sides of the membrane.

Compared to classical separation processes, MP have several advantages as they can work without phase change (in most MP) at low temperature and without the addition of chemicals. MP can achieve separations that are not possible with classical processes such as separation at the colloidal level and according to the size of molecules or particles. MP are energy effective. Higher temperature, for example, in evaporation and distillation, can destroy sensitive components of food and beverages. An important feature of MP is their modularity. Experimental data with relatively small pilot- or industrial-size membrane modules can be easily scaled up to industrial-size plant because the same modules are also used in large installations.

A brief overview of the applications of membrane filtration in the food industry, biotechnology, and related wastewater treatment together with the discussion of factors influencing the performance of MP will be presented, highlighting recent publications.

6.2 FOOD AND BIOTECHNOLOGY APPLICATIONS

As follows from typical cut-offs of membranes in pressure MP shown in Figure 6.1, microfiltration (MF) can be used, for example, in the clarification of liquids, the separation of microparticles and microorganisms and their concentration, in the cold sterilization of liquids, liquid foods, and beverages, and so on. Ultrafiltration (UF) can separate colloids, for example, milk fat globules, biopolymers, enzymes, and proteins, and can concentrate or fractionate them. Nanofiltration (NF) can be used in the separation and concentration of sugars, oligomers, peptides, amino acids, divalent ions, and so on. Reverse osmosis (RO) can effectively separate and concentrate smaller molecules such as sugars or organics responsible for COD/BOD (chemical/biological oxygen demand) in wastewaters and salts.

MP can be used alone or in combination with other separation processes, for example, centrifugation, filtration, extraction, adsorption, and so on to achieve the required effect effectively and economically. It can form an integrated system or downstream processing with individual processes working in series as used frequently in biotechnology. Integrated systems involving two or more MP and hybrid systems with MP will be discussed in Section 6.4.1.

6.2.1 FOOD AND BIOTECHNOLOGY APPLICATIONS

Reviews on the applications of membrane filtration in the food industry (Daufin et al. 2001; Charcosset 2006; Fillaudeau et al. 2008; Kulozik and Ripperger 2008; Echavarria et al. 2011, 2012; El Rayess et al. 2011) and in biotechnology (Rios et al. 2007; van Reis and Zydney 2007) are available. Examples on food and biotechnology applications of MP can be found in works shown in Table 6.1 and in Tables 6.4 and 6.5 for integrated and hybrid processes (Section 6.4).

The largest numbers of applications of MP are in dairy, beer, and wine-making industries (Daufin et al. 2001; Fillaudeau et al. 2006, 2008; Skrzypek and Burger 2010). MP have become important tools in the food processing industry over the last few decades. The share of MP applications in the food industry is 20–30% of the overall membrane market, the main food applications being in the dairy industry, close to 40% (Daufin et al. 2001).

The widely used MF of skim milk through ceramic membranes with mean pore sizes of 1.4 μm to remove bacteria and spores allow producing milk with an extended shelf life of 15–20 days. New ceramic membranes with uniform flux along the membrane have been developed with gradually decreasing thickness of the active layer of the membrane (Isoflux), which improved the operation properties of the membranes significantly (Skrzypek and Burger 2010). Other dairy applications are micellar casein separation from milk, regeneration of cheese brine, and cleaning in place (CIP) solutions (Daufin et al. 2001; Skrzypek and Burger 2010).

The recovery of beer from yeast suspensions of the tank bottoms is an important and widely used application with payback in less than 2 years (Fillaudeau et al. 2006, 2008). Sterile filtration and clarification of rough beer are advancing in full-scale applications but further improvements in fouling resistance and permeate flux are needed.

Stabilization and clarification of wine is widely used in the wine industry but further development of membranes and systems with higher flux is of interest (Fillaudeau et al. 2008; El Rayess et al. 2011). Severe fouling with wine components is not fully understood and its study should continue. A process comprising column adsorption with zirconium oxide and crossflow MF was applied to reduce the unstable proteins of wine and improve the permeate flux by 15–20% during wine MF. The phenolic composition of wine was not affected by adsorption with zirconium oxide or in the MF process (Salazar et al. 2007).

The most important applications of membrane processes in fruit juice treatment are connected with the clarification or the concentration of fruit juices. Substantial concentration of fruit juices is usually done in integrated processes with several membrane separations (Section 6.4.1). The trend in the utilization of waste fruit liquids and solids to produce nutraceuticals and food additives is visible from recent publications shown in Tables 6.2 through 6.5. More details can be found in reviews (Jiao et al. 2004; Echavarria et al. 2011, 2012).

TABLE 6.2

Selected Papers on Membrane Filtration in Food and Biotechnology Wastewater Treatment

Wastewater Type	MP	Membrane Type[a]	MP Purpose[b]	References
Dairy	MF, UF	C	Recirc.	Daufin et al. (2001); Skrzypek and Burger (2010)
	UF	P	Reco.	Chollangi and Hossain (2007)
	MBR/UF	P	Treat.	Farizoglu and Uzuner (2011)
	UF, NF,	P	Treat.	Kertesz et al. (2011)
	RO, NF	P	Treat.	Vourch et al. (2005); Luo and Ding (2011); Gong et al. (2012); Luo et al. (2012a); Riera et al. (2013)
	RO	P	Recirc.	Vourch et al. (2008)
Beer	MBR/UF	P	Treat.	Fillaudeau et al. (2006); Fillaudeau et al. (2008)
Wine and alcohol production	MBR/UF or MF	P	Treat.	Artiga et al. (2005); Artiga et al. (2007); Fillaudeau et al. (2008); Guglielmi et al. (2009)
Ethanol stillage	AnMBR	P	Treat	Dereli et al. (2012b)
Olive mill	UF	P	Treat.	Mameri et al. (2000); Cassano et al. (2011a,b)
	UF/DF	P	Treat., Reco.	Gkoutsidis et al. (2011)
	RO	P	Treat.	Ochando-Pulido et al. (2012)
Fish saline wastewater	MBR/UF or MF	P	Treat.	Artiga et al. (2008)
Various food wastewater	MBR/MF	P	Treat.	Chae and Shin (2007); Lesjean and Huisjes (2008); Meng et al. (2009); Lin et al. (2012)

[a] P, polymeric; C, ceramic.

[b] Conc., concentration; Reco., recovery; Recirc., recirculation; Sep., separation; Treat., treatment.

TABLE 6.3

Selected Papers on Membrane Filtration in Nutraceutical, Food Additive, and Therapeutic Component Applications

Application Area/Feed	MP	Membrane Type[a]	MP Purpose[b]	References
Antioxidants from				
Fruit waste	UF, NF	P	Sep.	Diaz-Reinoso et al. (2009); Diaz-Reinoso et al. (2010)
Olive mill wastewater	UF/DF	P	Pur., Conc.	Wei et al. (2010)
Olive mill wastewater	UF	P	Conc.	Mameri et al. (2000); Gkoutsidis et al. (2011); Zunin et al. (2011); Cassano et al. (2011a,b)
Plant extracts				
In organic solvents	NF	P	Sep., Conc.	Peshev et al. (2011); Achour et al. (2012)
In aqueous solvents	UF	P	Sep., Conc.	Nawaz et al. (2006); Diaz-Reinoso et al. (2011)
	NF	P		Paun et al. (2011)
Proteins				
Plant proteins	UF/DF	P	Sep., Pur.	Nabetani et al. (1995); Mondor et al. (2010); Aspelund and Glatz (2010a,b); Leberknight et al. (2011); Castel et al. (2012)
Extract from safflower	UF	P	Sep.	Ulloa et al. (2011)
Soybean extract	UF	P	Sep., Pur.	Campbell and Glatz (2010)
Corn extracts	UF	P	Clar.	Aspelund and Glatz (2010a,b)
Egg proteins, peptides	UF	P	Sep.	Ting et al. (2010)
Protein hydrolysates	NF	P	Sep.	Chabeaud et al. (2009)
Therapeutic proteins	UF	P	Sep., Pur.	Molek and Zydney (2007); van Reis and Zydney (2007); Zydney (2009)

Serum proteins	UF/DF	P	Sep., Pur.	van Reis et al. (1997); van Reis (1999)
Bovine serum	UF	P	Sep.	Causserand et al. (2001)
Oligosaccharides	NF, DF	P	Sep., Pur.	Goulas et al. (2002); Goulas et al. (2003); Yuan et al. (2004)
	NF/DF	P	Sep., Pur.	Zhang et al. (2011); Zhao et al. (2012a)
Peptides				
Whey	MF, UF	P	Sep., Conc.	Tavares et al. (2012)
Dipeptides, Meat	NF	P	Pur., Conc.	Nabetani et al. (2012)
Biopolymers				
xanthan	UF	P	Conc.	Lo et al. (1997); Torrestiana-Sanchez et al. (2007)
	EUF	P	Conc.	Gozke and Posten (2010)
Starch	UF	P	Sep., Conc.	Middlewood and Carson (2012)
Extracellular biopolymer	UF	P	Sep., Conc.	Li et al. (2011)

[a] P, polymeric; C, ceramic.

[b] Clar., clarification; Conc., concentration; Pur., purification; Sep., separation.

TABLE 6.4
Selected Papers on Integrated Processes with Two or More MP

Application Area/Feed	MP[a]	Membrane Type[b]	MP Purpose[c]	References
Whey protein	UF + UF	P	Fract.	Cheang and Zydney (2004)
	MF, UF, RO	P	Sep., Conc.	Tavares et al. (2012)
Oligosaccharides	UF + NF	P	Pur., Conc.	Kamada et al. (2002)
Organic acids	UF + NF/RO + FO[c]	P	Sep., Conc., MBR	Cho et al. (2012)
Fruit juices				
Apple	UF + MD	P	Conc.	Onsekizoglu et al. (2010)
Grape must	MF + RO	P	Clar., Conc.	Rektor et al. (2004); Rektor et al. (2007)
	MF + RO + MD	P		Rektor et al. (2006)
Blackcurrant	MD + RO + NF + MD	P	Conc.	Sotoft et al. (2012)
	MF + RO + MD	P	Conc.	Kozak et al. (2009)
Pomegranate	UF + MD	P	Clar., Conc.	Cassano et al. (2011a,b)
Plant extracts (aq.)	UF + NF	P	Sep., Conc.	Chhaya et al. (2012)
Sugar	RO + NF	P	Conc.	Nabetani (2012); Gul and Harasek (2012)
Dairy wastewater	UF + NF	P	Treat.	Gong et al. (2012)
	NF + RO	P	Treat.	Vourch et al. (2005)
Olive oil wastewater	MF + NF + MD	P	Rec.	Garcia-Castello et al. (2010)
	UF + NF + NF	P	Rec.	Zirehpour et al. (2012)

[a] FO, forward osmosis; MD, membrane distillation.
[b] P, polymeric; C, ceramic.
[c] Clar., clarification, Conc., concentration; Fract., fractionation, Pur., purification; Rec., recovery, Sep., separation, Treat., treatment.

TABLE 6.5
Selected Papers on Hybrid Processes Including Membrane Bioreactors

Application Area/Feed	MP	Membrane Type[a]	MP Purpose[b]	References
MBR with external module	MF	P	Sep.	Jung and Lovitt (2010); Dey and Pal (2012); Carstensen et al. (2012a)
MBR with submerged module	MF	C	Sep., Rem.	Wojciech et al. (2013)
	MF	P, C	Sep.	Navaratna et al. (2012)
	MF	P	Sep.	Ramchandran et al. (2012); Carstensen et al. (2012a); Carstensen et al. (2012b)
Organic acid production	MBR/UF + NF/RO + FO	P	Sep., Conc.,	Cho et al. (2012)
	MBR/MF + NF	P	Sep., Conc.	Dey and Pal (2012)
Bacteria production	MBR/UF	P	Sep.	Ramchandran et al. (2012)
Peptides	MBR + NF/DF	P	Biotran., Sep.	So et al. (2010)
	MBR/UF + UF	P	Biotran., Sep.	Das et al. (2012)
Oligosaccharides	MBR/UF	P	Biotran., Sep.	Kuroiwa et al. (2009); Hang et al. (2012a,b)
	Enzymatic MBR/UF + NF/DF	P	Biotran., Sep.	Das et al. (2011); Sen et al. (2011); Sen et al. (2012)
	MBR/UF enzyme in membrane	C	Biotran., Sep.	Nishizawa et al. (2000)
Biomass hydrolysates				
Oil seed meal	MBR/UF	P	Sep.	Das et al. (2009); Das et al. (2012)
Almond shells	MBR/UF	P	Sep.	Nabarlatz et al. (2007)
Saccharification of starch	MBR/MF	C	Sep., Rem.	Koutinas et al. (2001); Wojciech et al. (2013)
Biosurfactant production	MBR/MF	P	Aeration	Coutte et al. (2010)
Aglycon from olive mill waste	MBR/UF enzyme in membrane	P	Biotran., Sep.	Mazzei et al. (2009); Mazzei et al. (2010, 2012)
Wastewater treatment with activated sludge	MBR/MF	P	Treat.	Chae and Shin (2007); Lesjean and Huisjes (2008); Lin et al. (2012)

[a] P, polymeric; C, ceramic; FO, forward osmosis.

[b] Biotran., biotransformation; Conc., concentration; Clar., clarification; Desal., desalination; Rec., recovery; Recirc., recirculation; Rem., removal; Sep., separation; Treat., treatment.

The concentration and separation of a purified silage juice produced in a biorefinery pilot plant by nanofiltration to recover lactic acid and amino acid fractions was studied by Ecker et al. (2012). The separation of solutions in organic solvents has interesting applications in biotechnology. For example, the extract of the macrolide antibiotic spiramycin in butylacetate was concentrated by NF through a polyimide membrane, which showed excellent separation performance under test conditions. The permeation flux was over 20 L m^{-2} h^{-1}, and the rejection of spiramycin reached 99% (Shi et al. 2006).

6.2.2 WASTEWATER APPLICATIONS

Food and biotechnology wastewaters are highly polluted with high COD/BOD values. Their treatment is important from an ecological point of view. Membrane separations can contribute much in their treatment to achieve the required limits. Reviews on the application of membrane filtration in food wastewater treatment applications are available (Lesjean and Huisjes 2008; Lesjean et al. 2009; Lin et al. 2012; Meng et al. 2012). Selected references on the treatment of wastewaters from food and beverage technologies and biotechnologies are shown in Table 6.2.

The growing application of MP, especially membrane bioreactors (MBR), in wastewater treatment is an evidence. There were around 300 MBR plants for industrial wastewater treatment in Europe by 2006 with a capacity larger than 20 m^3/day and more than 50 new plants are supposed to be installed every year (Lesjean and Judd 2009).

Anaerobic membrane bioreactor (AnMBR) technology offers many advantages in the treatment of industrial wastewaters such as high organic matter removal efficiency, recovery of energy, and excess sludge reduction (van Lier et al. 2001; Stuckey 2012; Dereli et al. 2012a). AnMBRs present an attractive option for the treatment of industrial wastewaters at extreme conditions. So far, most of the research has been conducted at laboratory and pilot scales; however, a number of full-scale AnMBR systems are currently being operated worldwide. The performance of a pilot AnMBR with a volume of 12 m^3 for the treatment of ethanol stillage wastewater is presented by Dereli et al. (2012b).

6.3 APPLICATIONS IN NUTRACEUTICAL INDUSTRY

Nutraceutical is a food or food product that reportedly provides health and medical benefits, including the prevention and treatment of disease. Nutraceuticals are isolated from foods or prepared by biotechnological processes. There is an increasing demand for natural antioxidants, specific proteins and peptides, and so on in the food and pharmaceutical industries. They or their precursors are present in agricultural products and frequently also in liquid and solid wastes from the agro-food industry. To meet these needs, new methods of separation and concentration are being developed, frequently with the utilization of membrane separations, including membrane filtration. Overviews on the applications of MP in the production of functional foods and nutraceuticals can be found in Pinelo et al. (2009) and Akin et al. (2012). Additional data on the applications of MP in the production of food additives and nutraceuticals can be found in selected works shown in Table 6.3 and also in Tables 6.4 and 6.5 for integrated and hybrid processes (see the next section).

Not much data are available on the recovery of higher-value components from fruit by-products. In many cases, extracts from solid by-products will be treated to separate and concentrate target components as discussed in a review on the recovery of polyphenol antioxidants (Shi et al. 2005). Data on the concentration of extracts from grape seeds of direct grape pomace from must pressing (Nawaz et al. 2006) or from distilled grape pomace (Diaz-Reinoso et al. 2009; Diaz-Reinoso et al. 2010) are available. The pure ethanol extract of rosemary leaves was concentrated by nanofiltration through a solvent-resistant membrane and it was found that diafiltration improved the quality of the product (Peshev et al. 2011). The recovery of polyphenols from olive mill wastewater by UF was also studied (Cassano et al. 2011a).

The production of protein concentrate from amaranth seeds was examined in pilot plant experiments by two methods (Castel et al. 2012). Clarified aqueous alkaline extract of amaranth flour was isoelectrically precipitated (classical technology). Alternatively, acid pretreated flour was alkaline extracted and the clarified extract was concentrated by UF. Proteins produced by UF concentration had better quality.

The purification and concentration of high-value therapeutic proteins is applied industrially (van Reis and Zydney 2007; Zydney 2009). Further improvement of this process can be achieved by charged membranes, which are much less fouled (Rohani and Zydney 2012).

Egg-laying hens are discarded because the quality of their meat is low. To add further value to this waste, the purification and concentration processes for functional dipeptides have been studied (Nabetani et al. 2012). Hot water extract from whole chicken was further purified and concentrated by combining ion-exchange chromatography and NF. A concentrate of the antioxidant dipeptides anserine and carnosine has been recovered. On the basis of pilot plant experiments, a process that could treat 3.6 t of chicken carcasses per day and recover 7.4 kg of purified anserine and carnosine was designed.

6.4 INTEGRATED AND HYBRID PROCESSES

6.4.1 INTEGRATED PROCESSES

In some cases, there is a need to combine more MP to achieve the required separation and/or concentration. If those MP work in a series of consecutive operations, an integrated process is obtained. Integrated processes with two or more MP involved are applied in the examples shown in Table 6.4.

Pilot plant experiments on the manufacture of peptide concentrates, using integrated processes with membrane filtration techniques, have been reported (Tavares et al. 2012). Purified cow whey protein concentrate was separated by UF after enzymatic hydrolysis of α-lactoglobulin into peptide permeate and β-lactoglobulin retentate. The peptide fraction was concentrated by RO and the retentate was separated by UF into two peptide fractions with a biologically active fraction of the permeate through a 3-kDa membrane. Both the peptide and protein fractions are high-added-value products that can be used as nutritional and functional ingredients—thus yielding an economically viable alternative for the upgradation of whey.

The integrated membrane process for organic acid separation proposed by Cho et al. (2012) includes three steps: (1) clarification of the fermentation broth by UF, (2) organic acid separation and concentration by NF, and (3) further acid concentration by forward osmosis (FO). This process could achieve energy-efficient and environmentally friendly organic acid removal and recovery.

On the basis of laboratory and pilot data from several authors working with real blackcurrant juice, a full-scale plant with a capacity of 17283 t/year of concentrate was designed (Sotoft et al. 2012). From the filtered raw juice, aroma compounds are removed in the first stage by vacuum membrane distillation (MD) to obtain aroma concentrate after normal distillation. Juice is concentrated by RO and the retentate further concentrated by NF to 31.5% (w/w) of sugars; this retentate goes to direct-contact MD where it is further concentrated to a final 67% (w/w) of sugars. The aroma concentrate is subsequently added to obtain the final juice concentrate with full aroma. The permeate from NF containing 14.2% (w/w) sugars is returned to the feed stream for RO.

The sugar industry technology is energy intensive, mostly due to the evaporation of large amounts of water. The application of MP in this area was not very successful up to now because the osmotic pressure of saccharose required at higher concentrations is high. Two promising proposals for an integrated process combining RO with NF have been suggested recently. Thin sugar juice with an initial concentration of 15% (w/w) and a temperature of 80°C is concentrated by RO at a pressure of 3.2 MPa to about 25% (w/w) and fed to the second and third stages with NF membranes of different rejections. The final sugar concentration from membrane concentration will be about 50% (w/w) at 80°C and in subsequent multiple-effect evaporators, it will be increased to 65–72% (w/w) before crystallization (Gul and Harasek 2012). Also, 89% of water is removed by MP and that results in energy savings in evaporation of more than 80%. In the two-stage process suggested by Nabetani (2012), in the first RO stage, sugar concentration is increased from 10% to 30% (w/w) at 7 MPa and in the second NF stage to 40% (w/w). The permeate from NF with 20% (w/w) of sugar is recirculated to the RO feed.

6.4.2 Hybrid Processes

In a hybrid process, two or more processes proceed in one equipment or in two or more equipments working in parallel that are connected with circulation loop(s) and can influence each other. Membrane bioreactor (MBR) is a good example of a hybrid system where microorganisms or biocatalysts are kept in a bioreactor while the product is removed from it. This is important especially when product inhibition occurs. An alternative type is the MBR with catalyst on/in a membrane (Nishizawa et al. 2000; Mazzei et al. 2010, 2012; Sen et al. 2011, 2012). Reviews on MBR applications in wastewater treatment (Drews and Kraume 2005; Yang et al. 2006; Lesjean and Huisjes 2008; Lin et al. 2012) and in biotechnology (Rios et al. 2004; Carstensen et al. 2012a) are available.

Membranes or better membrane module(s) in MBR can be positioned directly in the reactor as submerged modules or as external membrane module(s) in an outer circulation loop. Configurations of membrane reactors are discussed in a review (Carstensen et al. 2012a). Owing to the lower operation costs as compared to external membrane modules, submerged modules and low-pressure filtration technologies will remain the standard for large MBRs in wastewater treatment with activated sludge applications in the near future (Lesjean et al. 2009). References to papers on hybrid processes with membranes are presented in Table 6.5 and also in Table 6.2 for MBR in water treatment.

An integrated process for the removal of organic acids from the fermentation broth by MP was examined by Cho et al. (2012). Microorganisms were separated from the broth by MF or UF and returned back to the fermenter. The clear broth was further separated by NF or RO, nutrients and salts were retained by the membrane and sent back to the reactor, while organic acids at lower pH passed through the membrane to the permeate and were further concentrated by forward osmosis (FO). Lower retention of nutrients in NF or higher rejection of acid in RO decreased the efficiency of the process.

Enhanced peptide synthesis was achieved in a reactor containing polyethyleneglycol (PEG) in the organic solvent dimethylformamide (DMF) coupled with a solvent-resistant nanofiltration membrane built inside the reactor, through which excess reagents were removed by diafiltration (So et al. 2010). A soluble polymeric support, methoxy–amino–PEG with molecular mass 5000 g mol^{-1} on which the peptide was assembled, enabled its easy separation from the reagents by the NF membrane. Thus, pure peptides with various sequences of amino acids can be produced, which are potential drugs for the treatment of diseases.

Edible oil industries suffer from the problem of seed meal utilization, which is a by-product of this technology. From the enzymatic hydrolysis of oil seed meals in MBR, peptide antioxidants can be produced (Das et al. 2012). Ultrafiltration with membranes of various cut-offs was used for the separation of peptides from the broth.

The continuous enzymatic production of difructose anhydride III (DFA III) from inulin as a substrate using an ultrafiltration membrane bioreactor system was investigated (Hang et al. 2012a,b). DFA III is a nondigestible oligosaccharide and has recently attracted interest as a prebiotic. The productivity of MBR was about five times higher than that of the batch reactor. In the batch reactor, each volume replacement required a recharge with fresh enzyme.

Olive mill wastewater is a rich source of a diverse range of biophenols such as oleuropein, which, together with hydroxytyrosol, have a wide array of biological activity. The isomer of oleuropein aglycon produced in the first step of the oleuropein hydrolysis catalyzed by β-glucosidase gained much interest in the last few years because it is responsible, together with hydroxytyrosol, for the antioxidant properties of olive oil. An interesting hybrid process has been designed for this application (Mazzei et al. 2010, 2012). The enzyme is captured in the outer sponge support structure of a polysulfone capillary membrane with an active UF layer on its inner surface. When the solution of oleuropein in buffer slowly flows from the shell to the capillary lumen, hydrolysis takes place in the membrane wall where the enzyme is entrapped and an isomer of oleuropein aglycon is formed. Further development of this reactor was to join this biocatalytic system with membrane emulgation and extraction of the product into the organic phase because aglycon is not stable in the aqueous solution. If an organic solvent, limonene in this case, flows in the capillary lumen, droplets will be formed at the pore mouths from which aglycon is extracted into the organic phase. The unreacted oleuropein remaining in the aqueous solution (droplets), because it is not extracted to the solvent, can be returned to the reactor after the separation of the aqueous phase from the emulsion. In this process, three operations take place in one equipment: enzymatic reaction, membrane emulgation, and solvent extraction of the product.

6.5 MEMBRANE FOULING AND FOULING CONTROL

6.5.1 MEMBRANE FOULING

The accumulation of rejected species at the membrane surface (concentration polarization) (Figure 6.2), adsorption, and deposition on/in the membrane results in fouling, which can greatly decrease the flux through the membrane up to several orders of magnitude. The performance of the membrane can change much with time as the fouling is progressing. Recent reviews on fouling in food industry applications are available (Mohammad et al. 2012; van der Sman et al. 2012; Koo et al. 2012; Giglia and Straeffer 2012). The mechanism of fouling, especially in the separation of very complex feed in food applications, is not simple and may vary from system to system and depends on the membrane material, hydrodynamic conditions, and operating mode. There are three main mechanisms of fouling:

a. Concentration polarization resulting in the formation of the filtration cake or the gel layer
b. Adsorption on the membrane surface or pore walls
c. Slow colloidal aggregation resulting in enhanced fouling (Maruyama et al. 2001; Aimar and Bacchin 2010) and changing structure of the cake with time

A specific phenomenon is the flux decline in the filtration of small molecules that are not rejected by the membrane. Surprisingly, remarkable flux decline with fast kinetics was observed in the MF of the model flavanoid catechin (m.w. 290.27 Da) and bitter acid (hop extract for beer) solutions through a ceramic membrane with a mean pore size of 500 nm (Stopka et al. 2000). The addition of fatty acids to a glycerine-rich aqueous solution caused significant flux decline, even though the molecular weight of glycerine and fatty acids are much smaller than the polymeric membrane cut-off (Amin et al. 2010). Flux decline in the MF of aqueous gallic acid solution (m.w. 170.12 Da) and diluted white and red grape must is shown in Figure 6.3 (Václavík 2011). It should be stressed that the rejection of the components of these solutions was below 9% but a steep flux decline was observed.

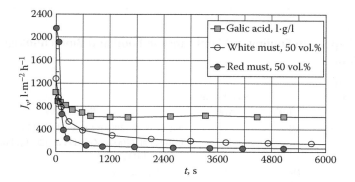

FIGURE 6.3 Permeate flux decline in the microfiltration of diluted clarified white and red grape musts and aqueous solution of gallic acid through a ceramic membrane (Mebralox) with a mean pore size of 100 nm at $\Delta p = 60$ kPa, 25°C, and a feed velocity of 1.26 m s^{-1}. (Adapted from Václavík, L. 2011. *Inst. Chem. Environ. Eng.*, Slovak University of Technology, Bratislava (SK).)

Data on fouling in food, biotechnology, and related applications or systems can be found in selected papers shown in Table 6.6.

Food liquids, fruit juices, fruit by-product solutions, and fermentation broths are quite complex in the number of components and their types, including polyphenols, flavonoids, polysaccharides, saccharides, oligosaccharides, proteins, enzymes, salts, various types of colloids, and so on. Their interaction with the membrane and also among themselves can greatly influence the separation through the fouling of the membrane and the permeate flux decline. To identify the main foulants present in each type of liquid foods, experiments with model solutions are of importance.

There are several models describing the flux decline due to the fouling of the membrane that can occur by the deposition of particles inside or on top of the membrane. They can not only describe the flux decline with time but also identify the fouling mechanism, which can help in finding methods for its prevention. The classical blocking model identifies four mechanistic models that are typically used to describe fouling in filtration (Hermia 1982), which have been modified for conditions in crossflow membrane filtration (Field et al. 1995). In many cases, this approach does not provide a good description of the situation. A combined pore blockage and cake filtration model, which can be used for a more detailed analysis of flux decline data in microfiltration, has been developed (Ho and Zydney 2000). Its simplified form is presented by van Reis and Zydney (2007). Physically less detailed combined models are suggested by Bolton et al. (2006).

6.5.2 Fouling Control and Improvement of Flux

The economic operation of MP requires the permeate flux to be as high as possible and the sustainable operation of the process. Both rely on the performance of the membrane, the driving force applied, and the interactions between solutes and

TABLE 6.6
Selected Papers on Fouling of Membranes in Food and Biotechnology Applications

Application Area/ Feed or Foulant Type	MP	Membrane Type[a]	References
Dairy applications	MF	P	Nigam et al. (2008); Luo et al. (2012a)
		C	Caric et al. (2000); Kuhnl et al. (2010); Popović and Tekić (2011)
	NF	P	Luo et al. (2012b)
Beer applications	MF	C	Stopka et al. (2000); Thomassen et al. (2005); Alicieo et al. (2007); Yazdanshenas et al. (2010)
		P	van der Sman et al. (2012)
Wine	MF	P	Ulbricht et al. (2009); Li et al. (2010); El Rayess et al. (2011); El Rayess et al. (2011)
Fruit juices	MF	P	de Oliveira et al. (2012)
		C	Mondor et al. (2000); de Oliveira et al. (2012)
	UF	P	Mondor et al. (2000); de Barros et al. (2003)
		C	Vladisavljevic et al. (2003); de Barros et al. (2003)
Membrane bioreactors (MBR)	MF	P	Drews et al. (2005); Carstensen et al. (2012a)
	UF, NF	P	Sen et al. (2012)
Proteins	MF	P	Ho and Zydney (1999, 2000, 2001, 2002); Zator et al. (2009)
	UF	P	Maruyama et al. (2001); Lim and Mohammad (2010); Leberknight et al. (2011)
Saccharose	UF	P	Saha et al. (2007, 2009)
	MF	C	Sim et al. (2009)
Oligosaccharides	UF	P	Nabarlatz et al. (2007); Sen et al. (2011); Sen et al. (2012)
Polysaccharides	UF	P	Ye et al. (2005); Susanto and Ulbricht (2006); Susanto et al. (2007); Saha et al. (2007); Susanto et al. (2008); de la Torre et al. (2009); Middlewood and Carson (2012)
	MF	C	Wojciech et al. (2013)
Extracellular polymeric substances	MF	P	Ye et al. (2005); Chen et al. (2006)
	UF	P	Sweity et al. (2011)
Natural organic matters (NOM)	MF	P	Kim et al. (2006); Xiao et al. (2009)
	UF	P	Zularisam et al. (2011)
Polyphenols	MF	C	Stopka et al. (2000); Zator et al. (2009); Wei et al. (2010); Václavík (2011)
		P	Stoller and Chianese (2010); Cassano et al. (2011a,b); Tsagaraki and Lazarides (2012)
Proteins	MF	P	Zator et al. (2009); Xiao et al. (2009)
Glycerine-rich fatty acid solutions	UF	P	Amin et al. (2010)

TABLE 6.6   (continued)
Selected Papers on Fouling of Membranes in Food and Biotechnology Applications

Application Area/ Feed or Foulant Type	MP	Membrane Type[a]	References
Enzymes	MF	C	Mercier-Bonin and Fonade (2002)
	UF	P	Enevoldsen et al. (2007a,b)
Microparticles and microorganisms	MF	C	Stopka et al. (2001); Mercier-Bonin and Fonade (2002)
	MF	P	Graves et al. (2006); Hwang and Chen (2007); Beier and Jonsson (2007); Rickman et al. (2012); Bilad et al. (2012b)
Small molecules	MF	P	Xiao et al. (2009); Leo et al. (2012)
		C	Stopka et al. (2000); Václavík (2011)
	UF	P	Kwon et al. (2009); Amin et al. (2010)
Dairy wastewater treatment	MF	P	Farizoglu and Uzuner (2011)
	UF/NF	P	Gong et al. (2012); Luo et al. (2012a); Luo et al. (2012b)
	NF/RO	P	Turan et al. (2002)
Olive mill wastewater	UF	P	Cassano et al. (2011a,b); Tsagaraki and Lazarides (2012)
	NF	P	Zirehpour et al. (2012)
Sewage water treatment in MBR	MF	P	Le-Clech et al. (2006); Chen et al. (2006); Meng et al. (2009); Johir et al. (2012); Lin et al. (2012)

[a] P, polymeric; C, ceramic.

membrane. The main factors that can decrease fouling and keep the permeate flux as high as possible are

1. Proper selection of the membrane and membrane module for a given application
2. Pretreatment of the feed if needed or using additives into the feed, thereby improving flux
3. Appropriate flux for sustainable operation of the system
4. Adequate hydrodynamic conditions at the membrane
5. Proper operation of the plant with well-designed cleaning cycles

Ad (1) Proper selection of the membrane is important not only from the point of view of separation (cut-off) properties but also of fouling properties. For example, earlier PVC–acrylonitrile membranes for protein concentration had one order of magnitude lower flux than recent regenerated cellulose ones (Zydney 2009). It was shown that charged composite cellulosic membranes have much better flux and selectivity in protein separation compared to classical ones (Mehta and Zydney 2005). Newly developed zwitterionic UF membranes show very low fouling behavior due to the highly hydrated structure of the zwitterions and could be the base for high-performance UF membranes for bioprocess applications (Rohani and Zydney 2012). A new antifouling NF membrane filled with nanoparticles is presented by Vatanpour et al. (2012).

Decreasing the TMP along the MF or UF membranes results in increased fouling at their entrance region, thereby decreasing the overall performance of the module. Several approaches were developed to solve this problem; among them was uniform flux ceramic membranes with gradually decreasing thickness of the active layer of the membrane in Isoflux membranes (Skrzypek and Burger 2010). An alternative approach was suggested by applying an impermeable coating on part of the external monolith surface so that uniform and low TMP along the membrane was achieved. This system was successfully tested on the MF of milk (Springer et al. 2011).

Ad (2) Enzymatic pretreatment of the feed can effectively improve the permeate flux as was shown in the MF of pineapple juice (Vaillant et al. 2001; Carneiro et al. 2002; de Carvalho et al. 2008; Laorko et al. 2010), the UF of apple juice (Youn et al. 2004) or other fruit juices in MF (Wang et al. 2005; Sin et al. 2006), or in the UF (Rai et al. 2007; Rai and De 2009; Rai et al. 2010). The pretreatment of the feed by adsorption was suggested (Causserand et al. 2001; Salazar et al. 2007; Bahceci 2012). The pH of the feed and the ionic strength of the solution can influence the membrane performance (Causserand et al. 2001; Rao and Zydney 2005; Graves et al. 2006). Specific additives can improve the flux and the separation selectivity, for example, small charged ligands in the separation of proteins by UF (Rao and Zydney 2006; Rao et al. 2007), powdered activated carbon and cationic polymer on biofouling mitigation in hybrid MBRs (Khan et al. 2012), and polymer addition (Zhang et al. 2010). Some additives may have a negative effect on the MBR performance (Iversen et al. 2009).

The presence of yeasts in the broth improved the flux in the enzyme MF (Mercier-Bonin and Fonade 2002). In the NF of the dairy effluent, flux decline was mitigated by the presence of casein micelles, and when the caseins were removed by the UF pretreatment, whey proteins would agglomerate through calcium bridges, resulting in a severe flux decline during the NF concentration in rotating and vibratory modules (Luo et al. 2012a).

Ad (3) The older concept of limiting the permeate flux, which is achieved at a transmembrane pressure (TMP) above which no increase of flux occurs, was connected with the suggestion to work under limiting flux conditions. This was exchanged by the critical flux concept up to which flux versus TMP dependence is linear (Howell 1995; Bacchin et al. 1995; Bacchin and Aimar 2005). A weak and a strong forms of the critical flux were defined (Bacchin et al. 2006), so as to work below the critical flux (Howell 1995). The value of the critical flux depends much on the method of estimation. The most frequently used is the step-up method (Bacchin et al. 2006; Beier and Jonsson 2010) but it is not much reliable. A better option is the step-up and step-down method but it also does not provide highly reproducible data (Beier and Jonsson 2010). Aimar and Bacchin (2010) suggest that the experimental determination of the critical flux is theoretically accessible for stable, medium-size, and large colloidal particles. The critical flux represents in these conditions a well-marked transition between no-fouling and fouling conditions. Macromolecules and nanoparticles, because they probably slowly aggregate or form gels, will make the measurement of a critical flux difficult and less accurate. In contrast, the critical flux concept could be a good base for the identification of sustainable flux regimes, for well-defined time period for sustainability, and for the identification of low-fouling, not zero-fouling region. There are studies for critical flux conditions in oenology (El Rayess et al. 2011), dairy (Rabiller-Baudry et al. 2009; Luo et al. 2012b), olive oil wastewater treatment (Stoller and Chianese 2006; Stoller and Bravi 2010), and in water treatment in MBR (Le-Clech et al. 2006; Brookes et al. 2006; Guglielmi et al. 2007, 2008).

Ad (4) There are several methods to maintain hydrodynamic conditions in the membrane aiming at reducing flux decline to values as low as possible. Increasing crossflow velocity in the membrane module has a positive effect on the permeate flux and control of fouling, especially when it is connected with the cake formation on the membrane. This was proved also in works on fruit liquids in MF (Ushikubo et al. 2007; Laorko et al. 2010) and UF (Cassano et al. 2008; Rai et al. 2010; Wei et al. 2010). Pulsations imposed on the retentate and permeate flow during the MF of yeast suspension improved the permeate flux by 25–220% (Olayiwola and Walzel 2009).

Another possibility is the introduction of disturbances into liquid flow at the membrane. Membranes with shaped surface can improve their performance as well. A ceramic membrane with a helically stamped surface has several advantages compared with a smooth membrane: flux, as well as limiting flux, are higher for the same velocity of the feed, and power consumption per unit volume of the permeate is lower for the stamped membrane and increases with increasing crossflow velocity of the feed (Stopka et al. 2001). The introduction of static mixers into the tubular membrane increased the permeate flux in MF (Krstic et al. 2006; Gaspar et al. 2011; Popović and Tekić 2011; Popović et al. 2013).

There is a potential for the intensification of the permeate flux in the equipment with increased share stress, for example, in rotating disk and vibrating modules as discussed in an overview (Jaffrin 2012) and showed in the MF of beer (Fillaudeau et al. 2007), the UF separation of polysaccharides (Brou et al. 2003) and the NF of milk products (Frappart et al. 2006). The production of galactosyl oligosaccharide by the hydrolysis of lactose by an enzyme immobilized on a membrane in a rotating disk membrane bioreactor was better than in a batch reactor (Sen et al. 2012). Magnetically induced membrane vibration in MBR for water treatment was also studied (Bilad et al. 2012a).

The backwashing of the membrane with the permeate can improve its performance as shown in the MF of beer (Stopka et al. 2000) and in the UF of fruit juice (Rai and De 2009). The dependence of the permeate flux on the backwash velocity goes through a maximum when the loss of the permeate starts to be larger than the increase of the flux (Stopka et al. 2000).

Gas sparging of modules significantly enhances the performance of membrane processes (Cui et al. 2003) and is widely used in MBRs for wastewater treatment. It keeps the membrane fouling at acceptable levels, but contributes to operation costs. Its modeling is suggested by Smith et al. (2006).

Ad (5) A proper operation of the plant is not limited to well-designed cleaning cycles, which are no doubt important. The proper selection of all parameters discussed in points 1–4 together also play an important role. Details such as not to start the operation of the plant at higher or even at the suggested pressure at once but increasing it gradually can be important as well. An interesting approach in maintaining fouling at low levels is using backwashing of the membrane with the feed (nutrient) solution (Carstensen et al. 2012b).

6.6   MEMBRANE CLEANING

When fouled membranes reach the lowest economically acceptable permeate flux, a proper cleaning procedure is required, the method and frequency of which depend much on the composition of the feed and the process parameters and should be designed for each system individually. An overview of the cleaning methods and strategies, which are very important for the successful operation of the process, is found in review papers (D'Souza and Mawson, 2005; Lin et al. 2010; Astudillo et al. 2010; Porcelli and Judd 2010). More details on the membrane cleaning in specific processes in dairy applications (D'Souza and Mawson, 2005; Nigam et al. 2008; Rabiller-Baudry et al. 2008; Astudillo et al. 2010; Madaeni et al. 2010, 2011), beer applications (Gan et al. 1999), wine MF (El Rayess et al. 2011), protein separation (Levitsky et al. 2012a,b), sugarcane juice (Sim et al. 2009), starch separation (Middlewood and Carson 2012), membrane emulsification (Trentin et al. 2012), and municipal wastewater treatment in MBR (Guglielmi et al. 2008) are available.

6.7   OUTLOOK

The applications of pressure-driven membrane separations discussed in this chapter and additional information presented in the references document the great potential for their wide application in industrial practice. There is still need for further development of membranes and membrane modules that are less sensitive to fouling in specific applications. Further development of technologies with membrane separations for the production of food and nutraceuticals, and in biotechnology as well as in the treatment of related waste streams will continue. Integrated and hybrid systems, especially membrane bioreactors, have a great application potential in the production of higher-value products. The application of MP other than pressure-driven ones will continue, as shown in a few examples in this review. MP utilizing the UF membranes but working with electric potential driving force such as electrodialysis (Bazinet et al. 2012) with UF membranes or electro-ultrafiltration (Sarkar et al. 2009; Sarkar and De 2010) also has the potential for development.

The complex nature of food and biotechnology feeds requires a deeper understanding of the interactions between its components and with membranes, especially in relation to fouling and its prevention. Unlike many industrial applications and published papers on the treatment of food industry liquids, fruit juices, and fermentation broths, there is unfortunately a lack of systematic studies on the performance of individual processes for well-defined liquids containing typical food components such as polyphenols and antioxidants in matrices close to real solutions of interest and under well-defined conditions. Not much data allowing the selection of process conditions for the sustainable operation of MP and related systems under well-defined conditions are available as well. Future effort has to fill this gap.

ACKNOWLEDGMENT

The support of the Slovak grant agency VEGA No. 1/1184/11 is acknowledged.

REFERENCES

Achour, S., Khelifi, E., Attia, Y., Ferjani, E., Hellal, A. N. 2012. Concentration of antioxidant polyphenols from *Thymus capitatus* extracts by membrane process technology. *J. Food Sci.* 77 (6):C703–C709.

Aimar, P., Bacchin, P. 2010. Slow colloidal aggregation and membrane fouling. *J. Membr. Sci.* 360 (1–2):70–76.

Akin, O., Temelli, F., Koseoglu, S. 2012. Membrane applications in functional foods and nutraceuticals. *Crit. Rev. Food Sci. Nutr.* 52 (4):347–371.

Alicieo, T. V. R., Mendes, E. S., Pereira, N. C., de Barros, S. T. D., Alves, J. A. 2007. Evaluation of fouling in beer microfiltration: A study of resistances. *Acta Sci.-Technol.* 29 (2):151–156.

Alicieo, T. V. R., Mendes, E. S., Pereira, N. C., de Barros, S. T. D., Innocenti, T. D., Alves, J. A. 2008. Analysis of the use of a 0.2 mu m ceramic membrane for beer clarification. *Acta Sci.-Technol.* 30 (2):181–186.

Alkhudhiri, A., Darwish, N., Hilal, N. 2012. Membrane distillation: A comprehensive review. *Desalination* 287:2–18.

Amin, I., Mohammad, A. W., Markom, M., Peng, L. C., Hilal, N. 2010. Flux decline study during ultrafiltration of glycerin-rich fatty acid solutions. *J. Membr. Sci.* 351 (1–2):75–86.

Artiga, P., Carballa, M., Garrido, J. M., Mendez, R. 2007. Treatment of winery wastewaters in a membrane submerged bioreactor. *Water Sci. Technol.* 56 (2):63–69.

Artiga, P., Ficara, E., Malpei, F., Garrido, J. M., Mendez, R. 2005. Treatment of two industrial wastewaters in a submerged membrane bioreactor. *Desalination* 179 (1–3):161–169.

Artiga, P., Garcia-Toriello, G., Mendez, R., Garrido, J. M. 2008. Use of a hybrid membrane bioreactor for the treatment of saline wastewater from a fish canning factory. *Desalination* 221 (1–3):518–525.

Aspelund, M. T., Glatz, C. E. 2010a. Clarification of aqueous corn extracts by tangential flow microfiltration. *J. Membr. Sci.* 365 (1–2):123–129.

Aspelund, M. T., Glatz, C. E.. 2010b. Purification of recombinant plant-made proteins from corn extracts by ultrafiltration. *J. Membr. Sci.* 353 (1–2):103–110.

Astudillo, C., Parra, J., Gonzalez, S., Cancino, B. 2010. A new parameter for membrane cleaning evaluation. *Sep. Purif. Technol.* 73 (2):286–293.

Bacchin, P., Aimar, P. 2005. Critical fouling conditions induced by colloidal surface interaction: From causes to consequences. *Desalination* 175 (1):21–27.

Bacchin, P., Aimar, P., Field, R. W. 2006. Critical and sustainable fluxes: Theory, experiments and applications. *J. Membr. Sci.* 281 (1–2):42.

Bacchin, P., Aimar, P., Sanchez, V. 1995. Model for colloidal fouling of membranes. *AIchE J.* 41 (2):368–376.

Bagger-Jorgensen, R., Meyer, A. S., Pinelo, M., Varming, C., Jonsson, G. 2011. Recovery of volatile fruit juice aroma compounds by membrane technology: Sweeping gas versus vacuum membrane distillation. *Innov. Food Sci. Emerg. Technol.* 12 (3):388–397.

Bahceci, K. S. 2012. Effects of pretreatment and various operating parameters on permeate flux and quality during ultrafiltration of apple juice. *Int. J. Food Sci. Technol* 47 (2): 315–324.

Baklouti, S., Ellouze-Ghorbel, R., Mokni, A., Chaabouni, S. 2012. Clarification of pomegranate juice by ultrafiltration: Study of juice quality and of the fouling mechanism. *Fruits* 67 (3):215–225.

Baldasso, C., Barros, T. C., Tessaro, I. C. 2011. Concentration and purification of whey proteins by ultrafiltration. *Desalination* 278 (1–3):381–386.

Bazinet, L. 2005. Electrodialytic phenomena and their applications in the dairy industry: A review. *Crit. Rev. Food Sci. Nutr.* 45 (4):307–326.

Bazinet, L., Brianceau, S., Dube, P., Desjardins, Y. 2012. Evolution of cranberry juice physicochemical parameters during phenolic antioxidant enrichment by electrodialysis with filtration membrane. *Sep. Purif. Technol.* 87:31–39.

Bazinet, L., Firdaous, L. 2009. Membrane processes and devices for separation of bioactive peptides. *Rec. Pat. Biotechnol.* 1:61–72.

Bazinet, L., Firdaous, L. 2011. Recent patented applications of ion-exchange membranes in the agrifood sector. *Rec. Pat. Biotechnol.* 4:207–216.

Beier, S. P., Jonsson, G. 2007. Separation of enzymes and yeast cells with a vibrating hollow fiber membrane module. *Sep. Purif. Technol.* 53 (1):111–118.

Beier, S. P., Jonsson, G. 2010. Critical flux determination by flux-stepping. *AIchE J.* 56 (7):1739–1747.

Belafi-Bako, K., Boor, A. 2011. Concentration of cornelian cherry fruit juice by membrane osmotic distillation. *Desalin. Water Treat.* 35 (1–3):271–274.

Bilad, M. R., Mezohegyi, G., Declerck, P., Vankelecom, I. F. J. 2012a. Novel magnetically induced membrane vibration (MMV) for fouling control in membrane bioreactors. *Water Res.* 46 (1):63–72.

Bilad, M. R., Vandamme, D., Foubert, I., Muylaert, K., Vankelecom, I. F. J. 2012b. Harvesting microalgal biomass using submerged microfiltration membranes. *Bioresour. Technol.* 111:343–352.

Bolton, G., LaCasse, D., Kuriyel, R. 2006. Combined models of membrane fouling: Development and application to microfiltration and ultrafiltration of biological fluids. *J. Membr. Sci.* 277 (1–2):75–84.

Brans, G., Schroen, C., van der Sman, R. G. M., Boom, R. M. 2004. Membrane fractionation of milk: State of the art and challenges. *J. Membr. Sci.* 243 (1–2):263–272.

Brookes, A., Jefferson, B., Guglielmi, G., Judd, S. J. 2006. Sustainable flux fouling in a membrane bioreactor: Impact of flux and MLSS. *Sep. Purif. Technol.* 41 (7):1279–1291.

Brou, A., Jaffrin, M. Y., Ding, L. H., Courtois, J. 2003. Microfiltration and ultrafiltration of polysaccharides produced by fermentation using a rotating disk dynamic filtration system. *Biotechnol. Bioeng.* 82 (4):429–437.

Bugan, S. G., Domeny, Z., Smogrovicova, D., Svitel, J., Schlosser, S., Stopka, J. 2000. Ceramic membrane cross-flow microfiltration for beer recovery from tank bottoms. *Mon. Schr. Brauwiss.* 53 (11–12):229–233.

Campbell, K. A., Glatz, C. E. 2010. Protein recovery from enzyme-assisted aqueous extraction of soybean. *Biotechnol. Prog.* 26 (2):488–495.

Caric, M. D., Milanovic, S. D., Krstic, D. M., Tekic, M. N. 2000. Fouling of inorganic membranes by adsorption of whey proteins. *J. Membr. Sci.* 165 (1):83–88.

Carneiro, L., Sa, I. D., Gomes, F. D., Matta, V. M., Cabral, L. M. C. 2002. Cold sterilization and clarification of pineapple juice by tangential microfiltration. *Desalination* 148 (1–3):93–98.

Carstensen, F., Apel, A., Wessling, M. 2012a. *In situ* product recovery: Submerged membranes versus external loop membranes. *J. Membr. Sci.* 394:1–36.

Carstensen, F., Marx, C., André, J., Melin, T., Wessling, M. 2012b. Reverse-flow diafiltration for continuous *in situ* product recovery. *J. Membr. Sci.* 421–422 (0):39–50.

Cassano, A., Conidi, C., Drioli, E. 2011a. Clarification and concentration of pomegranate juice (*Punica granatum* L.) using membrane processes. *J. Food Eng.* 107 (3–4): 366–373.

Cassano, A., Conidi, C., Drioli, E. 2011b. Comparison of the performance of UF membranes in olive mill wastewaters treatment. *Water Res.* 45 (10):3197–3204.

Cassano, A., Mecchia, A., Drioli, E. 2008. Analyses of hydrodynamic resistances and operating parameters in the ultrafiltration of grape must. *J. Food Eng.* 89 (2):171–177.

Castel, V., Andrich, O., Netto, F. M., Santiago, L. G., Carrara, C. R. 2012. Comparison between isoelectric precipitation and ultrafiltration processes to obtain *Amaranth mantegazzianus* protein concentrates at pilot plant scale. *J. Food Eng.* 112 (4):288–295.

Cath, T. Y., Childress, A. E., Elimelech, M. 2006. Forward osmosis: Principles, applications, and recent developments. *J. Membr. Sci.* 281 (1–2):70.

Causserand, C., Kara, Y., Aimar, P. 2001. Protein fractionation using selective adsorption on clay surface before filtration. *J. Membr. Sci.* 186 (2):165–181.

Chabeaud, A., Vandanjon, L., Bourseau, P., Jaouen, P., Guerard, F. 2009. Fractionation by ultrafiltration of a saithe protein hydrolysate (*Pollachius virens*): Effect of material and molecular weight cut-off on the membrane performances. *J. Food Eng.* 91 (3):408–414.

Chae, S. R., Shin, H. S. 2007. Effect of condensate of food waste (CFW) on nutrient removal and behaviours of intercellular materials in a vertical submerged membrane bioreactor (VSMBR). *Biores. Technol.* 98 (2):373.

Charcosset, C. 2006. Membrane processes in biotechnology: An overview. *Biotechnol. Adv.* 24 (5):482–492.

Cheang, B. L., Zydney, A. L. 2004. A two-stage ultrafiltration process for fractionation of whey protein isolate. *J. Membr. Sci.* 231 (1–2):159–167.

Chen, M. Y., Lee, D. J., Tay, J. H. 2006. Extracellular polymeric substances in fouling layer. *Sep. Sci. Technol.* 41 (7):1467–1474.

Chhaya, M. S., Majumdar, G. C., De, S. 2012. Clarifications of stevia extract using cross flow ultrafiltration and concentration by nanofiltration. *Sep. Purif. Technol.* 89:125–134.

Cho, Y. H., Lee, H. D., Park, H. B. 2012. Integrated membrane processes for separation and purification of organic acid from a biomass fermentation process. *Ind. Eng. Chem. Res.* 51 (30):10207–10219.

Chollangi, A., Hossain, M. M. 2007. Separation of proteins and lactose from dairy wastewater. *Chem. Eng. Process.* 46 (5):398–404.

Chung, T. S., Zhang, S., Wang, K. Y., Su, J. C., Ling, M. M. 2012. Forward osmosis processes: Yesterday, today and tomorrow. *Desalination* 287:78–81.

Coutte, F., Lecouturier, D., Yahia, S. A., Leclere, V., Bechet, M., Jacques, P., Dhulster, P. 2010. Production of surfactin and fengycin by *Bacillus subtilis* in a bubbleless membrane bioreactor. *Appl. Microbiol. Biotechnol.* 87 (2):499–507.

Cui, Z. F., Chang, S., Fane, A. G. 2003. The use of gas bubbling to enhance membrane processes. *J. Membr. Sci.* 221 (1–2):1–35.

D'Souza, N. M., Mawson, A. J. 2005. Membrane cleaning in the dairy industry: A review. *Crit. Rev. Food Sci. Nutr.* 45 (2):125–134.

Das, R., Bhattacherjee, C., Ghosh, S. 2009. Studies on membrane processing of sesame protein isolate and sesame protein hydrolysate using rotating disk module. *Sep. Sci. Technol.* 44 (1):131–150.

Das, R., Ghosh, S., Bhattacharjee, C. 2012. Enzyme membrane reactor in isolation of antioxidative peptides from oil industry waste: A comparison with non-peptidic antioxidants. *LWT-Food Sci. Technol.* 47 (2):238–245.

Das, R., Sen, D., Sarkar, A., Bhattacharyya, S., Bhattacharjee, C. 2011. A comparative study on the production of galacto-oligosaccharide from whey permeate in recycle membrane reactor and in enzymatic batch reactor. *Ind. Eng. Chem. Res.* 50 (2):806–816.

Daufin, G., Escudier, J. P., Carrere, H., Berot, S., Fillaudeau, L., Decloux, M. 2001. Recent and emerging applications of membrane processes in the food and dairy industry. *Food Bioprod. Process.* 79 (C2):89–102.

de Barros, S. T. D., Andrade, C. M. G., Mendes, E. S., Peres, L. 2003. Study of fouling mechanism in pineapple juice clarification by ultrafiltration. *J. Membr. Sci.* 215 (1–2):213–224.

de Carvalho, L. M. J., da Silva, C. A. B. 2010. Clarification of pineapple juice by microfiltration. *Cien. Technol. Aliment.* 30 (3):828–832.

de Carvalho, L. M. J., de Castro, I. M., da Silva, C. A. B. 2008. A study of retention of sugars in the process of clarification of pineapple juice (*Ananas comosus*, L. Merril) by micro- and ultra-filtration. *J. Food Eng.* 87 (4):447–454.

de la Torre, T., Harff, M., Lesjean, B., Drews, A., Kraume, M. 2009. Characterisation of polysaccharide fouling of an ultrafiltration membrane using model solutions. *Desalin. Water Treat.* 8 (1–3):17–23.

de Oliveira, R. C., Doce, R. C., de Barros, S. T. D. 2012. Clarification of passion fruit juice by microfiltration: Analyses of operating parameters, study of membrane fouling and juice quality. *J. Food Eng.* 111 (2):432–439.

Dereli, R. K., Ersahin, M. E., Ozgun, H., Ozturk, I., Jeison, D., van der Zee, F., van Lier, J. B. 2012a. Potentials of anaerobic membrane bioreactors to overcome treatment limitations induced by industrial wastewaters. *Bioresour. Technol.* 122:160–170.

Dereli, R. K., Urban, D. R., Heffernan, B., Jordan, J. A., Ewing, J., Rosenberger, G. T., Dunaev, T. I. 2012b. Performance evaluation of a pilot-scale anaerobic membrane bioreactor (AnMBR) treating ethanol thin stillage. *Environ. Technol.* 33 (13):1511–1516.

Dey, P., Pal, P. 2012. Direct production of L (+) lactic acid in a continuous and fully membrane-integrated hybrid reactor system under non-neutralizing conditions. *J. Membr. Sci.* 389:355–362.

Diaz-Reinoso, B., Gonzalez-Lopez, N., Moure, A., Dominguez, H., Parajo, J. C. 2010. Recovery of antioxidants from industrial waste liquors using membranes and polymeric resins. *J. Food Eng.* 96 (1):127–133.

Diaz-Reinoso, B., Moure, A., Dominguez, H., Parajo, J. C. 2009. Ultra- and nanofiltration of aqueous extracts from distilled fermented grape pomace. *J. Food Eng.* 91 (4):587–593.

Diaz-Reinoso, B., Moure, A., Dominguez, H., Parajo, J. C. 2011. Membrane concentration of antioxidants from *Castanea sativa* leaves aqueous extracts. *Chem. Eng. J.* 175:95–102.

Drews, A., Evenblij, H., Rosenberger, S. 2005. Potential and drawbacks of microbiology–membrane interaction in membrane bioreactors. *Environ. Prog.* 24 (4):426–433.

Drews, A., Kraume, M. 2005. Process improvement by application of membrane bioreactors. *Chem. Eng. Res. Des.* 83 (A3):276–284.

Drioli, E., Cassano, A. 2010. Advances in membrane-based concentration in the food and beverage industries: Direct osmosis and membrane contactors. In: Rizvi, S. S. H. *Separation, Extraction and Concentration Processes in the Food, Beverage and Nutraceutical Industries*, Cambridge: Woodhead Publishers Ltd.

Drioli, E., Giorno, L., eds. 2009. *Membrane Operations. Innovative Separations and Transformations.* Weinheim: Wiley-VCH Verlag.

Drioli, E., Giorno, L., eds. 2010. *Comprehensive Membrane Science and Engineering.* 4 vols: Elsevier.

Ecker, J., Raab, T., Harasek, M. 2012. Nanofiltration as key technology for the separation of LA and AA. *J. Membr. Sci.* 389:389–398.

Echavarria, A. P., Falguera, V., Torras, C., Berdun, C., Pagan, J., Ibarz, A. 2012. Ultrafiltration and reverse osmosis for clarification and concentration of fruit juices at pilot plant scale. *LWT-Food Sci. Technol.* 46 (1):189–195.

Echavarria, A. P., Torras, C., Pagan, J., Ibarz, A. 2011. Fruit juice processing and membrane technology application. *Food Eng. Rev.* 3 (3–4):136–158.

El Rayess, Y., Albasi, C., Bacchin, P., Taillandier, P., Mietton-Peuchot, M., Devatine, A. 2011. Cross-flow microfiltration of wine: Effect of colloids on critical fouling conditions. *J. Membr. Sci.* 385–386 (0):177–186.

El Rayess, Y., Albasi, C., Bacchin, P., Taillandier, P., Raynal, J., Mietton-Peuchot, M., Devatine, A. 2011. Cross-flow microfiltration applied to oenology: A review. *J. Membr. Sci.* 382 (1–2):1–19.

Enevoldsen, A. D., Hansen, E. B., Jonsson, G. 2007a. Electro-ultrafiltration of amylase enzymes: Process design and economy. *Chem. Eng. Sci.* 62 (23):6716–6725.

Enevoldsen, A. D., Hansen, E. B., Jonsson, G. 2007b. Electro-ultrafiltration of industrial enzyme solutions. *J. Membr. Sci.* 299 (1–2):28–37.

Farizoglu, B., Uzuner, S. 2011. The investigation of dairy industry wastewater treatment in a biological high performance membrane system. *Biochem. Eng. J.* 57:46–54.

Field, R. W., Wu, D., Howell, J. A., Gupta, B. B. 1995. Critical flux concept for microfiltration fouling. *J. Membr. Sci.* 100 (3):259–272.

Fikar, M., Kovács, Z., Czermak, P. 2010. Dynamic optimization of batch diafiltration processes. *J. Membr. Sci.* 355 (1–2):168–174.

Fillaudeau, L., Blanpain-Avet, P., Daufin, G. 2006. Water, wastewater and waste management in brewing industries. *J. Clean Prod.* 14 (5):463–471.

Fillaudeau, L., Boissier, B., Moreau, A., Blanpain-Avet, P., Ermolaev, S., Jitariouk, N., Gourdon, A. 2007. Investigation of rotating and vibrating filtration for clarification of rough beer. *J. Food Eng.* 80 (1):206–217.

Fillaudeau, L., Bories, A., Decloux, M. 2008. Brewing, winemaking and distilling: An overview of wastewater treatment and utilisation schemes. In: Klemeš, J., Smith, R., Kim, J. K. Vol. 160, *Handbook of Water and Energy Management in Food Processing*, Abington, UK: Woodhead Publ.

Frappart, M., Akoum, O., Ding, L. H., Jaffrin, M. Y. 2006. Treatment of dairy process waters modelled by diluted milk using dynamic nanofiltration with a rotating disk module. *J. Membr. Sci.* 282 (1–2):465–472.

Gan, Q. 2001. Beer clarification by cross-flow microfiltration—Effect of surface hydrodynamics and reversed membrane morphology. *Chem. Eng. Process.* 40 (5):413–419.

Gan, Q., Howell, J. A., Field, R. W., England, R., Bird, M. R., McKechinie, M. T. 1999. Synergetic cleaning procedure for a ceramic membrane fouled by beer microfiltration. *J. Membr. Sci.* 155 (2):277–289.

Garcia-Castello, E., Cassano, A., Criscuoli, A., Conidi, C., Drioli, E. 2010. Recovery and concentration of polyphenols from olive mill wastewaters by integrated membrane system. *Water Res.* 44 (13):3883–3892.

Gaspar, I., Koris, A., Bertalan, Z., Vatai, G. 2011. Comparison of ceramic capillary membrane and ceramic tubular membrane with inserted static mixer. *Chem. Pap.* 65 (5):596–602.

Gesan-Guiziou, G., Alvarez, N., Jacob, D., Daufin, G. 2007. Cleaning-in-place coupled with membrane regeneration for re-using caustic soda solutions. *Sep. Purif. Technol.* 54 (3):329–339.

Giglia, S., Straeffer, G. 2012. Combined mechanism fouling model and method for optimization of series microfiltration performance. *J. Membr. Sci.* 417:144–153.

Gkoutsidis, P. E., Petrotos, K. B., Kokkora, M. I., Tziortziou, A. D., Christodouloulis, K., Goulas, P. 2011. Olive mill waste water (OMWW) treatment by diafiltration. *Desalin. Water Treat.* 30 (1–3):237–246.

Gong, Y. W., Zhang, H. X., Cheng, X. N. 2012. Treatment of dairy wastewater by two-stage membrane operation with ultrafiltration and nanofiltration. *Water Sci. Technol.* 65 (5):915–919.

Goulas, A. K., Grandison, A. S., Rastall, R. A. 2003. Fractionation of oligosaccharides by nanofiltration. *J. Sci. Food Agric.* 83 (7):675–680.

Goulas, A. K., Kapasakalidis, P. G., Sinclair, H. R., Rastall, R. A., Grandison, A. S. 2002. Purification of oligosaccharides by nanofiltration. *J. Membr. Sci.* 209 (1):321–335.

Gozke, G., Posten, C. 2010. Electrofiltration of biopolymers. *Food Eng. Rev.* 2 (2):131–146.

Graves, K., Rozeboom, G., Heng, M., Glatz, C. 2006. Broth conditions determining specific cake resistance during microfiltration of *Bacillus subtilis*. *Biotechnol. Bioeng.* 94 (2):346–352.

Grzenia, D. L., Dong, R. W., Jasuja, H., Kipper, M. J., Qian, X., Ranil, W. S. 2012. Conditioning biomass hydrolysates by membrane extraction. *J. Membr. Sci.* 415–416 (0):75–84.

Guglielmi, G., Andreottola, G., Foladori, P., Ziglio, G. 2009. Membrane bioreactors for winery wastewater treatment: Case-studies at full scale. *Water Sci. Technol.* 60 (5):1201–1207.

Guglielmi, G., Chiarani, D., Judd, S. J., Andreottola, G. 2007. Flux criticality and sustainability in a hollow fibre submerged membrane bioreactor for municipal wastewater treatment. *J. Membr. Sci.* 289 (1–2):241–248.

Guglielmi, G., Chiarani, D., Saroj, D. P., Andreottola, G. 2008. Impact of chemical cleaning and air-sparging on the critical and sustainable flux in a flat sheet membrane bioreactor for municipal wastewater treatment. *Water Sci. Technol.* 57 (12):1873–1879.

Gul, S., Harasek, M. 2012. Energy saving in sugar manufacturing through the integration of environmental friendly new membrane processes for thin juice pre-concentration. *Appl. Therm. Eng.* 43:128–133.

Hang, H., Mu, W. M., Jiang, B., Zhao, M., Zhou, L. M., Zhang, T., Miao, M. 2012a. DFA III production from inulin with inulin fructotransferase in ultrafiltration membrane bioreactor. *J. Biosci. Bioeng.* 113 (1):55–57.

Hang, H., Mu, W. M., Jiang, B., Zhao, M., Zhou, L. M., Zhang, T., Miao, M. 2012b. Enzymatic hydrolysis of inulin in a bioreactor coupled with an ultrafiltration membrane. *Desalination* 284:309–315.

Hermia, J. 1982. Constant pressure blocking filtration laws—Application to power-law non-Newtonian fluids. *Trans. Inst. Chem. Eng.* 60:183–187.

Hinkova, A., Bubnik, Z., Kadlec, P., Pridal, J. 2002. Potentials of separation membranes in the sugar industry. *Sep. Purif. Technol.* 26 (1):101–110.

Ho, C. C., Zydney, A. L. 1999. Effect of membrane morphology on the initial rate of protein fouling during microfiltration. *J. Membr. Sci.* 155:261–275.

Ho, C. C., Zydney, A. L. 2000. A combined pore blockage and cake filtration model for protein fouling during microfiltration. *J. Colloid Interface Sci.* 232 (2):389–399.

Ho, C. C., Zydney, A. L. 2001. Protein fouling of asymmetric and composite microfiltration membranes. *Ind. Eng. Chem. Res.* 40 (5):1412–1421.

Ho, C. C., Zydney, A. L. 2002. Transmembrane pressure profiles during constant flux microfiltration of bovine serum albumin. *J. Membr. Sci.* 209 (2):363–377.

Howell, J. A. 1995. Sub-critcal flux of microfiltration. *J. Membr. Sci.* 107:165–171.

Hwang, K.-J., Chen, F.-F. 2007. Modeling of particle fouling and membrane blocking in submerged membrane filtration. *Sep. Sci. Technol.* 42 (12):2595–2614.

Iversen, V., Koseoglu, H., Yigit, N. O., Drews, A., Kitis, M., Lesjean, B., Kraume, M. 2009. Impacts of membrane flux enhancers on activated sludge respiration and nutrient removal in MBRs. *Water Res.* 43 (3):822–830.

Jaffrin, M. Y. 2012. Dynamic filtration with rotating disks, and rotating and vibrating membranes: An update. *Curr. Opin. Chem. Eng.* 1 (2):171–177.

Jaffrin, M. Y., Charrier, J. P. 1994. Optimization of ultrafiltration and diafiltration processes for albumin production. *J. Membr. Sci.* 97 (Dec):71–81.

Jiao, B., Cassano, A., Drioli, E. 2004. Recent advances on membrane processes for the concentration of fruit juices: A review. *J. Food Eng.* 63 (3):303–324.

Johir, M. A. H., Vigneswaran, S., Sathasivan, A., Kandasamy, J., Chang, C. Y. 2012. Effect of organic loading rate on organic matter and foulant characteristics in membrane bioreactor. *Bioresour. Technol.* 113:154–160.

Jung, I., Lovitt, R. W. 2010. A comparative study of the growth of lactic acid bacteria in a pilot scale membrane bioreactor. *J. Chem. Technol. Biotechnol.* 85 (9):1250–1259.

Kamada, T., Nakajima, M., Nabetani, H., Saglam, N., Iwamoto, S. 2002. Availability of membrane technology for purifying and concentrating oligosaccharides. *Eur. Food Res. Technol.* 214 (5):435–440.

Kelly, J., Kelly, P. 1995a. Desalination of acid casein whey by nanofiltration. *Int. Dairy J.* 5 (3):291–303.

Kelly, J., Kelly, P. 1995b. Nanofiltration of whey—Quality, environmental and economic-aspects. *J. Soc. Dairy Technol.* 48 (1):20–25.

Kertész, R., Schlosser, Š. 2005. Design and simulation of two phase hollow fiber contactors for simultaneous membrane based solvent extraction and stripping of organic acids and bases. *Sep. Purif. Technol.* 41 (3):275–287.

Kertesz, S., Laszlo, Z., Forgacs, E., Szabo, G., Hodur, C. 2011. Dairy wastewater purification by vibratory shear enhanced processing. *Desalin. Water Treat.* 35 (1–3):195–201.

Khan, S. J., Visvanathan, C., Jegatheesan, V. 2012. Effect of powdered activated carbon (PAC) and cationic polymer on biofouling mitigation in hybrid MBRs. *Bioresour. Technol.* 113:165–168.

Kim, H.-C., Hong, J.-H., Lee, S. 2006. Fouling of microfiltration membranes by natural organic matter after coagulation treatment: A comparison of different initial mixing conditions. *J. Membr. Sci.* 283 (1–2):266.

Koo, C. H., Mohammad, A. W., Suja, F., Talib, M. Z. M. 2012. Review of the effect of selected physicochemical factors on membrane fouling propensity based on fouling indices. *Desalination* 287:167–177.

Koutinas, A., Belafi-Bako, K., Kabiri-Badr, A., Toth, A., Gubicza, L., Webb, C. 2001. Enzymatic hydrolysis of polysaccharides—Hydrolysis of starch by an enzyme complex from fermentation by *Aspergillus awamori*. *Food Bioprod. Proc.* 79 (C1):41–45.

Kozak, A., Bekassy-Molnar, E., Vatai, G. 2009. Production of black-currant juice concentrate by using membrane distillation. *Desalination* 241 (1–3):309–314.

Krstic, D. M., Koris, A. K., Tekic, M. N. 2006. Do static turbulence promoters have potential in cross-flow membrane filtration applications? *Desalination* 191 (1–3):371–375.

Kuhnl, W., Piry, A., Kaufmann, V., Grein, T., Ripperger, S., Kulozik, U. 2010. Impact of colloidal interactions on the flux in cross-flow microfiltration of milk at different pH values: A surface energy approach. *J. Membr. Sci.* 352 (1–2):107–115.

Kulozik, U., Ripperger, S. 2008. Innovative inputs of membrane technology in the food technology. *Chem. Eng. Tech.* 80 (8):1045–1058.

Kuroiwa, T., Izuta, H., Nabetani, H., Nakajima, M., Sato, S., Mukataka, S., Ichikawa, S. 2009. Selective and stable production of physiologically active chitosan oligosaccharides using an enzymatic membrane bioreactor. *Process Biochem.* 44 (3):283–287.

Kwon, B., Shon, H. K., Cho, J. 2009. Investigating the relationship between model organic compounds and ultrafiltration membrane fouling. *Desalin. Water Treat.* 8 (1–3):177–187.

Laorko, A., Li, Z. Y., Tongchitpakdee, S., Chantachum, S., Youravong, W. 2010. Effect of membrane property and operating conditions on phytochemical properties and permeate flux during clarification of pineapple juice. *J. Food Eng.* 100 (3):514–521.

Le-Clech, P., Chen, V., Fane, T. A. G. 2006. Fouling in membrane bioreactors used in wastewater treatment. *J. Membr. Sci.* 284 (1–2):17–53.

Leberknight, J., Wielenga, B., Lee-Jewett, A., Menkhaus, T. J. 2011. Recovery of high value protein from a corn ethanol process by ultrafiltration and an exploration of the associated membrane fouling. *J. Membr. Sci.* 366 (1–2):405–412.

Leo, C. P., Lee, W. P. C., Ahmad, A. L., Mohammad, A. W. 2012. Polysulfone membranes blended with ZnO nanoparticles for reducing fouling by oleic acid. *Sep. Purif. Technol.* 89:51–56.

Lesjean, B., Ferre, V., Vonghia, E., Moeslang, H. 2009. Market and design considerations of the 37 larger MBR plants in Europe. *Desalin. Water Treat.* 6 (1–3):227–233.

Lesjean, B., Huisjes, E. H. 2008. Survey of the European MBR market: Trends and perspectives. *Desalination* 231 (1–3):71–81.

Lesjean, B., Judd, S. 2009. Facts on MBR-technology, http://www.mbr-network.eu/mbr-projects/index.php Accessed: October 2012.

Levitsky, I., Duek, A., Naim, R., Arkhangelsky, E., Gitis, V. 2012a. Cleaning UF membranes with simple and formulated solutions. *Chem. Eng. Sci.* 69 (1):679–683.

Levitsky, I., Naim, R., Duek, A., Gitis, V. 2012b. Effect of time in chemical cleaning of ultrafiltration membranes. *Chem. Eng. Technol.* 35 (5):941–946.

Li, H. F., Li, Z. W., Xiong, S. Q., Zhang, H. R., Li, N., Zhou, S. H., Liu, Y. M., Huang, Z. B. 2011. Pilot-scale isolation of bioactive extracellular polymeric substances from cellfree media of mass microalgal cultures using tangential-flow ultrafiltration. *Process Biochem.* 46 (5):1104–1109.

Li, M. S., Zhao, Y. J., Zhou, S. Y., Xing, W. H. 2010. Clarification of raw rice wine by ceramic microfiltration membranes and membrane fouling analysis. *Desalination* 256 (1–3):166–173.

Lim, Y. P., Mohammad, A. W. 2010. Effect of solution chemistry on flux decline during high concentration protein ultrafiltration through a hydrophilic membrane. *Chem. Eng. J.* 159 (1–3):91–97.

Lin, H. J., Gao, W. J., Meng, F. G., Liao, B. Q., Leung, K. T., Zhao, L. H., Chen, J. R., Hong, H. C. 2012. Membrane bioreactors for industrial wastewater treatment: A critical review. *Crit. Rev. Environ. Sci. Technol.* 42 (7):677–740.

Lin, J. C. T., Lee, D. J., Huang, C. P. 2010. Membrane fouling mitigation: Membrane cleaning. *Sep. Sci. Technol.* 45 (7):858–872.

Lo, Y. M., Yang, S. T., Min, D. B. 1997. Ultrafiltration of xanthan gum fermentation broth: Process and economic analyses. *J. Food Eng.* 31 (2):219–236.

Lopez, F., Pescador, P., Guell, C., Morales, M. L., Garcia-Parrilla, M. C., Troncoso, A. M. 2005. Industrial vinegar clarification by cross-flow microfiltration: Effect on colour and polyphenol content. *J. Food Eng.* 68 (1):133–136.

Luo, J., Cao, W., Ding, L., Zhu, Z., Wan, Y., Jaffrin, M. Y. 2012a. Treatment of dairy effluent by shear-enhanced membrane filtration: The role of foulants. *Sep. Purif. Technol.* 96:194–203.

Luo, J. Q., Ding, L. H. 2011. Influence of pH on treatment of dairy wastewater by nanofiltration using shear-enhanced filtration system. *Desalination* 278 (1–3):150–156.

Luo, J. Q., Ding, L. H., Wan, Y. H., Jaffrin, M. Y. 2012b. Threshold flux for shear-enhanced nanofiltration: Experimental observation in dairy wastewater treatment. *J. Membr. Sci.* 409:276–284.

Madaeni, S. S., Rostami, E., Rahimpour, A. 2010. Surfactant cleaning of ultrafiltration membranes fouled by whey. *Int. J. Dairy Technol.* 63 (2):273–283.

Madaeni, S. S., Tavakolian, H. R., Rahimpour, F. 2011. Cleaning optimization of microfiltration membrane employed for milk sterilization. *Separ. Sci. Technol.* 46 (4):571–580.

Mameri, N., Halet, F., Drouiche, M., Grib, H., Lounici, H., Pauss, A., Piron, D., Belhocine, D. 2000. Treatment of olive mill washing water by ultrafiltration. *Can. J. Chem. Eng.* 78 (3):590–595.

Maruyama, T., Katoh, S., Nakajima, M., Nabetani, H. 2001. Mechanism of bovine serum albumin aggregation during ultrafiltration. *Biotechnol. Bioeng.* 75 (2):233–238.

Mazzei, R., Drioli, E., Giorno, L. 2010. Biocatalytic membrane reactor and membrane emulsification concepts combined in a single unit to assist production and separation of water unstable reaction products. *J. Membr. Sci.* 352 (1–2):166–172.

Mazzei, R., Drioli, E., Giorno, L. 2012. Enzyme membrane reactor with heterogenized betaglucosidase to obtain phytotherapic compound: Optimization study. *J. Membr. Sci.* 390:121–129.

Mazzei, R., Giorno, L., Piacentini, E., Mazzuca, S., Drioli, E. 2009. Kinetic study of a biocatalytic membrane reactor containing immobilized beta-glucosidase for the hydrolysis of oleuropein. *J. Membr. Sci.* 339 (1–2):215–223.

Mehta, A., Zydney, A. L. 2005. Permeability and selectivity analysis for ultrafiltration membranes. *J. Membr. Sci.* 249:245–249.

Meng, F. G., Chae, S.-R., Shin, H.-S., Fenglin Y. F., Zhou, Z. 2012. Recent advances in membrane bioreactors: Configuration development, pollutant elimination, and sludge reduction. *Environ. Eng. Sci.* 29 (3):139–160.

Meng, F., Chae, S.-R., Drews, A., Kraume, M., Shin, H.-S., Yang, F. 2009. Recent advances in membrane bioreactors (MBRs): Membrane fouling and membrane material. *Water Res.* 43 (6):1489–1512.

Mercier-Bonin, M., Fonade, C. 2002. Air-sparged microfiltration of enzyme/yeast mixtures: Determination of optimal conditions for enzyme recovery. *Desalination* 148 (1–3):171–176.

Middlewood, P. G., Carson, J. K. 2012. Extraction of *Amaranth* starch from an aqueous medium using microfiltration: Membrane fouling and cleaning. *J. Membr. Sci.* 411:22–29.

Mihal', M., Vereš, R., Markoš, J., Štefuca, V. 2012. Intensification of 2-phenylethanol production in fed-batch hybrid bioreactor: Biotransformations and simulations. *Chem. Eng. Process.* 57–58:75–85.

Mohammad, A. W., Ng, C. Y., Lim, Y. P., Ng, G. H. 2012. Ultrafiltration in food processing industry: Review on application, membrane fouling, and fouling control. *Food Bioprocess Technol.* 5 (4):1143–1156.

Molek, J. R., Zydney, A. L. 2007. Separation of PEGylated alpha-lactalbumin from unreacted precursors and byproducts using ultrafiltration. *Biotechnol. Prog.* 23 (6):1417–1424.

Mondor, M., Ali, F., Ippersiel, D., Lamarche, F. 2010. Impact of ultrafiltration/diafiltration sequence on the production of soy protein isolate by membrane technologies. *Innov. Food Sci. Emerg. Technol.* 11 (3):491–497.

Mondor, M., Girard, B., Moresoli, C. 2000. Modeling flux behavior for membrane filtration of apple juice. *Food Res. Int.* 33 (7):539–548.

Muller, A., Chaufer, B., Merin, U., Daufin, G. 2003a. Prepurification of alpha-lactalbumin with ultrafiltration ceramic membranes from acid casein whey: Study of operating conditions. *Lait* 83 (2):111–129.

Muller, A., Chaufer, B., Merin, U., Daufin, G. 2003b. Purification of alpha-lactalbumin from a prepurified acid whey: Ultrafiltration or precipitation. *Lait* 83 (6):439–451.

Nabarlatz, D., Torras, C., Garcia-Valls, R., Montané, D. 2007. Purification of xylo-oligosaccharides from almond shells by ultrafiltration. *Sep. Purif. Technol.* 53 (3):235–243.

Nabetani, H. 2012. Optimization of membrane separation processes for liquid foods (in Japanese). *J. Jpn. Soc. Food Sci. Technol.-Nippon Shokuhin Kagaku Kogaku Kaishi* 59 (6):249–261.

Nabetani, H., Abbott, T. P., Kleiman, R. 1995. Optimal separation of jojoba protein using membrane processes. *Ind. Eng. Chem. Res.* 34 (5):1779–1788.

Nabetani, H., Hagiwara, S., Yanai, N., Shiotani, S., Baljinnyam, J., Nakajima, M. 2012. Purification and concentration of antioxidative dipeptides obtained from chicken extract and their application as functional food. *J. Food Drug Anal.* 20:179–183.

Nagaraj, N., Patil, B. S., Biradar, P. M. 2006. Osmotic membrane distillation—A brief review. *Int. J. Food Eng.* 2 (2) Article No. 5.

Navaratna, D., Shu, L., Baskaran, K., Jegatheesan, V. 2012. Model development and parameter estimation for a hybrid submerged membrane bioreactor treating *Ametryn*. *Bioresour. Technol.* 113:191–200.

Nawaz, H., Shi, J., Mittal, G. S., Kakuda, Y. 2006. Extraction of polyphenols from grape seeds and concentration by ultrafiltration. *Sep. Purif. Technol.* 48 (2):176–181.

Ndiaye, N., Pouliot, Y., Saucier, L., Beaulieu, L., Bazinet, L. 2010. Electroseparation of bovine lactoferrin from model and whey solutions. *Sep. Purif. Technol.* 74 (1):93–99.

Nigam, M. O., Bansal, B., Chen, X. D. 2008. Fouling and cleaning of whey protein concentrate fouled ultrafiltration membranes. *Desalination* 218 (1–3):313–322.

Nishizawa, K., Nakajima, M., Nabetani, H. 2000. A forced-flow membrane reactor for transfructosylation using ceramic membrane. *Biotechnol. Bioeng.* 68 (1):92–97.

Ochando-Pulido, J. M., Rodriguez-Vives, S., Hodaifa, G., Martinez-Ferez, A. 2012. Impacts of operating conditions on reverse osmosis performance of pretreated olive mill wastewater. *Water Res.* 46 (15):4621–4632.

Olayiwola, B., Walzel, P. 2009. Effects of in-phase oscillation of retentate and filtrate in crossflow filtration at low Reynolds number. *J. Membr. Sci.* 345 (1–2):36–46.

Onsekizoglu, P., Bahceci, K. S., Acar, M. J. 2010. Clarification and the concentration of apple juice using membrane processes: A comparative quality assessment. *J. Membr. Sci.* 352 (1–2):160–165.

Pabby, A. K., Rizvi, S. S. H., Sastre, A. M. 2009. *Handbook of Membrane Separations*, Boca Raton, FL: CRC Press.

Paulen, R., Fikar, M., Foley, G., Kovacs, Z., Czermak, P. 2012. Optimal feeding strategy of diafiltration buffer in batch membrane processes. *J. Membr. Sci.* 411:160–172.

Paun, G., Neagu, E., Tache, A., Radu, G. L., Parvulescu, V. 2011. Application of the nanofiltration process for concentration of polyphenolic compounds from *Geranium robertianum* and *Salvia officinalis* extracts. *Chem. Biochem. Eng. Q.* 25 (4):453–460.

Peshev, D., Peeva, L. G., Peev, G., Baptista, I. I. R., Boam, A. T. 2011. Application of organic solvent nanofiltration for concentration of antioxidant extracts of rosemary (*Rosmarinus officiallis* L.). *Chem. Eng. Res. Des.* 89 (3A):318–327.

Pinelo, M., Jonsson, G., Meyer, A. S. 2009. Membrane technology for purification of enzymatically produced oligosaccharides: Molecular and operational features affecting performance. *Sep. Purif. Technol.* 70 (1):1–11.

Piry, A., Heino, A., Kuhnl, W., Grein, T., Ripperger, S., Kulozik, U. 2012. Effect of membrane length, membrane resistance, and filtration conditions on the fractionation of milk proteins by microfiltration. *J. Dairy Sci.* 95 (4):1590–1602.

Piry, A., Kuhnl, W., Grein, T., Tolkach, A., Ripperger, S., Kulozik, U. 2008. Length dependency of flux and protein permeation in crossflow microfiltration of skimmed milk. *J. Membr. Sci.* 325 (2):887–894.

Popović, S., Jovičević, D., Muhadinović, M., Milanović, S., Tekić, M. N. 2013. Intensification of microfiltration using a blade-type turbulence promoter. *J. Membr. Sci.* 425–426 (0):113–120.

Popović, S., Tekić, M. N. 2011. Twisted tapes as turbulence promoters in the microfiltration of milk. *J. Membr. Sci.* 384 (1–2):97–106.

Porcelli, N., Judd, S. 2010. Chemical cleaning of potable water membranes: A review. *Sep. Purif. Technol.* 71 (2):137–143.

Rabiller-Baudry, M., Begoin, L., Delaunay, D., Paugam, L., Chaufer, B. 2008. A dual approach of membrane cleaning based on physico-chemistry and hydrodynamics application to PES membrane of dairy industry. *Chem. Eng. Process.* 47 (3):267–275.

Rabiller-Baudry, M., Bouzid, H., Chaufer, B., Paugam, L., Delaunay, D., Mekmene, O., Ahmad, S., Gaucheron, F. 2009. On the origin of flux dependence in pH-modified skim milk filtration. *Dairy Sci. Technol.* 89 (3–4):363–385.

Rafia, N., Aroujalian, A., Raisi, A. 2011. Pervaporative aroma compounds recovery from lemon juice using poly(octyl methyl siloxane) membrane. *J. Chem. Technol. Biotechnol.* 86 (4):534–540.

Rai, P., De, S. 2009. Clarification of pectin-containing juice using ultrafiltration. *Curr. Sci.* 96 (10):1361–1371.

Rai, P., Majumdar, G. C., Das Gupta, S., De, S. 2007. Effect of various pretreatment methods on permeate flux and quality during ultrafiltration of mosambi juice. *J. Food Eng.* 78 (2):561–568.

Rai, P., Majumdar, G. C., Dasgupta, S., De, S. 2010. Flux enhancement during ultrafiltration of depectinized mosambi (*Citrus sinensis* L Osbeck) juice *J. Food Process Eng.* 33 (3):554–567.

Ramchandran, L., Sanciolo, P., Vasiljevic, T., Broome, M., Powell, I., Duke, M. 2012. Improving cell yield and lactic acid production of *Lactococcus lactis* ssp cremoris by a novel submerged membrane fermentation process. *J. Membr. Sci.* 403:179–187.

Rao, S., Ager, K., Zydney, A. L. 2007. High performance tangential flow filtration using charged affinity ligands. *Separ. Sci. Technol.* 42 (11):2365–2385.

Rao, S., Zydney, A. L. 2005. Controlling protein transport in ultrafiltration using small charged ligands. *Biotechnol. Bioeng.* 91 (6):733–742.

Rao, S. M., Zydney, A. L. 2006. High resolution protein separations using affinity ultrafiltration with small charged ligands. *J. Membr. Sci.* 280 (1–2):781–789.

Rastogi, N. K., Nayak, C. A. 2011. Membranes for forward osmosis in industrial applications. In: Basile, A., Nunes, S. P. (Eds.), *Advanced Membrane Science and Technology for Sustainable Energy and Environmental Applications*, Abington, UK: Woodhead Publ.

Rektor, A., Kozak, A., Vatai, G., Bekassy-Molnar, E. 2007. Pilot plant RO-filtration of grape juice. *Sep. Purif. Technol.* 57 (3):473–475.

Rektor, A., Pap, N., Kokai, Z., Szabo, R., Vatai, G., Bekassy-Molinar, E. 2004. Application of membrane filtration methods for must processing and preservation. *Desalination* 162 (1–3):271–277.

Rektor, A., Vatai, G., Bekassy-Molnar, E. 2006. Multi-step membrane processes for the concentration of grape juice. *Desalination* 191 (1–3):446–453.

Rickman, M., Pellegrino, J., Davis, R. 2012. Fouling phenomena during membrane filtration of microalgae. *J. Membr. Sci.* 423–424:33–42.

Riera, F. A., Suárez, A., Muro, C. 2013. Nanofiltration of UHT flash cooler condensates from a dairy factory: Characterisation and water reuse potential. *Desalination* 309 (0):52–63.

Rios, G. M., Belleville, M. P., Paolucci-Jeanjean, D. 2007. Membrane engineering in biotechnology: Quo vamus? *Trends Biotechnol.* 25 (6):242–246.

Rios, G. M., Belleville, M. P., Paolucci, D., Sanchez, J. 2004. Progress in enzymatic membrane reactors—A review. *J. Membr. Sci.* 242 (1–2):189–196.

Rohani, M. M., Zydney, A. L. 2012. Protein transport through zwitterionic ultrafiltration membranes. *J. Membr. Sci.* 397:1–8.

Roman, A., Vatai, G., Ittzes, A., Kovacs, Z., Czermak, P. 2012. Modeling of diafiltration processes for demineralization of acid whey: An empirical approach. *J. Food Process Eng.* 35 (5):708–714.

Rombaut, R., Dejonckheere, V., Dewettinck, K. 2007. Filtration of milk fat globule membrane fragments from acid buttermilk cheese whey. *J. Dairy Sci.* 90 (4):1662–1673.

Ruby-Figueroa, R., Cassano, A., Drioli, E. 2012. Ultrafiltration of orange press liquor: Optimization of operating conditions for the recovery of antioxidant compounds by response surface methodology. *Sep. Purif. Technol.* 98 (0):255–261.

Saha, N. K., Balakrishnan, M., Ulbricht, M. 2007. Sugarcane juice ultrafiltration: FTIR and SEM analysis of polysaccharide fouling. *J. Membr. Sci.* 306 (1–2):287–297.

Saha, N. K., Balakrishnan, M., Ulbricht, M. 2009. Fouling control in sugarcane juice ultrafiltration with surface modified polysulfone and polyethersulfone membranes. *Desalination* 249 (3):1124–1131.

Salazar, F. N., de Bruijn, J. P. F., Seminario, L., Guell, C., Lopez, F. 2007. Improvement of wine crossflow microfiltration by a new hybrid process. *J. Food Eng.* 79 (4):1329–1336.

Santos, F. R., Catarino, I., Geraldes, V., De Pinho, M. N. 2008. Concentration and rectification of grape must by nanofiltration. *Am. J. Enol. Vitic.* 59 (4):446–450.

Sarkar, B., DasGupta, S., De, S. 2009. Flux decline during electric field-assisted cross-flow ultrafiltration of mosambi (*Citrus sinensis* (L.) Osbeck) juice. *J. Membr. Sci.* 331 (1–2):75–83.

Sarkar, B., De, S. 2010. Electric field enhanced gel controlled cross-flow ultrafiltration under turbulent flow conditions. *Sep. Purif. Technol.* 74 (1):73–82.

Sen, D., Sarkar, A., Das, S., Chowdhury, R., Bhattacharjee, C. 2012. Batch hydrolysis and rotating disk membrane bioreactor for the production of galacto-oligosaccharides: A comparative study. *Ind. Eng. Chem. Res.* 51 (32):10671–10681.

Sen, D., Sarkar, A., Gosling, A., Gras, S. L., Stevens, G. W., Kentish, S. E., Bhattacharya, P. K., Barber, A. R., Bhattacharjee, C. 2011. Feasibility study of enzyme immobilization on polymeric membrane: A case study with enzymatically galacto-oligosaccharides production from lactose. *J. Membr. Sci.* 378 (1–2):471–478.

Shi, D., Kong, Y., Yu, J., Wang, Y., Yang, J. 2006. Separation performance of polyimide nanofiltration membranes for concentrating spiramycin extract. *Desalination* 191 (1–3):309–317.

Shi, J., Nawaz, H., Pohorly, J., Mittal, G., Kakuda, Y., Jiang, Y. M. 2005. Extraction of polyphenolics from plant material for functional foods—Engineering and technology. *Food Rev. Int.* 21 (1):139–166.

Schlosser, S., Kertesz, R., Martak, J. 2005. Recovery and separation of organic acids by membrane-based solvent extraction and pertraction—An overview with a case study on recovery of MPCA. *Sep. Purif. Technol.* 41 (3):237–266.

Schlosser, Š. 2009. Extractive separations in contactors with one and two immobilized L/L interfaces: Applications and perspectives. In: Drioli, E., Giorno, L., *Membrane Operations. Innovative Separations and Transformations*, Weinheim: Wiley-VCH.

Sim, L., Shu, L., Jegatheesan, V., Phong, D. D. 2009. Effect of operating parameters and cleaning on the performance of ceramic membranes treating partially clarified sugar cane juice. *Sep. Sci. Technol.* 44 (15):3506–3537.

Sin, H. N., Yusof, S., Sheikh, N., Hamid, A., Rahman, R. A. 2006. Optimization of enzymatic clarification of sapodilla juice using response surface methodology. *J. Food Eng.* 73 (4):313–319.

Skrzypek, M., Burger, M. 2010. Isoflux (R) ceramic membranes—Practical experiences in dairy industry. *Desalination* 250 (3):1095–1100.

Smith, S. R., Field, R. W., Cui, Z. F. 2006. Predicting the performance of gas-sparged and nongas-sparged ultrafiltration. *Desalination* 191 (1–3):376–385.

So, S., Peeva, L. G., Tate, E. W., Leatherbarrow, R. J., Livingston, A. G. 2010. Membrane enhanced peptide synthesis. *Chem. Commun.* 46 (16):2808–2810.

Sotoft, L. F., Christensen, K. V., Andresen, R., Norddahl, B. 2012. Full scale plant with membrane based concentration of blackcurrant juice on the basis of laboratory and pilot scale tests. *Chem. Eng. Process.* 54:12–21.

Souchon, I., Pierre, F. X., Athes-Dutour, V., Marin, A. 2002. Pervaporation as a deodorization process applied to food industry effluents: Recovery and valorisation of aroma compounds from cauliflower blanching water. *Desalination* 148 (1–3):79–85.

Springer, F., Carretier, E., Veyret, D., Dhaler, D., Moulin, P. 2011. Numerical and experimental methodology for the development of a new membrane prototype intended to microfiltration bioprocesses. Application to milk filtration. *Chem. Eng. Process.* 50 (9): 904–915.

Stoller, M., Bravi, M. 2010. Critical flux analyses on differently pretreated olive vegetation waste water streams: Some case studies. *Desalination* 250 (2):578–582.

Stoller, M., Chianese, A. 2006. Optimization of membrane batch processes by means of the critical flux theory. *Desalination* 191 (1–3):62–70.

Stopka, J., Bugan, S. G., Broussous, L., Schlosser, S., Larbot, A. 2001. Microfiltration of beer yeast suspensions through stamped ceramic membranes. *Sep. Purif. Technol.* 25 (1–3):535–543.

Stopka, J., Schlosser, Š., Dömény, Z., Šmogrovičová, D. 2000. Flux decline in microfiltration of beer and related solutions of model foulants through ceramic membranes. *Pol. J. Environ. Stud.* 9 (1):65–69.

Stuckey, D. C. 2012. Recent developments in anaerobic membrane reactors. *Bioresour. Technol.* 122:137–148.

Susanto, H., Arafat, H., Janssen, E. M. L., Ulbricht, M. 2008. Ultrafiltration of polysaccharide–protein mixtures: Elucidation of fouling mechanisms and fouling control by membrane surface modification. *Sep. Purif. Technol.* 63 (3):558–565.

Susanto, H., Franzka, S., Ulbricht, M. 2007. Dextran fouling of polyethersulfone ultrafiltration membranes—Causes, extent and consequences. *J. Membr. Sci.* 296 (1–2):147–155.

Susanto, H., Ulbricht, M. 2006. Insights into polysaccharide fouling of ultrafiltration membranes. *Desalination* 200 (1–3):181–182.

Sweity, A., Ying, W., Belfer, S., Oron, G., Herzberg, M. 2011. pH effects on the adherence and fouling propensity of extracellular polymeric substances in a membrane bioreactor. *J. Membr. Sci.* 378 (1–2):186–193.

Tavares, T. G., Amorim, M., Comes, D., Pintado, M. E., Pereira, C. D., Malcata, F. X. 2012. Manufacture of bioactive peptide-rich concentrates from whey: Characterization of pilot process. *J. Food Eng.* 110 (4):547–552.

Thomassen, J. K., Faraday, D. B. F., Underwood, B. O., Cleaver, J. A. S. 2005. The effect of varying transmembrane pressure and crossflow velocity on the microfiltration fouling of a model beer. *Sep. Purif. Technol.* 41 (1):91–100.

Ting, B., Pouliot, Y., Gauthier, S. F., Mine, Y. 2010. Fractionation of egg proteins and peptides for nutraceutical applications. In: Rizvi, S. S. H., *Separation, Extraction and Concentration Processes in the Food, Beverage and Nutraceutical Industries*, Cambridge: Woodhead Publishers Ltd.

Tomasula, P. M., Mukhopadhyay, S., Datta, N., Porto-Fett, A., Call, J. E., Luchansky, J. B., Renye, J., Tunick, M. 2011. Pilot-scale crossflow–microfiltration and pasteurization to remove spores of *Bacillus anthracis* (Sterne) from milk. *J. Dairy Sci.* 94 (9):4277–4291.

Torrestiana-Sanchez, B., Balderas-Luna, L., Brito-De la Fuente, E., Lencki, R. W. 2007. The use of membrane-assisted precipitation for the concentration of xanthan gum. *J. Membr. Sci.* 294 (1–2):84–92.

Trentin, A., Guell, C., Gelaw, T., de Lamo, S., Ferrando, M. 2012. Cleaning protocols for organic microfiltration membranes used in premix membrane emulsification. *Sep. Purif. Technol.* 88:70–78.

Tsagaraki, E. V., Lazarides, H. N. 2012. Fouling analysis and performance of tubular ultrafiltration on pretreated olive mill waste water. *Food Bioprocess Technol.* 5 (2):584–592.

Turan, M., Ates, A., Inanc, B. 2002. Fouling of reverse osmosis and nanofiltration membranes by dairy industry effluents. *Water Sci. Technol.* 45 (12):355–360.

Ulbricht, M., Ansorge, W., Danielzik, I., Konig, M., Schuster, O. 2009. Fouling in microfiltration of wine: The influence of the membrane polymer on adsorption of polyphenols and polysaccharides. *Sep. Purif. Technol.* 68 (3):335–342.

Ulloa, J. A., Rosas-Ulloa, P., Ulloa-Rangel, B. E. 2011. Physicochemical and functional properties of a protein isolate produced from safflower (*Carthamus tinctorius* L.) meal by ultrafiltration. *J. Sci. Food Agric.* 91 (3):572–577.

Ushikubo, F. Y., Watanabe, A. P., Viotto, L. A. 2007. Microfiltration of umbu (*Spondias tuberosa* Arr. Cam.) juice. *J. Membr. Sci.* 288 (1–2):61–66.

Václavík, L. 2011. Membrane filtration of solutions of biomolecules (in Slovak). MSc thesis, *Inst. Chem. Environ. Eng.*, Slovak University of Technology, Bratislava (SK).

Vaillant, F., Millan, A., Dornier, M., Decloux, M., Reynes, M. 2001. Strategy for economical optimisation of the clarification of pulpy fruit juices using crossflow microfiltration. *J. Food Eng.* 48 (1):83–90.

van der Sman, R. G. M., Vollebregt, H. M., Mepschen, A., Noordman, T. R. 2012. Review of hypotheses for fouling during beer clarification using membranes. *J. Membr. Sci.* 396 (0):22–31.

van Lier, J. B., Tilche, A., Ahring, B. K., Macarie, H., Moletta, R., Dohanyos, M., Pol, L. W. H., Lens, P., Verstraete, W. 2001. New perspectives in anaerobic digestion. *Water Sci. Technol.* 43 (1):1–18.

van Reis, R. 1999. Protein ultrafiltration. In: Flickinger, M. C., Drew, S. W., *Encyclopedia of Bioprocess Technology. Fermentation, Biocatalyst and Bioseparation*, New York: A Wiley—Interscience Publication.

van Reis, R., Goodrich, E. M., Yson, C., Frautschy, L. N., Dzengelski, S., Lutz, H. 1997. Linear scale ultrafiltration. *Biotech. Bioeng.* 55 (5):737–746.

van Reis, R., Zydney, A. 2007. Bioprocess membrane technology. *J. Membr. Sci.* 297 (1–2):16–50.

Vatanpour, V., Madaeni, S. S., Moradian, R., Zinadini, S., Astinchap, B. 2012. Novel anti-bifouling nanofiltration polyethersulfone membrane fabricated from embedding TiO_2 coated multiwalled carbon nanotubes. *Sep. Purif. Technol.* 90:69–82.

Viladomat, F. G., Souchon, I., Pierre, F. X., Marin, M. 2006. Liquid–liquid and liquid–gas extraction of aroma compounds with hollow fibers. *AIchE J.* 52 (6):2079–2088.

Vincze, I., Vatai, G. 2004. Application of nanofiltration for coffee extract concentration. *Desalination* 162 (1–3):287–294.

Vladisavljevic, G. T., Vukosavljevic, P., Bukvic, B. 2003. Permeate flux and fouling resistance in ultrafiltration of depectinized apple juice using ceramic membranes. *J. Food Eng.* 60 (3):241–247.

Vourch, M., Balannec, B., Chaufer, B., Dorange, G. 2005. Nanofiltration and reverse osmosis of model process waters from the dairy industry to produce water for reuse. *Desalination* 172 (3):245–256.

Vourch, M., Balannec, B., Chaufer, B., Dorange, G. 2008. Treatment of dairy industry waste-water by reverse osmosis for water reuse. *Desalination* 219 (1–3):190–202.

Wang, B. J., Wei, T. C., Yu, Z. R. 2005. Effect of operating temperature on component distri-bution of West Indian cherry juice in a microfiltration system. *LWT-Food Sci. Technol.* 38 (6):683–689.

Wei, S., Hossain, M. M., Saleh, Z. S. 2010. Concentration of rutin model solutions from their mixtures with glucose using ultrafiltration. *Int. J. Mol. Sci.* 11 (2):672–690.

Wojciech, B., Celińska, E., Dembczyński, R., Szymanowska, D., Nowacka, M., Jesionowski, T., Grajek, W. 2013. Cross-flow microfiltration of fermentation broth containing native corn starch. *J. Membr. Sci.* 427 (0):118–128.

Xiao, K., Wang, X. M., Huang, X., Waite, T. D., Wen, X. H. 2009. Analysis of polysaccharide, protein and humic acid retention by microfiltration membranes using Thomas' dynamic adsorption model. *J. Membr. Sci.* 342 (1–2):22–34.

Yang, W. B., Cicek, N., Ilg, J. 2006. State-of-the-art of membrane bioreactors: Worldwide research and commercial applications in North America. *J. Membr. Sci.* 270 (1–2):201–211.

Yazdanshenas, M., Soltanieh, M., Nejad, S., Fillaudeau, L. 2010. Cross-flow microfiltration of rough non-alcoholic beer and diluted malt extract with tubular ceramic membranes: Investigation of fouling mechanisms. *J. Membr. Sci.* 362 (1–2):306–316.

Ye, H., Lei, J. J., Zhang, Y. Z., Li, H., Li, G. F., Cao, Z. P., Xie, L. P. 2011. Investigation of microfiltration for pretreatment of whey concentration. *Desalin. Water Treat.* 34 (1–3):173–178.

Ye, Y., Le Clech, P., Chen, V., Fane, A. G. 2005. Evolution of fouling during crossflow filtra-tion of model EPS solutions. *J. Membr. Sci.* 264 (1–2):190–199.

Youn, K. S., Hong, J. H., Bae, D. H., Kim, S. J., Kim, S. D. 2004. Effective clarifying pro-cess of reconstituted apple juice using membrane filtration with filter-aid pretreatment. *J. Membr. Sci.* 228 (2):179–186.

Yuan, Q. P., Zhang, H., Qian, Z. M., Yang, X. J. 2004. Pilot-plant production of xylo-oligosac-charides from corncob by steaming, enzymatic hydrolysis and nanofiltration. *J. Chem. Technol. Biotechnol.* 79 (10):1073–1079.

Zator, M., Ferrando, M., Lopez, F., Guell, C. 2009. Microfiltration of protein/dextran/poly-phenol solutions: Characterization of fouling and chemical cleaning efficiency using confocal microscopy. *J. Membr. Sci.* 344 (1–2):82–91.

Zhang, J. S., Zhou, J. T., Su, Y.-C., Fane, A. G. 2010. Transient performance of MBR with flux enhancing polymer addition. *Sep. Sci. Technol.* 45 (7):982–992.

Zhang, Z., Yang, R. J., Zhang, S., Zhao, H. F., Hua, X. 2011. Purification of lactulose syrup by using nanofiltration in a diafiltration mode. *J. Food Eng.* 105 (1):112–118.

Zhao, H. F., Hua, X., Yang, R. J., Zhao, L. M., Zhao, W., Zhang, Z. 2012a. Diafiltration process on xylo-oligosaccharides syrup using nanofiltration and its modelling. *Int. J. Food Sci. Technol.* 47 (1):32–39.

Zhao, S. F., Zou, L., Tang, C. Y. Y., Mulcahy, D. 2012b. Recent developments in forward osmosis: Opportunities and challenges. *J. Membr. Sci.* 396:1–21.

Zirehpour, A., Jahanshahi, M., Rahimpour, A. 2012. Unique membrane process integration for olive oil mill wastewater purification. *Sep. Purif. Technol.* 96 (0):124–131.

Zularisam, A. W., Ahmad, A., Sakinah, M., Ismail, A. F., Matsuura, T. 2011. Role of natural organic matter (NOM), colloidal particles and solution chemistry on ultrafiltratio performance, *Sep. Purif. Technol.* 78 (2):189–200.

Zunin, P., Fusella, G. C., Leardi, R., Boggia, R., Bottino, A., Capannelli, G. 2011. Effect of the addition of membrane processed olive mill waste water (OMWW) to extra virgin olive oil. *J. Am. Oil Chem. Soc.* 88 (11):1821–1829.

Zydney, A. L. 2009. Membrane technology for purification of therapeutic proteins. *Biotechnol. Bioeng.* 103 (2):227–230.

7 Chromatography

Katarzyna Wrzosek, Łukasz Wiśniewski,
Michal Gramblička, Monika Antošová,
and Milan Polakovič

CONTENTS

7.1   INTRODUCTION

Chromatography is a highly versatile separation technique used in the food industry mostly for the purification of saccharides and also proteins or chiral agents. The main downside of chromatography is its high cost. Therefore, for many applications, it is more viable to use less expensive techniques such as membrane filtration or crystallization. The advantage of chromatography is its remarkable selectivity, allowing the separation of very complex mixtures in a single step. This makes it a technique of choice for the separation of valuable products that are difficult to be obtained with demanded purity by other techniques.

Liquid chromatography is a type of adsorption-based process. Adsorption as a phenomenon is the adhesion of molecules dissolved in a liquid to a surface of a solid called the adsorbent. Chromatography is a fixed-bed fractional separation of a multicomponent mixture by different binding of its components. The separation occurs by simultaneous adsorption and desorption of components occurring in different zones of the fixed bed leading to their different mobility in a liquid phase. Adsorption is called a process, in which one or more components are bound to the adsorbent while the rest of the components are easily washed out from the system. The intrinsic adsorption occurs in a relatively narrow zone of the adsorbent fixed bed of which the frontal part is empty and the rear part is saturated. Adsorption, unlike chromatography, can also be carried out in batch, continuous stirred tank, as well as in expanded bed.

A brief overview of chromatographic and some adsorption processes applied in biotechnology is presented below. The fundamentals of process chromatography are also described. Owing to the development of new stationary phases and new technique improvements, mainly with regard to continuous chromatography, the application of chromatographic processes becomes more viable and common in biotechnology and the food industry.

7.2   BASIC THEORY

Chromatographic separation is a complicated process based on different properties of liquid and solid phases. The solid phase, also called the stationary phase, is an adsorbent usually in the form of particles, or alternatively in the form of a continuous bed (monolith) or membrane sheets. The liquid or mobile phase refers either to the extraparticle liquid or to the liquid in the membrane and monolith large throughpores. The stationary phase is typically composed of the carrier backbone and ligands that can interact with the adsorptive molecules introduced into the system in the mobile phase. In most typical porous materials, the solid phase includes both the components of the stationary phase and the stagnant fluid inside meso- and micropores. A choice of proper stationary and mobile phases is the first step in the design of chromatographic separation. The selected combination must ensure a good separation of components of interest and must be viable. Moreover, the appropriate solubility of a feedstock and high saturation capacity of the adsorbent are of importance.

The basic chromatographic parameter, characteristic for a given component, is its retention time (t_R) or retention volume (V_R), which are interrelated by the liquid flow rate. The retention time is the time from the moment of feeding of a component mixture into a column until the mean time of the appearance of a particular component in the column outlet (Figure 7.1). They characterize the migration rate of solute molecules in the column. While the retention time is a design parameter that often does not depend on the production capacity, the retention volume is proportional to the production capacity. For that reason, the retention volume is usually given in the relative form as the number of column/bed volumes. A wish of a designer of chromatographic process is to keep t_R and V_R relatively low to minimize the purification

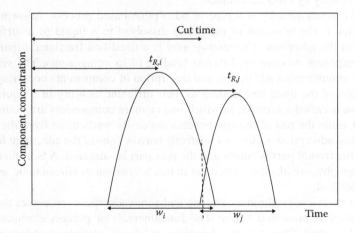

FIGURE 7.1   Illustration of a chromatogram, that is, the concentration profiles of separated components i and j in the column outlet stream.

time and the eluent consumption. Shorter purification times prevent losses of unstable separated molecules and lower eluent consumption results in less dilution that increases the costs for subsequent product concentration.

However, the requirement of short retention times is contradictory to the objectives of high adsorption capacity. The retention time is proportional to the ratio of the equilibrium of the solid- and liquid-phase concentrations; therefore, the more strongly adsorbing solutes have higher retention times (Figure 7.1). The retention times are also limited by another important adsorption equilibrium property—selectivity. Selectivity is a binary characteristic that determines the ratio of the retention times of two components. A prerequisite of the separation of two components is thus the value of selectivity different from 1. In a real column, the dispersion of component molecules occurs, which results in a concentration peak at the column outlet (Figure 7.1).

The peak width w is a measure of the dispersion and it is also another key parameter of chromatographic separation. As shown in Figure 7.1, the peaks in process chromatography may overlap, which means that the whole material cannot be collected in two pure fractions as it often happens in analytical applications. Chromatography does not have a thermodynamic limitation of other separation processes in which a perfect separation of components may happen only in an infinite number of stages. When the length of the chromatographic bed is increased, the difference between the retention times of two components increases more than the sum of their peak widths, and complete separation can be obtained in a real column. In process chromatography, however, this is uneconomic due to the prolonged cycle times of periodic operation, increased consumption of the mobile phase, and higher costs for the adsorbent and the column.

A typical industrial chromatographic process requires defining cut time(s) for collecting product fractions of desired purity (Figure 7.1). The main aim of process chromatography is the acquisition of a component of interest at high productivity, defined as the product amount per mass of adsorbent in a unit of time. The separation of this component from a complex mixture can be a highly complicated process; hence, a careful optimization is required to assure a high productivity at a desired level of product purity and yield. The purity is defined as the ratio of the amount of target compound to the amount of all components in the product preparation after a chromatographic separation step. The yield is the ratio of the amount of target compound to its amount in the feed. In many cases, the cost of the mobile phase is a critical part of the total separation cost. Therefore, the eluent consumption, defined as the amount of eluent needed for the purification of unit amount of the component of interest, has to be carefully considered during the design of a chromatographic separation process.

The basis for process optimization of chromatographic separations are adsorption isotherm equations and models of column dynamics composed of differential equations of the components of interest and describing their migration through the column. The following text presents only a very brief summary of the most significant isotherm equations and basic mathematical models used for the description of chromatographic processes pertinent to food biotechnology. More information can be found in numerous reviews and monographs (Bellot and Condoret 1993; Guiochon 2002; Guiochon and Lin 2003; Schmidt-Traub 2005; Guiochon et al. 2006).

The intrinsic adsorption and desorption of components on the adsorbent active surface are in most cases very fast and the thermodynamic equilibrium of adsorption at a constant temperature is described by adsorption isotherms, which are functions of the component concentrations in the solid and liquid phases, respectively. The simplest form of an isotherm is a linear isotherm with a single equilibrium constant (also called the Henry constant). The linear isotherm is useful in the conditions of a low coverage of the adsorbent surface or when size exclusion is the dominant sorption mechanism. For the latter reason, it has often been used to describe the adsorption equilibria of saccharides on cation-exchange resins even at concentrations as high as 400 g/L (Vente et al. 2005; Pedruzzi et al. 2008; Vaňková et al. 2010; Chilamkurthi et al. 2012). At very high saccharide concentrations, the isotherms become slightly convex. To properly describe such an equilibrium, a model based on the nonrandom two-liquid (NRTL) isotherm model for activity coefficients was applied (Abderafi and Bounahmidi 1994; Nowak et al. 2009).

The characteristic shape of nonlinear adsorption isotherms is a concave one, which is most frequently described by the well-known Langmuir equation. Its two parameters characterize the saturation capacity and affinity of a component for the adsorbent. The Langmuir isotherm is frequently used as a first approximation of adsorption equilibrium that provides a sufficient accuracy of the description of experimental data even in cases when some of its theoretical assumptions are not valid. A typical extension of this single-component Langmuir isotherm is the bi-Langmuir isotherm applied for cases of nonhomogenous adsorbent surface with two independent types of binding sites.

All cases of practical importance for process liquid chromatography involve multicomponent mixtures. The amount of each adsorbate then depends on the concentrations of several components as the adsorption has a competitive character. A simple approximation of multicomponent adsorption can be provided by the competitive Langmuir isotherm, which is however thermodynamically correct only for components with a similar affinity to the adsorbent.

Different adsorption isotherm equations may be required for systems with specific types of interactions. One example is the steric mass action model (SMA) developed for a proper description of ion-exchange chromatography of proteins (Brooks and Cramer 1992). The equation was derived from the assumption of stoichiometric reactions of both protein molecules and salt counterions competing for adsorbent binding sites. It also includes a parameter characterizing the size ratio of protein and ligand molecules. The model allows the description of equilibrium data at different ionic strength values. Another example is protein adsorption on hydrophobic adsorbents where the cosmotropic salt concentration has a strong effect on hydrophobicity being the basis of the adsorption mechanism. This has been best characterized by the modified Langmuir isotherm containing an exponential term describing the salt influence on the adsorption equilibrium (Antia and Horváth 1989).

A comprehensive description of the chromatographic process also requires a proper dynamic model besides an isotherm equation that accounts for the transport phenomenon. An ideal situation is a hypothetical column of finite length where all zone-spreading phenomena are absent. This hypothesis is a fundamental of ideal chromatography. It assumes that the shape of the zone profile at the column outlet is solely determined by the adsorption equilibrium. The modeling of such idealized separation is useful for process chromatography because it can estimate the approximate separation process course and can predict the maximum achievable yields and purities.

There are several models of nonideal chromatography considering finite column separation efficiency. Thermodynamic equilibrium is not reached instantaneously because the intrinsic adsorption process is limited by mass transfer kinetics. This results in concentration dispersion of individual adsorptives along the fixed bed and therewith in spreading of chromatographic zones. The typical phenomena responsible for zone broadening in chromatography are axial dispersion, film diffusion, pore or surface diffusion, and kinetic character of intrinsic adsorption. Each of these phenomena occurs at a different rate and the slowest one has the main influence on zone broadening. Proper identification of the contribution of the individual mechanism influences the complexity of a model needed for the description of the performance of the chromatographic bed.

The simplest model of nonideal chromatography is the equilibrium dispersive model (ED) (Table 7.1). Its basic form assumes that zone broadening is caused solely

---

TABLE 7.1

Mass Balance Equations and Initial and Boundary Conditions for Ideal, ED, and GRM Models of Chromatography

Model	Mass Balance Equation
Ideal	$\dfrac{\partial c_i}{\partial t} + \dfrac{(1-\varepsilon)}{\varepsilon}\dfrac{\partial q_i}{\partial t} + u\dfrac{\partial c_i}{\partial z} = 0$
Equilibrium dispersive	$\dfrac{\partial c_i}{\partial t} + \dfrac{(1-\varepsilon)}{\varepsilon}\dfrac{\partial q_i}{\partial t} + u\dfrac{\partial c_i}{\partial z} = D_{L,\text{app}}\dfrac{\partial^2 c_i}{\partial z^2}$
General rate	$\dfrac{\partial c_i}{\partial t} + u\dfrac{\partial c_i}{\partial z} + \dfrac{(1-\varepsilon)}{\varepsilon}\dfrac{3k_{\text{ext}}}{R_p(c_i - c_{p,i/r=R})} = D_L\dfrac{\partial^2 c_i}{\partial z^2}$
	(liquid phase)
	$\varepsilon_p\dfrac{\partial c_{p,i}}{\partial t} = D_{\text{eff},i}\left(\dfrac{\partial^2 c_{p,i}}{\partial r^2} + \dfrac{2}{r}\dfrac{\partial c_{p,i}}{\partial r}\right) - (1-\varepsilon_p)\dfrac{\partial q_i}{\partial t}$
	(solid phase)
Initial and boundary conditions (for GRM)	$c_{p,i}(r, z, t = 0) = 0$
	$\left(\dfrac{\partial c_{p,i}}{\partial t}\right)_{r=0} = 0$
	$\left(\dfrac{\partial c_{p,i}}{\partial t}\right)_{r=R_p} = \dfrac{3k_{\text{ext}}}{R_p D_{\text{eff},i}}(c_i - c_{p,i/r=R_p})$

Note: Symbols used: q_i, solute concentration in solid phase; c_i, solute concentration in liquid phase; t, time; ε, void fraction; z, axial coordinate; u, mobile phase velocity; R_P, particle radius; $c_{p,i}$, concentration in pores; r, radial coordinate; ε_p, particle porosity; $D_{L,\text{app}}$, apparent axial dispersion coefficient; D_L, axial dispersion coefficient; k_{ext}, external mass transfer coefficient; D_{eff}, effective pore diffusion coefficient.

by axial dispersion that arises from the radial distribution of velocity in the highly laminar flow of the mobile phase. In fact, the contribution of axial dispersion is often negligible in biotechnological applications of process chromatography. The exceptions are applications where size-exclusion effect is the key one such as separations of saccharides. Nonetheless, the usefulness of the ED model goes beyond these few applications. Van Deemter et al. (1956) showed that the ED model accurately describes a chromatographic process encompassing all mass transfer phenomena when the adsorption isotherm is linear. In this case, the single transport parameter of the model is the apparent axial dispersion coefficient, which lumps all mass transfer and adsorption kinetic parameters. Owing to its simplicity, the ED model is often used for the design and optimization of chromatographic processes with the nonlinear adsorption isotherm.

The most complete description of the chromatographic process can be provided by the general rate model (GRM) with three parameters accounting for various zone-broadening effects separately (Table 7.1): axial dispersion coefficient, external mass transfer coefficient, and the effective pore diffusion coefficient. It consists of the mass balance equations in the mobile and stationary phases. In each case, the mass balance equations have to be supplemented with isotherm equation(s) and initial and boundary conditions.

7.3   OPERATION MODES

The three basic modes of chromatographic operation are elution, frontal, and displacement. The first mode is the most common (Figure 7.2a). The composition of the eluent can be kept either constant (isocratic elution) or can be changed during separation (gradient elution). In process chromatography, gradient elution is carried out via one or several stepwise changes of pH or ionic strength of the liquid phase to control the target component retention so that products of desired purity are obtained. Gradient elution also results in the decrease of the process time and mobile phase consumption.

In displacement chromatography, a larger feed than in elution chromatography is introduced into the column at conditions allowing a significant retention of various components of the mixture (Figure 7.2b). The components adsorb along the column in order of their decreasing affinity to the stationary phase. In the next step, they are displaced using a mobile phase containing a displacer that has adsorption strength supreme to all elements of the feed. Each adsorbate is displaced by the more strongly bound one and the bands of the components move along the column due to repeated readsorption. They appear in the column outlet as a series of pure zones with some degree of mixing at their edges. An advantage of this technique is that it can concentrate on the individual components. Therefore, it can be used for the concentration of the product from a diluted solution or for the removal of trace impurities. However, it requires a thorough regeneration step to remove a displacer having very high affinity to the stationary phase, which significantly increases process costs.

Frontal chromatography is characteristic by a continuous supply of the feed (Figure 7.2c). The retention of components of the feed is competitive, resulting in the displacement of the least-retained components by the components with higher interaction strength. Only the least-retained component can be obtained in pure

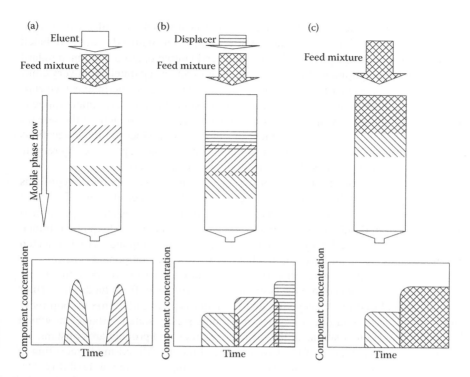

FIGURE 7.2   Illustration of chromatographic modes. (a) Elution, (b) displacement, and (c) frontal.

form; the remaining components leave the column in a mixture. In spite of this limitation, this technique found useful applications, for example, in the removal of nucleic acids or viruses in the production of monoclonal antibodies (Phillips et al. 2005; Wang et al. 2007).

Elution chromatography is performed in several steps when the column is operated periodically. First, the resin is equilibrated with a desired mobile phase selected to improve the interaction of a target component with the adsorbent. The mixture to be separated is loaded onto the column in the next step. An adequate loading volume has to be chosen with respect to the saturation capacity of the adsorbent to avoid excessive overload of the column. Moreover, the feed concentration and conditions have to be optimized for good adsorption. The adsorbent is subsequently washed to remove impurities with minor affinity to the stationary phase. Afterward, a mobile phase with higher elution strength is applied to desorb the target component. The conditions of elution should be optimized to assure the desired purity and minimal elution volume. The column bed has to be usually regenerated in the final step of the operational cycle. Regeneration allows the removal of strongly bound impurities and sanitation of the column. The choice of the regeneration procedure depends on the nature of the adsorbent and impurities and can involve, for example, washes with strong basic or acidic solutions.

This mode of periodic operation of elution chromatography is called batch chromatography although it is carried out in a flow regime. The main advantages of batch chromatography are fast process development and simple operation using standard chromatographic equipment. Multiple components can be separated from very complex mixtures in a single run. The disadvantages include the discontinuous feed supply and high solvent consumption increasing process costs. Batch chromatography is by far the most frequently applied mode. Efforts for the improvement of the separation performance led to the development of other modes, which found applications in process chromatography. The two most important types are represented by recycle and simulated moving bed (SMB) chromatography, respectively.

Recycling or recycle chromatography has a similar periodic operation regime as batch chromatography but it gives better efficiency (Charton et al. 1994). From the operational point of view, recycling in chromatographic processes may be performed in two ways. Both divide the eluate into three parts: raffinate, in which less retained components are collected; recycle, containing unresolved components to be injected into the column again; and extract, in which more retained components prevail. The recycle can be either returned directly to the column inlet or can be temporarily stored and then introduced into the column after mixing with fresh feed. The former type is called closed-mode recycling and the latter type is called mixed-loop recycling. After a certain number of cycles, a periodic steady state is attained at which the amounts and compositions of the raffinate, extract, and product are constant.

There are pros and cons of the two mentioned modes. Mixed-mode recycling is simpler in operation and easier in design, and gives a higher degree of flexibility. The cost of it is that the achieved separation in the preceding cycle is wasted when the recycle fraction is collected and mixed with the fresh feed before the next injection (Grill and Miller 1998). On the contrary, closed-loop recycling partially preserves the gained separation but its design is complicated. Compared to the mixed-mode recycling, closed-loop recycling gives an additional degree of freedom in operation, namely, the time point of introducing fresh feed with respect to the time of feeding of the recycle fraction.

It should be emphasized that recycling chromatography ought to be distinguished from the so-called rechromatography (Tarafder et al. 2008). Rechromatography is often employed in the industry to improve product recovery via a simple batch chromatography of the recycle. Since this additional step is carried out at the expense of processing fresh feed, it results in a significant decrease of productivity. On the contrary, recycling chromatography may improve both the yield and productivity and always significantly reduces the consumption of the eluent and reduces product dilution. It can be a good trade-off between the batch-mode best suited for high-resolution systems and continuous chromatography optimal for high-throughput, low-resolution systems offering the mentioned improvements with minor additional equipment costs (Schlinge et al. 2010).

A significant drawback of batch chromatography is that, in each moment, intrinsic separation occurs only in a short fraction of the column bed. An obvious inspiration for the improvement of traditional chromatographic processes was the design of industrial countercurrent continuous separation processes. A pursuit to perform chromatographic separation as a continuous process succeeded in 1961 when Universal Oil

Company patented a process called SMB chromatography (Broughton and Gerhold 1961). SMB chromatography was elaborated on the idea of a true moving bed (TMB) process with a countercurrent motion of solid and liquid phases (Figure 7.3), which is technically unfeasible on the industrial scale (Schmidt-Traub 2005).

The effect of the countercurrent movement is achieved in SMB chromatography by changing the positions of inlets and outlets of liquid streams (feed, eluent, extract, and raffinate). However, the change of the positions is not continuous but is periodic and synchronous with a switching time (tact time) defining the velocity of the stationary-phase motion. It does not occur along each axial position of a continuous bed but occurs between the inlets and outlets of a series of identical columns with liquid-phase recycle that are distributed in four sections. The number of columns in the individual sections does not have to be the same.

Figure 7.3 explains the role of each section (zone) of SMB chromatography on the basis of TMB theory with an example of the separation of the two-component feed mixture when component A is the less retained one and component B is the more retained one. Zones I and IV are regeneration zones for the regeneration of the adsorbent and the eluent, respectively. Zones II and III are separation zones providing raffinate and extract products. In zone 1, the residual component B is desorbed from the adsorbent while the regeneration of the eluent occurs in zone IV by the adsorption of the residual component A. The concentration of the more retained component B in zone II occurs by preferential desorption of A while the concentration of A in zone II is owing to the preferential adsorption of B.

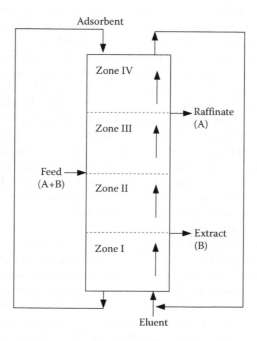

FIGURE 7.3 Principle of true moving bed chromatography.

Compared to batch chromatography, SMB chromatography provides several times higher productivity, reduced solvent consumption, and higher yield. These benefits come at the expense of higher equipment costs and more difficult process design, optimization, and control. It is also limited by the separation of feed only into two fractions, isocratic elution and absence of washing and regeneration phases. Several new concepts have appeared that remedy these limitations. The most important innovations include:

- Varicol system, where asynchronous inlet/outlet port shifts are implemented, providing a flexible use of each section length and an increase of the productivity value up to 30% (Toumi et al. 2003). This is the most successful innovation of conventional SMB.
- ModiCon system, which is based on the feed concentration variation within one switching period. This mode is limited only to nonlinear chromatographic processes (Schramm et al. 2003).
- PowerFeed system with the modulation of flow rates (Zhang et al. 2003).
- Partial feed (Zang and Wankat 2002).
- Two- and three-zone SMB improving the process economics by reduced equipment use (Lee 2003).

7.4 ADSORBENT TYPES

Chromatographic methods are capable of separating a broad spectrum of food components due to their ability to exploit marginal differences in solute–resin interactions. Therefore, all types of known chromatographic interactions are employed in food composition analysis (Nollet 2004; Pico 2012) and authentication (Sun 2008). However, the relatively high cost of preparative/process chromatography for lower added-value products considerably reduces the range of interaction modes applied (Rizvi 2010). The types of chromatography utilizing expensive materials (e.g., with affinity or chiral ligands) are seldom economically feasible. This section focuses on those types of chromatographic interactions that are relevant in the food processing.

The origin of solute retention lies in the reversible adsorptive interaction with the resin surface (electrostatic forces of various kinds) and/or in the geometric interaction with the porous structure (size exclusion). Multiple different mechanisms are present, to a certain degree, at each chromatographic separation. Thus, the interaction modes are approximately categorized into groups according to the prevailing retention mechanism assumed at the given separation.

Ion-exchange chromatography is the most widely used type of chromatographic separation in the food industry. The surface of the adsorbent contains ionic functional groups. The cation (for cation exchangers) or anion (for anion exchangers) of a functional group can be replaced by charged molecules of separated species. The most common functional groups used are carboxymethyl, orthophosphate, sulfoethyl, sulfopropyl, and sulfonate in cation exchangers and diethylaminoethyl, dimethylaminoethyl, trimethylaminoethyl, quaternary amino ethyl, and quaternary amine in anion exchangers (Janson and Rydén 1998). Adsorption equilibrium can be manipulated by pH and the concentration of counterions (ionic strength). The loading phase is performed at the conditions favoring the binding of the target species (positive mode) or of the contaminants (negative mode).

One group of compounds separated by chromatography using ion exchangers are saccharides, which are electroneutral in a wide pH range; therefore, they cannot displace counterions of the ion exchanger. The separation effect of ion exchangers is therefore assumed to stem from the selective complexation of saccharides with the counterion of the functional group (Goulding 1975). This also means that saccharide separations do not necessitate gradient elution and are performed in an isocratic mode (Vente et al. 2005). The said complexation is much weaker than ionic binding, resulting in a very low column-loading capacity. The process efficiency of saccharide chromatographic separations could be significantly enhanced by a continuous regime such as SMB (Azevedo and Rodrigues 2001; Vaňková and Polakovič 2012). On the contrary, the high binding capacity of proteins, peptides, and amino acids enables the separations performed sequentially in a single fixed bed (Rizvi 2010).

In the case of oligosaccharides, the primary separation effect was identified to be the size exclusion, even if ion exchangers are used (Gramblička and Polakovič 2007). A careful selection of the counterion for given saccharide separation is also of high importance, since it influences complexation as well as the adsorbent structure (Moravčík et al. 2012). In the case of di- and oligosaccharides, elevated temperatures ($\geq 60°C$) and the presence of strong cation exchangers might facilitate the acidic hydrolysis of saccharides (Vente 2004; Vaňková et al. 2010). This undesired effect strongly depends on the type of bonds between monosaccharide units. For instance, galactooligosaccharides are rather resistant to acid hydrolysis (Nakakuki 1993).

Hydrophobic interaction chromatography utilizes materials with low-polar surface. This feature stems either directly from the resin material (e.g., poly-(styrene–divinylbenzene), cross-linked aliphatic polymers, phenol–formaldehyde resins, and activated charcoal), or it is introduced into hydrophilic materials by chemical surface modification (e.g., silica or agarose functionalized by alkyl/aryl groups such as butyl, hexyl, octyl, decyl, octadecyl, or phenyl). The solutes are retained selectively according to their hydrophobicity or distribution of hydrophobic patches on their surface in the case of proteins. The molecules bind to the adsorbent surface via van der Waals and London dispersion forces (e.g., on activated charcoal) or due to salting-out effect promoted by water-structuring salts (ammonium sulfate is typically used for proteins). The adsorbed species are eluted by modifying the ionic strength and/or polarity of the mobile phase, elevated temperature, sodium hydroxide solution, or by the use of nitrogen or steam (for volatile compounds). In the case of contaminant capture by a low-cost adsorbent (e.g., activated charcoal), the material is typically not regenerated and it is disposed after saturation.

Size-exclusion chromatography is based on size- and shape-specific exclusion of solutes from parts of porous media. The components of the mobile phase selectively penetrate into the porous structure of the stationary phase. Small molecules have steric access to a relatively large volumetric fraction of the porous structure therefore their retention in the column is relatively high—close to the total liquid content of the bed. Larger solutes are less retained and elute sooner. The elution volume of the largest solutes that are entirely excluded from the porous structure equals the column's void volume. Thus, the pore size distribution of these materials is of vital importance. Other than steric interactions are generally not desired at this type of chromatography, since they might cause a retention not based entirely on the solute size. The absence of adsorptive interactions is the reason for the use of isocratic elution. Size-exclusion effect is the principal mechanism of the separation of prebiotic low-caloric oligosaccharides from high-caloric mono- and disaccharides, although ion exchangers are typically used for this process (Gramblička and Polakovič 2007).

A suitable column packing must assure both good permeability, that is, high flow velocity at low pressure drop, and mechanical strength to avoid the undesired changes of the stationary-phase volume, pore structure collapse, and adsorbent mechanical damage. The mechanical properties of porous adsorbent beds are significantly affected by the particle porosity, which has the values 0.5 and 0.9 for spherical monodisperse particles (Schmidt-Traub 2005). Another key parameter of the chromatographic column bed is void fraction that has typical values of 0.26–0.48 (Schmidt-Traub 2005). It influences the bed mechanical properties too but, together with the particle size and mobile-phase viscosity, it has a key effect on the bed permeability.

The choice of particle size is determined by the capacity of application and affordable product costs. Smaller particles have better mass transfer characteristics but invoke high pressure drop. On the other hand, large particles assure higher column permeability but have worse mass transfer properties. They must be packed into a longer bed to achieve the required separation efficiency and must have higher mechanical resistance. Softer, often gel-based, particles are used in smaller-scale applications, for example, therapeutic proteins or peptides production. These adsorbents have the particle diameter in the range of 10–50 μm and are almost monodisperse. Large-scale applications, such as high-fructose corn syrup production, employ particles of about 300 μm. The active surface area and the adsorption capacity of adsorbents for high-tech applications are often increased by the use of long polymer grafts, carrying multiple ligands and partially filling the pore space (DePhillips and Lenhoff 2001; Wrzosek et al. 2009).

Some limitations of the conventional particle stationary phases can be overcome by the use of monolith and membrane adsorbents. Monolith columns and membrane sheets are built of a single piece of a highly porous stationary phase instead of multiple individual particles. The structure of both types of adsorbents is based on large transport pores, with only very short micro- or mesopores, assuring a fast mass transfer. The transport pores of monoliths and membranes have sizes of several micrometers (Al-Bokari et al. 2002; Guiochon 2007; Tatárová et al. 2009). In both monoliths and membranes, the transport of components to the ligands takes place predominantly by convection and not by diffusion.

The overall mass transfer limitations are thus much smaller and the efficiency of the process becomes nearly independent of the mobile-phase velocity in a wide range of flow rates. This is a significant advantage compared to particle adsorbents for which efficiency is highly dependent on the flow rate. Therefore, both monoliths and membrane adsorbents can be operated at significantly higher flow velocities, which significantly shortens the process time. A high permeability, low mass transfer resistance, and short process times make them especially useful for the separation of large molecules with small diffusivities such as proteins, viruses, or entire cells (Arvidsson et al. 2002; Jungbauer and Hahn 2004; Charcosset 2006).

These materials however suffer from several drawbacks. The most significant drawback is that their separation capacity is limited due to their low specific surface area. It can be improved to some extent by using polymer grafts carrying multiple ligands. The grafts create hydrogel layers acting as an additional attainable volume for adsorbent–adsorbate interactions (Tatárová et al. 2009). Chromatographic membranes were successfully employed in flow-through polishing steps when a small amount of trace impurities had to be removed and when high adsorption capacity was not crucial (van Reis and Zydney 2001).

7.5 APPLICATIONS

Two types of adsorbents are utilized in chromatographic applications in the food industry—ion exchange and hydrophobic. The major applications of ion-exchange chromatography in the food industry are separations of saccharides and proteins, such as glucose–fructose separation (Viard and Lameloise 1992), separation of sucrose and betaine from molasses (Paananen and Kuisma 2000), demineralization and decolorization of materials such as raw sugar solutions, pectin juice, sweeteners (e.g., inulin), red/white grape most, and whey (Jensen 2007; Burkhardt et al. 2000), separation and fractionation of whey proteins from milk (Hahn et al. 1998), and separation of lysozyme from egg white (Chang and Chang 2006).

Adsorption or hydrophobic interaction chromatography is most often used for the removal of organic and protein impurities causing undesired coloration, off-flavors, and hazing. Examples of processed materials are starch hydrolyzates (removal of proteins, amino acids, foaming agents, iron complexes, tannins, and hydroxymethyl furfural) (Ammeraal and Delgado 1993; Orozco et al. 2012), organic acids such as citric and lactic (decolorization) (Margureanu and Gutmann 1989), fruit juices such as apple, grape, and pineapple (debittering, improvement of clarity, and color uniformity) (Wilson et al. 1989), glycerol and sweeteners (decolorization and odor removal) (Haselow 2010), and sugars (decolorization by the removal of caramel color bodies, melanoidines, and melanine–polyphenol–iron complexes) (Wilson et al. 1989).

High-fructose syrup (HFS) is extensively used mainly in the United States as the replacement for beet or cane sugar, while in the European Union, its production of HFS is limited by a production quota. The production of HFS is based on glucose syrup, which is obtained by the enzyme hydrolysis of maize starch. Glucose is then converted into fructose by the action of glucose isomerase enzyme. The resulting syrup, generally named HFS 42, is composed of fructose (~42% of dry weight), glucose (50%), and oligosaccharides (8%). The fructose content is then increased by the liquid chromatography of HFS 42 (Novasep 2010a), usually applying an SMB chromatographic system. In this way, a syrup containing approximately 90% of fructose is obtained (Novasep 2010a; Deng et al. 2010). HFS 90 is used either for the production of crystalline fructose of high purity or it is blended with HFS 42 to obtain HFS 55 (Packeus 1997). The application of the SMB system for glucose–fructose separation was patented by Rasche (1992). Sulfonated polystyrene–divinylbenzene resins in calcium form are commonly used in the above-mentioned processes (Dhingra and Pynnonen 1989; Katz 1983).

Molasses obtained as a by-product during the sugar refining process from sugar beet contains approximately 50% of sucrose, 10–12% of salts, up to 10% of proteins and amino acids, about 5% of betaine, about 2% of raffinose, and water. Sugar recovery in the conventional process is about 80% when its loss in molasses represents about 15% of the sucrose contained in sugar beet. Molasses desugaring by chromatographic process allows recovering up to 90% of sugar present in molasses and increases the recovery yield of sugar from beet to about 90% (Asadi 2005). Soft molasses (with hardness content less than 3.0 mEq per 100 g of dry substance) is the source feed. The key separation step is carried out by the chromatographic system, usually by continuous SMB or sequential SMB using sodium from ion-exchange resins based on styrene–divinylbenzene copolymers with the degree of cross-linking of 5–8%. Three fractions are collected in this process (Asadi 2005):

- Extract (high sugar fraction) with low color content contains more than 90% (on the dry substance basis) of sucrose, which is further crystallized after concentration of the extract to about 70% of the dry substance.
- Raffinate (low sugar fraction) contains about 15–22% (on the dry substance basis) of sucrose, salts, proteins, and amino acids, about 5% of trisaccharide raffinose, and other oligosaccharides. Concentrated raffinate is used as a valuable feed supplement.
- Betaine fraction consists (on the dry substance basis) of 50% of betaine, 10% of amino acids, 10% of organic anions, and other components. Betaine, used as a food and feed additive or in the cosmetic industry, represents additional revenue.

Besides the increase of sucrose recovery and the production of other valuable products, an additional advantage of the desugaring of molasses by chromatographic processes is lowering the production costs of sugar compared to the sugar made from beet, which enables a return of investments in less than 5 years (Asadi 2005). The mentioned process of molasses desugaring can be combined with other separation techniques such as crystallization, electrodialysis, or nanofiltration to improve the yield and purity of the obtained products (Paananen et al. 2006; Carter and Jensen 2007).

Pure raffinose can be produced from molasses too if a multicomponent separation SMB system is used for the separation. Raffinose is further double crystallized from a chromatographic fraction, whose dry mass contains 60–70% of raffinose, into raffinose crystals with a 98% purity (Inoue et al. 2001).

The prebiotics area is a fast evolving field that attracts significant interest by both academic and industrial communities. Prebiotic oligosaccharides are, in principle, produced in three ways: the first one is their isolation from natural sources or by-products or wastes of biotechnological productions, the second one is the acid or enzyme hydrolysis of polysaccharides, and the third one is the elongation of oligosaccharide chains from disaccharide substrates by the action of enzymes with glycosyltransferase activity. The last mentioned method is used in the production of fructooligosaccharides from sucrose and galactooligosaccharides from lactose.

A by-product of these transglycosylation reactions is glucose while a portion of unreacted disaccharide remains in the oligosaccharide mixture produced. Since mono- and disaccharides lack the prebiotic effect, their content decreases a prebiotic value of the resulting product. In this case, the use of a purification step is necessary to improve the quality of the oligosaccharide product. The chromatographic separation of oligosaccharides, disaccharides, and monosaccharides is well known and is frequently used in analytical applications.

Cation-exchange resins of the same type as for desugaring of molasses are mostly used for this purpose. The degree of copolymer cross-linking is 4–6% to provide the penetration of oligosaccharides into the pores with the degree of polymerization up to 10 and therewith achieve their good separation. The resins are converted into a convenient metal-cationic form; the most used are silver, sodium, and potassium cations, from which only resins containing sodium or potassium cations are compatible with food-grade requirements in industrial production.

Some industrial applications of chromatographic separations of oligosaccharides are documented in patent literature; for example, production of galactooligosaccharides with 90–97% purity using a cation-exchange resin in potassium form (Sawatzki 2003; Jeong et al. 2011) or fructooligosaccharide production patented by Meiji Seika Kaisha in 1996 using resins in sodium form (Matsumoto et al. 1996). The design of the industrial production of fructooligosaccharides, including their purification by an SMB chromatographic system, was described by Vaňková et al. (2008).

Lactose, a disaccharide found in milk, is traditionally produced from whey by crystallization when the lactose recovery yield from the whey permeate is about 65–70%. The incorporation of the chromatographic purification step in this production process can increase the lactose recovery yield up to 90–98% and lactose purity up to 99.5% (Novasep 2010b).

Raw bovine milk contains around 3–3.5% of proteins, which is a crucial source of important peptides in human diet. The two main groups of bovine milk proteins are caseins (80%) and whey proteins (20%). The second group contains several high-value proteins such as β-lactoglobulin (3–4 g/L), α-lactalbumin (1.2–1.5 g/L), bovine serum albumin (BSA) (0.3–0.6 g/L), immunoglobulin G, A, and M (0.6–0.9 g/L), lactoperoxidase (≤0.06 g/L), and lactoferrin (≤0.05 g/L) (Hahn et al. 1998). Their chromatographic separation from whey is economically justified because it is a cheap by-product of cheese manufacturing.

The fractionation of whey proteins has been mostly carried out by ion-exchange chromatography. Anion-exchange resins were found to be the most suitable for the separation of β-lactoglobulin from sweet whey mixtures (Gerberding and Byers 1998; Hahn et al. 1998). BSA and α-lactalbumin were successfully separated from whey using a strong anion-exchange membrane adsorber (Splitt et al. 1996). The separation of α-lactalbumin and β-lactoglobulin has recently been investigated by Garcia Rojas et al. (2011). The two most valuable whey proteins, lactoperoxidase and lactoferrin (antimicrobial agents), were purified in a multistep process, in which whey protein concentrate (whey devoid of fat and lactose) served as a feed for continuous chromatographic separation (Andersson and Mattiasson 2006).

Polyunsaturated fatty acids (especially ω-3 fatty acids) are known to reduce the risks of coronary heart disease, autoimmune disorders, and cancer (Espinosa et al.

2008). They have to be supplied in the diet since humans cannot synthesize them. The separation of these compounds is complex and countercurrent chromatography plays a prominent role here (Bousquet and Le Goffic 1995). Recently, SMB chromatography processes employing supercritical fluids (mainly carbon dioxide), which reduce consumptions of organic solvents, have been developed (Mazzotti et al. 1997; Perrut et al. 1998; Martínez Cristancho et al. 2012).

Lysozyme has been used in the pharmaceutical and food industries for many years due to its lytic activity toward Gram-positive microorganisms. Therefore, it is widely used as a food preservative and an antimicrobial agent (Cunningham et al. 1991). On the industrial scale, it is obtained from hen egg white using several separation steps, including ion-exchange chromatography (Dembczyński et al. 2010). Several types of macroporous cation exchangers have been applied for the separation of lysozyme with a good recovery yield and purity (Li-Chan et al. 1986; Bayramoğlu et al. 2007; Yan et al. 2011).

Several chromatographic modes are employed in amino acid separation. This comes from the versatile nature and diversity of these compounds. SMB chromatography using a polymeric resin was exploited by Molnár et al. (2005) to separate glycine from aqueous solutions of L-phenylalanine. An excellent yield and purity of 99% were obtained. A more advanced concept of SMB reactor/separator employing a weak cation exchanger was introduced for the production of threonine from glycine and acetaldehyde by Makart et al. (2008). Kostova and Bart (2007a, b) presented the application of chiral ligand exchange chromatography for the separation of enantiomers of racemic mixtures of amino acids.

REFERENCES

Abderafi, S., and Bounahmidi, T. 1994. Measurement and modeling of atmospheric pressure vapor–liquid equilibrium data for binary, ternary and quaternary mixtures of sucrose, glucose, fructose and water components. *Fluid Phase Equilibria* 93:337–351.

Al-Bokari, M., Cherrak, D., and Guiochon, G. 2002. Determination of the porosities of monolithic columns by inverse size-exclusion chromatography. *Journal of Chromatography A* 975:275–284.

Ammeraal, R., and Delgado, G. 1993. Fractionating starch hydrolysates. US Patent 5,194,094.

Andersson, J., and Mattiasson, B. 2006. Simulated moving bed technology with a simplified approach for protein purification: Separation of lactoperoxidase and lactoferrin from whey protein concentrate. *Journal of Chromatography A* 1107:88–95.

Antia, F. D., and Horváth, C. 1989. Gradient elution in non-linear preparative liquid chromatography. *Journal of Chromatography A* 484:1–27.

Arvidsson, P., Plieva, F. M., Savina, I. N., Lozinsky, V. I., Fexby, S., Bülow, L., Yu Galaev, I., and Mattiasson, B. 2002. Chromatography of microbial cells using continuous supermacroporous affinity and ion-exchange columns. *Journal of Chromatography A* 977:27–38.

Asadi, M. 2005. Molasses-desugaring process. In *Beet-Sugar Handbook*, John Wiley & Sons, Inc., Hoboken, NJ, pp. 517–546.

Azevedo, D. C. S., and Rodrigues, A. E. 2001. Fructose–glucose separation in a SMB pilot unit: Modeling, simulation, design, and operation. *AIChE Journal* 47:2042–2051.

Bayramoğlu, G., Ekici, G., Beşirli, N., and Arica, M. Y. 2007. Preparation of ion-exchange beads based on poly(methacrylic acid) brush grafted chitosan beads: Isolation of lysozyme from egg white in batch system. *Colloids and Surfaces A: Physicochemical and Engineering Aspects* 310:68–77.

Bellot, J. C., and Condoret, J. S. 1993. Modelling of liquid chromatography equilibria. *Process Biochemistry* 28:365–376.

Bousquet, O., and Le Goffic, F. 1995. Counter-current chromatographic separation of polyunsaturated fatty acids. *Journal of Chromatography A* 704:211–216.

Brooks, C. A., and Cramer, S. M. 1992. Steric mass-action ion exchange: Displacement profiles and induced salt gradients. *AIChE Journal* 38:1969–1978.

Broughton, D. B., and Gerhold, C.G. 1961. Continuous sorption process employing fixed bed of sorbent and moving inlets and outlets. United States Patent 2985589.

Burkhardt, M. O., Schick, R., and Freudenberg, T. 2000. Continuous thin juice decalcification with weakly acidic cation exchangers in Pfeifer & Langen's Appeldorn factory. *Zuckerindustrie* 125:673–682.

Carter, M., and Jensen, J. P. 2007. Process for the recovery of sucrose and/or non-sucrose components.US Patent 2007/0169,772.

Chang, Y.-K., and Chang, I.-P. 2006. Method development for direct recovery of lysozyme from highly crude chicken egg white by stirred fluidized bed technique. *Biochemical Engineering Journal* 30:63–75.

Charcosset, C. 2006. Membrane processes in biotechnology: An overview. *Biotechnology Advances* 24:482–492.

Charton, F., Bailly, M., and Guiochon, G. 1994. Recycling in preparative liquid chromatography. *Journal of Chromatography A* 687:13–31.

Chilamkurthi, S., Willemsen, J.-H., van der Wielen, L. A. M., Poiesz, E., and Ottens, M. 2012. High-throughput determination of adsorption equilibria for chromatographic oligosaccharide separations. *Journal of Chromatography A* 1239:22–34.

Cunningham, F. E., Proctor, V. A., and Goetsch, S. J. 1991. Egg-white lysozyme as a food preservative: An overview. *World's Poultry Science Journal* 47:141–163.

Dembczyński, R., Białas, W., Regulski, K., and Jankowski, T. 2010. Lysozyme extraction from hen egg white in an aqueous two-phase system composed of ethylene oxide–propylene oxide thermoseparating copolymer and potassium phosphate. *Process Biochemistry* 45:369–374.

Deng, Y., Li, R., Shang, H., Xue, P., Zhang, X., and Zhang, Y. 2010. Method for preparing high fructose corn syrup. CN Patent 101,766,289.

DePhillips, P., and Lenhoff, A. M. 2001. Determinants of protein retention characteristics on cation-exchange adsorbents. *Journal of Chromatography A* 933:57–72.

Dhingra, Y. R., and Pynnonen, B. W. 1989. Chromatographic separations using ion-exchange resins. EP Patent 0,327,400.

Espinosa, S., Diaz, M. S., and Brignole, E. A. 2008. Food additives obtained by supercritical extraction from natural sources. *The Journal of Supercritical Fluids* 45:213–219.

Garcia Rojas, E. E., Coimbra, J. S. R., Saraiva, S. H., and Vicente, A. A. 2011. Modeling of the α-lactalbumin and β-lactoglobulin protein separation. *Chemical Engineering Research and Design* 89:156–163.

Gerberding, S., and Byers, C. 1998. Preparative ion-exchange chromatography of proteins from dairy whey. *Journal of Chromatography A* 808:141–151.

Goulding, R. W. 1975. Liquid chromatography of sugars and related polyhydric alcohols on cation exchangers: The effect of cation variation. *Journal of Chromatography A* 103:229–239.

Gramblička, M., and Polakovič, M. 2007. Adsorption equilibria of glucose, fructose, sucrose, and fructooligosaccharides on cation exchange resins. *Journal of Chemical & Engineering Data* 52:345–350.

Grill, C. M., and Miller, L. 1998. Separation of a racemic pharmaceutical intermediate using closed-loop steady state recycling. *Journal of Chromatography A* 827:359–371.

Guiochon, G. 2002. Preparative liquid chromatography. *Journal of Chromatography A* 965:129–161.

Guiochon, G. 2007. Monolithic columns in high-performance liquid chromatography. *Journal of Chromatography A* 1168:101–168.

Guiochon, G., Felinger, A., and Shirazi, D. G. G. 2006. *Fundamentals of Preparative and Nonlinear Chromatography*, 2nd edition. Academic Press, Amsterdam, the Netherlands.

Guiochon, G., and Lin, B. 2003. *Modeling for Preparative Chromatography*. Academic Press.

Hahn, R., Schulz, P. M., Schaupp, C., and Jungbauer, A. 1998. Bovine whey fractionation based on cation-exchange chromatography. *Journal of Chromatography A* 795:277–287.

Haselow, J. 2010. Crude glycerol purification process. WIPO Patent Application WO/2010/033817.

Inoue, H., Semba, Y., Suda, O., and Ohwada, Y. 2001. Method for preparing raffinose crystals and equipment for the same. US Patent 6,224,684.

Janson, J.-C., and Rydén, L. 1998. *Protein Purification: Principles, High-Resolution Methods, and Applications*. Wiley-VCH, New York.

Jensen, C. R. C. 2007. Direct white sugar manufacture in the cane sugar industry via membrane filtration and continuous ion-exchange demineralization. *Zuckerindustrie* 132:446–452.

Jeong, H.-S., Kwon, H.-K. and Choi, J.-S. 2011. High-purity galactooligosaccharides and uses thereof. US Patent 2011/0189,342.

Jungbauer, A. and Hahn, R. 2004. Monoliths for fast bioseparation and bioconversion and their applications in biotechnology. *Journal of Separation Science* 27:767–778.

Katz, E., Davis, H. S., and Scallet, B. 1983. High fructose syrup and process for making same. US Patent 4,395,292.

Kostova, A., and Bart, H.-J. 2007a. Preparative chromatographic separation of amino acid racemic mixtures: I. Adsorption isotherms. *Separation and Purification Technology* 54:340–348.

Kostova, A., and Bart, H.-J. 2007b. Preparative chromatographic separation of amino acid racemic mixtures: II. Modelling of the separation process. *Separation and Purification Technology* 54:315–321.

Lee, K. 2003. Continuous separation of glucose and fructose at high concentration using two-section simulated moving bed process. *Korean Journal of Chemical Engineering* 20:532–537.

Li-Chan, E., Nakai, S., Sim, J., Bragg, D. B., and Lo, K. V. 1986. Lysozyme separation from egg white by cation exchange column chromatography. *Journal of Food Science* 51:1032–1036.

Makart, S., Bechtold, M., and Panke, S. 2008. Separation of amino acids by simulated moving bed under solvent constrained conditions for the integration of continuous chromatography and biotransformation. *Chemical Engineering Science* 63:5347–5355.

Margureanu, G., and Gutmann, F. 1989. Process for producing citric acid. US Patent 4855494.

Martínez Cristancho, C. A., Peper, S., and Johannsen, M. 2012. Supercritical fluid simulated moving bed chromatography for the separation of ethyl linoleate and ethyl oleate. *The Journal of Supercritical Fluids* 66:129–136.

Matsumoto, H., Nishizawa, K., Kawakami, T., Hirayama, M., and Adachi, T. 1996. Production of high purity oligosaccharides from highly concentrated mixed saccharide solution. JP Patent 8,140,691.

Mazzotti, M., Storti, G., and Morbidelli, M. 1997. Supercritical fluid simulated moving bed chromatography. *Journal of Chromatography A* 786:309–320.

Molnár, Z., Nagy, M., Aranyi, A., Hanák, L., Argyelán, J., Pencz, I., and Szánya, T. 2005. Separation of amino acids with simulated moving bed chromatography. *Journal of Chromatography A* 1075:77–86.

Moravčík, J., Gramblička, M., Wiśniewski, Ł., Vaňková, K., and Polakovič, M. 2012. Influence of the ionic form of a cation-exchange adsorbent on chromatographic separation of galactooligosaccharides. *Chemical Papers* 66:583–588.

Nakakuki, T. (Ed.) 1993. *Oligosaccharides: Production, Properties and Applications.* Gordon and Breach Science Publishers, Amsterdam, the Netherlands.

Nollet, L. M. L. 2004. *Handbook of Food Analysis*, 2nd edition—*3 Volume Set*. CRC Press, Boca Raton, FL.

Novasep, 2010a. Starch sweeteners production. Application data sheet. http://www.novasep.com/biomolecules/food-ingredients/Applications/sugar-and-sweeteners-production.asp

Novasep, 2010b. Purification solutions for lactose. Application data sheet. http://www.novasep.com/biomolecules/food-ingredients/Applications/Lactose-purification.asp

Nowak, J., Poplewska, I., Antos, D., and Seidel-Morgenstern, A. 2009. Adsorption behaviour of sugars versus their activity in single and multicomponent liquid solutions. *Journal of Chromatography A* 1216:8697–8704.

Orozco, R. L., Redwood, M. D., Leeke, G. A., Bahari, A., Santos, R. C. D., and Macaskie, L. E. 2012. Hydrothermal hydrolysis of starch with CO_2 and detoxification of the hydrolysates with activated carbon for bio-hydrogen fermentation. *International Journal of Hydrogen Energy* 37:6545–6553.

Paananen, H., Heikkila, H., Puuppo, O., Koivikko, H., Monten, K.-E., Manttari, M., and Nystrom, M. 2006. Process for recovering betaine. US Patent 7,009,076.

Paananen, H., and Kuisma, J. 2000. Chromatographic separation of molasses components. *Zuckerindustrie* 125:978–981.

Packeus, L.W. 1997. Integrated process for producing crystalline fructose and a high-fructose, liquid phase sweetener. US Patent 5,656,094.

Pedruzzi, I., da Silva, E. A. B., and Rodrigues, A. E. 2008. Selection of resins, equilibrium and sorption kinetics of lactobionic acid, fructose, lactose and sorbitol. *Separation and Purification Technology* 63:600–611.

Perrut, M., Nicoud, R.-M., and Breivik, H. 1998. Processes for chromatographic fractionation of fatty acids and their derivatives. US Patent 5719302.

Phillips, M., Cormier, J., Ferrence, J., Dowd, C., Kiss, R., Lutz, H., and Carter, J. 2005. Performance of a membrane adsorber for trace impurity removal in biotechnology manufacturing. *Journal of Chromatography A* 1078:74–82.

Pico, Y. 2012. *Chemical Analysis of Food: Techniques and Applications.* Academic Press, Boston, MA.

Rasche, J. F. 1992. Simulated moving bed chromatographic separation. US Patent 5,122,275.

Rizvi, S. 2010. *Separation, Extraction and Concentration Processes in the Food, Beverage and Nutraceutical Industries.* Woodhead Publishing, Oxford, UK.

Sawatzki, G. 2003. Process for continuous production of galacto-oligosaccharides. EP Patent 1,352,967.

Schlinge, D., Scherpian, P., and Schembecker, G. 2010. Comparison of process concepts for preparative chromatography. *Chemical Engineering Science* 65:5373–5381.

Schmidt-Traub, H. 2005. *Preparative Chromatography: Of Fine Chemicals and Pharmaceutical Agents.* Wiley-VCH, Weinheim, Germany.

Schramm, H., Kienle, A., Kaspereit, M., and Seidel-Morgenstern, A. 2003. Improved operation of simulated moving bed processes through cyclic modulation of feed flow and feed concentration. *Chemical Engineering Science* 58:5217–5227.

Splitt, H., Mackenstedt, I., and Freitag, R. 1996. Preparative membrane adsorber chromatography for the isolation of cow milk components. *Journal of Chromatography A* 729:87–97.

Sun, D.-W. 2008. *Modern Techniques for Food Authentication.* Academic Press, San Diego, CA.

Tarafder, A., Aumann, L., and Morbidelli, M. 2008. Improvement in industrial re-chromatography (recycling) procedure in solvent gradient bio-separation processes. *Journal of Chromatography A* 1195: 67–77.

Tatárová, I., Fáber, R., Denoyel, R., and Polakovič, M. 2009. Characterization of pore structure of a strong anion-exchange membrane adsorbent under different buffer and salt concentration conditions. *Journal of Chromatography A* 1216:941–947.

Toumi, A., Engell, S., Ludemann-Hombourger, O., Nicoud, R. M., and Bailly, M. 2003. Optimization of simulated moving bed and Varicol processes. *Journal of Chromatography A* 1006:15–31.

Van Deemter, J. J., Zuiderweg, F. J., and Klinkenberg, A. 1956. Longitudinal diffusion and resistance to mass transfer as causes of nonideality in chromatography. *Chemical Engineering Science* 5:271–289.

Van Reis, R., and Zydney, A. 2001. Membrane separations in biotechnology. *Current Opinion in Biotechnology* 12:208–211.

Vaňková, K., Ačai, P., and Polakovič, M. 2010. Modelling of fixed-bed adsorption of mono-, di-, and fructooligosaccharides on a cation-exchange resin. *Biochemical Engineering Journal* 49:84–88.

Vaňková, K., Onderková, Z., Antošová, M., and Polakovič, M. 2008. Design and economics of industrial production of fructooligosaccharides. *Chemical Papers* 62:375–381.

Vaňková, K., and Polakovič, M. 2012. Design of fructooligosaccharide separation using simulated moving-bed chromatography. *Chemical Engineering and Technology* 35:161–168.

Vente, J. A. 2004. *Adsorbent Functionality in Relation to Selectivity and Capacity in Oligosaccharide Separations.* PhD Thesis, University of Twente, The Netherlands.

Vente, J. A., Bosch, H., de Haan, A. B., and Bussmann, P. J. T. 2005. Evaluation of sugar sorption isotherm measurement by frontal analysis under industrial processing conditions. *Journal of Chromatography A* 1066:71–79.

Viard, V., and Lameloise, M.-L. 1992. Modelling glucose–fructose separation by adsorption chromatography on ion exchange resins. *Journal of Food Engineering* 17:29–48.

Wang, C., Soice, N. P., Ramaswamy, S., Gagnon, B. A., Umana, J., Cotoni, K. A., Bian, N., and Cheng, K.-S. C. 2007. Cored anion-exchange chromatography media for antibody flow-through purification. *Journal of Chromatography A* 1155:74–84.

Wilson, C. W., Wagner, C. J., and Shaw, P. E. 1989. Reduction of bitter components in grapefruit and navel orange juices with.beta.-cyclodextrin polymers or XAD resins in a fluidized bed process. *Journal of Agricultural and Food Chemistry* 37:14–18.

Wrzosek, K., Gramblička, M., and Polakovič, M. 2009. Influence of ligand density on antibody binding capacity of cation-exchange adsorbents. *Journal of Chromatography A* 1216:5039–5044.

Yan, L., Shen, S., Yun, J., and Yao, K. 2011. Isolation of lysozyme from chicken egg white using polyacrylamide-based cation-exchange cryogel. *Chinese Journal of Chemical Engineering* 19:876–880.

Zang, Y., and Wankat, P. C. 2002. SMB operation strategy—Partial feed. *Industrial and Engineering Chemistry Research* 41:2504–2511.

Zhang, Z., Mazzotti, M., and Morbidelli, M. 2003. PowerFeed operation of simulated moving bed units: Changing flow-rates during the switching interval. *Journal of Chromatography A* 1006:87–99.

8 Crystallization

António Ferreira and Fernando Alberto Rocha

CONTENTS

8.1 INTRODUCTION

Crystallization is the conversion of a substance or several substances from an amorphous solid, liquid, or gaseous state into the crystalline state [1]. The crystallization process is one of the oldest and most important unit operations that can be found in different industries. From the chemical to food industries, crystallization is a widely implemented process with an enormous economic importance. Although a generally efficient process for purification, it is often a complex operation with many factors that can influence the final product. In the food industry, the crystallization control is a very difficult task potentiated by the production process and, in most of the cases, by the restrictions due to the used ingredients.

The crystalline structure of foods is important for the product quality, texture, and stability; however, in some cases, the avoidance of the solid forms is desired, for example, the oleic acid precipitation in olive oil during its storage. So, the crystallization control is crucial not only during the product production but also during its storage. Understanding the phase equilibrium of the system as well as the nucleation and growth kinetics is fundamental to obtain the final product with the desired characteristics. In this chapter, the authors discuss the previous phenomena, presenting some basic principles of crystallization and opening the window for a more deep knowledge by indicating some useful bibliography.

8.2 PHASE EQUILIBRIUM

In the food industry, crystallization control is a key factor in quality as it relates to texture, with some foods requiring the promotion of crystallization and others requiring its prevention [2]. Table 8.1 shows the importance of crystallization control in some characteristics of food products. For example, to make a smooth texture of ice cream, many small crystals must be formed during processing and, for that, the proper crystallization conditions need to be found. In the same way, in fondant processing, the temperature at which the syrup is nucleated is a critical parameter; if nucleation is induced at a temperature other than the optimal temperature, fewer crystals will be formed and the texture of the fondant will be unsatisfactory [3].

In the food industry, the crystallization is normally obtained by decreasing the temperature (cooling crystallization), rather than by solvent evaporation. The main differences of these types of crystallization are related to the crystal growth rate and the metastable zone (range between the solubility concentration and the supersaturation concentration after which the 3D homogeneous nucleation of the solute starts). To understand these two parameters, some basic fundamentals need to be known.

Some questions arise when we are dealing with crystallization: in what conditions does a crystal grow or dissolve, or be in equilibrium with the solution? When does

TABLE 8.1
Characteristics of Food Products for Which Crystallization Control Is Important

Control Number, Size, Shape, and Polymorph		
Product	Ingredient	Desired Characteristic
Caramel	Sucrose or lactose	Smooth and short textures
Fondant	Sucrose	Smooth texture
Panned confections	Sucrose	Brittle texture (hard panned)
		Soft texture (soft panned)
Cereal coating	Sucrose	Appearance
Refined sugars	Sucrose, lactose, and so on	Size distribution for separation
Refined salt	NaCl	Size distribution for separation
Organic acids	Citric acid and so on	Size distribution for separation
Bread, baked products	Starch	Texture
Frozen foods	Ice	Thawed quality
Frozen desserts	Ice	Smooth texture
Freeze concentration	Ice	Size distribution for separation
Chocolate	Cocoa butter	Texture (snap) and gloss
		Shelf stability for blooming
Butter	Milk fat	Hardness, spreadability
Margarine	Vegetable fats	Hardness, spreadability
Peanut butter	Vegetable fats	Texture, spreadability
Fat fractionation	Palm oil, tallow, and milk fat	Size distribution for separation

Source: Adapted from R.W. Hartel, *Crystallization in Foods*, Gaithersburg, MD: Springer, 2001.

a solution start to nucleate? These are some questions, beyond others, which can be answered if some fundamental principles of phase equilibrium are known. One of them is the solubility.

8.2.1 SOLUBILITY

Solubility can be defined as the maximum weight of the anhydrous solute that will dissolve in 100 g of the solvent; in this condition, a solution is said to be saturated. In the food industry, the solvent is generally water.

Figure 8.1, adapted from Charles [4] and Gharsallaoui et al. [5], shows the sucrose solubility in water, which is a function of the temperature. In most of the cases, the solubility increases with the temperature as can be observed here. However, in the food industry, particular attention needs to be given to milk and its by-products, since calcium phosphate compounds present in the milk are characterized by their solubility decreasing with the temperature. For example, this characteristic increases the problem related to fouling, caused by milk heat treatment [6–9]. To reduce this problem, it is important to understand the precipitation and growth processes of calcium phosphate compounds and, for that, their solubility at different temperatures and system pHs (the solubility of calcium phosphate compounds changes drastically with the system pH [10,11]).

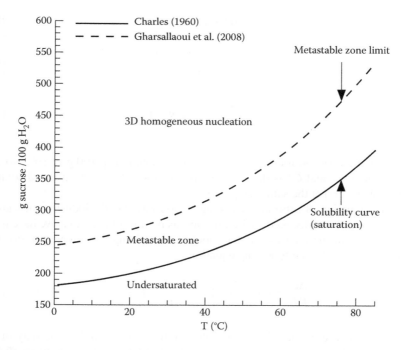

FIGURE 8.1   Solubility and metastable zone width of sucrose in aqueous solution. (Adapted from D.F. Charles, *International Sugar Journal*, 62, 1960, 125; A. Gharsallaoui, B. Rogé, and M. Mathlouthi, *Food Chemistry*, 106, 2008, 1329–1339.)

The solubility curve depends on the solution properties and also on the different types and the amount of impurities present in the system. Thus, to know the solubility of a certain compound in a system, experimental work needs to be done. If the system is not very complex, some thermodynamic tools can be useful to predict the compound solubility in a given system. UNIQUAC (universal quasichemical) and UNIFAC (UNIQUAC functional-group activity coefficients) are two thermodynamic models that can be found in the literature with high applicability [12]. On the basis of UNIQUAC and UNIFAC principles, it is possible to predict the solubility of a certain compound in specific conditions. In Figure 8.2, adapted from Ferreira [13] using the UNIQUAC and UNIFAC thermodynamic parameters obtained by Peres and Macedo [14,15], the comparison between the experimental and predicted sucrose solubility in pure and impure systems is presented. A good agreement is obtained, in this case, using the UNIQUAC model.

8.2.2 Undersaturation and Supersaturation

A solution is said to be *undersaturated* when it has a lower concentration than a saturated solution under the same conditions. A solution is said to be *supersaturated* when it has a higher concentration than a saturated solution under the same conditions. The most common expressions of supersaturation are the concentration driving force (ΔC), the supersaturation ratio (S), and the absolute or relative supersaturation (σ) defined by

$$\Delta C = C - C^* \tag{8.1}$$

$$S = \frac{C}{C^*} \tag{8.2}$$

$$\sigma = \frac{C - C^*}{C^*} = S - 1 \tag{8.3}$$

where C is the concentration of the solute in the solution (g/100 g of solvent) at a given temperature and C^* is the concentration of the solute in a saturated solution (g/100 g of solvent) at the same temperature.

Despite Equations 8.1 through 8.3 being the most used in the industry, their application in scientific works needs to be cautious, as in most of the cases, we have nonideal solutions. In these cases, the thermodynamic driving force for crystallization ($\Delta\mu/RT$) is defined by the following equation:

$$\frac{\Delta\mu}{RT} = \ln\left(\frac{a}{a^*}\right) = \ln\frac{\gamma \cdot m}{\gamma^* \cdot m^*} = \ln(\sigma + 1) \tag{8.4}$$

where a and a^* are the activity of the solute in the solution and the activity of the solute in the solution at saturation, respectively. γ and m are the activity coefficient and the molality, respectively. The asterisk (*) refers to the saturated solution. The activity coefficients can be calculated using the UNIQUAC and UNIFAC models.

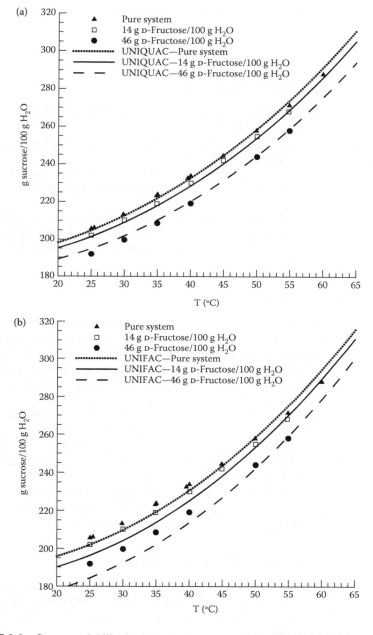

FIGURE 8.2 Sucrose solubility in the ternary system, sucrose/H_2O/D-fructose: experimental (symbols) and predicted values using (a) UNIQUAC and (b) UNIFAC models. (Adapted from A. Ferreira, *Caracterização fenomenológica dos processos de transferência de impurezas para os cristais*, Faculdade de Engenharia da Universidade do Porto, Portugal, 2008; A.M. Peres and E.A. Macedo, *Fluid Phase Equilibria*, 123, Aug. 1996, 71–95; A.M. Peres and E.A. Macedo, *Fluid Phase Equilibria*, 139, 1997, 47–74.)

According to the previous equation, σ is now defined as

$$\sigma = \frac{a - a^*}{a^*} \qquad (8.5)$$

Thus, a crystalline solid in contact with a solution

- *Dissolves*, if the activity of the solute in the solution is lower than its activity in the saturated solution, $(\Delta\mu/RT) < 0$
- *Grows*, if the activity of the solute in the solution is higher than its activity in the saturated solution, $(\Delta\mu/RT) > 0$
- *Remains in equilibrium with the solution*, if the activity of the solute in the solution is equal to its activity in the saturated solution, $(\Delta\mu/RT) = 0$

In short, according to Figure 8.1, if one has seeds (small crystals of the solute) in a solution with a concentration below the saturation line at a certain temperature, these will dissolve. In contrast, if one is above the saturation line, but inside the metastable zone, the seeds will grow.

8.2.3   METASTABILITY AND NUCLEATION

The metastable zone is characterized, in the crystallization process, by the region between the saturation and the supersaturation curves (defined in Figure 8.1 by the metastable zone limit). In this region, aggregates of solute molecules (called clusters) are formed but are then dispersed again and they will not grow. If seed crystals are added to the system, they will act as centers of crystallization, and then the solute molecules will be deposited on them, leading to their growth. To allow the development of the metastable clusters to three-dimensional (3D) stable nuclei, the system needs to achieve a required degree of supersaturation. The minimum degree of this supersaturation corresponds to the metastable zone limit, presented in Figure 8.1. Above this line, crystals form spontaneously and rapidly, without an external initiating action, this being called spontaneous nucleation.

It is important to note that the metastable zone limit depends on several variables such as the type and amount of impurities present in the system, temperature, agitation speed, crystallizer type, and cooling velocity, among others.

8.3   MECHANISMS OF CRYSTAL GROWTH

Once the energy barrier that opposes the formation of stable nuclei is surpassed, the supersaturated system develops, by nuclei growing, until the saturation. The growth of crystals in the supersaturated solution is a very complex process and its complete knowledge remains an open area.

8.3.1   TWO-STEP MODEL

It is generally considered that crystal growth from the solution is the result of the existence of two steps in series: bulk or volume diffusion through a hypothetical stagnant film (diffusional step), followed by the integration of growth elements into the crystal lattice ("kinetic" step or surface "reaction" step). The slower step determines the global process rate. These two stages, occurring under the influence of different concentration driving forces, can be represented by the following equations:

$$\frac{dm}{dt} = k_d A(C - C_i) \qquad \text{(diffusion)} \tag{8.6}$$

and

$$\frac{dm}{dt} = k_r A(C_i - C^*)^r \qquad \text{(diffusion)} \tag{8.7}$$

where m is the mass of the solid deposited in time t, A is the surface area of the crystal, k_d is the coefficient of mass transfer by diffusion, k_r is the rate constant for surface reaction (integration) process, r is the kinetic order, and C_i is the solute concentration in the solution at crystal–solution interface. In Figure 8.3, a representation of these two stages is shown.

As Equations 8.6 and 8.7 are difficult to apply, because of the C_i term (very difficult to measure), a semiempirical overall growth rate equation is frequently used:

$$\frac{dm}{dt} = K_G A(C - C^*)^g \tag{8.8}$$

where K_G is an overall crystal growth coefficient and g is the order of the overall crystal growth process.

Thus, the overall growth rate (R_G) can be written as

$$R_G = \frac{1}{A} \cdot \frac{dm}{dt} = K_G(C - C^*)^g \tag{8.9}$$

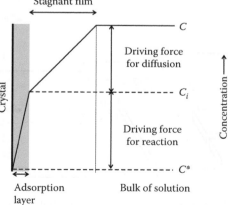

FIGURE 8.3 Concentration driving forces in crystallization from the solution.

More information about the applicability of the previous equation to the systems controlled by diffusion and surface integration processes can be found in Mullin [16].

Despite the large-scale application of the presented model, difficulties in the scale-up of the crystallizers are often reported due to uncertainties in the prediction of the growth rates. To eliminate the limitations and inconsistencies of the two-step model, Martins and Rocha [17] recently presented a new model, where the integration step is preceded by two other steps (solute diffusion and adsorption) occurring concurrently over the adsorption layer.

8.3.2   DIFFUSION-CONTROLLED CRYSTAL GROWTH

When the integration of growth elements into the crystal lattice ("kinetic" step or surface "reaction" step) is too fast, the crystallization process is controlled by the diffusive/convective transport of units. In this case, the physical properties of the system, such as the viscosity and the velocity of the liquid comparative to that of the crystal (on increasing the velocity, the stagnant film thickness decreases, increasing, by this way, the crystal growth rate), play an important role in the crystallization process. In contrast, if the system is controlled by the integration process, the crystal surface properties become an important variable in the crystallization process.

8.3.3   INTEGRATION-CONTROLLED CRYSTAL GROWTH

When the units of the crystallizing substance arrive at the crystal surface, they are not immediately integrated into the lattice, but lose one degree of freedom (during this process, the solute molecules also lose some of the solvation solvent molecules and heat, Figure 8.4, step 2) and are free to migrate over the crystal face (surface diffusion, Figure 8.4, step 3). This migration will continue until the units reach a position where the attractive forces are the greatest, that is, at the "active centers." At this point, the crystallizing units lose the remainder of the solvation molecules and heat, and are integrated into the lattice (Figure 8.4, step 4). The solvent molecules initiate the

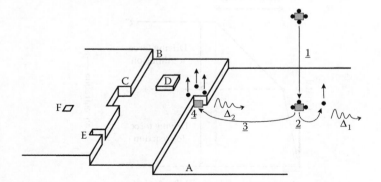

FIGURE 8.4 Modified Kossel's model of a growing crystal surface showing (a) flat surfaces, (b) steps, (c) kinks, (d) surface-adsorbed growth units, (e) edge vacancies, and (f) surface vacancies. The heat is released in steps 2 (Δ_1) and 4 (Δ_2).

counterdiffusion through the adsorption layer, reducing the growth velocity by decreasing a local supersaturation. Under ideal conditions, the integration of growth units into the lattice will continue until the whole plane face is complete. To facilitate the process visualization, it is often used by Kossel's model (Figure 8.4). This model considers that the apparently flat crystal surface is in fact made up of moving layers (*steps*) of monatomic height, which may contain one or more *kinks*. In addition, there will be loosely adsorbed growth units on the crystal surface and vacancies on the surfaces and steps. Growth units are most easily incorporated into the crystal at a kink; the kink moves along the step and the face is eventually completed. A fresh step could be created by surface nucleation and this frequently starts at the corners [16]. To understand the mechanisms involved in the formation of a fresh step, several models can be found in the literature, the most known being the birth and spread (B + S) model (Volmer, 1939, [16]), the spiral growth model or the Burton–Cabrera–Frank (BCF) model [18], and, more recently, the spiral nucleation model (SNM) proposed by Martins and Rocha [19].

As mentioned previously, during the units' integration in the crystal lattice, heat is released ($\Delta_1 + \Delta_2$) (Figure 8.4, steps 2 and 4); this heat is called the heat of crystallization. For most food materials, the heat of crystallization is negative, that is, heat is released during the crystallization process.

8.4 INFLUENCE OF TEMPERATURE

The crystallization kinetics is strongly influenced by the temperature. In some cases, the temperature changes can lead to an alteration in the mechanism controlling the crystallization process. For example, in sucrose crystallization, if the crystallization occurs at temperatures lower than 45–50°C, the crystallization process is controlled by the integration process. If one works at higher temperatures, the diffusion step starts to be the limiting factor of crystallization. This information is crucial when the main objective is to control the crystallization process. Knowing the mechanism that is controlling the crystallization permits to decide what kind of variables are needed to control, if the variables are related to the solution properties (agitation speed, viscosity, etc.) or the others are related to the crystal surface (impurities presence, growth rate history, surface roughness, etc.).

8.5 INFLUENCE OF IMPURITIES

The impurities are, nowadays, one of the most investigated variables in the crystallization processes. According to their physical and chemical characteristics, the impurities have higher or lower impact on the crystal growth rate. Some impurities act on the properties of the solutions whereas others act on the surface of the crystals. The impurity action on the crystal surface is face dependent, as the molecule orientation on the crystal lattice exhibits different functional groups in the different faces, as one can see in Figure 8.5 for sucrose crystal. On the basis of this property, it is possible to change the crystal shape. Choosing the correct impurity, it is possible to block the growth rate of a specific face and, by this way, modify the crystal shape. In Figure 8.6, obtained from Khaddour [20], it is possible to see the dextran effect on sucrose crystal growth. In this figure, it is clearly seen that the step advancement

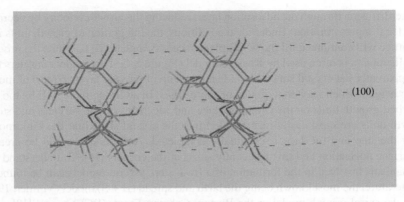

FIGURE 8.5 Sucrose molecule orientation on the crystal lattice showing the available groups (OH) on the face (100) of the sucrose crystal.

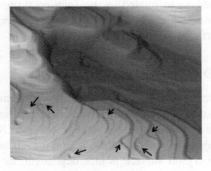

FIGURE 8.6 Detail of the face [100] of a sucrose crystal grown in the presence of dextran. The cavities (indicated by arrows) are expected to represent the locations where the dextran molecules are incorporated. The dimensions of the image are approximately 3×3.5 μm. (From I. Khaddour, A. Ferreira, L. Bento, and F. Rocha, Sucrose crystal growth in the presence of dextran of different molecular weights, *Journal of Crystal Growth*, 355, Sep. 2012, 17–25.)

is reduced at the places (cavities) where the incorporation of the dextran molecules occurred. The result of that is the decrease of the face growth rate and, consequently, the change in the sucrose crystal shape.

In the food industry, the importance of this property is related to the crystallization control. Using some specific additives, it is possible to control the crystallization, to obtain the product with the desired properties.

REFERENCES

1. A. Mersmann, *Crystallization Technology Handbook*, New York: Marcel Dekker, Inc., 1995.
2. R.W. Hartel, *Crystallization in Foods*, Gaithersburg, MD: Springer, 2001.
3. R.L. Earle and M.D. Earle, *Unit Operations in Food Processing*, Web Edition: The New Zealand Institute of Food Science & Technology (Inc.), 2004.
4. D.F. Charles, The solubility of pure sucrose in water, *International Sugar Journal*, 62, 1960, 125.
5. A. Gharsallaoui, B. Rogé, and M. Mathlouthi, Water–disaccharides interactions in saturated solution and the crystallisation conditions, *Food Chemistry*, 106, 2008, 1329–1339.
6. R. Rosmaninho and L.F. Melo, Calcium phosphate deposition from simulated milk ultrafiltrate on different stainless steel-based surfaces, *International Dairy Journal*, 16, Jan. 2006, 81–87.
7. R. Rosmaninho, F. Rocha, G. Rizzo, H. Müller-Steinhagen, and L.F. Melo, Calcium phosphate fouling on TiN-coated stainless steel surfaces: Role of ions and particles, *Chemical Engineering Science*, 62, Jul. 2007, 3821–3831.
8. R. Rosmaninho, G. Rizzo, H. Müller-Steinhagen, and L.F. Melo, Deposition from a milk mineral solution on novel heat transfer surfaces under turbulent flow conditions, *Journal of Food Engineering*, 85, Mar. 2008, 29–41.
9. T.J.M. Jeurnink, P. Walstra, and C.G. de Kruif, Mechanism of fouling in dairy processing, *Netherlands Milk and Dairy Journal*, 50, 1996, 407–426.
10. A. Ferreira, C. Oliveira, and F. Rocha, The different phases in the precipitation of dicalcium phosphate dihydrate, *Journal of Crystal Growth*, 252, May. 2003, 599–611.
11. J. Elliot, *Structure and Chemistry of the Apatites and Other Calcium Orthophosphates*, Amsterdam: Elsevier, 1994.
12. J.M. Smith, H.C.V. Ness, and M.M. Abbott, *Introduction to Chemical Engineering Thermodynamics*, New York and Singapore: McGraw-Hill, 1996.
13. A. Ferreira, *Caracterização fenomenológica dos processos de transferência de impurezas para os cristais*, Faculdade de Engenharia da Universidade do Porto, Portugal, 2008.
14. A.M. Peres and E. A. Macedo, Thermodynamic properties of sugars in aqueous solutions: Correlation and prediction using a modified UNIQUAC model, *Fluid Phase Equilibria*, 123, Aug. 1996, 71–95.
15. A.M. Peres and E.A. Macedo, A modified UNIFAC model for the calculation of thermodynamic properties of aqueous and non-aqueous solutions containing sugars, *Fluid Phase Equilibria*, 139, 1997, 47–74.
16. J.W. Mullin, *Crystallization*, Oxford: Butterworth-Heinemann, 1992.
17. P.M. Martins and F. Rocha, A new theoretical approach to model crystal growth from solution, *Chemical Engineering Science*, 61, Sep. 2006, 5696–5703.
18. W. Burton, N. Cabrera, and F. Frank, The growth of crystals and the equilibrium structure of their surfaces, *Philosophical Transactions of the Royal Society*, A243, 1951, 299.
19. P.M. Martins and F. Rocha, Characterization of crystal growth using a spiral nucleation model, *Surface Science*, 601, Aug. 2007, 3400–3408.
20 I. Khaddour, A. Ferreira, L. Bento, and F. Rocha, Sucrose crystal growth in the presence of dextran of different molecular weights, *Journal of Crystal Growth*, 355, Sep. 2012, 17–25.

9 Supercritical Extraction

M. Gabriela Bernardo-Gil

CONTENTS

9.1 INTRODUCTION

Over the past few decades, several remarkable consumer trends have emerged, namely on the quality and safety of foods and medicines, as well as on the toxicity and nutritive levels. Besides, some people believe that "natural" is good in opposition to synthetics, thus contributing to the growth of the production of bioactive compounds of high value, especially health-promoting foods and ingredients for the natural products industries, such as functional foods or nutraceuticals, pharmaceuticals, flavorings, and perfumery.

Several problems arise when traditional extraction is used to obtain products for the food industry, due to the use of huge amounts of toxic solvents, their subsequent recuperation and separation of extract and solvent, as well as the presence of solvent residues in the final product. Many studies have been performed to develop alternative methods of extraction to overcome the inconveniences of the conventional methods. In this context, looking for environmentally friendly solvents, the use of supercritical fluids (SCFs) has had a large increase, as these fluids present a great potential in the demand of "natural" products free from solvents and obtained by clean separation technologies (Verbeke, 2005) that contribute to food quality and safety requirements.

One of the main characteristics of SCFs is the possibility of modifying their density by changing the pressure and/or the temperature. As the solvent power of SCFs significantly depends on the density, the solvent-extracting ability of SCFs increases with the increase of pressure, at a given temperature, or with the decrease of temperature, at a given pressure, or both. However, with increasing pressure, more compounds are extracted, thus reducing the selectivity. So, by varying the temperature and/or the pressure, it is possible to manipulate the solubility of compounds in the SCFs, according to the needs of the process.

The densities of SCFs are similar to those of liquids, and the viscosities are similar to those of gases. SCFs also have very low surface tensions, and moderately high diffusion coefficients, favoring the diffusion through solid materials. Table 9.1 presents properties of SCFs when compared to those of gases and liquids.

Carbon dioxide (SCCO2) is the most commonly used solvent in operations involving supercritical fluids due to its low values of critical pressure and temperature (7.38 MPa, 31.1°C), allowing its use in the extraction of thermolabile compounds (Mendiola et al., 2007); it is nontoxic and nonflammable, exists in abundance, is practically chemically inert, and is cheap with high purity when compared with other solvents. Besides, as CO_2 is a gas at room temperature, at the end of the extraction, by decompression, a solvent-free extract is achieved. Carbon dioxide is nonpolar, making it very useful for the extraction of nonpolar compounds such as hydrocarbons, but its large quadrupole moment also enables it to dissolve some moderately polar compounds such as alcohols, esters, aldehydes, and ketones (Lang and Wai, 2001). Problems arise when extracting polar compounds. However, with the addition of a polar cosolvent (modifier or entrainer), it is possible to alter the solvent power and almost all the available compounds useful for the food, cosmetics, and pharmaceutical industries can be extracted. Indeed, by selecting a liquid modifier, its addition in a small amount to the process can significantly enhance the extraction efficiency, reducing the extraction time. Many studies have been published showing the effect of the use of a modifier. As an example, Hawthorne et al. (1993) verified that the addition of only 0.5 mL of dichloromethane to a 500 mg sample would reduce the extraction time from 90 to 30 min, in the extraction of essential oils from

TABLE 9.1
Comparison of Physical Properties of Supercritical Fluids, Gases, and Liquids

Physical Properties	Liquids	SCF	Gases
Density ρ (kg m^{-3})	600–1600	100–1000	0.6–2
Dynamic viscosity μ (mPa s)	0.2–1200	0.01–0.03	0.01–0.3
Kinematic viscosity ν (10^6 m^2 s^{-1})	0.1–5	0.2–0.1	5–500
Thermal conductivity λ (W/mK)	0.1–0.2	Max[a]	0.01–0.025
Diffusion coefficient D (10^6 m^2 s^{-1})	0.0002–0.002	0.010.1	1–40
Surface tension σ (dyn/cm^2)	20–40	—	—

Source: Adapted from Pereda, S., S. B. Bottini, and E. A. Brignole. 2008. *Supercritical Fluid Extraction of Nutraceuticals and Bioactive Compounds*, ed. J. L. Martinez, 1–23, Boca Raton, FL: CRC Press, Taylor & Francis Group; Span, R. and W. Wagner, 1996. *J. Phys. Chem. Ref. Data* 25:1509–1596.

[a] Thermal conductivity presents maximum values in the near-critical region, highly dependent on temperature.

aromatic plants, with recoveries that agreed well with the results of hydrodistillation performed during 4 h. Also, Bernardo-Gil et al. (2011) showed that the addition of a mixture of ethanol/water (80:20, v/v) enhanced the extraction of compounds with antioxidant activity from carob kibbles.

For the SCFs, many applications have been suggested including extraction and fractionation (Mukhopadhyay, 2000), material processing (Liu et al., 2012); supercritical antisolvent (SAS) (Catchpole et al., 2004; Sierra-Pallares et al., 2012; Erriguible et al., 2012); micronization (Park et al., 2008; Pham et al., 2012); rapid expansion supercritical solutions (RESS) (Keshavarz et al., 2012); coprecipitation (Subra-Paternault et al., 2012; Montes et al., 2012); cell microencapsulation (Della Porta et al., 2012; Vergara-Mendoza et al., 2012); inactivation of microorganisms (Kim et al., 2007; Ortuño et al., 2012); reactions (Jessop and Leitner, 1999; Brunner, 2004; Doll and Erhan 2005; Lubary et al., 2010; Elbashir et al., 2010; Natalia et al., 2012) cleaning and drying (Mohamed and Mansoori, 2002; Abbas et al., 2008; Pereira and Meireles, 2010; Brown et al., 2010). Some works were published on the removal of undesirable compounds from foodstuff, such as organic pollutants (Antunes et al., 2003; Kim et al., 2011), pesticides (Pearce, 1997; Fiddler et al., 1999; Aguilera et al., 2005), and herbicides (Pensabene, 2000).

9.2   SUPERCRITICAL EXTRACTION OF BIOACTIVE COMPOUNDS

Supercritical fluid extraction (SFE) is the most common application of SCFs, and has received huge attention as an alternative extraction method due to the advantages that this technique presents over the conventional ones, namely the absence of toxic residue in the final product, the easy process of the separation of the solute/solvent, and the recovery of the solvent. However, SFE has a major drawback related to its high initial capital investment (CI), when compared to those of conventional ones (Reverchon and de Marco, 2008). Nevertheless, Rosa and Meireles (2005) for the SFE of clove bud oil and ginger oleoresin, and Leal et al. (2008), for the SFE of sweet basil, showed that the principal factor contributing to the cost of manufacture was the raw material cost. Pereira and Meireles (2007) compared the cost of extracts of rosemary, fennel, and anise obtained by SFE with the cost of essential oils of the same plants obtained by hydrodistillation and verified that the manufacturing costs of extracts produced by SFE were lower than those of essential oils.

An increasing interest has been registered in the extraction of high-added-value compounds such as oils that show antioxidant and pharmaceutical properties (Reverchon and de Marco, 2006; Cossuta et al., 2008; Tang et al., 2011). Several works on the extraction of natural products have been published, such as lipids from olive husk (Esquível and Bernardo-Gil, 1993), walnut (Oliveira et al., 2002), hazelnut (Bernardo-Gil et al., 2002), *Cucurbita ficifolia* seeds (Bernardo-Gil and Lopes, 2004), acorn fruit of *Quercus rotundifolia* L. (Lopes and Bernardo-Gil, 2005; Bernardo-Gil et al., 2007), *Hibiscus cannabinus* L. seed (Chan and Ismail, 2009), coconut (Norulaini et al., 2009b), chia (*Salvia hispanica* L.) (Ixtaina et al., 2010; 2011), coriander seeds (Mhemdi et al., 2011), and grape seeds (Prado et al., 2012). Volatile oils were obtained from *Thymus zygis* L. subsp. *sylvestris* (Moldão-Martins et al., 2000), *Melissa officinalis* L. (Ribeiro et al., 2001), bark of *Cinnamomum zeylanicum* (Marongiu et al., 2007), clove buds (Guan et al., 2007), Italian coriander seeds (Grosso et al., 2008), coriander, winter savory, cotton lavender, and thyme (Grosso et al., 2010), myrtle (Pereira et al., 2012), *Thymus vulgaris* (Grosso et al., 2010), berries and needles of Estonian juniper (*Juniperus communis* L.) (Orav et al., 2010), and coriander seeds (Mhemdi et al., 2011). Phenolic compounds from sweet grass (*Hierochloe odorata*) (Grigonis et al., 2005), lotus germ oil (Li et al., 2009), olive oil mill waste (Lafka et al., 2011), and carob kibbles (*Ceratonia siliqua* L.) (Bernardo-Gil et al., 2011); alkaloids from *Lupinus mutabilis* and *L. albus* (Nossack et al., 2000), and Robusta coffee (*Coffea canephora* var. *Robusta*) husks (Tello et al., 2011); caffeine and theobromine from dry leaves of mate (*Ilex paraguariensis*) (Cardozo et al., 2007); allelochemicals from sunflower leaves (Marsnia et al., 2011); carotenoids from red pepper (*Capsicum frutescens* L.) (Duarte et al., 2004); other colorants from *Alkanna tinctoria* (Akgun, 2011); squalene and phytosterols contained in the deodorization distillate of rice bran oil (Sugihara et al., 2010); tocopherol-enriched oil from Kalahari melon (*Citrullus lanatus*) and roselle (*Hibiscus sabdariffa*) seeds (Nyam et al., 2010); and β-carotene, flavonoids, vitamin A, and α-tocopherol from the microalga *Spirulina platensis* (Wang et al., 2007) were also obtained.

Extracts with high content of carotol were obtained by the SCCO2 extraction of carrot fruit (*Daucus carota* L., cultivar "Chanteney") at optimal conditions of 10 MPa and 40°C (Glišić et al., 2007), as well as from *Cynanchum paniculatum* (Bge.) kitagawa at 15 MPa and 55°C, containing 72% of paeonol at 99.4% purity (Sun et al., 2008). High contents of zerumbone plus α-caryophyllene and camphene were found in SCCO2 extracts obtained at 30°C and 55 MPa from ginger (*Zingiber zerumbet* (L.) Smith) (Norulaini et al., 2009a). Lycopene-rich extracts were obtained from tomato skins by SCCO2 at optimal conditions of 40 MPa and 100°C (Yi et al., 2009). Oil with high antioxidant activity was extracted by SCCO2 at 47 MPa, 46.5°C, and a CO_2 flow rate of 10 kg/h from *Opuntia dillenii* Haw seeds (Liu et al., 2009). Cardanol and phenol from bio-oils (Patel et al., 2011); γ-linolenic acid at high levels from *Microula sikkimensis* seeds oil (Lina et al., 2010); wedelolactone from *Elipta alba* (Savita and Prakashchandra, 2011); hernandulcin from *Lippia dulcis* Trev (Oliveira et al., 2012); and bioactive compounds of amaranth seeds, *Sophora japonica* flower buds, *Stephania rotunda* stems, and *Stevia rebaudiana* leaves (Ikonnikov et al., 2010) were also obtained.

Edible oil with good antioxidant activity was extracted by using SCCO2 at 32 MPa and 45°C from tea (*Camellia sinensis* L.) seeds, the yield being higher than that obtained by Soxhlet extraction (Wang et al., 2011). The *Lepidium apetalum* seed oil extracted by SCCO2 at 30 MPa, 70°C, and about 26 L/h of flow rate presents high antioxidant activity (Xu et al., 2011). Fish oil from different parts of Indian mackerel (*Rastrelliger kanagurta*), rich in ω-3 fatty acids, especially eicosapentaenoic acid (EPA) and docosahexaenoic acid (DHA), was extracted by SCCO2 at 35 MPa and 60°C (Sahena et al., 2010).

Phytosterols from sea buckthorn (*Hippophae rhamnoides* L.) seeds were obtained by SCCO2, the maximum yield of β-sitosterol being found at 15 MPa, 40°C. Both yield and phytosterol concentration in the extract were higher than that obtained with hexane (Sajfrtova et al., 2010). Phytosterols were also extracted from *Typhonium giganteum* Engl. tubers by using SCCO2, β-sitosterol, β-sitosterol-D-glucoside, DL-inositol, and cerebroside being the main components (Li et al., 2011). Uquiche et al. (2012) studied the possibility of fractionating minor lipids (sterols, tocopherols, carotenoids) from cold-pressed rapeseed cake by using SCCO2, having verified an increase in the antioxidant activity of the oil in the second half of the extraction, as compared to the first half, suggesting that SCCO2 extraction could be used to isolate vegetable oil fractions with different functional values.

Lutein and β-carotene were extracted by SCCO2 from a freeze-dried powder of the marine microalga, *Scenedesmus almeriensis*, but a pressure of 40 MPa and a temperature of 60°C were necessary to obtain a significant yield of pigments (Macías-Sánchez et al., 2010). Lutein and zeaxanthin from daylily (*Hemerocallis disticha*) flowers could be obtained by SCCO2 extraction at 80°C and 60 MPa, the extracts presenting strong antioxidant activity (Hsu et al., 2011). Lu et al. (2011) recovered the aroma compounds from Zhenjiang aromatic vinegar, highly prized in China, by SCCO2 extraction at 35 MPa and 50°C. The essential oil of *Ferulago angulata*, which is used as food preservative, was extracted by SCCO2, the highest yield being obtained at 19 MPa and 35°C (Sodeifian and Ansari, 2011). Grevenstuk et al. (2012) studied the possibility of extracting high-purity plumbagin from micropropagated *Drosera intermedia* plants, having verified that, although the highest extraction yield was obtained with SCCO2 extraction, the amount of plumbagin recovered (2.5 ± 0.1 mg/g plant) was inferior to the one obtained with ultrasonic-assisted extraction (2.74 mg of plumbagin/g plant), meaning that other undesired compounds were coextracted. Costa et al. (2012) obtained extracts from *Thymus lotocephalus* with high antioxidant activity by using SCCO2 at 40°C and 12 and 18 MPa. Extracts from the leaves of *Baccharis dracunculifolia*, showing high antioxidant activity, and high content of Artepillin C, were obtained by SCCO2 at 40 MPa and 60°C (Martinez-Correa et al., 2012). An extract with high yield, high antioxidant activity and high thymoquinone quantity was obtained by SCCO2 at 35 MPa and 60°C from *Nigella sativa* L. seeds (Solati et al., 2012). The filamentous fungi *Mortierella alpine*, which is a potent source of arachidonic acid (ARA), an important polyunsaturated fatty acid (PUFA) of the n − 6 series, was extracted by SCCO2, the highest extraction rate being obtained at 25 MPa and 50°C (Nisha et al., 2012).

Brunner (2005) states that some products that are part of our daily diet can be produced by SCF technology. Examples of those products are: decaffeinated coffee and tea; flavor-enhanced orange juice; vitamin and antioxidant additives; dealcoholized wine and beer; defatted meat, potato chips, and French fries; or flavor-enhanced distillates.

9.3 SUPERCRITICAL EXTRACTION PROCESSING

In the SFE of a solid, four steps can be considered: compression and temperature adjustment of the CFC; extraction; expansion; and extract/solvent separation. When the extraction occurs, the solvent penetrates into the cells, dissolves the compounds that are transferred to the solid surface by diffusion; the solvent with the extract flow to the exterior of the extractor, and the separation occurs by changing the pressure and/or temperature conditions.

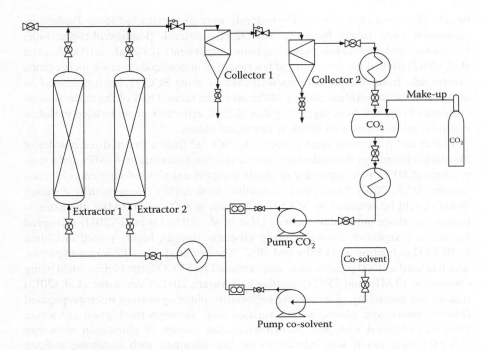

FIGURE 9.1   SFE process flow diagram.

A schematic representation of an SFE process flow diagram is illustrated in Figure 9.1. The process starts with the adjustment of the pressure and the temperature to the operating conditions, by using a solvent pump and a heat exchanger, the supercritical carbon dioxide flows through the bed with the substratum and the components are extracted, the mixture of SCCO2 and extract being subjected to rapid decompression in one or more collectors, which can be at different values of temperature and pressure. The decompression causes the separation of the extract, and the CO_2 is subjected to cooling and compression that is recirculated. When a cosolvent is used, another pump exists to introduce it in the process. The existence of two or more collectors operating in series, at different pressures and temperatures, allows the fractionation of the SCFs extracts in fractions of different composition by setting adequate temperature and pressure (Senorans et al., 2000; Gaspar et al., 2003; Bensebia et al., 2009). SCCO2 extracts from plants of the *Lamiaceae* family, namely, oregano (*Origanum vulgare*), thyme (*Thymus zygis*), sage (*Salvia officinalis*), and rosemary (*Rosmarinus officinalis*) were collected in two separators (Fornari et al., 2012). Oregano extract was mainly recovered in the first collector, sage and thyme extracts in the second separator, and similar amounts of rosemary extract were recovered in both separators.

To promote the fractional separation of compounds from the raw material, it is also possible to perform extractions in successive steps with varying pressure and/ or temperature (Ibáñez et al., 1999; Gaspar et al., 2003; Smith et al., 2003; Ramirez et al., 2004). This procedure is useful to avoid the co-extraction of compounds of the same family that have different solubilities in SCCO2, or when different mass transfer resistances occur in the solid matrix. For example, it is possible to perform a first extraction operating at low CO_2 density (e.g., 290 kg m^{-3}, 9 MPa, 50°C) followed by a second extraction step at high CO_2 density (e.g., 870 kg m^{-3}, 30 MPa, 50°C). The most-soluble compounds (such as the volatile oils) are extracted during the first step, whereas the less soluble compounds (lipids, antioxidants, pigments) are extracted in the second step. Marzouki et al. (2008) used two steps to extract oils from the berries of *Laurus nobilis* L. by SCCO2 at 40°C: in the first step, at 9 MPa, the volatile oil was obtained; after, in the second step, at 25 MPa, the extract content was mainly triacylglycerols. A similar process was used by Silva et al. (2009) for the extraction of compounds from winter savory (*Satureja montana*); the volatiles being extracted at 9 MPa, and the nonvolatile fractions, at 25 MPa, showed a high content of (+)-catechin; and chlorogenic, vanillic, and protocatechuic acids. Limonoids and naringin were extracted from grapefruit (*Citrus paradise* Macf.) seeds by SCCO2 extraction in two steps: limonin being extracted at 48.3 MPa and 50°C; the highest yield of limonin-17-β-D-glucopyranoside being achieved at 41.4 MPa, 60°C, and using 30% ethanol as cosolvent; and the naringin obtained at 41.4 MPa, 50°C, and using 20% ethanol (Yu et al., 2007). Rosemary (*Rosmarinus officinalis* L.) was subjected to a three-step SCCO2 extraction, the first step at 40°C and 10 MPa, the second step at 20.7 MPa and 40°C, and the third step at 19.3 MPa, and 55°C, with 5% ethanol. The antioxidant activity of the first two extracts was identical, but the highest total phenolics content was observed in the oil extracted by using ethanol as a modifier (Irmak et al., 2010).

Grigonis et al. (2005) compared different extraction techniques for the isolation of antioxidants from sweet grass (*Hierochloë odorata*), and concluded that a two-step SFE was the preferable method. The first step was performed from dried herbs with SCCO2 at 35 MPa, 40°C, and 20% of ethanol, and the crude extract was submitted to SFE, in a second step, at 25 MPa and 40°C.

9.4   EFFECTS OF TECHNOLOGICAL PARAMETERS

The technological parameters of SFE are pressure and temperature, which determine the SCF density, and the solubility of desired compounds in the fluid; the solvent flow rate; extraction time; relation height/diameter of the extractor; structure of the matrix, and its pretreatment; and the use of modifiers. Several authors, such as Bernardo-Gil and Lopes (2004), Ma et al. (2008), Li et al. (2010), Mariod et al. (2010), Herrero et al. (2010), Xu et al. (2011), Ghoreishi and Bataghva (2011), and Bimakr et al. (2012), have used experimental design techniques to optimize some of these variables.

Operating at high pressures, the solvent power of the SCFs is enhanced, but its selectivity is low. The compromise between these two variables should be taken into account to extract compounds of interest, reducing the coextraction of the undesired compounds (Reverchon, 1997). In addition, the solubility of solid compounds in supercritical fluid could be influenced by the repulsive solute–fluid interaction (Ma et al., 2008). With the increase of the SCFs densities due to the rise of pressure, solubility generally increases. However, as the pressure continues to rise, the repulsive solute–fluid interaction keeps increasing, surpassing the increase of solubility. This interaction thus contributes to the effective lowering in the solubility of the compounds and, therefore, to the decrease of the extraction yields. So, the solubility of a solute in SCFs depends on a complex balance among fluid density, solute vapor pressure, and the repulsive solute–fluid interaction, which are controlled by the temperature and the pressure.

The increase of temperature at a fixed pressure reduces the SCF solvent power but increases the vapor pressure of the compounds to be extracted, which favors its passage to the fluid phase. The improvement of solubility with temperature is dependent on which effect is more important. If the density effect is predominant, the solubility of compounds in the supercritical phase should decrease at higher temperatures. In case the vapor pressure is overwhelming, the solubility would increase with an increase in the temperature (Erkucuk et al., 2009). Some care must be taken with the increase of temperature, since the bioactivity of natural extracts may decrease when subjected to high processing temperatures. When the compounds are thermolabile, the temperature must be fixed between 35°C and 60°C to avoid degradation (Reverchon and de Marco, 2008).

Solvent flow rate and extraction time are also important since these parameters influence the extraction rate, and the percentage of extracted material, being connected to the kinetics of the extraction, which depends on mass transfer resistances, and the solubility of material in the SCFs. Nobre et al. (2009) show that the content of *trans*-lycopene obtained by SFE of Portuguese tomato waste increased with an increase of the solvent flow rate. Lee et al. (2010) also verified that the extraction rate of nobiletin and tangeretin from *Citrus depressa* hayata, increased by increasing the CO_2 flow rate, using SCCO2 with ethanol as the modifier. The influence of the solvent flow rate on the extraction of hazelnut oil was investigated by Bernardo-Gil et al. (2002), and optimum value was found, for the range of the studied conditions. Similar conclusions were drawn by Nyam et al. (2011) for the SFE of Kalahari melon seeds oil, the flow rate being the most significant parameter, this factor having no significant effect on the phytosterol content in the oil. Wang et al. (2012) verified that the flow rate of SCCO2 had great influence on the extraction yield of volatile oil from *Cyperus rotundus* L.

Particle size of the raw material is an important parameter mainly when the concentrations of the desired compounds are low, and also when internal mass transfer controls the extraction. Smaller particles increase the contact area, improving the diffusion of SCFs. As examples, the case of SFE of paprika (Nagy and Simándi, 2008), or the recovery of *trans*-lycopene from Portuguese tomato industrial waste (Nobre et al., 2009), can be mentioned. However, if the particles are too small, they can cause channeling problems inside the extraction bed, the fluid does not contact all the material to be extracted, and the efficiency can be poor, causing a decrease in yield. This effect was observed by Oliveira et al. (2002) in the SFE of walnut oil. The grinding of plant material to very small particles can also lead to a loss of volatile components. The most usual pretreatment is milling to improve the cellular breakdown, and also hydration to promote the swelling of the material (e.g., decaffeination of coffee); dehydration can also be used, depending on the structure of the substratum and the location of the compounds to be extracted, which influence the mass transfer resistances.

The quantity of water in the raw material is also a parameter to be taken into account, since, if it is high, it can lead to coalescence of particles of the milled material, producing particles of different sizes along the bed, which can cause irregular extractions. Similar to grinding, drying can also promote the loss of volatile components. However, water can have the effect of a cosolvent. Ge et al. (2002) verified that the increase of the water content to about 5% in wheat germ promotes the increase in the yield of vitamin E in the oil, but with an increase in the moisture content above that value, a decrease was observed. Nagy and Simándi (2008) observed that the increase of the water content in paprika had a negligible effect on the extractability of the oil, but, when the moisture content was above 18%, the extraction efficiency decreased. Nobre et al. (2009) showed that an increase in the moisture content of the samples of Portuguese tomato industrial wastes from 4.6% to 22.8% led to a rise in the extraction yield and to a decrease in the recovery of *trans*-lycopene, but, at higher moisture contents, both yield and recovery decreased. The increase in the moisture content of *Helichrysum italicum* flowers enhanced the solubility of solute in SCCO2, enabling a higher rate of extraction and thus decrease in SCCO2 consumption necessary to achieving the desired extraction yield (Ivanovic et al., 2011). Sun et al. (2002) verified that the alkylamide yields from fresh roots of *Echinacea angustifolia* (moisture content 75.3%) were significantly lower than those from dried roots, while air-dried roots (moisture content 8.4%) and freeze-dried roots (moisture content 4.9%) yielded similar amounts of extracts. Similar conclusions were obtained by Snyder et al. (1984) on the extraction of soybeans at moisture levels of 3% and 12%, and by Dunford et al. (1997) on the extraction of oil from Atlantic mackerel, in which they found that the oil yields were similar, for mackerel containing either 10.2% moisture, or 3.8% moisture. Tello et al. (2011) recovered caffeine of at least 94% purity, from robusta coffee husks, by using SCCO2 at 30 MPa and 100°C, after wetting the coffee husks. The results indicated that increasing the amount of water up to 32% resulted in higher extraction yields, but exceeding this percentage was harmful. Araújo and Sandi (2007) verified that it was possible to obtain valuable oil from green coffee beans (humidity 9.98%), with high levels of diterpenes, namely cafestol and kahweol, for cosmetic uses, extracted with SCCO2. From roasted coffee beans with a humidity of about 2.4%, healthy coffee oil with low levels of diterpenes can be obtained for the food industry. The swelling effect on the extraction of several plant families was studied by Stamenic et al. (2010).

9.5 SUB- AND SUPERCRITICAL SOLVENTS

Although CO_2 is the most used solvent on unit operations involving supercritical fluids, many other compounds have been used as SCFs (Table 9.2).

Besides its high cost, ethane was studied for application as SFC due to its critical temperature (32.3°C) close to that of CO_2 and critical pressure (4.87 MPa) lower that of CO_2. Mendes et al. (1999) studied the solubility of β-carotene in SCCO2 and in ethane. Raeissi et al. (2002) studied the possibility of the use of ethane to obtain linalool as part of deterpination of citrus essential oils. Mohamed et al. (2002) extracted caffeine, theobromine, and cocoa butter from cocoa beans using SCCO2 and ethane.

TABLE 9.2

Physical Properties of Some Gases Used as Supercritical Fluids

Solvent	Molecular Weight (g mol⁻¹)	Normal Boiling Point (°C)	Critical Temperature (°C)	Critical Pressure (MPa)	Critical Density (kg m⁻³)
Carbon dioxide	44.01	−78.5	31.1	7.38	469
Acetone	58.08	56.0	235.1	4.70	278
Ammonia	17.03	−33.4	132.5	11.3	240
Benzene	78.11	80.1	289.0	4.89	302
Ethane	30.07	−88.0	32.2	4.87	203
Ethanol	46.07	78.4	240.9	6.14	276
Ethylene	28.05	−103.7	9.3	5.04	215
Methane	16.04	−161.2	−82.7	4.60	162
Methanol	32.04	64.8	239.6	8.09	272
Nitrous oxide	44.02	−89.0	36.5	7.10	457
Propane	44.09	−44.5	96.7	4.25	217
Propylene	42.08	−47.7	91.9	4.60	232
Toluene	92.13	110.0	318.6	4.11	290
Water	18.015	100.0	374.2	22.06	322

Source: Adapted from Klesper, E. 1980. In *Extraction with Supercritical Gases,* Verlag Chemie, Weinheim, Germany.

At the same conditions of temperature and pressure, the amount of cocoa butter extracted with ethane was much larger than the one extracted with SCCO2.

Extractions of whole fruit, seeds, and peel of rosehip were studied with propane, CO_2, and propane+ CO_2, as solvents under super- and subcritical conditions, biologically active compounds such as carotenoids and tocopherols were preferentially extracted by SCCO2 alone or with propane (Illés et al., 1997). Propane and CO_2, under sub- and supercritical conditions were used to produce extracts from coriander seeds, a complete oil recovery being attained with propane or propane-rich solvents at 25°C and 5, 8, and 10 MPa; while supercritical conditions are essentially required for the complete extraction of essential oils, mild subcritical conditions with CO_2, CO_2+ propane, and propane are sufficient to extract tocopherols and triglycerides from coriander seeds (Illés et al., 2000). Gnayfeed et al. (2001) compared the extraction of pungent paprika by using SCCO2 and subcritical propane, the yield of paprika extract being affected by the extraction conditions with SCCO2, but fairly constant at different conditions with subcritical propane. Catchpole et al. (2003) studied the extraction of chili, black pepper, and ginger with near-critical CO_2, propane, and dimethyl ether; subcritical dimethyl ether being effective at extracting the pungent principles from the spices as SCCO2, although a substantial amount of water was also extracted; whereas subcritical propane was the least effective solvent. The yields of capsaicins obtained by SCCO2 and dimethyl ether were similar and approximately double of the one produced by propane extraction; and the yield of piperines obtained from black pepper by extraction using propane was about 10% of the one achieved with dimethyl ether and SCCO2. Hamdan et al. (2008) extracted oil from cardamom by using CO_2 at supercritical conditions (35–55°C and 10–30 MPa), as well as subcritical CO_2 (25°C and 8–10 MPa), and subcritical propane (2–5 MPa and 25°C), the maximum yield (~7.2%) being obtained with propane at 5 MPa and 25°C, higher than the one obtained with SCCO2 (~6.6%) at 30 MPa and 35°C, and no significant increase in the yield was observed with the addition of ethanol (25%) to SCCO2. Carrara et al. (2012) compared the pyrrolidine alkaloid content extracted by SCCO2 and by compressed propane from *Piper amalago* L. leaves and verified that SCCO2 at 40°C and 12.55 MPa showed the highest selectivity.

Denery et al. (2004) studied carotenoid extraction from the microalgae, *Haematococcus pluvialis* and *Dunaliella salina*, using pressurized ethanol, and the extraction yield was shown to be similar to those obtained using traditional extraction techniques. Compounds from *Phormidium* microalgae species were obtained by using pressurized hexane, ethanol, and water; hexane and ethanol extracts showing higher antioxidant capacity, which was mainly attributed to carotenoids (Rodríguez-Meizoso et al., 2008). A pressurized mixture of ethanol and ethyl lactate was used to obtain γ-linolenic acid from *Spirulina—Arthrospira platensis* (Golmakani et al., 2012).

Catchpole and Proells (2001) measured the solubilities of squalene, oleic acid, soya oil, and deep-sea shark liver oil in subcritical R134a, the results showing that the solubilities increased almost linearly with increasing temperature at fixed pressure, and increased logarithmically with increasing pressure at fixed temperature. They concluded that R134a could be used as an alternative to supercritical CO_2. Mustapa et al. (2011) studied the extraction of β-carotene from palm oil mesocarp using subcritical R134a, the palm oil yield increasing with pressure and temperature, and the maximum solubility of β-carotene was obtained at 10 MPa and 60°C, while the lowest solubility occurred at 8 MPa and 40°C.

9.5.1 SUBCRITICAL WATER

Subcritical water extraction, also known as high-pressure water (HPW) extraction, or superheated water extraction, has recently emerged as a promising technology, not only for the concentration of bioactive compounds such as antioxidants, lignans, and anthocyanin from natural materials, but also in sample preparation to extract organic contaminants from foodstuffs, for food safety analysis, and solids/sediments for environmental monitoring purposes (Shi et al., 2012). HPW is used at temperatures between 100°C and 374°C (the latter being the water critical temperature) and under pressures high enough to maintain the liquid state (usually from 1 to 6 MPa). With HPW, it is possible to selectively extract different classes of compounds, depending on the temperature used, with the more polar compounds extracted at lower temperatures, and the less polar compounds extracted at higher temperatures (Shi et al., 2012). The most important factor to consider in this type of extraction procedure is the variability of the dielectric constant with temperature. Water at room temperature is a very polar solvent, with a dielectric constant close to 80. However, this value can be significantly decreased to values close to 27 when water is heated up to 250°C, while maintaining its liquid state, by applying the appropriate pressure.

In general, the use of subcritical water extraction provides a number of advantages over traditional extraction techniques (i.e., hydrodistillation, organic solvents, solid–liquid extraction). These are, mainly, low extraction times, higher quality of the extracts (mostly for essential oils), lower costs of the extracting agent, and it is an environmentally compatible technique (Herrero et al., 2006). HPW was used for the extraction of antioxidant compounds, such as eugenol and eugenyl acetate from clove (*Syzygium aromaticum*) (Rovio et al., 1999 and Clifford et al., 1999); rosmarinic and carnosic acids, carnosol, and methyl carnosate from sage (Ollanketo et al., 2002); carnosol, rosmanol, carnosic acid, methyl carnosate, and some flavonoids such as cirsimaritin and genkwanin from rosemary leaves (Ibañez et al., 2003); gallic acid, ellagic acid, and corilagin from *Terminalia chebula* Retz fruits (Rangsriwong et al., 2009); quercetin-3-galactosidase, kaempferol, and isorhamnetin, from seabuckthorm (Kumar, 2011), and curcumin from turmeric rhizomes (Euterpio et al., 2011). Antioxidants from canola meal (Hassas-Roudsari et al., 2009), and from pomegranate (*Punica granatum* L.) seed residues (He et al., 2012) were also extracted by using HPW, which also proved to be efficient on the extraction of anthocyanins from the freeze-dried skin of a highly pigmented red wine grape (Ju and Howard, 2003); for the extraction of thermally labile and reasonably polar components such as berberine in coptidis rhizoma, glycyrrhizin in radix glycyrrhizae/liquorice, and baicalein in scutellariae radix (Ong and Len, 2003); and for the extraction of lactones from a kava (*Piper methysticum*) root (Kuvátová et al., 2001b). Antioxidants were also extracted by HPW from the microalga *Spirulina platensis* (Herrero et al., 2004), carotenoids from *Haematococcus pluvialis* and *Dunaliella salina*, kavalactones from *Piper methysticum* (Denery et al., 2004), and short-chain fatty acids (responsible for antimicrobial activity), vitamin E together with simple phenols (responsible for antioxidant activity) from *Haematococcus pluvialis* microalga in red phase (Rodríguez-Meizoso et al., 2010).

Aroma compounds from *R. officinalis* (Basile et al., 1998), from laurel leaves (Fernández-Pérez et al., 2000), from fennel (*Foeniculum vulgare*) (Gámiz-Gracía and Luque de Castro, 2000), from oregano (*Lippia graveolens*) leaves (Ayala and Luque de Castro, 2001), from *Satureja hortensis* and *Mentha piperita* (Kuvátová et al., 2001a), from *Thymbra spicata* (Ozel et al., 2003), from *Zataria multiflora* Boiss (Khajenoori et al., 2009), and from *Bunium persicum* Boiss (Mortazavi et al., 2010) were also obtained by extraction with subcritical water.

Damnacanthal, an anthraquinone that has been proven to have antiviral, antibacterial, anticancer activities, was extracted from the dried root of *Morinda citrifolia*, by using HPW, best results being obtained at 4 MPa and 170°C (Anekpankul et al., 2007). Saponins and cyclopeptides were extracted from cow cockle seeds by using pressurized water (Güçlü-Üstündağ et al., 2007; Güçlü-Üstündağ and Mazza, 2008). Lignans, proteins, and carbohydrates from defatted flaxseed meal (Ho et al., 2007), and pectin contained in the flavedo of *Citrus junos* fruits (Ueno et al., 2008) were also extracted by pressurized low-polarity water. Polysaccharides were obtained by pressurized water extraction from *Lentinula edodes* with about 90% recovery (Lo et al., 2007).

Buranov and Mazza (2009) verified that pressurized low-polarity water, pressurized aqueous ethanol, and pressurized aqueous ammonia were efficient in the extraction of ferulic acid and vanillin from flax shives, wheat bran, and corn bran. Phenolic compounds from flax shives were also obtained by pH-controlled pressurized low-polarity water (Kim and Mazza, 2009). Abdelmoez et al. (2011) used subcritical water for the extraction of cottonseed oil, the results showing that the optimum temperature, mean particle size, water:seed ratio, and extraction time were 270°C, <0.5 mm, 2:1, and 30 min, respectively, and the extracted oil was identical to that extracted using the traditional hexane method.

Pawlowski and Poole (1998) used HPW to extract two fungicides, thiabendazole (TBZ) and carbendazim (MBC), from agricultural commodities including banana pulp, whole lemons, orange pulp, mushrooms, and rice at extraction temperatures below 100°C and an extraction pressure of 5 MPa.

The addition of organic solvent such as ethanol into the extraction with PHWE—pressurized hot water extraction—will result in higher recoveries for the compounds in medicinal plants (Ong et al., 2006). Curren and King (2001) showed that the use of subcritical water with 30% ethanol, as modifier, was adequate for the complete removal of atrazine from beef kidney.

9.6 EFFECTS OF THE USE OF COSOLVENTS ON THE SFE OF BIOACTIVE COMPOUNDS

As mentioned, the solubility of the compounds to be extracted is one of the parameters that influence the extraction yield the most. There are many data on the solubility of compounds in SCCO2, but data on the solubility of natural mixtures, which are multicomponent systems, are scarce. However, the intermolecular interactions between components can significantly alter the selectivity of the SCCO2. The performance of a modifier can be based on this concept. In reality, the addition of a cosolvent to SCCO2 resulted in an enhancement of the solubility of the targeted component. Several compounds have been studied to be applied as cosolvents or modifiers (Modey et al., 1996; Jeong and Chesney, 1999). The addition of those compounds to SCCO2 can improve the extraction efficiency by increasing the solubility of solutes especially for highly polar compounds, or for compounds with high molecular mass. As an example, the case of the extraction of β-carotene from freeze-dried carrots can be mentioned, in which the optimum extraction conditions were obtained with the use of 10% ethanol as cosolvent. Although β-carotene is a nonpolar compound, its solubility in CO_2 is small, because it is a large molecule (Barth et al., 1995).

The type of cosolvent to be used is determined by taking into account the affinity between the SCF, the cosolvent, the substratum, and the compounds to be extracted. The amount of cosolvents involved in extraction is also an important factor for the optimization of the operating conditions, as the cosolvent in excess may cause either negative or uneconomical effects (Shi et al., 2009). In general, the amount of cosolvent to be added varies from 1% to 15% (Pereira and Meireles, 2010).

In some cases, up to a certain value, the addition of cosolvent has no further effect on the extraction of different bioactive compounds, but in others, an optimum value exists. Berna et al. (2001) showed that the solubility of cathechin was enhanced with the increase in the amount of ethanol as cosolvent, at different pressures. The same effect was observed by Berna et al. (2001) and Cháfer et al. (2002) for the extraction of epicathechin, and by Cháfer et al. (2004) for the extraction of quercetin. The addition of small amounts of ethanol (2 vol%) to SCCO2 enhanced the extraction yield of jojoba seed oil from the seed matrix as compared to SCCO2 extraction, but higher concentrations of ethanol has no effect (Salgin, 2007). The amounts of genistein and daidzein (soy isoflavones) increased with increasing the amount of ethanol, as cosolvent, in the SFE of okara oil (Quitain et al., 2006). The yield of gallic acid extracted by SFE with ethanol, at 40°C, has a small effect at pressures of upto 30 MPa, but presents a highly increased value at 40 MPa (Cháfer et al., 2007). The SFE of *Ginkgo biloba* leaves was studied by Yang et al. (2002), which verified that the yields of total flavonoids and total terpenoids of ginkgo biloba extracts have an optimum value at 30 MPa and 60°C, with about 5% of ethanol as cosolvent. The mixture of ethanol and water (80:20, vol%) was used as cosolvent by Bernardo-Gil et al. (2011) for the SFE of carob kibbles.

Aqueous ethanol and aqueous methanol were used as modifiers to extract poly-methoxylflavones from *Citrus depressa* hayata peel by Lee et al. (2010), their results showing that, with aqueous methanol, the yield of nobiletin and tangeretin was relatively lower than the amount obtained when aqueous ethanol was used; besides, they have verified that the yield of nobiletin and tangeretin increased with the increasing percentage of water in ethanol of up to 15% (v/v), and decreased again with the increasing amount of water. A similar effect was found by Li et al. (2010) on the SFE of kaempferol glycosides from tea seed cake by using mixtures of ethanol:water as modifiers; having an optimal concentration of ethanol between 60% and 70%, and a rather sharp drop in yield as the ethanol concentration approaches either 50% or 90% (v/v). Ashraf-Khorassani et al. (1995) have showed that methanol was more effective as cosolvent than acetonitrile, acetone, water, or dichloromethane in the SFE of phenolic acids. Chun et al. (1996) studied the influence of methanol, ethyl acetate, dichloromethane, diethyl ether, and dichloromethane, as modifiers, on the SFE of paclitaxel (taxol) and baccatin III from the needles of *Taxus cuspidate*, and it was found that dichloromethane was the most effective one for the extraction of taxol, but for the extraction of baccatin III, diethyl ether was the best. Supercritical extraction of glycosides from grapes was studied by using methanol and ethyl acetate as modifiers: the recovery was 16% when pure SCCO2 was used, whereas the recovery increased to 58.8% if 15% methanol was added; and with ethyl acetate, the recovery was threefold lower than with methanol (Palma et al., 2000). The triterpenes gin-senosides were extracted by Wood et al. (2006) from the root of North American ginseng, using SCCO2 and methanol or dimethylsulfoxide (DMSO) as entrainers; up to 90% of the total gensenosides extracted by methanol, in Soxhlet, were extracted at pressures of 20.7–48.3 MPa using SCCO2 and methanol. Wang et al. (2001) have studied the SFE of Korean ginseng root hair before by using SCCO2 and aqueous ethanol as cosolvent, having verified that 55% of the ginsenosides normally extracted by conventional Soxhlet during 12 h were recovered after 4 h of SCCO2 extraction at 31.2 MPa, 60°C, with 6% of ethanol:water mixture.

Isopropanol was used as cosolvent with SCCO2 for the recovery of tagatose from mixtures with galactose, and a recovery of about 75% tagatose with purity >90% was obtained at 30 MPa, 80°C and 25% (v:v) of isopropanol (Montanés et al., 2006). The use of isopropyl alcohol and ethanol, as cosolvents, showed to be efficient on the extraction of antioxidant compounds from guava (*Psidium guajava*), when compared with conventional extraction processes (Moura et al., 2012).

Ollanketo et al. (2001) studied the SCCO2 extraction of lycopene from tomato skins, by using acetone, methanol, hexane, ethanol, dichloromethane, and water, as modifiers; methanol and acetone achieving better results, 100% of total lycopene being recovered at 110°C, 40 MPa, in 50 min, but 94% of recovery was achieved in just 15 min with acetone.

Choi et al. (1999) verified that hyoscyamine and scopolamine, which are important tropane alkaloids acting as parasympatholytic substances, were soluble in SCCO2, but it was not possible to extract them from *Scopolia japonica* Maxim. The extraction with SCCO2 modified with methanol, water, 10% (v/v) diethylamine in methanol, and 10% (v/v) diethylamine in water was then studied at 60°C and 34.0 MPa: methanol (1–10%) proved to be the best option for the extraction of hyoscyamine; however, to obtain higher amounts of scopolamine, the best option was water (5–10%). The results of SCCO2 extraction of volatile components from *Salvia mirzayanii* showed that, under pressure of 35.5 MPa, temperature of 35°C, 6% methanol, the linalyl acetate was selectively extracted (Yamini et al., 2008). Extracts of *Aloe vera* leaf skin obtained by SCCO2 at 50°C, 35 MPa and 20% of methanol, presented high antioxidant activity (Hu et al., 2005). Coumarin from *Cuscuta reflexa* was extracted by using SCCO2 associated with a small amount of methanol (Mitra et al., 2011). Carminic acid was obtained by SFE from cochineal (*Dactylopius coccus*), by using, as cosolvents, mixtures of methanol/water and ethanol/water at 100 and 150°C, respectively (Borges et al., 2012). Ghasemi et al. (2011) proved that the SFE using SCCO2 and methanol, as modifier, is more selective than the conventional hydrodistillation for the extraction of essential oils from *Myrtus communis* L. leaves. Similar results were obtained by Khajeh et al. (2010) for the extraction of essential oil of *Nepeta persica*, by using SCO2 and methanol (1.5% v/v) at 20.3 MPa and 45°C. The SFE of essential oil from *Satureja hortensis* confirmed that the use of methanol (8.6% v/v), as cosolvent of SCCO2 (35 MPa, and 72.6°C) required shorter extraction time (15 min) when compared to hydrodistillation (4 h) (Khajeh, 2011). SCCO2 and hydromethanolic cosolvent 10% (v/v) were used to extract lipids from two strains of bifidobacteria—*Bifidobacterium longum* and *B. angulatum* at 25 MPa and 45°C (Izhyk et al., 2012). Methanol, as a modifier of SCCO2, was also used for the determination of nitrosamines in sausages at pressures of 12.9–36 MPa with recoveries, respectively, of 21–82% (Sanches Filho et al., 2007).

Methanol is the solvent most commonly used as a modifier for several matrices, because it is up to 20% miscible with CO_2 (Lang and Wai, 2001). However, it is toxic, and though ethanol is not as polar as methanol, can be a good choice in the SFE of natural products (Pereira and Meireles, 2010). However, other foodgrade solvents such as water and lipids, beyond ethanol, can be used as cosolvents to increase solubility and enhance extract yield.

Water was used as a cosolvent of SCCO2 to obtain extracts from *Ocimum basilicum* (sweet basil), in which eugenol, germacrene-D, malic acid, tartaric acid, ramnose, kaempferol, caffeoylquinic, caffeic, and quinic acids were identified (Leal et al., 2008). Water was also used as a cosolvent for obtaining biocompounds from sunflower leaves (Casas et al., 2008). Huang et al. (2007) studied the influence of the addition of water to SCCO2 to remove caffeine and retaining catechins from green tea powder and concluded that, when SCCO2 was used at 60°C and 30 MPa, 91.5% of the caffeine was removed and 80.8% of catechins were retained in the tea. Identical study was made by Park et al. (2007), who verified that, when the dry green tea leaves were extracted with SCCO2 modified with 95% (v/v) ethanol, at 30 MPa and 70°C, for 120 min, the caffeine content in the decaffeinated green tea leaves was reduced to 2.6% of the initial content, but a substantial loss of epigallocatechin gallate, as much as 37.8% of original content, proved unavoidable.

Coniferyl ferulate from *Angelica sinensis* was obtained by SCCO2 at 35 MPa, 40°C. Among the three solvents used (aqueous ethanol, ethanol, and ethyl acetate), ethyl acetate was selected as the modifier because, with it, the highest extraction efficiency was obtained (Xie et al., 2009). Ethyl acetate, as entrainer, was also used in the SCCO2 extraction of trace amounts of amphenicols (chloramphenicol, florfenicol, and thiamphenicol) in shrimp (Liu et al., 2010). SCCO2 modified with ethyl acetate or ethanol was used to obtain the phenolic fraction from guava seeds (*Psidium guajava* L.), the antioxidant potential of the extracts indicating ethanol (10% w/w) as the best cosolvent in SFE at 50°C and 30 MPa (Castro-Vargas et al., 2010). Volatile compounds of *Mentha spicata* L. were extracted by using SCCO2 at 50°C and 30 MPa, the highest yield (2.38% w/w) was obtained when ethanol was used as a cosolvent, as compared to the one obtained with the use of ethyl acetate (Almeida et al., 2012).

Flavonoids and phenolics from *Ampelopsis grossedentata* stems were obtained by SFE at best conditions of 25 MPa, 40°C, and using as modifier mixtures of methanol/ethanol (1:3, v/v), and (1:1, v/v) for the extraction of flavonoids and phenolics, respectively (Wang et al., 2011).

Rajaei et al. (2005) showed that the yield of tea seed oil obtained using SCCO2 and ethanol was similar to or higher than the ones obtained by other methods, such as Soxhlet or ultrasonic extraction. Tanaka et al. (2004) used SCCO2+ethanol to extract phospholipids from salmon fish roe, being verified that the amount of extracted phospholipids increased with raising the addition of ethanol; more than 80% of the phospholipids were recovered when the extraction was performed at 17.7 MPa, 33°C, and 20% ethanol. Bensebia et al. (2009) studied the effect of ethanol on the SFE of rosemary oil. Leal et al. (2003) studied the SFE of ginger (*Zingiber officinalis* Roscoe), rosemary (*R. officinalis* L.), and turmeric (*Curcuma longa* L.), by using SCCO2 without and with ethanol or isopropyl alcohol as the cosolvent, the extracts presenting antioxidant, anticancer, and antimycobacterial activities, rosemary extract exhibiting the strongest antioxidant activity, followed by the ginger and turmeric extracts. Carotenoids, namely β-carotene, were obtained by SCCO2 and ethanol from freeze-dried fruits of *Rosa canina* L. (Tozzi et al., 2008). Ethanol was added (5 wt%) as a cosolvent to increase the polarity of the SCCO2, thus favoring the extraction of xanthones present in mangosteen (*Garcinia mangostana* L.) fruit pericarp, the maximum extraction yield (15 wt%) being achieved at 28 MPa, 50°C, and an extraction time of 8 h, while without cosolvent, the yield was 7.5 wt% (Zarena et al., 2012). Studies demonstrated that the SCCO2 extraction of limonoid glucosides from grapefruit molasses at 48.3 MPa, and 50°C, with 10% ethanol as entrainer, has practical significance for commercial production (Yu et al., 2006).

Cajaninstilbene acid and pinostrobin from pigeonpea (*Cajanus cajan* L. Mill sp.) leaves were extracted by using SCCO2 at 30 MPa, 60°C, and 80% ethanol (Kong et al., 2009). Flavonoids from *Pueraria lobata* (Wang et al., 2008), and phenolic compounds from Macela (*Achyrocline satureioides*) (Takeuchi et al., 2010) were obtained by SFE using mixtures of CO_2 and ethanol. The bioactive flavonoid compounds (catechin, epicatechin, rutin, luteolin, myricetin, apigenin, and naringenin) of spearmint (*Mentha spicata* L.) leaves were extracted by using SCCO2 and ethanol at the optimal conditions of 20.9 MPa, 50°C, and 7.39 g/min cosolvent (Bimakr et al. 2012). Cerón et al. (2012) obtained antioxidant compounds from Andes berry fruits (*Rubus glaucus* Benth), by using SCCO2 and ethanol, the anthocyanin pigments yields improved up to 59.3% when compared to the yields obtained by the traditional process. Alkylamides, mainly spilanthol, were obtained from jambú (*Spilanthes acmella*) flowers, leaves, and stems by using SCCO2 and ethanol, water, and their mixtures as solvent enhancers (Dias et al., 2012). Felfoldi-Gava et al. (2012) studied the influence of the use of ethanol as cosolvent for the extraction of phytochemicals from *Alnus glutinosa* L., the optimum conditions found at 30 MPa, 60°C, and 10% of ethanol; although the highest yield was obtained by using ethanolic Soxhlet extraction, the concentration of triterpenes in the extracts was very low due to the simultaneous extraction of undesirable substances. Extracts from propolis were obtained by SCCO2, the highest yield being obtained at 15 MPa, 40°C, using 5% of ethanol (Biscaia and Ferreira, 2009). SCCO2 at 18 MPa and 40°C with ethanol as cosolvent was used to produce extracts from wormwood (*Artemisa absinthium*), which presented high activity against insect pests (Martín et al., 2011a,b). SCCO2 with 15% ethanol was used to selectively extract different carotenoids from a marine strain of the Cyanobacterium *Synechococcus* sp. (Cyanophyceae), the carotenoids being extracted are: zeaxanthin, β-carotene, β-cryptoxanthin, and equinenone; chlorophyll was poorly extracted, and myxoxanthophyll, another major carotenoid, was not extracted under any experimental condition (Montero et al., 2005). Hegel et al. (2011) used SCCO2 to extract lipids from the yeast *Saccharomyces cerevisiae* at 20 MPa and 40°C, using 9%, (w/w) ethanol, which allowed more efficient extraction of triglycerides, and also an extraction/fractionation of phospholipids. Astaxanthin (80.6%) could be extracted from the microalga *Haematococcus pluvialis* at moderate pressure and temperature, by using ethanol as entrainer, whereas without ethanol, the highest astaxanthin extracted by SCCO2 was 12.3% at higher pressure and temperature (Machmudah et al., 2006). Ethanol was also used as an entrainer in the SCCO2 extraction of γ-linolenic acid and other lipids from the *Arthrospira* (*Spirulina*) *maxima*, yields being compared to the ones obtained by solvent extraction (Mendes et al., 2006). Dejoye et al. (2011) verified that the yields and fatty acid contents obtained by using SCCO2 from freeze-dried *Chlorella vulgaris* were higher if, prior to SFE, the microalgae were submitted to a microwave pretreatment.

Vasapollo et al. (2004) studied the effect of the use of different vegetable oils (almond, peanut, hazelnut, and sunflower seed oil) as cosolvents on the SCCO2 extraction of lycopene from tomato; higher extraction yields (60% of the total amount of extractable oil) were obtained when hazelnut oil (10%) was used. Shi et al. (2009) used ethanol, water, and canola oil, as modifiers, on the extraction of lycopene from tomato skins, and verified that the extraction efficiency was improved by the addition of any of the three modifiers, increasing with the increase in the modifier's amounts (from 5% to 10%), but the rate of yield increase was lower when the proportion of ethanol concentration was rising from 10% to 15%. When canola oil was added to SCCO2 in the extraction of carotenoids from carrot samples, α-carotene and β-carotene yields were improved more than twice, and lutein yield was more than four times higher when compared to those obtained with SCCO2 extraction alone (Sun and Temelli, 2006). Saldaña et al. (2010) studied the extraction of lycopene and β-carotene from freeze-dried tomatoes (skin+pulp) at 40°C and 40 MPa by using pure SCCO2 or SCCO2+5% (w/w) cosolvent (ethanol or canola oil), having verified that the apparent solubility of those compounds was higher when canola oil was used. SCCO2 extraction of lutein esters from marigold (*Tagetes erect* L.) petals with soybean oil as cosolvent was studied by Ma et al. (2008), under the extraction condition of 30 MPa, 52.5°C, 10 L/h CO_2, flow rate, 6.1% soybean oil, the lutein yield was more than twice higher than the one obtained without the cosolvent. Soybean and olive oils were used as entrainers of SCCO2 by Krichnavaruk et al. (2008), for the SCCO2 extraction of astaxanthin from *Haematococcus pluvialis*, being verified that the efficiency was enhanced by about 30% in the presence of 10% soybean oil, but with 10% of olive oil in SCCO2, the efficiency was comparable to that obtained when ethanol was used as a cosolvent.

9.6.1 RECOVERY OF WASTES AND BY-PRODUCTS BY USING SFE: SOME EXAMPLES

Isoflavones from soybean cake were extracted by using SCCO2 and a mixture of ethanol:water (70:30, v/v) as the modifier, a large yield of malonylglucoside and glucoside being produced at 60°C and 35 MPa, while a high amount of acetylglucoside and aglycone was formed at 80°C and 35 MPa (Kao et al., 2008).

Seabra et al. (2012) studied the SFE for the valorization of tara (*Caesalpinia spinosa*) fruit seed coat by using mixtures of SCCO2, water, and ethanol, and verified that, with higher quantity of water, high extraction yields were obtained, but maximum phenolic contents were produced with ethanol-rich mixtures.

Distilled white grape pomace (*Vitis vinifera* var. Garnacha) was subjected to SCCO2, the addition of ethanol as modifier (8%) favoring the release of phenolic compounds, doubling those obtained by solid–liquid extraction employing 96% ethanol and water (Pinelo et al., 2007).

Mezzomo et al. (2010) obtained an oil of high quality, rich in oleic acid and flavonoids, by SCCO2 extraction (27 MPa, 40°C) from peach (*Prunus persica*) kernels, which are industrial residues from the peach processing, having verified that the use of 5% ethanol as a cosolvent did not result in a significant effect on yield and total phenolic content.

Carob pulp kibbles, a by-product of carob bean gum production, were subjected to SCCO2, the optimal values of extraction yields and antioxidant activity being found at 22 MPa, 40°C, 0.27 mm particle size, flow rate 0.29 kg h^{-1}, and 12.4% of a mixture of ethanol and water (80:20, v/v) as the cosolvent. The extract presented high antioxidant activity with compounds such as naringenin, caffeic, ferulic, and gallic acids, tricetin-3',5'-dimethyl ether, and myricetin glucoside; the solid residue, yet with antioxidant activity, proved to be a dietary fiber that can be compared with Caromax™, a carob fiber commercialized by Nutrinova Inc. (Bernardo-Gil et al., 2011).

Vitamin E and provitamin A (β-carotene) were recovered from red pepper (*Capsicum annum* L.) by-products, by using SCCO2, the highest extract yield being found at 24 MPa and 60°C, without the addition of a modifier (Romo-Hualde et al., 2012).

High-value phenolic compounds were obtained from maritime pine bark by using SCCO2 at 25 MPa and 30°C and ethanol as entrainer, the highest contents of total phenolics and procyanidins found when ethanol (70% v/v) was used (Seabra et al., 2012).

By using SCCO2 at 20 MPa, 60°C, and 40% (w/w) of ethanol, it was possible to reduce to approximately 70% the volatile compounds from a wine-making inactive dry yeast that may be released into the wines and therefore affecting their sensory characteristics (Pozo-Bayón, 2010).

Phospholipids were also extracted and fractionated with more than 5% ethanol in SCCO2 (17.7 MPa and 33°C) from tuna shavings, a by-product of factory-processed tuna (Tanaka and Sakaki, 2005). Ethanol at 3% (v/v) was used as an entrainer in SCCO2 for removing lipids and bad flavor from tuna viscera (Kang et al., 2005).

Sanchez-Camargo et al. (2012) studied the effect of ethanol addition to SCCO2 on the extraction of astaxanthin and ω-3 fatty acid (EPA+DHA, mainly) from red-spotted shrimp waste (*Farfantepenaeus paulensis*). An increase of 136% was reached in the total lipid extraction yield when the proportion of ethanol was increased from 5% to 15% wt, with maximum recoveries of 93.8% and 65.2% for lipids and astaxanthin, respectively, occurring when the proportion of 15% wt. of ethanol in the SCCO2/ethanol mixture was used; the best results for the recovery of EPA and DHA were also obtained under these conditions.

SCCO2 of Northern shrimp (*Pandelus borealis* Kreyer) by-products at 35 MPa and 40°C generated a deep red oil, rich in ω-3 PUFAs, specifically 7.8% EPA and 8.0% DHA (Amiguet et al., 2012).

Rubio-Rodríguez et al. (2012) verified that SCCO2 extraction at moderate conditions (25 MPa and 40°C) was an advantageous process for obtaining oil from freeze-dried, low-fat, fish by-products, reducing fish oil oxidation, especially when fish oil is rich in ω-3.

9.7 MODELING AND SCALE-UP OF SUPERCRITICAL FLUID EXTRACTION

Finding a model that represents the SFE process is very important since, besides enabling the interpolation of data, it allows developing scaling-up procedures from the laboratory stage to the industrial-scale production.

In the SCF extraction of a solid, the solid particles at an adequate particle size to provide a large surface area stay in a fixed bed inside a batch extractor, the SCF flowing continually through it, and pressure and temperature within the extractor being maintained constant at values providing adequate solubility of the compounds to be extracted. The concentration of solutes in the SCF, known as loading, has the highest value at the beginning of the extraction, remaining constant for some time, and falling afterwards. Supercritical fluid extraction involves the following sequential steps: (1) penetration and diffusion of the CO_2 in the solid matrix; (2) solubilization of the components; (3) transport of solutes through the solid, with formation of a thin liquid film around the solid particles; (4) convective transport of solutes from the external surface of the solid to the bulk of the fluid; and (5) desorption of solutes from the solid matrix. Depending on the mechanism of mass transfer, there may be three different regimes, and a typical extraction curve shows a first stage of constant extraction rate controlled by solubility, followed by a stage of decreasing extraction rate controlled mainly by diffusion of solutes in the solid particle, and a final stage where the rate-limiting step is internal diffusion due to desorption of solutes. The constant rate period depends on the flow rate of the solvent as well as on the characteristics of the substrate, namely its content in the solid matrix. When the substrate concentration is low, the constant rate period may not exist at all.

Considering the various mechanisms that can exist, several approaches have been proposed for modeling of SFE: empirical models; models based on heat and mass transfer analogy (Fick's second law of diffusion); and models based on differential mass balances along the extractor bed. Empirical models (Nguyen et al., 1991; Goodrum et al., 1996; Esquível et al., 1999; Zekovic et al., 2003; Quispe-Condori et al., 2005) use exponential or hyperbolic equations, and can be useful when information on mass transfer mechanisms and on equilibrium relationships is missing. However, they are not adequate for scaling-up, since they do not provide a description of underlying mass transfer phenomena that occurs also in industrial-size extraction vessels (Del Valle and de la Fuente, 2006).

Mass transfer models concerning the heat-transfer analogy, based on a simple geometry, were reported (Del Valle et al., 2000; Gaspar et al., 2003). Some authors consider only the fluid-phase resistance (Reverchon et al., 1993, 1994); when the major mass transfer resistance is within the solid phase, models must also consider solute transport within the solid particles (Bartle et al., 1990; Lojkova et al., 1997; Zancan et al., 2002).

Sovová (1994) developed a model which is an extension of the Lack's plug flow model for application to SFE that has proved to describe very well the SFE of several products, such as grape oil (Sovová et al., 1994); black pepper essential oil (Ferreira and Meireles, 2002); aniseed oil (Rodrigues et al., 2003); apricot kernel oil (Ozkal et al., 2005); walnuts and hazelnuts oils (Bernardo-Gil and Casquilho, 2007); acorn oil (Bernardo-Gil et al. 2007); and fig leaf gourd seeds oil (Bernardo-Gil et al., 2009). Sovová's mass transfer model takes into account the solute solubility in the solvent phase and the mass transfer coefficient both in the fluid and in the solid phases. The model assumes pseudo-steady state and plug flow, with temperature, pressure, and solvent velocity being kept constant throughout the operation. The model also assumes that the bed is homogeneous regarding particle size distribution and the initial solute distribution in the bed. Axial dispersion and solute accumulation in the fluid phase are assumed to be negligible.

Reis-Vasco et al. (2000) developed a model that takes into account the desorption of pennyroyal essential oil located near the leaf surface, and the mass transfer resistance to the extraction of essential oil contained in the internal part of the vegetable structure, considering the concept of broken and intact cells (Reverchon et al., 2000; Reverchon and Marrone, 2001). Axial dispersion was also considered.

Sovová (2005) presented a new model for the modeling of SFE of natural products. It is based on the concept of broken and intact cells, and it is suited to fit experimental data, as it almost independently simulates two extraction periods, the first one governed by phase equilibrium and the second one governed by internal diffusion in particles, having taken into account different types of phase equilibrium and solvent flow patterns. This model was applied to SFE of hyssop oil (Langa et al., 2009); propolis (Biscaia and Ferreira, 2009); rosemary (Bensebia et al., 2009), and wormwood (*Artemisia absinthium* L.) (Martín et al., 2011b).

Rahimi et al. (2011) presented a model for SFE of chamomile that accounts for both particle and fluid phase, the distribution coefficient of chamomile extract between solid and solvent being determined using genetic algorithm method.

Revisions on mathematical models used for SFE can be recommended, such as: Del Valle et al. (2005); Sovová (2005); Díaz-Reinoso et al. (2006); Del Valle and de la Fuente (2006); Reverchon and De Marco (2008).

REFERENCES

Abbas, K.A., A. Mohamed, A.S. Abdulamir, and H.A. Abas. 2008. A review on supercritical fluid extraction as new analytical method. *Am. J. Biochem. Biotechnol.* 4:345–353.

Abdelmoez, W., R. Abdelfatah, A. Tayeb, and H. Yoshida. 2011. Extraction of cottonseed oil using subcritical water technology. *AIChE J.* 57:2353–2359.

Aguilera, A., M. Rodriguez, M. Brotons, M. Boulaid, and A. Valverde. 2005. Evaluation of supercritical fluid extraction/aminopropyl solid-phase "in-line" cleanup for analysis of pesticide residues in rice. *J. Agric. Food Chem.* 53:9374–9382.

Akgun, I. H., A. Erkucuk, M. Pilavtepe, and O. Yesil-Celiktas. 2011. Optimization of total alkannin yields of *Alkanna tinctoria* by using sub- and supercritical carbon dioxide extraction. *J. Supercrit. Fluids* 57:31–37.

Almeida, P.P., N. Mezzomo, and S. R. S. Ferreira. 2012. Extraction of *Mentha spicata* L. volatile compounds: Evaluation of process parameters and extract composition. *Food Bioprocess Technol.* 5:548–559.

Amiguet, V. T., K. L. Kramp, J. Q. Mao, C. McRae, A. Goulah, L. E. Kimpe, J. M. Blais, and J. T. Arnason. 2012. Supercritical carbon dioxide extraction of polyunsaturated fatty acids from Northern shrimp (*Pandalus borealis* Kreyer) processing by-products. *Food Chem.* 130: 853–858.

Anekpankul, T., M. Goto, M. Sasaki, P. Pavasant, and A. Shotipruk. 2007. Extraction of anti-cancer damnacanthal from roots of *Morinda citrifolia* by subcritical water. *Sep. Purif. Technol.* 55:343–349.

Antunes, P., O. Gil, and M. G. Bernardo-Gil. 2003. Supercritical fluid extraction of organochlorines from fish muscle with different sample preparation. *J. Supercritic. Fluids* 25:135–142.

Araújo, J. M. A. and D. Sandi. 2007. Extraction of coffee diterpenes and coffee oil using supercritical carbon dioxide. *Food Chem.* 101:1087–1094.

Ashraf-Khorassani, M., S. Gidanian, and Y. Yamini. 1995. Effect of pressure, temperature, modifier concentration and sample matrix on supercritical fluid extraction efficiency of different phenolic compounds. *J. Chromatogr. Sci.* 33:658–662.

Ayala, R. S. and M. D. Luque de Castro. 2001. Continuous subcritical water extraction as a useful tool for isolation of edible essential oils. *Food Chem.* 75:109–113.

Bartle, K. D., A. A. Clifford, S. B. Hawthorne, J. J. Langenfeld, D. J. Miller, and R. Robinson. 1990. A model for dynamic extraction using a supercritical fluid. *J. Supercrit. Fluids* 3:143–149.

Barth, M. M., C. Zhou, K. M. Kute, and G. A. Rosenthals. 1995. Determination of optimum conditions for supercritical fluid extraction of carotenoids from carrot (*Daucus carota* L.) tissue. *J. Agric. Food Chem.* 43:2876–2878.

Basile, A., M. M. Jimeénez-Carmona, and A. A. Clifford. 1998. Extraction of rosemary by superheated water. *J. Agric. Food Chem.* 46:5205–5209.

Bensebia, O., D. Barth, B. Bensebia, and A. Dahmani. 2009. Supercritical CO_2 extraction of rosemary: Effect of extraction parameters and modeling. *J. Supercrit. Fluid* 49:161–166.

Berna, A., A. Cháfer, J. B. Montón, and S. Subirats. 2001. High-pressure solubility data of system ethanol (1) + catechin (2) + CO_2(3). *J. Supercrit Fluid* 20:157–162.

Bernardo-Gil, M.G., J. Grenha, J. Santos, and P. Cardoso. 2002. Supercritical fluid extraction and characterisation of hazelnut oil. *Eur. J. Lipid Sci. Technol.* 104:402–409.

Bernardo-Gil, M. G. and L. M. C. Lopes. 2004. Supercritical fluid extraction of *Cucurbita ficifolia* seeds oil. *Eur. Food Res. Technol.* 219:593–597.

Bernardo-Gil, M.G. and M. Casquilho. 2007. Modeling the supercritical fluid extraction of hazelnut and walnut oils. *AIChE J.* 53:2980–2985.

Bernardo-Gil, M. G., M. Casquilho, M. M. Esquível, and M. A. Ribeiro. 2009. Supercritical fluid extraction of fig leaf gourd seeds oil. Fatty acids composition and extraction kinetics. *J. Supercritic. Fluids* 49:32–36.

Bernardo-Gil, M. G., I. M. G. Lopes, M. Casquilho, M. A. Ribeiro, M. M. Esquível, and J. Empis. 2007. Supercritical carbon dioxide extraction of acorn oil. *J. Supercrit. Fluids* 40: 344–348.

Bernardo-Gil, M. G., R. Roque, L. B. Roseiro, L. C. Duarte, F. Girio, and P. Esteves. 2011. Supercritical extraction of carob kibbles (*Ceratonia siliqua* L.). *J. Supercrit. Fluids* 59:36–42.

Bimakr, M., R. A. Rahman, A. Ganjloo, F. S. Taip, L. M. Salleh, and M. Z. I. Sarker. 2012. Optimization of supercritical carbon dioxide extraction of bioactive flavonoid compounds from spearmint (*Mentha spicata* L.) leaves by using response surface methodology. *Food Bioprocess Technol.* 5:912–920.

Biscaia, D. and S. R.S. Ferreira. 2009. Propolis extracts obtained by low pressure methods and supercritical fluid extraction. *J. Supercrit. Fluids* 51:17–23.

Borges, M. E., R. L. Tejera, L. Díaz, P. Esparza, and E. Ibáñez. 2012. Natural dyes extraction from cochineal (*Dactylopius coccus*). New extraction methods. *Food Chem.* 132:1855–1860.

Brown, Z. K., P. J. Fryer, I. T. Norton, and R. H. Bridson. 2010. Drying of agar gels using supercritical carbon dioxide. *J. Supercrit. Fluids* 54: 89–95.

Brunner, G. 2004. *Supercritical Fluids as Solvents and Reaction Media*. Amsterdam: Elsevier.

Buranov, A. U. and G. Mazza. 2009. Extraction and purification of ferulic acid from flax shives, wheat and corn bran by alkaline hydrolysis and pressurised solvents. *Food Chem.* 115: 1542–1548.

Cardozo Jr., E. L., L. Cardozo-Filho, O. Ferrarese-Filho, and E. F. Zanoelo. 2007. Selective liquid CO_2 extraction of purine alkaloids in different *Ilex paraguariensis* progenies grown under environmental influences. *J. Agric. Food Chem.* 55:6835–6841.

Carrara, V. S., L. Z. Serra, L. Cardozo-Filho, E. F. Cunha-Júnior, E. C. Torres-Santos, and D. A. G. Cortez. 2012. HPLC analysis of supercritical carbon dioxide and compressed propane extracts from Piper amalago L. with antileishmanial activity. *Molecules* 17: 15–33.

Casas, L., C. Mantell, M. Rodríguez, A. Torres, F. A. Macías, and E. J. Martínez de la Ossa. 2008. Supercritical fluid extraction of bioactive compounds from sunflower leaves with carbon dioxide and water on a pilot plant scale. *J. Supercrit. Fluids* 45:37–42.

Castro-Vargas, H. I., L. I. Rodríguez-Varela, S. R. S. Ferreira, and F. Parada-Alfonso. 2010. Extraction of phenolic fraction from guava seeds (*Psidium guajava* L.) using supercritical carbon dioxide and co-solvents. *J. Supercrit. Fluids* 51:319–324.

Catchpole, O. J., J. B. Grey, K. A. Mitchell, and J. S. Lan. 2004. Supercritical antisolvent fractionation of propolis tincture. *J. Supercrit. Fluids* 29:97–106.

Catchpole, O. J., J. B. Grey, N. B. Perry, E. J. Burgess, W. A. Redmond, and N. G. Porter. 2003. Extraction of chili, black pepper, and ginger with near-critical CO_2, propane, and dimethyl ether: Analysis of the extracts by quantitative nuclear magnetic resonance. *J. Agric. Food Chem.* 51:4853–4860.

Catchpole, O. J. and K. Proells. 2001. Solubility of squalene, oleic acid, soya oil, and deep sea shark liver oil in subcritical R134a from 303 to 353 K. *Ind. Eng. Chem. Res.* 40:965–972.

Cerón, I. X., J. C. Higuita, and C. A. Cardona. 2012. Design and analysis of antioxidant compounds from Andes berry fruits (*Rubus glaucus* Benth) using an enhanced-fluidity liquid extraction process with CO_2 and ethanol. *J. Supercrit. Fluids* 62:96–101.

Cháfer, A., A. Berna, J. B. Montón, and R. Muñoz. 2002. High-pressure solubility data of system ethanol(1)+epicatechin (2)+CO_2 (3). *J. Supercrit. Fluid* 24:103–109.

Cháfer, A., T. Fornari, A. Berna, and R. P. Stateva. 2004. Solubility of quercetin in supercritical CO_2+ethanol as a modifier: Measurements and thermodynamic modeling. *J. Supercrit. Fluid* 32:89–96.

Cháfer, A., T. Fornari, R. P. Stateva, A. Berna, and J. García-Reverter. 2007. Solubility of the natural antioxidant gallic acid in supercritical CO_2+ethanol as a co-solvent. *J. Chem. Eng. Data* 52:116–121.

Chan, K. W. and M. Ismail. 2009. Supercritical carbon dioxide fluid extraction of *Hibiscus cannabinus* L. seed oil: A potential solvent-free and high antioxidative edible oil. *Food Chem.* 114: 970–975.

Choi, Y. H., Y. W. Chin, J. Kim, S. H. Jeon, and K. P. Yoo. 1999. Strategies for supercritical fluid extraction of hyoscyamine and scopolamine salts using basified modifiers. *J. Chromatogr. A* 863:47–55.

Chun, M. K., H. W. Shin, and H. Lee. 1996. Supercritical fluid extraction of paclitaxel and Baccatin III from needles of *Taxus cuspidate*. *J. Supercrit. Fluids* 9:192–198.

Clifford, A. A., A. Basile, and S. H. Al-Saidi. 1999. A comparison of the extraction of clove buds with supercritical carbon dioxide and superheated water. *Fresenius J. Anal. Chem.* 364:635–637.

Cossuta, D., B. Simándi, E. Vági, J. Hohmann, A. Prechl, É. Lemberkovics, Á. Kéry, and T. Keve. 2008. Supercritical fluid extraction of *Vitex agnus castus* fruit. *J. Supercrit. Fluids* 47:188–194.

Costa, P., S. Gonçalves, C. Grosso, P. B. Andrade, P. Valentão, M. G. Bernardo-Gil, and A. Romano. 2012. Chemical profiling and biological screening of *Thymus lotocephalus* extracts obtained by supercritical fluid extraction and hydrodistillation. *Ind. Crops Prod.* 36:246–256.

Curren, M. S. S., and J. W. King. 2001. Ethanol-modified subcritical water extraction combined with solid-phase microextraction for determining atrazine in beef kidney. *J. Agric. Food Chem.* 49:2175–2180.

Dejoye, C., M. A. Vian, G. Lumia, C. Bouscarle, F. Charton, and F. Chemat. 2011. Combined extraction processes of lipid from *Chlorella vulgaris* microalgae: Microwave prior to supercritical carbon dioxide extraction. *Int. J. Mol. Sci.* 12:9332–9341.

Del Valle, J. M. and J. C. de la Fuente. 2006. Supercritical CO_2 extraction of oilseeds: Review of kinetic and equilibrium models. *Crit. Rev. Food Sci. Nutr.* 46:131–160.

Del Valle, J. M., J. C. de la Fuente, and D. A. Cardarelli. 2005. Contributions to supercritical extraction of vegetable substrates in Latin America. *J. Food Eng.* 67: 35–57.

Del Valle, J. M., P. Napolitano, and N. Fuentes. 2000. Estimation of relevant mass transfer parameters for the extraction of packed substrate beds using supercritical fluids. *Ind. Eng. Chem. Res.* 39:4720–4728.

Della Porta, G., F. Castaldo, M. Scognamiglio, L. Paciello, P. Parascandola, and E. Reverchon. 2012. Bacteria microencapsulation in PLGA microdevices by supercritical emulsion. *J. Supercrit. Fluids* 63:1–7.

Denery, J. R., K. Dragull, C. S. Tang, and Q. X. Li. 2004. Pressurized fluid extraction of carotenoids from *Haematococcus pluvialis* and *Dunaliella salina* and kavalactones from *Piper methysticum*. *Anal. Chim. Acta* 501:175–181.

Dias, A. M. A., P. Santos, I. J. Seabra, R. N. C. Júnior, M. E. M. Braga, and H. C. de Sousa. 2012. Spilanthol from *Spilanthes acmella* flowers, leaves and stems obtained by selective supercritical carbon dioxide extraction. *J. Supercrit. Fluids* 61:62–70.

Díaz-Reinoso, B., A. Moure, H. Domínguez, and J. C. Parajó. 2006. Supercritical CO_2 extraction and purification of compounds with antioxidant activity. *J. Agric. Food Chem.* 54:2441–2469.

Doll, K. M. and S. Z. Erhan. 2005. Synthesis of carbonated fatty methyl esters using supercritical carbon dioxide. *J. Agric. Food Chem.* 53:9608–9614.

Duarte, C., M. Moldão-Martins, A. F. Gouveia, S. Beirão-da-Costa, A. E. Leitão, and M. G. Bernardo-Gil. 2004. Supercritical fluid extraction of red pepper (*Capsicum frutescens* L.). *J. Supercrit. Fluids* 30:155–161.

Dunford, N. T., F. Temelli, and E. LeBlanc. 1997. Supercritical CO_2 extraction of oil and residual proteins from Atlantic mackerel (*Socomber scombrus*) as affected by moisture content. *J. Food Sci.* 62:289–294.

Elbashir, N. O., D. B. Bukur, E. Durham, and C. B. Roberts. 2010. Advancement of Fischer–Tropsch synthesis via utilization of supercritical fluid reaction media. *AIChE J.* 56:997–1015.

Erkucuk, A., I. H. Akgun, and O. Yesil-Celiktas. 2009. Supercritical CO_2 extraction of glycosides from *Stevia rebaudiana* leaves: Identification and optimization. *J. Supercrit. Fluids* 51:29–35.

Erriguible, A., S. Vincent, and P. Subra-Paternault. 2012. Numerical investigations of liquid jet breakup in pressurized carbon dioxide: Conditions of two-phase flow in supercritical antisolvent process. *J. Supercrit. Fluids* 63:16–24.

Esquível, M. M. and M. G. Bernardo-Gil. 1993. Extraction of olive husk oil with compressed carbon dioxide. *J. Supercrit. Fluids* 6:91–94.

Esquível, M. M., M. G. Bernardo-Gil, and M. B. King. 1999. Mathematical models for supercritical extraction of olive husk oil. *J. Supercrit. Fluids* 16:43–58.

Euterpio, M. A., C. Cavaliere, A. L. Capriotti, and C. Crescenzi. 2011. Extending the applicability of pressurized hot water extraction to compounds exhibiting limited water solubility by pH control: Curcumin from the turmeric rhizome. *Anal. Bioanal. Chem.* 401:2977–2985.

Felfoldi-Gava, A., S. Szarka, B. Simándi, B. Blazics, B. Simon, and A. Kery. 2012. Supercritical fluid extraction of *Alnus glutinosa* (L.) Gaertn. *J. Supercrit. Fluids* 61:55–61.

Fernández-Pérez, V., M. M. Jiménez-Carmona, and M. D. Luque de Castro. 2000. An approach to the static–dynamic subcritical water extraction of laurel essential oil: Comparison with conventional techniques. *Analyst* 125: 481–485.

Ferreira, S. R. S. and M. A. A. Meireles. 2002. Modeling the supercritical fluid extraction of black pepper (*Piper nigrum* L.) essential oil. *J. Food Eng.* 54:263–269.

Fiddler, W., J. W. Pensabene, R. A. Gates, and D. J. Donoghue. 1999. Supercritical fluid extraction of organochlorine pesticides in eggs. *J. Agric. Food Chem.* 47:206–211.

Fornari, T., A. Ruiz-Rodriguez, G. Vicente, E. Vázquez, M. R. García-Risco, and G. Reglero. 2012. Kinetic study of the supercritical CO_2 extraction of different plants from *Lamiaceae* family. *J. Supercrit. Fluids* 64:1–8.

Gámiz-Gracía, L., and M. D. Luque de Castro. 2000. Continuous subcritical water extraction of medicinal plant essential oil: Comparison with conventional techniques. *Talanta* 51: 1179–1185.

Gaspar, F., T. Lu, R. Santos, and B. Al-Duri. 2003. Modelling the extraction of essential oils with compressed carbon dioxide. *J. Supercrit. Fluids* 25:247–260.

Ge, Y., H. Yan, B. Hui, Y. Ni, S. Wang, and T. Cai. 2002. Extraction of natural vitamin E from wheat germ by supercritical carbon dioxide. *J. Agric. Food Chem.* 50:685–689.

Ghasemi, E., F. Raofie, and N. M. Najafi. 2011. Application of response surface methodology and central composite design for the optimisation of supercritical fluid extraction of essential oils from *Myrtus communis* L. leaves. *Food Chem.* 126:1449–1453.

Ghoreishi, S. M. and E. Bataghva. 2011. Supercritical extraction of evening primrose oil: Experimental optimization via response surface methodology. *AIChE J.* 57:3378–3384.

Glišić, S. B., D. R. Mišić, M. D. Stamenić, I. T. Zizovic, R. M. Ašanin, and D. U. Skala. 2007. Supercritical carbon dioxide extraction of carrot fruit essential oil: Chemical composition and antimicrobial activity. *Food Chem.*105: 346–352.

Gnayfeed, M. H., H. G. Daood, V. Illés, and P. A. Biacs. 2001. Supercritical CO_2 and subcritical propane extraction of pungent paprika and quantification of carotenoids, tocopherols, and capsaicinoids. *J. Agric. Food Chem.* 49:2761–2766.

Golmakani, M. T., J. A. Mendiola, K. Rezaei, and E. Ibáñez. 2012. Expanded ethanol with CO_2 and pressurized ethyl lactate to obtain fractions enriched in γ-linolenic acid from *Arthrospira platensis* (Spirulina). *J. Supercrit. Fluids* 62:109–115.

Goodrum, J. W., M. B. Kilgo, and C. R. Santerre. 1996. Oil solubility and extraction modeling. In *Supercritical Fluid Technology in Oil and Lipid Chemistry*, eds. J. W. King, and G. R. List, 101–131. AOCS Press, Champaign, IL.

Grevenstuk, T., S. Gonçalves, J. M. F. Nogueira, M. G. Bernardo-Gil, and A. Romano. 2012. Recovery of high purity plumbagin from *Drosera intermedia. Ind. CropsProd.* 35:257–260.

Grigonis, D., P. R. Venskutonis, B. Sivik, M. Sandahl, and C. S. Eskilsson. 2005. Comparison of different extraction techniques for isolation of antioxidants from sweet grass (*Hierochloe odorata*). *J. Supercrit. Fluids* 33:223–233.

Grosso, C., J. A. Coelho, J. S. Urieta, A. M. F. Palavra, and J. G. Barroso. 2010. Herbicidal activity of volatiles from coriander, winter savory, cotton lavender, and thyme isolated by hydrodistillation and supercritical fluid extraction. *J. Agric. Food Chem.* 58:11007–11013.

Grosso, C., V. Ferraro, A. C. Figueiredo., J. G. Barroso, J. A., Coelho, and A. M. Palavra. 2008. Supercritical carbon dioxide extraction of volatile oil from Italian coriander seeds. *Food Chem.* 111:197–203.

Grosso, C., A. C. Figueiredo, J. Burillo, A. M. Mainar, J. S. Urieta, J. G. Barroso, J. A. Coelho, and A. M. F. Palavra. 2010. Composition and antioxidant activity of *Thymus vulgaris* volatiles: Comparison between supercritical fluid extraction and hydrodistillation. *J. Sep. Sci.* 33:2211–2218.

Guan, W., S. Li, R. Yan, S. Tang, and C. Quant. 2007. Comparison of essential oils of clove buds extracted with supercritical carbon dioxide and other three traditional extraction methods. *Food Chem.* 101: 1558–1564.

Güçlü-Üstündağ, Ö., J. Balsevich, and G. Mazza. 2007. Pressurized low polarity water extraction of saponins from cow cockle seed. *J. Food Eng.* 80:619–630.

Güçlü-Üstündağ, Ö. and G. Mazza. 2008. Extraction of saponins and cyclopeptides from cow cockle seed with pressurized low polarity water. *LWT – Food Sci. Technol.* 9:1600–1606.

Hamdan, S., H. G. Daood, M. Toth-Markus, and V. Illés. 2008. Extraction of cardamom oil by supercritical carbon dioxide and sub-critical propane. *J. Supercrit. Fluids* 44:25–30.

Hassas-Roudsari, M., P. R. Chang, R. B. Pegg, and R. T. Tyler. 2009. Antioxidant capacity of bioactives extracted from canola meal by subcritical water, ethanolic and hot water extraction. *Food Chem.* 114:717–726.

Hawthorne, S. B., M. L. Riekkola, K. Serenius, Y. Holm, R. Hiltunen, and K. Hartonen. 1993. Comparison of hydrodistillation and supercritical fluid extraction for the determination of essential oils in aromatic plants. *J. Chromatogr. A* 634:297–308.

He, L., X. Zhang, H. Xu, C. Xu, F. Yuan, Z. Knez, Z. Novak, and Y. Gao. 2012. Subcritical water extraction of phenolic compounds from pomegranate (*Punica granatum* L.) seed residues and investigation into their antioxidant activities with HPLC–ABTS•+ assay. *Food Bioprod. Process* 90:215–223.

Hegel, P. E., S. Camy, P. Destrac, and J. S. Condoret. 2011. Influence of pretreatments for extraction of lipids from yeast by using supercritical carbon dioxide and ethanol as cosolvent. *J. Supercrit. Fluids* 58:68–78.

Herrero, M., A. Cifuentes, and E. Ibañez. 2006. Sub- and supercritical fluid extraction of functional ingredients from different natural sources: Plants, food-by-products, algae and microalgae: A review. *Food Chem.* 98:136–148.

Herrero, M., E. Ibáñez, J. Señoráns, and A. Cifuentes. 2004. Pressurized liquid extracts from *Spirulina platensis* microalga: Determination of their antioxidant activity and preliminary analysis by micellar electrokinetic chromatography. *J. Chromatogr. A* 1047:195–203.

Herrero, M., J. A. Mendiola, A. Cifuentes, and E. Ibañez. 2010. Supercritical fluid extraction: Recent advances and applications. *J. Chromatogr. A* 1217:2495–2511.

Ho, C. H. L., J. E. Cacace, and G. Mazza. 2007. Extraction of lignans, proteins and carbohydrates from flaxseed meal with pressurized low polarity water. *LWT – Food Sci. Technol.* 40:1637–1647.

Hsu, Y. W., C. F. Tsai, W. K., Y. C. C. Ho, and F. J. Lu. 2011. Determination of lutein and zeaxanthin and antioxidant capacity of supercritical carbon dioxide extract from daylily (*Hemerocallis disticha*). *Food Chem.* 129:1813–1818.

Hu, Q., Y. Hu and J. Xu. 2005. Free radical-scavenging activity of Aloe vera (Aloe barbadensis Miller) extracts by supercritical carbon dioxide extraction. *Food Chem.* 91: 85–90.

Huang, K. J., J. J. Wu, Y. H. Chiu, C. Y. Lai, and C. M. J. Chang. 2007. Designed polar cosolvent-modified supercritical CO_2 removing caffeine from and retaining catechins in green tea powder using response surface methodology. *J. Agric. Food Chem.* 55:9014–9020.

Ibañez, E., A. Kubátová, F. J. Señoráns, S. Cavero, G. Reglero, and S. B. Hawthorne. 2003. Subcritical water extraction of antioxidant compounds from rosemary plants. *J. Agric. Food Chem.* 51:375–382.

Ibañez, E., A. Oca, G. Murga, S. Lopez-Sebastian, J. Tabera, and G. Reglero. 1999. Supercritical fluid extraction and fractionation of different preprocessed rosemary plants. *J. Agric. Food Chem.* 47:1400–1404.

Ikonnikov, V. K., P. A. Egoyants, S. A. Sirotin, and T. T. Hieu. 2010. The determination of the parameters of sub- and supercritical extraction of plant raw materials. *Russ. J. Phys. Chem. B.* 4:1265–1271.

Illés, V., H. G. Daood, S. Perneczki, L. Szokonya, and M. Then. 2000. Extraction of coriander seed oil by CO_2 and propane at super- and subcritical conditions. *J. Supercrit. Fluids* 17:177–186.

Illés, V., O. Szalai, M. Then, H. Daood, and S. Perneczki. 1997. Extraction of hiprose fruit by supercritical CO_2 and propane. *J. Supercrit. Fluids* 10:209–218.

Irmak, S., K. Solakyildirim, A. Hesenov, and O. Erbatur. 2010. Study on the stability of supercritical fluid extracted rosemary (*Rosmarinus offcinalis* L.) essential oil. *J. Analyt. Chem.* 65:899–906.

Ixtaina, V. Y., F. Mattea, D. A. Cardarelli, M. A. Mattea, S. M. Nolasco, and M. C. Tomás. 2011. Supercritical carbon dioxide extraction and characterization of Argentinean chia seed oil. *J. Am. Oil Chem. Soc.* 88:289–298.

Ixtaina, V. Y., A. Vega, S. M. Nolasco, M. C. Tomás, M. Gimeno, E. Bárzana, and A. Tecante. 2010. Supercritical carbon dioxide extraction of oil from Mexican chia seed (*Salvia hispanica* L.): Characterization and process optimization. *J. Supercrit. Fluids* 55:192–199.

Ivanovic, J., M. Ristic, and D. Skala. 2011. Supercritical CO_2 extraction of *Helichrysum italicum*: Influence of CO_2 density and moisture content of plant material. *J. Supercrit. Fluids* 57:129–136.

Izhyk, A., G. Novik, and E. S. Dey. 2012. Extraction of polar lipids from *Bifidobacteria* by supercritical carbon dioxide (scCO$_2$). *J. Supercrit. Fluids* 62:149–154.

Jeong, M. L. and D. J. Chesney. 1999. Investigation of modifier effects in supercritical CO$_2$ extraction from various solid matrices. *J. Supercrit. Fluid.* 16:33–42.

Jessop, P. G. and W. Leitner. 1999. *Chemical Synthesis Using Supercritical Fluids.* Wiley, VCH, New York.

Ju, Z. Y. and L. R. Howard. 2003. Effects of solvent and temperature on pressurized liquid extraction of anthocyanins and total phenolics from dried red grape skin. *J. Agric. Food Chem.* 51:5207–5213.

Kang, K. Y., D. H. Ahn, S. M. Jung, D. H. Kim, and B. S. Chun. 2005. Separation of protein and fatty acids from tuna viscera using supercritical carbon dioxide. *Biotechnol. Bioprocess Eng.* 10:315–321.

Kao, T. H., J. T. Chien, and B. H. Chen. 2008. Extraction yield of isoflavones from soybean cake as affected by solvent and supercritical carbon dioxide. *Food Chem.* 107:1728–1736.

Keshavarz, A., J. Karimi-Sabet, A. Fattahi, A. Golzary, M. Rafiee-Tehrani, and F. A. Dorkoosh. 2012. Preparation and characterization of raloxifene nanoparticles using rapid expansion of supercritical solution (RESS). *J. Supercrit. Fluids* 63:169–179.

Khajeh, M., Y. Yamini, and S. Shariati. 2010. Comparison of essential oils compositions of *Nepeta persica* obtained by supercritical carbon dioxide extraction and steam distillation methods. *Food Bioproducts Proc.* 88: 227–232.

Khajeh, M. 2011. Optimization of process variables for essential oil components from *Satureja hortensis* by supercritical fluid extraction using Box–Behnken experimental design. *J. Supercrit. Fluids* 55: 944–948.

Khajenoori, M., A. H. Asl, and F. Hormozi. 2009. Proposed models for subcritical water extraction of essential oils. *Sep. Sci. Eng. —Chin. J. Chem. Eng.* 17:359–365.

Kim, K., K. S. Kim, S. H. Son, J. Cho, and Y. C. Kim. 2011. Supercritical water oxidation of transformer oil contaminated with PCBs. A road to commercial plant from bench-scale facility. *J. Supercrit. Fluids* 58: 121–130.

Kim, J. W. and G. Mazza. 2009. Extraction and separation of carbohydrates and phenolic compounds in flax shives with pH-controlled pressurized low polarity water. *J. Agric. Food Chem.* 57:1805–1813.

Kim, S. R., M. S. Rhee, B. C. Kim, H. Lee, and K. H. Kim. 2007. Modeling of the inactivation of Salmonella typhimurium by supercritical carbon dioxide in physiological saline and phosphate-buffered saline. *J. Microbiolog. Meth.* 70:132–141.

Klesper, E. 1980. Chromatography with supercritical fluids. In *Extraction with Supercritical Gases,* eds. G. M. Schneider, E. Stahl, and G. Wilke, Weinheim, Germany: Verlag Chemie.

Kong, Y., Y. J. Fu, Y. G. Zu, W. Liu, W. Wang, X. Hua, and M. Yang. 2009. Ethanol modified supercritical fluid extraction and antioxidant activity of cajaninstilbene acid and pinostrobin from pigeonpea (*Cajanus cajan* [L.] Millsp.) leaves. *Food Chem.* 117:152–159.

Krichnavaruk, S., A. Shotipruk, M. Goto, and P. Pavasant. 2008. Supercritical carbon dioxide extraction of astaxanthin from *Haematococcus pluvialis* with vegetable oils as co-solvent. *Bioresour. Technol.* 99:5556–5560.

Kumar, M. S. Y., R. Dutta, D. Prasad, and K. Misra. 2011. Subcritical water extraction of antioxidant compounds from Seabuckthorn (*Hippophae rhamnoides*) leaves for the comparative evaluation of antioxidant activity. *Food Chem.* 127:1309–1316.

Kuvátová, A., A. J. M. Lagadec, D. J. Miller, and S. B. Hawthorne. 2001a. Selective extraction of oxygenates from savory and peppermint using subcritical water. *FlavourFragr. J.* 16: 64–73.

Kuvátová, A., D. J. Miller, and S. B. Hawthorne. 2001b. Comparison of subcritical water and organic solvents for extracting kava lactones from kava root. *J. Chromatogr. A* 923:187–194.

Lafka, T. I., A. E. Lazou, V. J. Sinanoglou, and E. S. Lazos. 2011. Phenolic and antioxidant potential of olive oil mill wastes. *Food Chem.* 125: 92–98.

Lang, Q. and C. M. Wai. 2001. Supercritical fluid extraction in herbal and natural product studies—A practical review. *Talanta* 53:771–782.

Langa, E., J. Cacho, A. M. F. Palavra, J. Burillo, A. M. Mainar, and J. S. Urieta. 2009. The evolution of hyssop oil composition in the supercritical extraction curve: Modelling of the oil extraction process. *J. Supercrit. Fluids* 49:37–44.

Leal, P. F., M. E. M. Braga, D. N. Sato, J. E. Carvalho, M. O. M. Marques, and M. A. A. Meireles. 2003. Functional properties of spice extracts obtained via supercritical fluid extraction. *J. Agric. Food Chem.* 51:2520–2525.

Leal, P. F., N. B. Maia, Q. A. C. Carmello, R. R. Catharino, M. N. Eberlin, and A. A. Meireles. 2008. Sweet basil (*Ocimum basilicum*) extracts obtained by supercritical fluid extraction (SFE): Global yields, chemical composition, antioxidant activity, and estimation of the cost of manufacturing. *Food Bioprocess Technol.* 1:326–338.

Lee, Y. H., A. L. Charles, H. F. Kung, C. T. Hod, and T. C. Huang. 2010. Extraction of nobiletin and tangeretin from *Citrus depressa* Hayata by supercritical carbon dioxide with ethanol as modifier. *Ind. CropsProd.* 31:59–64.

Li, J., M. Zhang, and T. Zheng. 2009. The *in vitro* antioxidant activity of lotus germ oil from supercritical fluid carbon dioxide extraction. *Food Chem.* 115:939–944.

Li, B., Y. Xu, Y. X. Jin, Y. Y. Wu, and Y. Y. Tu. 2010. Response surface optimization of super-critical fluid extraction of kaempferol glycosides from tea seed cake. *Ind. Crops Prod.* 32:123–128.

Li, Q., C. Jiang, Y. Zu, Z. Song, B. Zhang, X. Meng, W. Qiu, and L. Zhang. 2011. SFE–CO$_2$ extract from *Typhonium giganteum* Engl. Tubers, induces apoptosis in human hepatoma SMMC-7721 cells involvement of a ROS-mediated mitochondrial pathway. *Molecules* 16:8228–8243.

Lina, S., R. Fei, Z. Xudong, D. Yangong, and H. Fa. 2010. Supercritical carbon dioxide extraction of *Microula sikkimensis* seed oil. *J. Am. Oil Chem. Soc.* 87:1221–1226.

Liu, W., Y. J. Fu, Y. G. Zu, M. H. Tong, N. Wu, X. L. Liu, and S. Zhang. 2009. Supercritical carbon dioxide extraction of seed oil from *Opuntia dillenii* Haw. and its antioxidant activity. *Food Chem.* 114:334–339.

Liu, C., G. Hu, and H. Gao. 2012. Preparation of few-layer and single-layer graphene by exfoliation of expandable graphite in supercritical *N,N*-dimethylformamide. *J. Supercrit. Fluids* 63:99–104.

Liu, W. L., R. J. Lee, and M. R. Lee. 2010. Supercritical fluid extraction *in situ* derivatization for simultaneous determination of chloramphenicol, florfenicol and thiamphenicol in shrimp. *Food Chem.* 121:797–802.

Lo, T. C. T., H. H. Tsao, A. Y. Wang, and C. A. Chang. 2007. Pressurized water extraction of polysaccharides as secondary metabolites from *Lentinula edodes*. *J. Agric. Food Chem.* 55:4196–4201.

Lojková, L., J. Slanina, M. Mikesova, E. Taborska, and J. Vejrosta. 1997. Supercritical fluid extraction of lignans from seeds and leaves of *Schizandra chinensis*. *Phytochem. Anal.* 8:261–265.

Lopes, I. M. G. and M. G. Bernardo-Gil. 2005. Characterisation of acorn oils extracted by hexane and by supercritical carbon dioxide. *Eur. J. Lipid Sci. Technol.* 107:12–19.

Lu, Z. M., W. Xu, N. H. Yu, T. Zhou, G. Q. Li, J. S. Shi, and Z. H. Xu. 2011. Recovery of aroma compounds from Zhenjiang aromatic vinegar by supercritical fluid extraction. *Int. J. Food Sci. Technol.* 46: 1508–1514.

Lubary, M., P. J. Jansens, J. H. Horst, and G. W. Hofland. 2010. Integrated synthesis and extraction of short-chain fatty acid esters by supercritical carbon dioxide. *AIChE J.* 56:1080–1089.

Ma, Q., X. Xu, Y. Gao, Q. Wang, and J. Zhao. 2008. Optimization of supercritical carbon dioxide extraction of lutein esters from marigold (*Tagetes erect* L.) with soybean oil as a co-solvent. *Int. J. Food Sci. Tech.* 43:1763–1769.

Machmudah, S., A. Shotipruk, M. Goto, M. Sasaki, and T. Hirose. 2006. Extraction of astaxanthin from *Haematococcus pluvialis* using supercritical CO_2 and ethanol as entrainer. *Ind. Eng. Chem. Res.* 45:3652–3657.

Macías-Sánchez, M. D., J. M. Fernandez-Sevilla, F. G. A. Fernández, M. C. C. García, and E. M. Grima. 2010. Supercritical fluid extraction of carotenoids from *Scenedesmus almeriensis*. *Food Chem.* 123:928–935.

Mariod, A. A., S. I. Abdelwahab, A. Gedi, and Z. Solati. 2010. Supercritical carbon dioxide extraction of sorghum bug (*Agonoscelis pubescens*) oil using response surface methodology. *J. Am. Oil Chem. Soc.* 87:849–856.

Marongiu, B., A. Piras, S. Porcedda, E. Tuveri, E. Sanjust, M. Meli, F. Sollai, P. Zucca, and A. Rescigno. 2007. Supercritical CO_2 extract of *Cinnamomum zeylanicum*: Chemical characterization and antityrosinase activity. *J. Agric. Food Chem.* 55:10022–10027.

Marsnia, Z. E., L. Casas, C. Mantella, M. Rodríguez, A. Torres, F. A. Macias, E. J. M. Ossa, J. M. G. Molinillo, and R. M. Varela. 2011. Potential allelopathic of the fractions obtained from sunflower leaves using supercritical carbon dioxide. *J. Supercrit. Fluids* 60:28–37.

Martín, L., L. F. Julio, J. Burillo, J. Sanz, A. M. Mainar, and A. González-Coloma. 2011a. Comparative chemistry and insect antifeedant action of traditional (Clevenger and Soxhlet) and supercritical extracts (CO_2) of two cultivated wormwood (*Artemisia absinthium* L.) populations. *Ind. Crops Prod.* 34:1615–1621.

Martín, L., A. M. Mainar, A. González-Coloma, J. Burillo, and J. S. Urieta. 2011b. Supercritical fluid extraction of wormwood (*Artemisia absinthium* L.). *J. Supercrit. Fluids* 56:64–71.

Martinez-Correa, H. A., F. A. Cabral, P. M. Magalhães, C. L. Queiroga, A. T. Godoy, A. P. Sánchez-Camargo, and L. C. Paviani. 2012. Extracts from the leaves of *Baccharis dracunculifolia* obtained by a combination of extraction processes with supercritical CO_2, ethanol and water. *J. Supercritic. Fluids* 63:31–39.

Marzouki, H., A. Piras, B. Marongiu, A. Rosa, and M. A. Dessì. 2008. Extraction and separation of volatile and fixed oils from berries of *Laurus nobilis* L. by supercritical CO_2 *Molecules* 13:1702–1711.

Mendes, R. L., A. D. Reis, and A. A. Palavra. 2006. Supercritical CO_2 extraction of γ-linolenic acid and other lipids from *Arthrospira* (*Spirulina*) *maxima*: Comparison with organic solvent extraction. *Food Chem.* 99:57–63.

Mendes, R. L., B. P. Nobre, J. P. Coelho, and A. A. Palavra. 1999. Solubility of β-carotene in supercritical carbon dioxide and ethane. *J. Supercrit. Fluids* 16:99–106.

Mendiola, J. A., M. Herrero, A. Cifuentes, and E. Ibáñez. 2007. Use of compressed fluids for sample preparation: Food applications. *J. Chromatogr. A* 1152:234–246.

Mezzomo, N., B. R. Mileo, M. T. Friedrich, J. Martínez, and S. R. S. Ferreira. 2010. Supercritical fluid extraction of peach (*Prunus persica*) almond oil: Process yield and extract composition. *Bioresour. Technol.* 101:5622–5632.

Mhemdi, H., E. Rodier, N. Kechaou, and J. Fages. 2011. A supercritical tuneable process for the selective extraction of fats and essential oil from coriander seeds. *J. Food Eng.*105:609–616.

Mitra, P., P. C. Barman, and K. S. Chang. 2011. Coumarin extraction from *Cuscuta reflexa* using supercritical fluid carbon dioxide and development of an artificial neural network model to predict the coumarin yield. *Food Bioprocess Technol.* 4:737–744.

Modey, W. K., D. A. Mulholland, and M. W. Raynor. 1996. Analytical supercritical fluid extraction of natural products. *Phytochem. Anal.* 7: 1–15.

Engineering Aspects of Food Biotechnology

Mohamed, R. S. and G. A. Mansoori. 2002. The use of supercritical fluid extraction technology in food processing. Featured Article—*Food Technology Magazine,* June. The World Markets Research Center, London, UK.

Mohamed, R. M., M. D. A. Saldana, and P. Mazzafera. 2002. Extraction of caffeine, theobromine, and cocoa butter from Brazilian cocoa beans using supercritical CO_2 and ethane. *Ind. Eng. Chem. Res.* 41:6751–6758.

Moldão-Martins, M., A. Palavra, M. L. Beirão-da-Costa, and M. G. Bernardo-Gil. 2000. Supercritical CO_2 extraction of *Thymus zygis* L. subsp. *sylvestris* aroma. *J. Supercrit. Fluids* 18:25–34.

Montanés, F., T. Fornari, P. J. Martín-Álvarez, N. Corzo, A. Olano, and E. Ibáñez. 2006. Selective recovery of tagatose from mixtures with galactose by direct extraction with supercritical CO_2 and different cosolvents. *J. Agric. Food Chem.* 54:8340–8345.

Montero, O., M. D. Macías-Sánchez, C. M. Lama, L. M. Lubián, C. Mantell, M. Rodríguez, and E. M. de la Ossa. 2005. Supercritical CO_2 extraction of â-carotene from a marine strain of the cyanobacterium *Synechococcus* species. *J. Agric. Food Chem.* 53:9701–9707.

Montes, A., M. D. Gordillo, C. Pereyra, and E. J. Martínez de la Ossa. 2012. Polymer and ampicillin co-precipitation by supercritical antisolvent process. *J. Supercrit. Fluids* 63:92–98.

Mortazavi, S. V., M. H. Eikani, H. Mirzaei, M. Jafari, and F. Golmohammad. 2010. Extraction of essential oils from *Bunium persicum* Boiss using superheated water. *Food Bioprod. Process* 88:222–226.

Moura, P. M., G. H. C. Prado, M. A. A. Meireles, and C. G. Pereira. 2012. Supercritical fluid extraction from guava (*Psidium guajava*) leaves: Global yield, composition and kinetic data. *J. Supercrit. Fluids* 62:116–122.

Mukhopadhyay, M. 2000. *Natural Extracts Using Supercritical Carbon Dioxide.* CRC Press, Boca Raton, FL: Taylor & Francis.

Mustapa, A. N., Z. A. Manan, C. Y. M. Azizi, W. B. Setianto, and A. K. M. Omar. 2011. Extraction of β-carotenes from palm oil mesocarp using sub-critical R134a. *Food Chem.* 125:262–267.

Nagy, B. and B. Simándi. 2008. Effects of particle size distribution, moisture content, and initial oil content on the supercritical fluid extraction of paprika. *J. Supercrit. Fluids* 46:293–298.

Natalia, D., L. Greiner, W. Leitner, and M. B. Ansorge-Schumacher. 2012. Stability, activity, and selectivity of benzaldehyde lyase in supercritical fluids. *J. Supercrit. Fluids* 62:173–177.

Nguyen, K., P. Barton, and J. S. Spencer. 1991. Supercritical carbon dioxide extraction of vanilla. *J. Supercrit. Fluids* 4:40–46.

Nisha, A. K. U. Sankar, and G. Venkateswaran. 2012. Supercritical CO_2 extraction of *Mortierella alpine* single cell oil: Comparison with organic solvent extraction. *Food Chem.* 133:220–226.

Nobre, B. P., A. F. Palavra, F. L. P. Pessoa, and R. L. Mendes. 2009. Supercritical CO_2 extraction of trans-lycopene from Portuguese tomato industrial waste. *Food Chem.* 116:680–685.

Norulaini, N. A. N., O. Anuar, A. F. M. AlKarkhi, W. B. Setianto, M. O. Fatehah, F. Sahena, and I. S. M. Zaidul. 2009a. Optimization of SC–CO_2 extraction of zerumbone from ginger *Zingiber zerumbet* (L) Smith. *Food Chem.* 114:702–705.

Norulaini, N. A. N., W. B. Setianto, I. S. M. Zaidul, A. H. Nawi, C. Y. M. Azizi, and A. K. M. Omar. 2009b. Effects of supercritical carbon dioxide extraction parameters on virgin coconut oil yield and medium-chain triglyceride content. *Food Chem.*116:193–197.

Nossack, A. C., J. H. Y. Vilegas, D. Von Baer, and F. M. Lanças. 2000. Supercritical fluid extraction and chromatographic analysis (HRGC–FID and HRGC–MS) of *Lupinus* spp. alkaloids. *J. Braz. Chem. Soc.* 11:495–501.

Nyam, K. L., C. P. Tan, R. Karim, O. M. Lai, K. Long, and Y. B. C. Man. 2010. Extraction of tocopherol-enriched oils from Kalahari melon and Roselle seeds by supercritical fluid extraction (SFE–CO_2). *Food Chem.* 119:1278–1283.

Nyam, K. L., C. P. Tan, O. M. Lai, K. Long, and Y. B. C. Man. 2011. Optimization of supercritical CO_2 extraction of phytosterol-enriched oil from Kalahari melon seeds. *Food Bioprocess Technol.* 4:1432–1441.

Oliveira R., M. F. Rodrigues, and M. G. Bernardo-Gil. 2002. Characterization and supercritical carbon dioxide extraction of walnut oil. *J. Am. Oil Chem. Soc.* 79:225–230.

Oliveira, P. F., R. A. F. Machado, A. Bolzan, and D. Barth. 2012. Supercritical fluid extraction of hernandulcin from *Lippia dulcis* Trev. *J. Supercrit. Fluids* 63:161–168.

Ollanketo, M., A. Peltoketo, K. Hartonen, R. Hiltunen, and M. L. Riekkola. 2002. Extraction of sage (*Salvia officinalis* L.) by pressurized hot water and conventional methods: Antioxidant activity of the extracts. *Eur. Food Res. Technol.* 215:158–163.

Ollanketo, M., K. Hartonen, M. L. Riekkola, Y. Holm, and R. Hiltunen. 2001. Supercritical carbon dioxide extraction of lycopene in tomato skins. *Eur. Food Res. Technol.* 212:561–565.

Ong, E. S., J. S. H. Cheong, and D. Goh. 2006. Pressurized hot water extraction of bioactive or marker compounds in botanicals and medicinal plant materials. *J. Chromatogr. A* 1112:92–102.

Ong, E. S. and S. M. Len. 2003. Pressurized hot water extraction of berberine, baicalein and glycyrrhizin in medicinal plants. *Anal. Chim. Acta* 482:81–89.

Orav, A., M. Koel, T. Kailas, and M. Müürisepp. 2010. Comparative analysis of the composition of essential oils and supercritical carbon dioxide extracts from the berries and needles of Estonian juniper (*Juniperus communis* L.). *Procedia Chem.* 2:161–167.

Ortuño, M., M. T. Martínez-Pastor, A. Mulet, and J. Benedito. 2012. Supercritical carbon dioxide inactivation of *Escherichia coli* and *Saccharomyces cerevisiae* in different growth stages. *J. Supercritic.* 63:8–15.

Ozel, M. Z., F. Gogus, and A. C. Lewis. 2003. Subcritical water extraction of essential oil from *Thymbra spicata. Food Chem.* 82:381–386.

Ozkal, S. G., M. E. Yener, and L. Bayindirli. 2005. Mass transfer modelling of apricot kernel oil extraction with supercritical carbon dioxide. *J. Supercrit. Fluids* 35:119–127.

Palma, M., L. T. Taylor, B. W. Zoecklein, and L. S. Douglas. 2000. Supercritical fluid extraction of grape glycosides. *J. Agric. Food Chem.* 48:775–779.

Park, H. S., H. J. Lee, M. H. Shin, K. W. Lee, H. Lee, Y. S. Kim, K. O. Kim, and K. H. Kim. 2007. Effects of cosolvents on the decaffeination of green tea by supercritical carbon dioxide. *Food Chem.* 105:1011–1017.

Park, C. I., M. S. Shin, and H. Kim. 2008. Micronization of arbutine using supercritical antisolvent. *Kor. J. Chem. Eng.* 25:581–584.

Patel, R. N., S. Bandyopadhyay, and A. Ganesh. 2011. Extraction of cardanol and phenol from bio-oils obtained through vacuum pyrolysis of biomass using supercritical fluid extraction. *Energy* 36:1535–1542.

Pawlowski, T. M. and C. F. Poole. 1998. Extraction of thiabendazole and carbendazim from foods using pressurized hot (subcritical) water for extraction: A feasibility study. *J. Agric. Food Chem.*, 46:3124–3132.

Pearce, K. L., C. Trenerry, and S. Were. 1997. Supercritical fluid extraction of pesticide residues from strawberries. *J. Agric. Food Chem.* 45:153–157.

Pensabene, J. W., W. Fiddler, and D. J. Donoghue. 2000. Supercritical fluid extraction of atrazine and other triazine herbicides from fortified and incurred eggs. *J. Agric. Food Chem.* 48:1668–1672.

Pereda, S., S. B. Bottini, and E. A. Brignole. 2008. Fundamentals of supercritical fluid technology, in *Supercritical Fluid Extraction of Nutraceuticals and Bioactive Compounds*, ed. J. L. Martinez, 1–23, Boca Raton, FL: CRC Press, Taylor & Francis Group.

Pereira, C. G. and M. A. A. Meireles. 2007. Economical analysis of rosemary, fennel and anise essential oils obtained by supercritical fluid extraction. *Flavour Fragr. J.* 22:407–413.

Pereira, C. G. and M. A. A. Meireles. 2010. Supercritical fluid extraction compounds: Fundamentals, applications and economic perspectives. *Food Bioprocess Technol.* 3:340–372.

Pereira, P., M. J. Cebola, and M. J.Bernardo-Gil. 2012. Comparison of antioxidant activity in extracts of *Myrtus communis* L. obtained by SFE vs. solvent extraction. *J. Environm. Sci. Eng. A* 1:115–120.

Pham, M., S. Pollak and M. Petermann. 2012. Micronisation of poly(ethylene oxide) solutions and separation of water by PGSS-drying. *J. Supercritic. Fluids* 64:19–24.

Pinelo, M., A. Ruiz-Rodríguez, J. Sineiro, F. J. Senorans, G. Reglero, and M. J. Nunez. 2007. Supercritical fluid and solid–liquid extraction of phenolic antioxidants from grape pomace: A comparative study. *Eur. Food Res. Technol.* 226:199–205.

Pozo-Bayón, M. A., I. Andújar-Ortiz, J. A. Mendiola, E. Ibánez, and M. V. Moreno-Arribas. 2010. Application of supercritical CO_2 extraction for the elimination of odorant volatile compounds from winemaking inactive dry yeast preparation. *J. Agric. Food Chem.* 58:3772–3778.

Prado, J.M., I. Dalmolin, N. D. D. Carareto, R. C. Basso, A. J. A. Meirelles, V. Oliveira, E. A. C. Batista, and A. A. Meireles. 2012. Supercritical fluid extraction of grape seed: Process scale-up, extract chemical composition and economic evaluation. *J. Food Eng.* 109:249–257.

Quispe-Condori, S., D. Sánchez, M. A. Foglio, P. T. V. Rosa, C. Zetzl, G. Brunner, and M. A. A. Meireles. 2005. Global yield isotherms and kinetic of artemisinin extraction from *Artemisia annua* L. leaves using supercritical carbon dioxide. *J. Supercrit. Fluids* 36:40–48.

Quitain, A. T., K. Oro, S. Katoh, and T. Moriyoshi. 2006. Recovery of oil components of okara by ethanol-modified supercritical carbon dioxide extraction. *Bioresour. Technol.* 97:1509–1514.

Raeissi, S., J. C. Asensi, and C. J. Peters. 2002. Phase behavior of the binary system ethane+linalool. *J. Supercrit. Fluids* 24:111–121.

Rahimi, E., J. M. Prado, G. Zahedi, and M. A. A. Meireles. 2011. Chamomile extraction with supercritical carbon dioxide: Mathematical modeling and optimization. *J. Supercritic. Fluids* 56:80–88.

Rajaei, A., M. Barzegar, and Y. Yamini. 2005. Supercritical fluid extraction of tea seed oil and its comparison with solvent extraction. *Eur. Food Res. Technol.* 220:401–405.

Ramirez, P., F. J. Senorans, E. Ibanez, and G. Reglero. 2004. Separation of rosemary antioxidant compounds by supercritical fluid chromatography on coated packed capillary columns. *J. Chromatogr. A* 1057:241–245.

Rangsriwong, P., N. Rangkadilok, J. Satayavivad, M. Goto, and A. Shotipruk. 2009. Subcritical water extraction of polyphenolic compounds from *Terminalia chebula* Retz. fruits. *Separ. Purif. Technol.* 66:51–56.

Reis-Vasco, E. M. C., J. A. P. Coelho, A. M. F. Palavra, C. Marrone, and E. Reverchon. 2000. Mathematical modelling and simulation of pennyroyal essential oil supercritical extraction. *Chem. Eng. Sci.* 55:2917–2922.

Reverchon, E. 1997. Supercritical fluid extraction and fractionation of essential oil and products, *J. Supercrit. Fluids* 10:1–37.

Reverchon, E. and I. De Marco. 2006. Supercritical fluid extraction and fractionation of natural matter—A review. *J. Supercrit. Fluids* 38:146–166.

Reverchon, E. and I. De Marco. 2008. Essential oils extraction and fractionation using supercritical fluids, in *Supercritical Fluid Extraction of Nutraceuticals and Bioactive Compounds*, ed. J. L. Martinez, 305–335. Boca Raton, FL: CRC Press, Taylor & Francis Group.

Reverchon, E., G. Donsi, and L. S. Osseo. 1993.Modeling of supercritical fluid extraction from herbaceous matrices. *Ind. Eng. Chem. Res.* 32:2721–2726.

Reverchon, E., A. Kaziunas, and C. Marrone. 2000. Supercritical CO_2 extraction of hiprose seed oil: Experiments and mathematical modeling. *Chem. Eng. Sci.* 55: 2195–2201.

Reverchon, E. and C. Marrone. 2001. Modeling and simulation of the supercritical CO_2 extraction of vegetable oils. *J. Supercrit. Fluids* 19:161–175.

Reverchon, E., L. S. Osseo, and D. Gorgoglione. 1994. Supercritical CO_2 extraction of basil oil: Characterization of products and process modeling. *J. Supercrit. Fluids* 7:185–190.

Ribeiro, M. A., M. G. Bernardo-Gil, and M. M. Esquível. 2001. *Melissa officinalis*, L.: Study of antioxidant activity in supercritical residues. *J. Supercrit. Fluids* 21:51–60.

Rodrigues, V. M. R., P. T.V. Rosa, M. O. M. Marques, A. J. Petenate, and M. A. A. Meireles. 2003. Supercritical extraction of essential oil from aniseed (*Pimpinella anisum* L.) using CO_2: Solubility, kinetics, and composition data. *J. Agric. Food Chem.* 51:1518–1523.

Rodríguez-Meizoso, I., L. Jaime, S. Santoyo, A. Cifuentes, G. García-Blairsy Reina, F. J. Señoráns, and E. Ibáñez. 2008. Pressurized fluid extraction of bioactive compounds from *Phormidium* species. *J. Agric. Food Chem.* 56:3517–3523.

Rodríguez-Meizoso, I., L. Jaime, S. Santoyo, F. J. Señoráns, A. Cifuentes, and E. Ibáñez. 2010. Subcritical water extraction and characterization of bioactive compounds from *Haematococcus pluvialis* microalga. *J. Pharm. Biomed. Anal.* 51:456–463.

Romo-Hualde, A., A. I. Yetano-Cunchillos, C. González-Ferrero, M. J. Sáiz-Abajo, and C. J. González-Navarro. 2012. Supercritical fluid extraction and microencapsulation of bioactive compounds from red pepper (*Capsicum annum* L.) by-products. *Food Chem.* 133:1045–1049.

Rosa, P. T. V. and M. A. A. Meireles. 2005. Rapid estimation of the manufacturing cost of extracts obtained by supercritical fluid extraction. *J. Food Eng.* 67:235–240.

Rovio, S., K. Hartonen, Y. Holm, R. Hiltunen, and M. L. Riekkola. 1999. Extraction of clove using pressurized hot water. *FlavourFragr. J.* 14:399–404.

Rubio-Rodríguez, N., S. M. de Diego, S. Beltrán, I. Jaime, M. T. Sanz, and J. Rovira. 2012. Supercritical fluid extraction of fish oil from fish by-products: A comparison with other extraction methods. *J. Food Eng.* 109:238–248.

Sahena, F., I. S. M. Zaidul, S. Jinap, A. M. Yazid, A. Khatib, and N. A. N. Norulaini. 2010. Fatty acid compositions of fish oil extracted from different parts of Indian mackerel (*Rastrelliger kanagurta*) using various techniques of supercritical CO_2 extraction. *Food Chem.* 120:879–885.

Sajfrtova, M., I. Licková, M. Wimmerova, H. Sovová, and Z. Wimmer. 2010. β-Sitosterol: Supercritical carbon dioxide extraction from sea Buckthorn (*Hippophae rhamnoides* L.) seeds. *Int. J. Mol. Sci.* 11:1842–1850.

Saldaña, M. D. A., F. Temelli, S. E. Guigard, B. Tomberli, and C. G. Gray. 2010. Apparent solubility of lycopene and β-carotene in supercritical CO_2, CO_2+ethanol and CO_2+canola oil using dynamic extraction of tomatoes. *J. Food Eng.* 99:1–8.

Salgin, U. 2007. Extraction of jojoba seed oil using supercritical CO2+ethanol mixture in green and high-tech separation process. *J. Supercrit. Fluids* 39:330–337.

Sanches Filho, P. J., A. Rios, M. Valcarcel, M. I. S. Melecchi, and E. B. Camarao. 2007. Method of determination of nitrosamines in sausages by CO_2 supercritical fluid extraction (SFE) and micellar electrokinetic chromatography (MEKC). *J. Agric. Food Chem.* 55:603–607.

Sanchez-Camargo, A. P., M. A. A. Meireles, A. L. K. Ferreira, E. Saito, and F. A. Cabral. 2012. Extraction of ω-3 fatty acids and astaxanthin from Brazilian redspotted shrimp waste using supercritical CO_2+ethanol mixtures. *J. Supercrit. Fluids* 61:71–77.

Savita, K. and K. Prakashchandra. 2011. Optimization of extraction conditions and development of a sensitive HPTLC method for estimation of wedelolactone in different extracts of *Eclipta alba. Int. J. Pharm. Sci. Drug Res.* 3:56–61.

Seabra, I. J., A. M. A. Dias, M. E. M. Braga, and H. C. de Sousa. 2012. High pressure solvent extraction of maritime pine bark: Study of fractionation, solvent flow rate and solvent composition. *J. Supercrit. Fluids* 62:135–148.

Seabra, I. J., M. E. M. Braga, and H. C. de Sousa. 2012. Statistical mixture design investigation of CO_2–ethanol–H_2O pressurized solvent extractions from tara seed coat. *J. Supercrit. Fluids* 64:9–18.

Senorans, F. J., E. Ibanez, S. Cavero, J. Tabera, and G. Reglero. 2000. Liquid chromatographic–mass spectrometric analysis of supercritical fluid extracts of rosemary plants. *J. Chromatogr. A* 870:491–499.

Shi, J., C. Yi, J. Xue, Y. Jiang, Y. Ma, and D. Li. 2009. Effects of modifier on lycopene extract profile from tomato skin using supercritical–CO_2 fluid. *J. Food Eng.* 93:431–436.

Shi, J., S. J. Xue, Y. Ma, Y. Jiang, X. Ye, and D. Yu. 2012. Green separation technologies in food processing: Supercritical–CO_2 fluid and subcritical water extraction. In *Green Technologies in Food Production and Processing*, eds. J. I. Boye, and Y. Arcand, 273–294, Springer, London.

Sierra-Pallares, J., D. L. Marchisio, M. T. Parra-Santos, J. García-Serna, F. Castro, and M. J. Cocero. 2012. A computational fluid dynamics study of supercritical antisolvent precipitation: Mixing effects on particle size. *AICHE J.* 58:385–398.

Silva, F. V. M., A. Martins, J. Salta, N. R. Neng, J. M. F. Nogueira, D. Mira, N. Gaspar, J. Justino, C. Grosso, J. S. Urieta, A. M. S. Palavra, and A. P. Rauter. 2009. Phytochemical profile and anticholinesterase and antimicrobial activities of supercritical versus conventional extracts of *Satureja montana*. *J. Agric. Food Chem.* 57:11557–11563.

Smith, R. L. Jr., R. M. Malaluan, W. B. Setianto, H. Inomata, and K. Arai. 2003. Separation of cashew (*Anacardium occidentale* L.) nut shell liquid with supercritical carbon dioxide. *Bioresour. Technol.* 88: 1–7.

Snyder, J. M., J. P. Friedrich, and D. D. Christianson. 1984. Effect of moisture and particle size on the extractability of oils from seeds with supercritical CO_2. *J. Am. Oil Chem. Soc.* 61:1851–1856.

Sodeifian, G. and K. Ansari. 2011. Optimization of *Ferulago Angulata* oil extraction with supercritical carbon dioxide. *J. Supercrit. Fluids* 57:38–43.

Solati, Z., B. S. Baharin, and H. Bagheri. 2012. Supercritical carbon dioxide (SC–CO_2) extraction of *Nigella sativa* L. oil using full factorial design. *Ind. Crops Prod.* 36:519–523.

Sovová, H. 1994. Rate of the vegetable oil extract with supercritical CO_2—I. Modeling of extraction curves. *Chem. Eng. Sci.* 49:409–414.

Sovová, H. 2005. Mathematical model for supercritical fluid extraction of natural products and extraction curve evaluation. *J. Supercrit. Fluids* 33:35–52.

Sovová, H., J. Kucera, and J. Jez. 1994. Rate of the vegetable oil extraction with supercritical CO_2—II. Extraction of grape oil. *Chem. Eng. Sci.* 49:415–420.

Span, R. and W. Wagner. 1996. A new equation of state for carbon dioxide covering the fluid region from the triple-point temperature to 1100 K at pressures up to 800 MPa. *J. Phys. Chem. Ref. Data* 25:1509–1596.

Stamenic, M., I. Zizovic, R. Eggers, P. Jaeger, H. Heinrich, E. Rój, J. Ivanovic, and D. Skala. 2010. Swelling of plant material in supercritical carbon dioxide. *J. Supercrit. Fluids* 52:125–133.

Subra-Paternault, P., D. Vrel, and C. Roy. 2012. Coprecipitation on slurry to prepare drug–silica–polymer formulations by compressed antisolvent. *J. Supercritic. Fluids* 63:69–80.

Sugihara, N., A. Kanda, T. Nakano, T. Nakamura, H. Igusa, and S. Hara. 2010. Novel fractionation method for squalene and phytosterols contained in the deodorization distillate of rice bran oil. *J. Oleo Sci.* 59:65–70.

Sun, Y., Z. Liu, J. Wang, W. Tian, H. Zhou, L. Zhu, and C. Zhang. 2008. Supercritical fluid extraction of paeonol from *Cynanchum paniculatum* (Bge.) Kitagawa and subsequent isolation by high-speed counter-current chromatography coupled with high-performance liquid chromatography–photodiode array detector. *Separ. Purif. Technol.* 64:221–226.

Sun, L., K. A., Rezaei, F. Temelli, and B. Ooraikul. 2002. Supercritical fluid extraction of alkylamides from *Echinacea angustifolia*. *J. Agric. Food Chem.* 50:3947–3953.

Sun, M. and F. Temelli. 2006. Supercritical carbon dioxide extraction of carotenoids from carrot using canola oil as a continuous co-solvent. *J. Supercrit. Fluids* 37:397–408.

Takeuchi, T., M. L. Rubano, and M. A. A. Meireles. 2010. Characterization and functional properties of Macela (*Achyrocline satureioides*) extracts obtained by supercritical fluid extraction using mixtures of CO_2 plus ethanol. *Food Bioprocess Technol.* 3:804–812.

Tanaka, Y. and I. Sakaki. 2005. Extraction of phospholipids from unused natural resources with supercritical carbon dioxide and an entrainer. *J. Oleo Sci.* 54:569–576.

Tanaka, Y., I. Sakaki, and T. Ohkubo. 2004. Extraction of phospholipids from *Salmon roe* with supercritical carbon dioxide and an entrainer. *J. Oleo Sci.* 53:417–423.

Tang, S., C. Qin, H. Wang, S. Li, and S. Tian. 2011. Study on supercritical extraction of lipids and enrichment of DHA from oil-rich. *J. Supercrit. Fluids* 57:44–49.

Tello, J., M. Viguera, and L. Calvo. 2011. Extraction of caffeine from Robusta coffee (*Coffea canephora* var. *Robusta*) husks using supercritical carbon dioxide. *J. Supercrit. Fluids* 59:53–60.

Tozzi, R., N. Mulinacci, K. Storlikken, I. Pasquali, F. F. Vincieri, and R. Bettini. 2008. Supercritical extraction of carotenoids from *Rosa canina* L. Hips and their formulation with β-cyclodextrin. *Pharm. Sci. Tech* 9:693–700.

Ueno, H., M. Tanaka, M. Hosino, M. Sasaki, and M. Goto. 2008. Extraction of valuable compounds from the flavedo of *Citrus junos* using subcritical water. *Sep. Purif. Technol.* 62:513–516.

Uquiche, E., X. Fica, K. Salazar, and J. M. Del Valle. 2012. Time fractionation of minor lipids from cold-pressed rapeseed cake using supercritical CO_2. *J. Am. Oil Chem. Soc.* 89:1135–1144.

Vasapollo, G., L. Longo, L. Rescio, and L. Ciurlia. 2004. Innovative supercritical CO_2 extraction of lycopene from tomato in the presence of vegetable oil as cosolvent. *J. Supercritic. Fluids* 29:87–96.

Verbeke, W. 2005. Consumer acceptance of functional foods: Socio-demographic, cognitive and attitudinal determinants. *Food Qual. Preference* 16:45–57.

Vergara-Mendoza, M. S., C. H. Ortiz-Estrada, J. González-Martínez, and J. A. Quezada-Gallo. 2012. Microencapsulation of coenzyme Q_{10} in poly(ethylene glycol) and poly(lactic acid) with supercritical carbon dioxide. *Ind. Eng. Chem. Res.* 51:5840–5846.

Wang, H., C. Chen, and C. J. Chang. 2001. Carbon dioxide extraction of ginseng root hair oil and ginsenosides. *Food Chem.* 72:505–509.

Wang, H., Y. Liu, S. Wei, and Z. Yan. 2012. Application of response surface methodology to optimise supercritical carbon dioxide extraction of essential oil from *Cyperus rotundus* Linn. *Food Chem.* 132:582–587.

Wang, L., B. Pan, J. Sheng, J. Xu, and Q. Hu. 2007. Antioxidant activity of *Spirulina platensis* extracts by supercritical carbon dioxide extraction. *Food Chem.* 105:36–41.

Wang, Y., D. Sun, H. Chen, L. Qian, and P. Xu. 2011. Fatty acid composition and antioxidant activity of tea (*Camellia sinensis* L.) seed oil extracted by optimized supercritical carbon dioxide. *Int. J. Mol. Sci.* 12: 7708–7719.

Wang, L., B. Yang, X. Du, and C. Yi. 2008. Optimisation of supercritical fluid extraction of flavonoids from Pueraria lobata. *Food Chem.* 108:737–741.

Wood, J. A., M. A. Bernards, W. K. Wan, and P. A. Charpentier. 2006. Extraction of ginsenosides from North American ginseng using modified supercritical carbondioxide. *J. Supercrit. Fluids* 39:40–47.

Xie, J. J., J. Lu, Z. M. Qian, Y. Yu, J. A. Duan, and S. P. Li. 2009. Optimization and comparison of five methods for extraction of coniferyl ferulate from *Angelica sinensis*. *Molecules* 14:555–565.

Xu, W., K. Chu, H. Li, L. Chen, Y. Zhang, and X. Tang. 2011. Extraction of *Lepidium apetalum* seed oil using supercritical carbon dioxide and anti-oxidant activity of the extracted oil. *Molecules* 16:10029–10045.

Yamini, Y., M. Khajeh, E. Ghasemi, M. Mirza, and K. Javidnia. 2008. Comparison of essential oil compositions of *Salvia mirzayanii* obtained by supercritical carbon dioxide extraction and hydrodistillation methods. *Food Chem.* 108: 341–346.

Yang, C., Y. R. Xu, and W. X. Yao. 2002. Extraction of pharmaceutical components from *Ginkgo biloba* leaves using supercritical carbon dioxide. *J. Agric. Food Chem.* 50:846–849.

Yi, C., J. Shi, S. J. Xue, Y. Jiang, and D. Li. 2009. Effects of supercritical fluid extraction parameters on lycopene yield and antioxidant activity. *Food Chem.* 113:1088–1094.

Yu, J., D. V. Dandekar, R. T. Toledo, R. K. Singh, and B. S. Patil. 2006. Supercritical fluid extraction of limonoid glucosides from grapefruit molasses. *J. Agric. Food Chem.* 54:6041–6045.

Yu, J., D. V. Dandekar, R. T. Toledo, R. K. Singh, and B. S. Patil. 2007. Supercritical fluid extraction of limonoids and naringin from grapefruit (*Citrus paradise* Macf.) seeds. *Food Chem.* 105:1026–1031.

Zancan, K. C., M. O. M. Marques, A. J. Petenate, and M. A. A. Meireles. 2002. Extraction of ginger (*Zingiber officinale* Roscoe) oleoresin with CO_2 and cosolvents: A study of the antioxidant action of the extracts. *J. Supercrit. Fluids* 24:57–76.

Zarena, A. S., N. M. Sachindra, and K. U. Sankar. 2012. Optimisation of ethanol modified supercritical carbon dioxide on the extract yield and antioxidant activity from *Garcinia mangostana* L. *Food Chem.* 130:203–208.

Zekovic, Z., Z. Lepojevic, and A. Tolic. 2003. Modeling of the thymesupercritical carbon dioxide extraction system. II. The influence of extraction time and carbon dioxide pressure. *Sep. Sci. Technol.* 38:541–552.

10 Innovative Unit Operations

Jane Selia dos Reis Coimbra, João Paulo Martins, and Luiza Helena Meller da Silva

CONTENTS

10.1 INTRODUCTION

The multicomponent constitution of biotechnological fluids in downstream processing demands innovative separation technologies in the liquid phase aiming at extracting, concentrating, and purifying target biomolecules, especially for food industry applications. A biotechnology-based compound intended to be used as a food ingredient must be handled at gentle processing conditions to preserve its structure and bioactivity, and consequently, to maintain the related functionality and nutritional quality in the added foodstuff. The liquid–liquid extraction unit operation with its variants (1) liquid extraction using aqueous two-phase system (ATPS); (2) liquid extraction using ionic liquids (ILs); and (3) liquid extraction using deep eutectic solvent (DEP) meets the criteria of mild, energy cost-reduced, and safe processing conditions. Owing to the increasing concerns about environmental effects and also to obtain a marketable and competitive bioproduct, the organic solvent in these variants is replaced by a fluid phase, which should be "green" and environmentally safe, bearing the features of being nonthermolabile, readily biodegradable, biorenewable, having low biotoxicity and low cost, and being largely marketable. The following sections focus on process conditions needed to apply the approach of "green" liquid extraction to bioseparations.

10.2   EXTRACTION USING AQUEOUS TWO-PHASE SYSTEM

Traditionally, the majority of liquid–liquid extraction processes use organic compounds as solvents. The use of organic solvents provides a variety of combinations to purify several kinds of solutes of industrial and scientific interest. However, this traditional method of extraction has presented many environmental and health problems because it uses organic solvents that degrade biomolecules and are volatile, flammable, and toxic to humans and the environment.

An alternative to conventional liquid–liquid extraction using economically viable organic solvents to replace the traditional methods of separation is known as the aqueous two-phase system. The ATPS was discovered in 1896 by Beijerink. He observed that the mixture of gelatine + agar + water or soluble starch + gelatine + water under certain temperature and concentration conditions separated into two phases with different intensive thermodynamic properties. The chemical analysis of ATPS phases showed that the lower phase was rich in starch or agar and the upper phase was enriched with gelatine. However, the first application of ATPS for the separation of biological material was reported only in 1956 (Albertsson, 1986).

ATPS will form when two types of water-soluble polymers or a water-soluble polymer and a low-molecular-weight substance (inorganic salt in general) dissolve in an aqueous solution above the critical concentrations. Each system can be characterized by its unique phase diagram, which contains the equilibrium phase compositions required for the formation of two phases. ATPS are used because they provide different physical and chemical environments for the partitioning of biomolecules, thus making possible its application in biomolecule separation and purification. Biomolecule partitioning in ATPS is influenced by many variables such as salt type and concentration, temperature, and the structural properties of the biomolecule.

The implementation of ATPS separation processes requires that, apart from the efficient separation/purification of the target molecule, the following aspects are considered: to reduce component cost, to increase the reused components' performance, and to develop recycling of the components.

10.2.1   PHASE DIAGRAMS

The graphical representation of ATPS is expressed by a phase diagram, which may be triangular or rectangular. Figure 10.1 shows a typical rectangular phase diagram formed by a polymer + salt + water system. The concentration of components can be observed in the axes (vertical and horizontal). Water quantities in each phase, in % mass, are calculated from the mass balance in each phase. In the same figure, a classical behavior of a polymer–salt ATPS (mutual exclusion between polymer and salt) can be observed. The salt has a tendency to concentrate in the lower phase and the polymer prefers the upper phase. The curve that divides the biphasic area from the monophasic area is called the binodal curve. The biphasic area is localized above the binodal curve and under the binodal curve the system has just one phase. Thus,

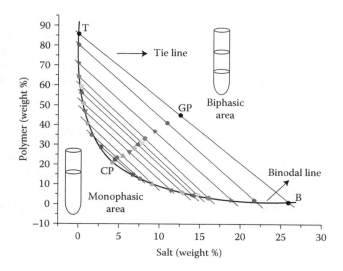

FIGURE 10.1 Rectangular phase diagram for a polymer–salt ATPS.

the biphasic region must be used for compound extraction using two liquid phases. The global composition or global point (GB) in liquid–liquid extraction is the over-all mixture composition, which form two immiscible phases.

All the global points are inserted in a line named tie line (TL), which is formed by the points that represent the composition of the top phase (T), global point (GP), and bottom phase (B). Each tie line has its own intensive properties, such as density, refractive index, interfacial tension, composition, and other physicochemical properties (Zaslavsky, 1995). In theory, there are infinite GBs for just one TL, but the extensive properties, such as volume, can be modified if the GB changes in the same TL. In general, in the middle of the tie line, the volume of top and bottom phases is the same. A particular composition of an ATPS is represented in Figure 10.1 as the critical point (CP). In the CP, the two phases have the same properties and it is not possible to distinguish between both the phases.

A parameter that expresses the difference in intensive thermodynamic properties of each phase of the ATPS is the tie line length (TLL), which can be described by Equation 10.1:

$$TLL = \left\{ (W_P^T - W_P^B)^2 + (W_s^T - W_s^B)^2 \right\}^{1/2} \tag{10.1}$$

where W_P and W_S are the polymer and salt concentrations, and the superscripts T and B designate the top and bottom phases, respectively.

Another important parameter is the slope of tie line (STL) described by Equation 10.2:

$$STL = (W_P^T - W_P^B)/(W_s^T - W_s^B) \tag{10.2}$$

where the variables have the same meanings as before.

These systems can be used for the purification or separation of chemical compounds such as metallic complexes; amino acids and proteins; biological materials, such as viruses, antibodies, and cells; and natural dyes (Martins et al., 2010; Rosa et al., 2011; Ruiz-Ruiz et al., 2012). The partitioning behavior of chemical substances will depend on the physical–chemical properties of ATPS and specific thermodynamic conditions.

10.2.2 PARTITIONING COEFFICIENT AND THERMODYNAMIC PARAMETERS

The partition coefficient (K_P) is given by the ratio of the activities (a_i) of the solute in the upper and lower phases, respectively. ATPS are appropriate to perform the separation of solutes present at very low concentrations in solutions, or equally at "infinite dilution." Based on this behavior, the activity coefficient (γ_i^α) obeys the limit law represented by Equation 10.3:

$$\lim_{C \to 0^+} \gamma_i^\alpha = 1 \tag{10.3}$$

The result expressed by Equation 10.3 implies that the partition coefficient can be represented by Equation 10.4:

$$K_P = \frac{[\text{solute}]_i^{FS}}{[\text{solute}]_i^{FI}} \tag{10.4}$$

where $[\text{solute}]_i^{FS}$ and $[\text{solute}]_i^{FI}$ are the equilibrium concentrations of the partitioned biomaterials or any solute in the top and bottom phase, respectively.

The thermodynamic parameter partition coefficient, Gibbs free energy, enthalpy, and entropy can be used to understand the partitioning behavior of biomaterials or other solutes in ATPS. These parameters can be obtained by using the classical equations of thermodynamics.

If the partition coefficient is determined at infinite dilution, and at different temperatures in the same TLL, the nonlinear Van't Hoff equation can be applied to obtain the transfer enthalpy energy associated with the solute ($\Delta_{tr}H^o$):

$$\Delta_{tr}H^o = -R\left[b + 2c\left(\frac{1}{T}\right) + 3d\left(\frac{1}{T}\right)^2 + \cdots \right] \tag{10.5}$$

where R is the universal gas constant, T is the temperature, and b, c, and d are the coefficients of Equation 10.5.

The transfer free energy change ($\Delta_{tr}G^o$) is determined from the equation

$$\Delta_{tr}G^o = -RT \ln K \tag{10.6}$$

and the corresponding entropy change from the equation

$$\Delta_{tr}S^o = \frac{(\Delta_{tr}H^o - \Delta_{tr}G^o)}{T} \qquad (10.7)$$

Thermodynamically, $\Delta_{tr}G^o$ energy is the result of two contributions, one enthalpic, $\Delta_{tr}H^o$, and another entropic, $\Delta_{tr}S^o$. These thermodynamic parameters are useful in a qualitative interpretation of the driving forces dominating the spontaneous transference of a solute between the phases.

10.2.3 EMPLOYMENT OF AQUEOUS TWO-PHASE SYSTEMS

ATPS has a great potential to be used by the industry to obtain a product with high purity and with little impact on the environment. A large number of substances can be separated or purified using ATPS (Rosa et al., 2011; Mageste et al., 2012; Ruiz-Ruiz et al., 2012).

One of the most studied types of molecules in ATPS partitioning is protein. As an example, the partitioning behavior of the proteins α-La and β-Lg in aqueous two-phase systems and salts can be mentioned (Silva and Meirelles, 2000, 2001). Boaglio et al. (2006) studied the effect of polymer, pH, and molar mass on the partitioning of these proteins in ATPS formed by PEG + sodium citrate. Rodrigues et al. (2001) partitioned α-La and β-Lg in ATPS formed by PEG + ammonium sulfate and their results show that it is possible to separate these proteins, once α-La (with a partitioning coefficient of 12.8) prefers to stay in the polymer-rich phase and β-Lg (with a partitioning coefficient of 0.34) prefers to stay in the salt-rich phase. Also, the partitioning of norbixin, from natural dye, was studied in different ATPS with a partitioning coefficient of 130 (Mageste et al., 2012).

In the last decade, a variant of ATPS formed by ionic liquids + salt + water was introduced (Gutowski et al., 2003; Liu et al., 2011; Freire et al., 2012).

10.3 EXTRACTION USING IONIC LIQUIDS

An IL is a salt with a low melting point and, by definition, exists at the liquid phase at temperatures lower than the boiling point of water, usually below 100°C (Hallet and Welton, 2011). ILs are not salts dissolved in liquids. They are a phase composed solely of ions—cations and anions. An IL melts without decomposing or vaporizing. ILs are also called fused salts, liquid salts, molten salts, liquid electrolytes, syrup solvent, ionic melts, ionic fluids, and ionic glasses. Molten salts at room temperature are named room-temperature ionic liquids (RTILs).

In 1888, the first IL, ethanolammonium nitrate, with a melting point of 55°C, was reported. In 1914, the first room-temperature IL, ethylammonium nitrate (melting point of 12°C) was synthesized. Also, IL 1-butyl-3-methyl imidazolium tetrafluoroborate has a melting point as low as −82°C and IL mono-6-butyl imidazolium iodide, as high as 242°C. For comparison, the melting point of sodium chloride, a typical

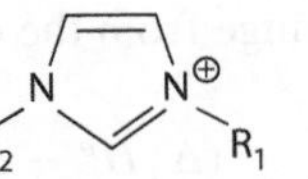

FIGURE 10.2   Structure of imidazolium cation.

ionic solid, is 803°C (Earle and Seddon, 2000; Kokorin, 2011). ILs include a great variety of compounds, in the range of 10^4 (Chiappe and Pieraccini, 2005).

ILs generally consist of three parts: (1) the charged group of the cation (an organic head group); (2) the anion (an organic or inorganic group); and (3) the side chain. Usually, the cation is larger than the anion. In IL structure, there is at least one organic ion and one ion with a delocalized charge. The ions are also generally poorly coordinated as the head group (cation) has a low degree of symmetry due to the difference in the size of the ions. This peculiarity provides conformational flexibility (large entropy changes) to the ions, which diminishes the lattice energy (small lattice enthalpies) of the ILs, favoring the system's liquid condition. Consequently, the melting point is low and the formation of stable crystal lattices is hindered (Hallet and Welton, 2011). Imidazolium is a cation head group much studied with the advent of ILs (Figure 10.2).

Other cation groups are ammonium, guanidinium, phosphonium, piperidinium, pyridinium, pyrrolidinium, sulfonium, and xanthinium, among many others. Examples of alkyl side chains of the cation group are methyl, ethyl, propyl to tetradecyl, and various chains with greater carbon numbers. Some anions used in ILs are acetate, alkylphosphate, bistriflimide, chloride, halide, hydrogen sulfate, methylsulfonate, thiocyanate, tetrafluoroborate, and tetrahalogenoaluminate (Freire et al., 2012).

10.3.1   SIMPLIFIED SYNTHESIS OF IONIC LIQUIDS

The typical procedure for the preparation of ILs can be explained by using cations based upon alkylated amines: (1) alkylation of an amine to produce an intermediate salt (imidazole materials and allyl chloride (or chloroethanol) are added into a round bottom flask); (2) the reaction is conducted under refluxing for 8–48 h. After reaction, the mixture is extracted (with ether, for example) for several times to remove the residual imidazole. The IL is obtained after drying at 80°C under vacuum (0.08 MPa, for example) for 24 h. The structures of synthesized ILs must be confirmed by characterization techniques, such as spectroscopic (FT-IR, NIR, 1HNMR, IR), chromatographic, and thermal analyses (such as TGA), among others (Hallett and Welton, 2011; Kokorin, 2011).

10.3.2   IONIC LIQUID FEATURES

ILs can act as solvents for synthesis, separation, and analysis in different chemical processes, as well as catalysts. ILs for industrial applications have the following advantages (Kokorin, 2011):

- Low vapor pressure (as low as 1×10^{-10} Pa, at 25°C): The absence of volatility is one of the most important benefits of ILs as compared to low-boiling-point organic solvents and results in ILs' reduced air emission into the environment, nonflammability, and nonexplosiveness. They are nonvolatile and thus do not emit vapors. As a consequence, ILs are easy to handle and store. Exception should be noted for some ILs that can be combustible and require careful handling (Smiglak et al., 2006).
- Reduced toxicity: ILs offer lower toxicity compared to volatile organic chemicals (VOCs), which makes them ideal replacements for volatile and nonsafe organic solvents.
- High thermal stability: ILs show thermal stability in a wide range, from −40°C to 200°C, which is larger than for water and organic solvents such as acetone, benzene, or methanol. Some ILs can be submitted to vacuum distillation at about 300°C, producing vapor-containing ion pairs, can generate flammable gases on thermal decomposition, and have short-term thermal stability (few minutes).
- Solubility: ILs have the behavior of anion (proton acceptor) or cation (proton donor) and therefore act as acceptor or donor substances. The miscibility of ILs in water is influenced by the type of anion present and by the cation structure. ILs can serve as a reaction medium for aqueous and nonaqueous reactions. There are also ILs immiscible in many organic solvents. Extraction processes demand solubility data to predict their behavior with other solvents and also the solute partitioning into the ILs.
- Solvents of different inorganic and organic solutes: ILs also exhibit the ability to dissolve a wide variety of compounds with varying polarity. This, coupled with the relatively inexpensive production costs, would make them ideal candidates for use as cleaners in household and industrial environments. ILs dissolve materials such as limescale, bleaches, metal tarnishes, fats, proteins, and surfactants, which can then be used in the solubilization of oils, sugars, and polysaccharides.
- Good electrical stability: ILs have wide electrochemical windows (that is, a large range of electrochemical potential in which the electrolyte is neither reduced nor oxidized at an electrode) making them suitable for various electrochemical applications such as rechargeable batteries and electroplating of base metals.

According to Marsh et al. (2004): (1) melting point of ILs is affected by the charge distribution on the ions, hydrogen bonding ability, Van der Waals interactions, and the symmetry of the ions; (2) most of the known ILs are denser than water, with values between 1 and 1.6 g/cm³. The density decreases with increasing length of the alkyl chain; (3) IL data of surface tension are scarce. Values ranging between 33.8 and 54.7 N/m were measured. They are lower than that of water (72.7 N/m, at 20°C), and higher than that of n-alkane (25.6 N/m for dodecane, at 20°C). As the alkyl chain length increases, the surface tension value decreases.

Most ILs are highly viscous (viscosity similar to that of oils), with values between 10 and 500 cP at room temperature. Water viscosity is 0.89 cP at 25°C.

High viscosity values negatively affect mass transfer and require high power for mixing heterogeneous liquid–liquid systems (Kokorin, 2011).

ILs could be hydrophobic or hydrophilic depending on the properties of their ions. IL 1-butyl-3-methylimidazolium *tetrafluoroborate* is hydrophilic while 1-butyl-3-methylimidazolium *hexafluorophosphate* is hydrophobic. The difference between both ILs lies in the anion group type.

Density, viscosity, and solubility (crucial in any extraction procedure) are among the more tunable IL features, and could be adjusted by changing the side chain (alkyl substituent), the symmetry, and the length of alkyl substituent of the cation head group, the presence of hydrophobic groups, and the degree of anion charge delocalization of the anion group. To optimize the use of ILs and to design desirable ILs, knowledge of their physical and chemical properties is crucial for their industrial implementation.

IL data on structure, physical and chemical properties, synthesis procedures, and measurement methods are compiled in the database on the web: IUPAC Ionic Liquids Database (IL-THERMO/National Institute of Standards and Technology—NIST; Standard Reference Database #147; USA); and Dortmund Data Bank (Data Bank of Ionic Liquids, DECHEMA, DETHERM, Germany), among others.

The characteristics listed previously indicate that ILs are environmentally benign. However, currently, the incipient knowledge on the biodegradability and the (eco) toxicological behavior of ILs still hinders their classification as sustainable solvents (Pham et al., 2010).

10.3.3 IONIC LIQUID SAFETY CONSIDERATIONS

The literature has reported the increase in the biodegradability of ILs with the elongation in the side chain length, and thus in the lipophilic character. On the other hand, the increment in the (eco)toxicity of ionic liquids with the increase on the side chain is also seen (Jastorff et al., 2003).

The (eco)toxicological behavior is associated with the impact of disposable substances on the aquatic and terrestrial ecosystems. Screening of IL solvents is carried out by using tests comprising different levels of biological complexity, such as enzymes, organisms, and multispecies systems. These types of biological tests pointed out the hazardous potential of ILs as mostly guided by the hydrophobicity level of the side chains bounded to the cationic head group. That is, a reduction in the side chain will decrease the (eco)toxicity of ILs. However, smaller side chains can increment the IL sorption to organic matter and mineral clay and diminish the biodegradability (Morrissey et al., 2009).

Thereby, both statements, one for maximizing the biodegradability and the other for minimizing the (eco)toxicity, create a conflict in designing sustainable ILs (Stolte et al., 2007).

However, ILs are represented by structures of cation and anion that could be designed to change their constitutive ions to produce a more environmentally friendly compound, especially when polar side chains and a nontoxic anionic group are combined. This indicates the possibility of manipulating IL structures to design sustainable solvents.

For example, from the viewpoints of both (eco)toxicity and biodegradability: (1) the pyridinium cationic group has been found to be more environmentally friendly than the imidazolium cationic group; (2) ether or polyether side chains showed a reduction in toxicity compared with the long-chain alkyl-substituted derivatives; (3) an ester group is preferred over an amide; (4) the anions methyl sulfate, or octyl sulfate, methyl sulfonates, and salts of organic acids (acetate or lactate) are safer than anions containing fluorine ([(CF3SO2)2N]–, [(C2F5)2PO2]–, [(CF3SO2)3C]–). [(CF3SO2)3C]– fluorine anion is persistent in the environment and should be avoided. Summaries of toxicological and ecotoxicological data for ILs, their cations, and different types of anions as a function of values for enzyme inhibition, cell viability, inhibition of the bioluminescence of marine life can be found, for example, in Pham et al. (2010), Morrissey et al. (2009), Ranke et al. (2007), and Rogers et al. (2003).

Currently, there are some restrictions to the application of ILs such as hygroscopy (can lead to the hydrolysis of the anion group); impurity (changes the expected properties of ILs); instability (from the decomposition of ILs due to sample preparations using thermal, ultrasound, or microwave energy); and alteration of the structure of dissolved target solute (Pham et al., 2010).

If nonsafe ILs must be used in a step of a certain technological process, a risk management study is needed to prevent contamination by the hazardous compounds. The use of membrane filtration in the waste treatment is a powerful regenerative method.

10.3.4 GENERAL APPLICATIONS AND PERSPECTIVES OF IONIC LIQUIDS

The first industrial application of an IL, in 2003, was the biphasic acid scavenging using ionic liquid (BASIL) by BASF industry, in which 1-alkylimidazole removes acid (HCl) during the production of alkoxyphenylphosphines. The IL produced is easily removed from the reaction mixture by decantation, showing advantages of cost and ease compared to filtration (Seddon, 2003).

Attesting the versatility of ILs, some applications of RTILs in several subjects are

- As catalyst or support for esterification and transesterification processes of vegetable oils and animal fats for producing biodiesel (Andreani and Rocha, 2012).
- As algae cell wall rupture agent aiming at releasing lipids and sugars from cell contents that can be converted into biofuel (biodiesel and ethanol).
- As polymerization media in several types of polymerization processes; ILs facilitate the separation of the polymer from the residual catalyst and reduce the extent of side reactions (Malhotra, 2010).
- As solvent for biomass. Biomass dissolved in ILs is more susceptible to chemical attack by external reagents and catalysts allow for the development of new processes and intensify existing ones to replace traditional solvents with ILs, such as for cellulose processing (Rogers and Seddon, 2003).
- As a novel solvent for separation using extraction and azeotropic distillation. Some soluble compounds in ionic liquids may be recycled together with the ionic liquid, after extraction with water and the nonpolar organic

solvent used for product separation. The catalyst and IL may be recycled several times.

- As a capture agent in the separation of gases, in particular, stripping of CO_2 and critical gas components from power plant flue gases (Brennecke and Maginn, 2001).
- In the synthesis of nanomaterials (Kokorin, 2011).
- As a reagent in Diels–Alder reactions (Hallett and Welton, 2011) and in Heck reactions.
- As pharmaceutical dual-active ILs in which the actions of two drugs are combined using a pharmaceutically active cation and a pharmaceutically active anion (Malhotra, 2010).
- In ionic liquid-in-oil microemulsions, in which an IL is dispersed in an oil-continuous phase by suitable surfactants (Yu et al., 2008).
- As a heat transfer and storage media in solar thermal energy systems (Malhotra, 2010).
- As an assistant for microwave-absorbing materials.
- As waste-recycling adjuvants for the solubilization and separation of synthetic goods, plastics, and metals, which could help avoid incinerating plastics or dumping them in landfills.
- In the synthesis of chemical and petrochemical compounds, as solvents, reagents, or catalyst promoters. A sustainable chemical/petrochemical industry uses sustainable raw materials and technologies for the manufacture of end products without the generation of hazardous metabolites.
- In biotechnology on biocatalysis, biomass dissolution and characterization, in membrane separation processes, in synthetic reactions, as reactive media, and as antimicrobial compounds, among others (Malhotra, 2010).

10.3.5 APPLICATIONS AND PERSPECTIVES OF ILS IN FOODS

Safer ILs when used in the food industry behave as follows:

- As an alternative reaction media to conventional organic solvents for sugar dissolution and lipase-catalyzed sugar ester synthesis. Sugar fatty acid esters are nonionic biosurfactants used in the food and pharmaceutical industries (Yang and Huang, 2012).
- As a solvent for the total solubilization of fruit pulps allowing for the study of fruit ripening by NMR (Fort et al., 2006).
- As a selective solvent to extract specific compounds from plants for pharmaceutical, nutritional, and cosmetic applications (Lapkin et al., 2006).
- As a functional component to produce new polysaccharide-based functional materials aiming at improving the solubility, processability, and feasibility of naturally occurring polysaccharides without inactivation of enzymes. ILs show particularly good affinities for natural polysaccharides (Malhotra, 2010).
- As renewable materials-based ILs such as choline chloride/citric acid to participate in the conversion of fructose into 5-hydroxymethylfurfural (Hu et al., 2008).

- As low-cost reaction media for the hydrolysis of carbohydrates into mono-saccharides, with expected reduced process cost (Binder and Raines, 2010). This is relevant, as a plant biomass has the potential to become a sustainable source of fuels and chemicals (Stark, 2011).

Additional reports on the use of ILs in food systems include the work of Park and Kazlauskas (2003), who concluded that ILs do not inactivate enzymes (as observed when using polar organic solvents): when suspended in ILs, enzymes remain stable and catalytically active as opposed to their behavior with organic solvents. Also, the stability of enzymes in ILs was higher than in organic solvents; thus, these solvents permit enzyme-catalyzed reactions on polar substrates in nonaqueous media. Han and Row (2010) listed several bioactive compounds extracted from natural plants using ILs, such as polyphenolic compounds, essential oils, tocopherol homologs, and para-red and sudan dyes.

10.4 AQUEOUS TWO-PHASE SYSTEMS BASED ON IONIC LIQUIDS

In fact, ATPS is a simpler and environmentally friendly unit operation as compared to traditional liquid extraction. However, polymer–salt and polymer–polymer ATPS work in a limited range of difference in polarities due to reactant characteristics, making many solute partitions unfeasible in such systems. The difference in polarities between the two phases of an ATPS drives solute partitioning. Assuming that ILs can work in a wide hydrophilicity–hydrophobicity range as a result of their properties, they are potential candidates to be used as reagents for phase forming of new aqueous biphasic systems (Freire et al., 2012).

In 2003, a first report on a new type of ATPS consisting of 1-butyl-3-methylimidazolium chloride ([C4mim]Cl) and K_3PO_4 and their corresponding phase diagrams was presented (Rogers and Seddon, 2003). Some properties of IL aqueous two-phase systems (ILATPS) were favored with the association between IL and ATPS, such as reduced viscosity, little emulsion formation, quick phase separation, and gentle biocompatible environment (Tan et al., 2012).

ILATPS are formed by adding the phase-forming reactant IL, inorganic salt, and water. Above the critical concentrations of the IL and salt in aqueous solution, the phase splitting occurs with the formation of an IL-rich top phase and a salt-rich bottom phase. ILATPS were used in the extraction of proteins, enzymes, and spices. Hydrophobic interactions, electrostatic interactions, and salting-out effects are important for the transfer of the proteins (Liu et al., 2011; Tan et al., 2012).

The critical review of Freire et al. (2012) on ILATPS presents discussions on physicochemical properties of these systems; on data of phase equilibrium, tie line length, and tie line slope; on the behavior of ATPS composed by IL/inorganic salt, IL/amino acids, IL/carbohydrates, IL/polymer, IL/inorganic salt/polymer as affected by IL type, type of salt, pH, and temperature; on the applications of ILATPS to extract several solutes; and the recovery of ILs from aqueous media.

10.5 CONCLUSION

Aqueous two-phase systems are an innovative, economically viable, and a low environmental risk extraction process. The excellent experimental results obtained on the separation/purification of chemical compounds show that ATPS can greatly contribute to the advancement of industrial and food processes.

It is also clear that ionic liquids can be applied in various situations involving liquid–liquid extraction, and new extraction processes can be developed based on the synthesis of new ILs and its association with ATPS.

ACKNOWLEDGMENTS

The authors express their gratitude to the CAPES, CNPq, FAPEMIG, and PROPESP-UFPA Brazilian agencies for financial support.

REFERENCES

Albertsson P. A. *Partition of Cell and Macromolecules*. New York: John Wiley, 346 p. 1986.

Andreani L., Rocha J. D. Use of ionic liquids in biodiesel production: A review. *Braz. J. Chem. Eng.*, 29(1), 1–13, 2012.

Binder J. B., Raines R. T. Fermentable sugars by chemical hydrolysis of biomass. *Proc. Nat. Acad. Sci. USA*, 107(10), 4516–4521, 2010.

Boaglio A., Bassani G., Picó G., Nerli B. Features of the milk whey protein partitioning in polyethyleneglycol-sodium citrate aqueous two-phase systems with the goal of isolating human alpha-1 antitrypsin expressed in bovine milk. *J. Chromatogr. B.*, 837, 18–23, 2006.

Brennecke J. F., Maginn E. J. Ionic liquids: Innovative fluids for chemical processing. *AIChE J.*, 47, 2384–2389, 2001.

Chiappe C., Pieraccini D. Ionic liquids: Solvent properties and organic reactivity. *J. Phys. Org. Chem.*, 18, 4, 275–297. 2005.

Earle M. J., Seddon K. R. Ionic liquids. Green solvents for the future. *IUPAC Pure Appl. Chem.*, 72(7), 1391–1398, 2000.

Fort D. A., Swatloski R. P., Moyna P., Rogers R. D., Moyna G. Use of ionic liquids in the study of fruit ripening by high-resolution C^{13} NMR spectroscopy: "Green" solvents meet green bananas. *Chem. Commun.*, 714–716, 2006.

Freire M. G., Claudio A. F. M., Araujo J. M. M., Coutinho J. A. P., Marrucho I. M., Lopesac J. N. C., Rebelo L. P. N. Aqueous biphasic systems: A boost brought about by using ionic liquids. *Chem. Soc. Rev.*, 41, 4966–4995, 2012.

Gutowski K. E., Broker G. A., Willauer H. D., Huddleston J. G., Swatloski R. P., Holbrey J. D., Rogers R. D., Controlling the aqueous miscibility of ionic liquids: Aqueous biphasic systems of water-miscible ionic liquids and water-structuring salts for recycle, metathesis, and separations. *J. Am. Chem. Soc.*, 125, 3, 6632–6633, 2003.

Hallett J. P., Welton T. Room-temperature ionic liquids: Solvents for synthesis and catalysis. Part 2. *Chem. Rev.*, 111, 3508–3576, 2011.

Han D., Row K. O. Recent applications of ionic liquids in separation technology. *Molecules*, 15, 2405–2426, 2010.

Hu S., Zhang Z., Zhou Y., Han B., Fan H., Li W., Song J., Xie Y. Conversion of fructose to 5-hydroxymethylfurfural using ionic liquids prepared from renewable materials. *Green Chem.*, 10, 1280–1283, 2008.

Jastorff B., Störmann R., Ranke J., Mölter K., Stock F., Oberheitmann B., Hoffmann W., Hoffmann J., Nüchter M., Ondruschka B., Filser J. How hazardous are ionic liquids? Structure activity relationships and biological testing as important elements for sustainability evaluation. *Green Chem.* 5, 136–142, 2003.

Lapkin A., Plucinski P. K., Cutler M. Comparative assessment of technologies for extraction of artemisinin. *J. Nat. Prod.*, 11, 69, 1653–1664, 2006.

Liu Y., Yu L., Chen Z., Xiao X. Advances in aqueous two-phase systems and applications in protein separation and purification. *Can. J. Chem. Eng. Tech.*, 2, 1–7, 2011.

Kokorin A. *Ionic Liquids: Applications and Perspectives.* InTech, Croatia, 2011.

Malhotra S. *Ionic Liquid Applications: Pharmaceuticals, Therapeutics, and Biotechnology, ACS Symp. Series; ACS: Washington, DC*, Org. Div., 236th Nat. Meet. Am. Chem. Society, Pennsylvania, 2010.

Mageste A. B., Senra T. D. A., Da Silva M. C. H., Bonomo R. C. F., da Silva L. H. M. Thermodynamics and optimization of norbixin transfer processes in aqueous biphasic systems formed by polymers and organic salts. *Sep. Purif. Technol.*, 98, 69–77, 2012.

Marsh, K. N., Boxall, J. A., Lichtenthaer, R., Room temperature ionic liquids and their mixtures—A review. *Fluid Phase Equil.*, 219, 1, 93–98, 2004.

Martins J. P., Silva, M. C. H., Silva L. H. M., Senra T. D. A., Ferreira G. M. D., Coimbra J. S. R., Minim LA. Liquid liquid phase equilibrium of triblock copolymer F68, poly(ethylene oxide), poly (propylene oxide) and poly(ethylene oxide), with sulfate salts. *J. Chem. Eng. Data*, 55, 1618–1622, 2010.

Morrissey S., Pegot B., Coleman D., Garcia M. T., Ferguson D., Quilty B., Gathergood N. Biodegradable, non-bactericidal oxygen-functionalised imidazoliumesters: A step towards "greener" ionic liquids. *Green Chem.*, 11, 475–483, 2009.

Park S., Kazlauskas R. J. Biocatalysis in ionic liquids—Advantages beyond green technology. *Cur. Opin. Biotechnol.*, 14, 432–437, 2003.

Pham T. P. T., Cho C.-W., Yun Y.-S. Environmental fate and toxicity of ionic liquids: A review. *Water Res.*, 44, 352–372, 2010.

Ranke J., Stolte S., Störmann R., Arning J., Jastorff B. Design of sustainable chemical products the example of ionic liquids. *Chem. Rev.*, 107, 6, 2183–2206, 2007.

Rodrigues L. R., Venâncio A., Teixeira J. A. Partitioning and separation of α-lactalbumin and β-lactoglobulin in polyethylene glycol/ammonium sulphate aqueous two-phase systems. *Biotechnol. Lett.*, 23, 1893–1897, 2001.

Rogers R. D., Seddon K. R. Ionic liquids—Solvents of the future. *Science*, 302, 5646, 792–793, 2003.

Rosa P. A. J., Azevedo A. M., Sommerfeld S., Bäcker W., Aires-Barros M. R. Aqueous two-phase extraction as a platform in the biomanufacturing industry: Economical and environmental sustainability. *Biotechnol. Adv.*, 29, 559–567, 2011.

Ruiz-Ruiz, F., Benavides, J., Aguilar O., Rito-Palomares, M. Aqueous two-phase affinity partitioning systems: Current applications and trends. *J. Chromatogr. A*, 1244, 1– 13, 2012.

Seddon K. R. Ionic liquids: A taste of the future. *Nat. Mat.*, 2, 6, 363–365, 2003.

Silva, L. H. M., Meirelles, A. J. A. Bovine serum albumin, a-lactoalbumin and b-lactoglobulin partitioning in polyethylene glycol/maltodextrin aqueous two-phase systems. *Carbohyd. Polym.*, 42(3), 279–282, 2000.

Silva, L. H. M., Meirelles, A. J. A. PEG + potassium phosphate + urea aqueous two-phase systems: Phase equilibrium and protein partitioning. *J. Chem. Eng. Data*, 46(2), 251–255, 2001.

Smiglak M., Reichert W., Holbrey J., Wilkes J., Sun L., Thrasher, Kirichenko K., Singh S., Katritzky A., Rogers R. D. Combustible ionic liquids by design: Is laboratory safety another ionic liquid myth. *Chem. Commun.*, 24, 2554–2556, 2006.

Stark A. Ionic liquids in the biorefinery: A critical assessment of their potential, energy environ. *Science*, 4, 19, 2011.

Stolte S., Matzke M., Arning J., Böschen A., Pitner W-R., Welz-Biermann U., Jastorff B., Ranke J. Effects of different head groups and functionalised side chains on the aquatic toxicity of ionic liquids. *Green Chem.*, 9, 1170–1179, 2007.

Tan Z., Li F., Xu X. Isolation and purification of aloe anthraquinones based on an ionic liquid/salt aqueous two-phase system. *Sep. Purific. Technol.*, 98, 150–157, 2012.

Yang Z., Huang Z-L. Enzymatic synthesis of sugar fatty acid esters in ionic liquids. *Catal. Sci. Technol.*, 2, 1767–1775, 2012.

Yu S., Yan F., Zhang X., You J.,Wu P., Lu J., Xu Q., Xia X., Ma G. Polymerization of ionic liquid-based microemulsions: A versatile method for the synthesis of polymer electrolytes. *Macromolecules*, 41(10), 3389–3392, 2008.

Zaslavsky B. Y. *Aqueous Two Phase Partitioning*. New York: Marcel Dekker, 1995.

11 Process Analytical Technology

José A. Teixeira, António A. Vicente, Fernando Filipe Macieira da Silva, João Sérgio Azevedo Lima da Silva, and Rui Miguel da Costa Martins

CONTENTS

11.1 INTRODUCTION

The term "process analytical technology" (PAT) has been used to describe the application of analytical chemistry and process chemistry combined with the multivariate tools for process understanding. PAT is a wide and complex theme, making it difficult to capture all the features in this chapter.

This chapter is presented from a food technology perspective. The food industry is aiming at better-controlled, faster, safer, more sustainable, and more automated processes to achieve an efficient management of products' quality. PAT allows achieving process control and continuous product quality management by integrating specific sensors in the streamline. These sensors are connected to an intelligent and dynamic informatics platform, allowing a better process understanding in all stages.

Chemometrics aims at extracting relevant information from a (bio)chemical process by applying computational statistical techniques, being crucial for the integration of PAT in industrial processes. With the present computation development allied to sensor development, the multivariate ability of chemometrics will allow real-time managing of industrial food processes at different levels.

11.2 INTRODUCTION TO PAT

After World War II, the reindustrialization that took place in Europe brought the inherent process control with process analyzers (PAs). PAs were used to monitor temperature, pressure, or other simple process parameters. However, when the monitoring of complex parameters was needed (e.g., product composition), PAs were noneffective. Bakeev (2010) reported that PAT was adopted in the early 1930s, when infrared (IR) photometers, oxygen, and conductivity sensors were being used in the petrochemical and chemical industry for refining processes (Bakeev, 2010). The modern developments in analytical technologies are currently able to provide chemical and analytical insights for all types of chemical reactions and process monitoring such as drying, distillation, crystallization, and hydrogenation among others.

The food industry is becoming more and more sophisticated and strict policies for quality control and monitoring of specific process phases are being adopted. The U.S. Food and Drug Administration (FDA) created a guidance document for the industry, called "PAT—A Framework for Innovative Pharmaceutical Development, Manufacturing, and Quality Assurance." This FDA initiative invokes the use of in-line sensors for very accurately monitoring and documenting the key parameters of the stream process. This automation demand will improve operator safety and, simultaneously, reduce the human errors when processing. The aim of this PAT framework was to encourage industrial manufacturers to implement efficient innovative approaches for process quality and control, thus improving the overall process understanding (FDA, 2004).

PAT can be implemented by using specific tools that are necessary to understand a certain process. These can be (a) multivariate tools for design, data acquisition, and analysis; (b) process analyzers; (c) process control tools; and (d) continuous improvement and knowledge management tools. Different combinations of these PAT tools in the industry will be able to supply different types of information enabling process understanding.

PAT has been used to describe the application of analytical chemistry and process chemistry sensor signals with chemometric multivariate tools that allow a specific process understanding. This initiative of the FDA is a collaborative effort with the industry to use automation and real-time parameters (FDA, 2004; Ganguly and Vogel, 2006).

To implement PAT effectively, automation needs to support the effort of continuously and automatically collecting data not just from the sensors directly associated with the process, but from all other factors that could influence the results. Figure 11.1 shows how a PAT system could operate.

Real-time data generated from PAs provide relevant information, such as temperature, pressure, dissolved oxygen, and pH; these parameters are being frequently obtained online (see Section 11.2.3). PA can be controlled by a network for communication and monitoring, giving a more complete picture of the process (Brosilow and Joseph, 2002).

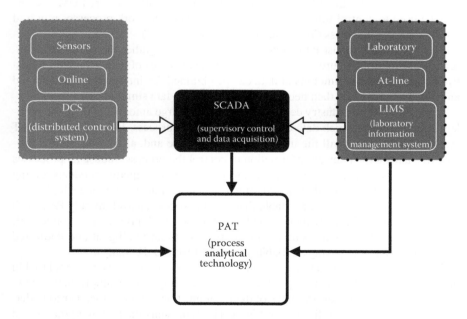

FIGURE 11.1 Process analytical technology integration.

In contrast, the data may be obtained at-line, meaning that it was necessary to analyze a sample in a different physical place, such as a laboratory; this type of data (e.g., HPLC, GC, MS, and microbiology) can be inserted in the laboratory information management system (LIMS) platform (Gibbon, 1996). Figure 11.1 clearly shows the integration of different sources of information to build a robust PAT system. Here, the supervisory control and data acquisition (SCADA) acts as a database, compiling information from LIMS and distributed control system (DCS). Thus, PAT can be very useful by using all process data to transform them into relevant information for interpretation and further process understanding (Andreas, 1997).

One relevant advantage of PAT is the reduction of product variability. Monitoring and control in industrial processes are often univariate, where process variables are analyzed individually. However, this approach offers a shallow and inefficient understanding of the process, where control strategies are more difficult to implement. This chapter enhances the use of monitoring and control strategies using a multivariate approach to understand the process variables relation in a food-based process. Section 11.4 presents PAT applications in scientific studies related to the food and beverage industry.

11.2.1 INDUSTRIAL PAT

Industrial environments are often complex, each company having its own data acquisition methods and processes, making the implementation of a PAT system difficult. There are solutions to integrate the data from different sources inside a company, called product development management (PDM).

Figure 11.2 shows the different areas inside an organization, where PDM systems may be used to control information, files, documents, and work processes required for the company business (Tony Liu and William Xu, 2001).

To implement a robust PAT system, it is necessary to gather the available meaningful information. Figure 11.3 shows the different sources of data generated industrially, where massive amounts of data can be obtained. To interpret and understand an industrial process is then necessary to analyze the data simultaneously.

Computation and industry certification generates large amounts of data, the problem being to select which data explains an industrial process. Ideally, a robust PAT system should process all the information in real time and, as a result, should provide the analyst the relevant information to control the process. Industrial processes require high precision, accuracy, and control to obtain high-quality products, giving satisfaction to customers, thus demanding a careful monitorization of each phase of the process. A few tools are available that enable process control, mainly PAs, such as temperature probes, pH meters, and densimeters. PAT promises to connect all relevant parameter signals obtained in the process to an intelligent and dedicated informatics platform, giving a reliable, real-time result to the analyst.

In terms of the food and beverage industry, PAT systems can be a powerful tool in quality improvement and in the product development stage. For example, in the brewing industry, there are several types of data that must be analyzed together to understand and act in the overall process. Figure 11.4 represents the brewing process and Table 11.1 shows the different parameters that must be monitored during this process.

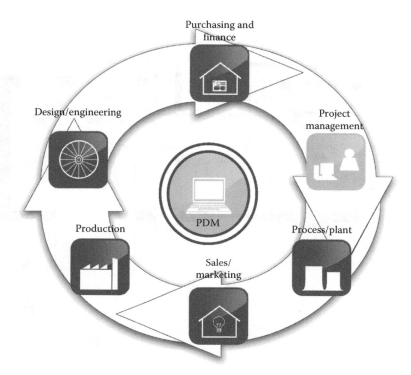

FIGURE 11.2 Application of PDM in a company.

Table 11.1 displays the different parameters susceptible of being monitored online during the brewing process. Monitoring is essential to address the end-product quality (beer, in this case). Generally, industrial processes have several stages that can only be monitored by a few online sensors; in the brewing industry, it is possible to monitor pH, temperature, pressure, dissolved oxygen, and conductivity (Figure 11.5).

At-line sampling occurs when samples are taken from the process and analyzed in a dedicated laboratory. In contrast, online sampling is obtained by probes placed directly in the process to measure specific parameters, as shown in Figure 11.5.

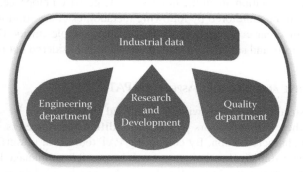

FIGURE 11.3 Data generated by companies.

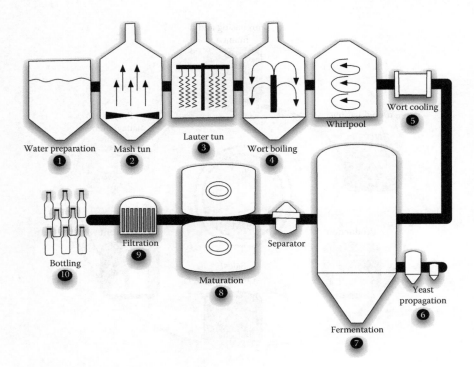

FIGURE 11.4   Brewing process.

The ability to monitor and control food bioprocesses is essential in terms of optimization. Concentration monitoring of products and substrates can be implemented by different means. Figure 11.6 shows how PAT systems can operate with online and at-line sampling, and how the output can help the food-based industry to control the quality of their products. Here, ethanol and sugar concentration profiles are shown where the data were obtained by online or at-line sampling. The obtained data must be processed and visualized to obtain robust multivariate models such as principal component analysis (PCA) or partial least squares (PLS) (see Section 11.3.2), providing detailed information about the bioprocess. Time-based bioprocesses, such as fermentations, can be monitored by following metabolites concentration (e.g., ethanol production or sugar consumption). In Figure 11.6, samples inside the circle are within control limits and samples outside the circle are considered out of control.

11.2.2   Positive and Negative Aspects of PAT

In future, industrial bioprocesses will be monitored and controlled by sophisticated PAT systems where online sensors will measure different parameters in real time, providing their characterization. By combining PAT tools with powerful computational multivariate statistical techniques, it will be possible to manage bioprocesses automatically and remotely. PAT will enhance the transfer of data generated during the development phase to the streamline, making the new product development

TABLE 11.1

Parameters Measured On-Line during the Brewing Process

Process Stage	Parameters
1. Water preparation	pH sensor
	Conductivity sensor
2. Mash tun	pH sensor
	Thermometer
	pH sensor
3. Lauter tun	Dissolved oxygen sensor
	Temperature sensor
	pH sensor
4. Wort boiling	Temperature sensor
	pH sensor
5. Wort cooling	Temperature sensor
	Conductivity sensor
	Dissolved oxygen sensor
	pH sensor
6. Yeast propagation	Dissolved oxygen sensor
	Temperature sensor
	pH sensor
7. Fermentation	Conductivity sensor
	Dissolved oxygen sensor
	Temperature sensor
	Dissolved oxygen sensor
8. Maturation	Temperature sensor
	Pressure
9. Filtration	Dissolved oxygen sensor
10. Bottling	pH sensor

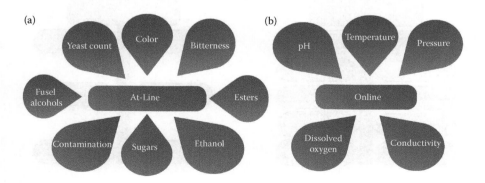

FIGURE 11.5   (a) At-line versus (b) online parameters obtained during the brewing process.

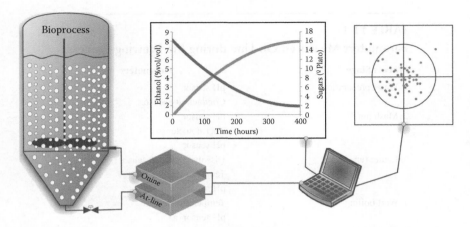

FIGURE 11.6 PAT system in a bioprocess, where online and at-line samples are taken from a bioprocess and control charts may be used to monitor the process.

easier and faster. Process analytical technologies applied to the food industry will improve the uniformity of the process due to maximum information analysis. The ability to store process information is an important feature, supporting continuous quality improvement. Figure 11.7 highlights the positive and negative aspects of implementing a PAT system that are schematized.

PAT may provide a basis for identifying and understanding relationships among various critical formulations and process factors and for developing effective risk mitigation strategies (e.g., product specifications, process controls, and training). The data and information to help understand these relationships can be leveraged through preformulation programs, development, and scale-up studies (FDA, 2004).

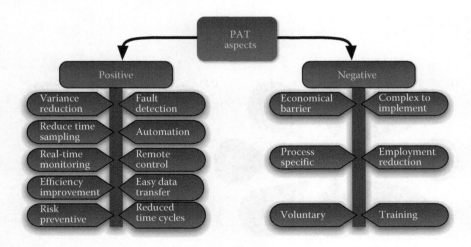

FIGURE 11.7 Positive and negative aspects of PAT implementation.

The PAT implementation in the food industry being an FDA initiative, making its implementation voluntary, it is difficult to create generic certification policies for quality control strategies. There is also an economical barrier associated with the usage of new systems for data acquisition, storage, preprocessing software, and streamline changes. By implementing PAT, process automation and remote monitoring can be possible in real time; thus, employment issues may be generated. The other negative aspect of PAT is process specificity, being necessary to adapt PAT tools for each purpose.

11.2.3 ECONOMICAL IMPORTANCE OF PAT

The world population is growing everyday, with more than seven billion people worldwide (Worldometers, 2012). Food demand is increasing at the same rate, and the importance that is being given by manufacturers is the minimization of food waste (Rosegrant et al., 2001). Food waste occurs at different points in the food supply chain, in particular, in the processing phase. Food and drink waste is estimated to be approximately 14 megatonnes in the United Kingdom, of which 20% is associated with food processing, distribution, and retail (WRAP, 2010).

Food is not a simple commodity, and the world demand has been changing dramatically over the last two decades. Consumers have been shifting their consumption patterns away from grains toward livestock and meat products (Yu et al., 2004). These changes were imposed by the changing market prices, for example, wheat and maize prices in the mid-1996 were 50% higher than a year before, while rice prices were 20% above the 1994 levels. The growing demand for food caused by expanding populations and shifting consumption habits will necessitate future food production increases, but unexploited, available arable land is limited, placing the burden for these increases on technologically driven yield improvements, where PAT systems will play an important role (Rosegrant et al., 2001). Therefore, the food and agricultural industry trend is to use systems able to achieve better control and monitoring toward avoiding food waste, leading to billions of savings yearly (Wang et al., 2006). Therefore, PAT systems will enable sustainability in different phases of the food chain supply.

11.2.4 CHEMOMETRICS

The emergence of chemometrics came from the realization that traditional univariate statistics was not sufficient to describe and model chemical experiments (Miller and Miller, 2005). Chemometrics appeared as a response to new technologies: (a) instrumentation giving multivariate responses for each sample analyzed and (b) the availability of computers (Geladi, 2003). Succinctly, chemometrics can be described as the extraction of information from multivariate data using statistics and mathematics. In Figure 11.8, the main areas involved in chemometrics sciences are shown.

Statistical techniques that can directly correlate quality parameters or physical properties with analytical data make chemometrics a crucial field for PAT purposes. Chemometrics can extract relevant patterns from the available data and can transform them into mathematical models. These models can be routinely used to predict

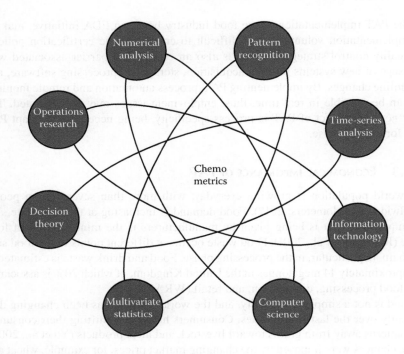

FIGURE 11.8 Main areas involved in chemometrics.

process performance, decreasing the fault in the streamline (Wold et al., 2001). Efficient laboratory practices, automated quality control systems, and the study in real time are some applications of chemometrics that lead to process understanding.

11.3 MONITORING AND CONTROL SYSTEMS

There is a substantial opportunity to enhance product quality and to optimize the manufacturing process by implementing systems and solutions for online monitoring, analysis, and control. The application of PAT systems in the industry opens the door for improvements in the product by enhanced process monitoring based on knowledge-based data analysis, "quality in design" concepts, and through feedback control (Gnoth et al., 2007; Hinz, 2006). The main difficulty in the implementation of PAT systems in a production process is the unavailability of tools and methods through which a PAT system can be systematically implemented (Singh et al., 2009).

11.3.1 INTEGRATION OF PAT

The design of a system of monitoring and analysis is a process divided into several steps. First, the critical process variables are selected, followed by the selection and placement of the control and analysis equipment. Finally, coupling of the monitoring information with analysis tools takes place, creating a control system capable of ensuring that the critical process variables can be controlled. Figure 11.9 shows the

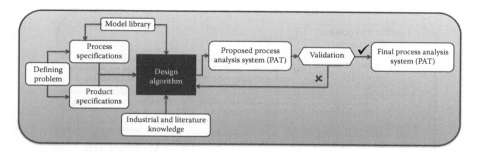

FIGURE 11.9 PAT overview.

design methodology, starting with the problem definition in terms of specifications of the process and product quality, which can be supplied by the manufacturer or system designer. In addition, it is possible to create a template library and a base of theoretical and practical knowledge, acting as tools to support the establishment of the PAT system. To generate the PAT system, the algorithm design must relate product and process specifications with the available support tools. If the PAT system created satisfies all requirements, it can be validated. System validation is performed by comparing the simulated performance of the process with the known process specifications (Singh et al., 2009; Wu et al., 2011). If validation is not satisfactory, the project steps are repeated until a valid project is obtained (Figure 11.9).

Theoretical and industrial knowledge must present useful information for the development of PAT systems, which must be built based on an extensive literature and industry search. This knowledge should cover the widest range of industrial processes and should contain information of the involved unit processes. Process variables, manipulated variables, and measurement equipment should be described extensively (type of equipment, accuracy, precision, operating range, response time, resolution, drift, cost, etc.) (Lopes et al., 2004; Singh et al., 2009).

The model library should contain a set of mathematical models able to characterize different types of process units, sensors, and controllers. The models must be consistent with the analysis process and must allow the generation of additional or missing data, which are necessary for the development of the PAT (Singh et al., 2009; Skibsted and Engelsen, 2010).

11.3.2 Design Methodology

The design of a PAT system is divided into three principal stages: (a) process analysis, (b) sensitivity analysis, and (c) analysis of monitoring tools, as shown in Figure 11.10.

In the process analysis stage, all the variables involved in the process and points to be monitored are listed. This list is generated based on product specifications and process, complemented with all the theoretical and industrial knowledge (Chen et al., 2011; Singh et al., 2009).

Sensitivity analysis identifies the critical process variables from the list of process variables. This identification is made based on process data and model libraries. First, the operational objectives of the process should be taken into account and the

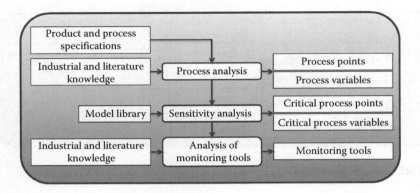

FIGURE 11.10   PAT system design methodology.

variables that violate the operational limit that have a great effect on product quality are considered as the critical process variables. These are the variables that must be monitored and controlled throughout the process (Rosas et al., 2012).

In the analysis of monitoring tools phase, the appropriate control tools for each critical variable in the process are selected. The measuring equipment adequate for each critical variable is then selected. It should also be possible to compare the performance of measurement equipments available for each critical variable. Usually, the equipment selection is based on the performance of one or more criteria: accuracy, precision and resolution, sensor drift, response time and cost, and operating range (Menezes, 2011; Singh et al., 2009).

11.3.3   Sensors

The main objective of PAT is to obtain quantitative and qualitative information about chemical, physical, and biological processes.

Sensors are devices that can continuously provide information of a chemical or physical nature, converting this information into an electrical signal. Physical sensors measure the properties of physical nature, such as viscosity, temperature, refractive index, and so on. Chemical sensors recognize chemical constituents and may be used to measure, for example, the concentration of a substance. Sensors for molecular identification based on biochemical processes (or reactions) are conventionally called biosensors and can be considered as a subgroup of chemical sensors (Pons, 1992).

In process control, measurements must be fast enough to allow a counteraction of the system. When the analyzer system is fast enough to allow the controller to reset the process variables, returning to normal, we call these measurements as analysis in real time (Shinskey, 1994).

According to the measuring system, process analyzers can be classified into five types: "off-line," "at-line," "online," "in-line," and "noninvasive" and can be distributed at strategic points of a process as shown in Figure 11.11 (Trevisan and Poppi, 2006).

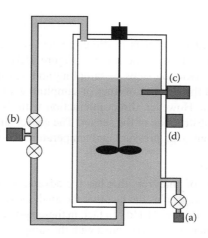

FIGURE 11.11 Different types of sensors applied in a reactor: (a) at-line, (b) online, (c) in-line, and (d) noninvasive.

11.3.3.1   At-Line and Off-Line

These analyzers employ manual sampling over sampling valves (Figure 11.11a) and transport samples to a laboratory. In "off-line" sensors, the sample is analyzed with sophisticated instruments, which are generally automated. The advantages of "off-line" systems include the more extensive use of the equipment, the availability of a technical specialist as a consultant, and the ease of methodology development and maintenance. The disadvantages include the delay between sample submission and reporting of results, which has led to methods "at-line." In these systems, an instrument is positioned close to the sampling points. The advantages include higher speed in obtaining the results, a more efficient control in the conditioning of the sample, and use of instrumentation that is low cost, requires simple maintenance, and is easy to use. Owing to the time-consuming operation, these analyzers are rarely used in controlling and monitoring processes, and are instead used for measuring the technical specifications of raw materials and quality control of final products (Bluma et al., 2010).

11.3.3.2   Online

In this type of analyzer, an automated system is used to extract the sample through a pipeline (Figure 11.11b), condition and measure, collect, and process the data. This class can be subdivided into two categories: intermittent systems where the transfer of a portion of the process flow for an analyzer instrument (e.g., chromatographic methods) takes place, and continuous systems, where the sample is continuously passing through a measuring cell, returning back to the process. The major disadvantage of these systems is the sampling process, which consists of the separation of an analytical line. As in other cases, sampling must ensure that the sample is representative and keeps its properties. These conditions should be valid throughout the sample line, until the sample is analyzed by the sensor; however, before the actual measurement, the sample may need to be prepared. These disadvantages have contributed to the development of analyzers "in-line" (Henrikson, 1995).

11.3.4 IN-LINE

In this case, the analytical sensor is in direct contact with the line of the process (*in situ*), interacting directly with the sample (Figure 11.11c). This system presents an extreme advantage, avoiding steps of sampling and getting more representative information because of the absence of lines of sampling while performing measurements in a shorter time. However, direct interaction with the process analyzer can cause exhaustion and obstruction of the sensor. The sensor should be able to operate under extreme conditions (e.g., pressure and temperature) (Vojinović et al., 2006).

11.3.4.1 Noninvasive

This is the newest class of analyzers that has the advantages of sensors "in-line" by not requiring sampling steps, plus the fact that the analysis system does not enter into contact with the process (Figure 11.11d). In fact, in these systems, the analyzer does not destroy and does not enter into direct contact with the sample, thus not causing changes in its composition and minimizing the risks of contamination (Tewari et al., 2010). These properties make these sensors the most appropriate for analyzing processes.

11.3.4.2 Sampling

Most problems related to the implementation of sensors and processes are related to the sampling steps, which have contributed to the preference for sensors "in-line" (Koch, 1999). Sampling techniques depend on the nature of the process, the matrix, the information required, and the type of analyzer used.

In "online" analyzers, sampling consists of transferring a sample representative of the process to the analyzer. In "in-line" and "noninvasive" systems, despite the removal of the sample being not necessary, an appropriate interface with the sensor inside the process is needed. Obtaining a representative sample of the process is the key step to obtain efficient measures of an industrial process.

Today, it is possible to find a wide variety of process analyzers for the most varied applications. However, the incompatibility and variability of these devices' characteristics make it difficult to choose the appropriate sensor (e.g., slow and very precise analysis by another less precise, but faster, instrument) (Beutel and Henkel, 2011).

11.3.5 ONLINE VERSUS LABORATORY SYSTEMS

In the laboratory, samples are handled under strict conditions of control and can be pretreated to promote an increase in selectivity and/or sensitivity. In addition, the instruments are not exposed to corrosive environments or samples, allowing the use of classical analytical techniques. Many measures in analytical laboratories are used only to ensure product quality, but not to control the process (Vojinović et al., 2006).

Process sensors, here defined as analytical devices implemented in the line of the process, must be resistant to the chemical environment of the plants, support changes in temperature and humidity, and be able to sample and analyze materials under extreme conditions (Cimander et al., 2002).

Sampling, sample handling, measurement, data collection, and processing should be automated. The purpose of the process analysis is to eliminate or reduce the causes of variability in the production line, increasing quality, productivity, and competitiveness of the product (Ganguly and Vogel, 2006).

The increased yields expected with the implementation of in-line sensors are due to the increase in the optimization and process control; so, it is always necessary to maintain an analytical laboratory to perform measurements, calibrations, and maintenance of the sensors distributed throughout the process.

11.3.6 INFORMATICS PLATFORM

To increase the productivity in the analysis of industry data and design of sensors and systems for monitoring and control, it is necessary to use the appropriate software (e.g., "OpenPAT").

"OpenPAT" is an open source software of PAT. It quantifies and diagnoses the main effects that affect a process or product, allowing to build evaluation tools for industrial applications. This software shows a potential in different industries such as food, chemical, biotechnology, and pharmaceutical.

"OpenPat" is structured to make the most of the analysis of industrial data. The software guides the user through the analysis of data in a simple and structured way: (i) preprocessing, (ii) data analysis, (iii) models of monitoring, and (iv) models for process control.

11.4 PAT WORKFLOW: FROM DATA TO PROCESS

11.4.1 PREPROCESSING

In data analysis, the first and most important step is the preprocessing of data. The importance of this step derives from the fact that it influences the success of subsequent steps.

Owing to several factors inherent to these data, it is always necessary to make their preprocessing (Zeaiter et al., 2005). The most important factors are

- Not all data are on the same scale of units
- Not all data have the same degree of variability
- It is necessary to diagnose the presence of outliers

11.4.1.1 Filtering

Filtering data corresponds to the first step of preprocessing. This procedure is applied to identify and correct data failure, attribute failure, or aggregate data. Filtering the data also allows eliminating the presence of errors and outliers (noise) and correcting the discrepancies in data points' names or codes (Jung, 2004).

11.4.1.2 Data Integration

The integration of data comes from the fact that there are data from different sources with the same information, which cause inconsistency and redundancy in the information. This comes from various errors: the same concept with different names, the same value expressed in different ways, and repeated variables in different tables (Cozzolino et al., 2011).

11.4.1.3 Relevant Data

Not all data are important to explain a process; so, it should be removed in the data analysis, leaving only the variables that are really important for the study. Therefore, it is necessary to perform an exploratory analysis to first know which are the important variables for the study of the process (Gurden et al., 1998).

11.4.1.4 Normalization

To prevent the excessive weight of variables with higher ranges, data normalization can be applied. Normalization of data is the reduction of different variables to comparable scales, for example, keeping all the variables in intervals between [0, 1] and [−1, 1] (Bylesjö et al., 2009).

11.4.1.5 Missing Data

In most cases, complete information of a data system under study is not available. It is necessary to find a consistent estimator of these data on failure, to complete these data sets. Examples of possible estimators are mean (distortions in the sample space), linear regression (collinearity in the data set), and PCA models (the use of principal components [PCs] that describe the variability of the data set to estimate the values in failure). The preprocessing of data is in most cases the limiting step in the time needed for data analysis (Arteaga and Ferrer-Riquelme, 2009).

11.4.2 MULTIVARIATE DATA ANALYSIS

The real world is multivariate; there is always more than one factor influencing each real phenomenon. Multivariate data analysis consists of mathematical and statistical techniques using two or more variables to provide a result that takes into account their relationship (Esbensen et al., 2002). Nowadays, analytical instrumentation generates enormous amounts of data (variables) for a large number of samples that can be analyzed in a relatively short time. Therefore, sample analysis requires multivariate data matrices that demand the use of mathematical and statistical procedures to efficiently extract the maximum useful information from the data (Berrueta et al., 2007).

In food chemistry, the data used for multivariate treatments are generally physical or chemical data, such as (a) conductivity; (b) pH; (c) moisture; (d) total acidity; (e) chemical concentrations measured by analytical techniques (HPLC or GC); (f) fingerprinting data such as chromatograms or spectroscopic measurements (specific signals or complete spectra) obtained by IR, nuclear magnetic resonance (NMR), mass spectrometry (MS), ultraviolet–visible (UV–VIS), or fluorescence spectrophotometry; (g) signals from sensor arrays such as electronic noses or tongues; and (h) data from sensorial analysis of samples (Arvanitoyannis et al., 1999; Sohn et al., 2005). There are publications dealing with chemometric applications in food analysis, such as general reviews on food analysis (Tzouros and Arvanitoyannis, 2001) or reviews on particular foods, such as fish (Arvanitoyannis et al., 2005b), meat (Arvanitoyannis and van Houwelingen-Koukaliaroglou, 2003), wine (Cozzolino et al., 2004; 2008a,b; Korenovska and Suhaj, 2005), beer (Siebert, 2001), or honey (Arvanitoyannis et al., 2005a).

In this section, focus will be given to the key topics of multivariate analysis involved in PAT systems. These are mainly (a) PCA, (b) pattern recognition,

(c) partial least-squares regression (PLSR), and (d) model validation. These are imperative matters to understand how PAT systems can be used to control or monitor a food or a biotechnological process in general.

11.4.2.1 Principal Component Analysis

PCA is a technique that reduces the original data dimensionality, capturing as much information as possible, and allows its visualization (Esbensen et al., 2002; Berrueta et al., 2007). PCA converts the original variables into new uncorrelated variables called PCs. Each PC is a linear combination of the original measured variables. This technique affords a group of orthogonal axes that represent the directions of higher variance in these data. PCA is a standard technique for visualizing high-dimension data and is often used for data preprocessing. PCA reduces the dimensionality, the number of variables of a data set, by maintaining as much variance as possible. The first principal component (PC1) accounts for the maximum of the total variance, whereas the second principal component (PC2) is uncorrelated with the first and accounts for the maximum of the residual variance. For practical reasons, it is sufficient to retain only those components (PC1 and PC2) that account for a considerable percentage of the total variance. The linear coefficients of the inverse relation of linear combinations are called the component loadings, being the correlation coefficients between the original variables and the PCs. The values characterizing samples in the space defined by the PCs are known as component scores. Scores can also be used as an input to other multivariate techniques, instead of using the original variables (Berrueta et al., 2007).

This makes PCA a very important multivariate technique, well suited for data visualization and interpretation (Wold et al., 1987; Daszykowski et al., 2007). In some scenarios, PCA is part of a specific chemometric technique itself, when data preprocessing is often performed (Wold et al., 1987; Jackson, 1991). PAT systems may use PCA to reduce the system dimensionality, offering fast and reliable information of the process.

11.4.2.2 Pattern Recognition

Pattern recognition can be described as an artificial intelligence computational algorithm that enhances the variance between data sets. Several types of pattern recognition methods have been used in food science studies. Two types of methods are commonly distinguished: those focused on discrimination among classes, such as linear discriminant analysis (LDA), k-nearest neighbors (kNN), classification and regression trees (CART), partial least- squares discriminant analysis (PLS-DA), and artificial neural networks (ANN), and those oriented toward modeling classes, such as soft independent modeling of class analogy (SIMCA) and unequal dispersed classes (UNEQ) (Brereton, 2003).

The variety of food studies by pattern recognition methods is extensive (e.g., wine, honey, dairy products, meat, fruits, beverages, cereals, and fish). The goal of these studies is to develop models that are able to classify foods in terms of (a) geographical, animal, or botanical origins; (b) technological processes; (c) quality state; (d) detection of adulteration; and others. Table 11.2 lists food samples analyzed by pattern recognition techniques.

TABLE 11.2

Pattern Recognition Techniques Applied to Food Sciences

Sample	Type of Data	Pattern Recognition Techniques		Classification Criteria	Reference
		Supervised	Unsupervised		
Alcoholic beverages	NIR spectra	PCA	SIMCA	Genuine or adulterated	Pontes et al. (2006)
Orange juice, milk, and tonic	e-Tongue	PCA	PCA + kNN, PCA + ANN, and ANN	Orange cultivar, milk brand, and tonic brand	Ciosek et al. (2006)
Beer	Fluorescence spectra	PCA	kNN, LDA	Storage conditions	Sikorska et al. (2006)
Starch gels	FT–Raman spectra	PCA	PCA + CVA, PLS + CVA, and SIMCA	Irradiated/nonirradiated, extent of irradiation	Kizil and Irudayaraj (2006)
Infant formulae	13 elements by AAS and AES	CA, PCA	LDA	Type of infant formula	Sola-Larranaga and Navarro-Blasco (2006)
Cereal-based foods	Raman spectra	PCA	SIMCA, PLS-DA	Specific nutritional contents high or low	Sohn et al. (2005)
Cooked foods	Visible spectra	PCA	PCA + kNN, PCA + ANN	Status of cooking	O'Farrell et al. (2005)
Wheat flour/grain	Eight HPLC peaks	PCA, CA	PCA + CVA	Wheat cultivar	Ordaz-Ortiz et al. (2005)
Chocolate	NIR/fluorescence spectra	PCA	PCA + PLS-DA	Under-, over-, or well-tempered	Svenstrup et al. (2005)
Fish sauces	NIR spectra	—	LDA, FA + LDA, SIMCA, and kNN	Three groups of total nitrogen content	Ritthiruangdej et al. (2005)
Wheat	Vis–NIR (one or two wavelengths)	—	LDA	Normal or scab-damaged	Delwiche and Gaines (2005)
Aniseed drinks	Seven elements by ICP-AES	PCA, CA	LDA, SIMCA	Brand	Jurado et al. (2005)
Food processed in ovens	Visible spectra	PCA	ANN	State of cooking	O'Farrell et al. (2005)
Onions	20 elements	No	LDA, SIMCA	Geographical origin	Ariyama et al. (2004)
Alcoholic beverages	Schlieren effect	PCA	SIMCA	Genuine or adulterated	da Costa et al. (2004)

Sample	Data	Exploratory	Classification	Purpose	Reference
Alcoholic beverages	Five VOCs, three organic acids, 14 polyphenols, and 11 metals	CA, PCA	PLS-DA, kNN	Type of beverage	Cardoso et al. (2004)
Fish	e-Nose (16 sensors)	PCA	kNN, PLS-DA, and LDA	State of conservation	Dodd et al. (2004)
Clams	Morphometrical data	No	CVA	Geographical origin	Palmer et al. (2004)
Alcoholic beverages	Three metals by AAS	No	LDA, QDA, and ANN	Geographical origin, genuine, or adulterated	Hernández-Caraballo et al. (2003)
Alcoholic beverages	Four general parameters, six metals, 10 VOCs, three sugars, and three NMR signals	CA, PCA	RDA, CART	Type of beverage	Kokkinofta and Theocharis (2005)
Vinegars	NIR spectra	CA, PCA	PCA + LDA, QDA, kNN, SIMCA, and UNEQ	Origin of the raw material, elaboration process	Casale et al. (2006)
Beers	53 variables: VOCs, organic acids, and routine parameters	CA, PCA	LDA, kNN, and SIMCA	Brand	Siebert and Lynn (2005)

11.4.2.3 Partial Least-Squares Regression

PLSR is a multivariate analysis tool that enables parameter estimation. The PLSR model can be developed from a training set of N observations (objects, cases, compounds, and process time points) with KX variables denoted by x_k ($k = 1,...,K$) and MY variables denoted by y_m ($m = 1,2,...,M$). These training data form the two matrices X and Y of dimensions ($N \times K$) and ($N \times M$) (Geladi and Kowalski, 1986; Wold et al., 2001; Abdi, 2003). X corresponds to the process variable data and Y corresponds to the product quality data, where the goal is to extract latent variables that explain the variation in the process data and in the respective data quality. The PLSR technique works by successively extracting factors from both X and Y such that the covariance between the extracted factors is maximized (MacGregor and Kourti, 1995).

11.4.2.4 Model Validation

Any model needs to be validated before it is used for understanding or for predicting new events such as the biological activity of new compounds or the yield and impurities in other different process conditions. The best validation of a model is that it consistently and precisely predicts the Y values of observations with the new X values. But an independent and representative validation set is rare. Model validation is necessary to ensure the quality of the analytical results. Although constructing the model involves the use of validation techniques that allow some basic characteristics of the model to be established, a set of samples not employed in the calibration process is required for prediction to confirm model robustness. Such samples can be selected from the initial data set and should possess similar properties as those in the calibration set. In the absence of a real validation set, two reasonable ways of model validation are given by cross-validation (CV), which simulates how well the model predicts new data, and model reestimation after data randomization, which estimates the chance (probability) to get a good fit with random response data (Wold et al., 2001). The quality of the result can be assessed in terms of parameters such as the relative standard error of prediction (RSEP) or the root mean square error of prediction (RMSEP) (Berrueta et al., 2007; Esbensen et al., 2002; Krzanowski, 2000).

11.4.3 Monitoring and Control Models

11.4.3.1 Model Structure

In most cases, the monitoring and control is considered as a single variable, with an easy interpretation, which does not require statistical treatment (register in control charts). The single-varied monitoring is less efficient in terms of information obtainable from a process, and therefore is used as a way to develop control strategies.

Multivariate models focus more on the extraction of relationships between variables than qualitative models. These models are based on the theoretical knowledge of processes and the direct use of real data processes, which facilitates the implementation of models for control and monitoring.

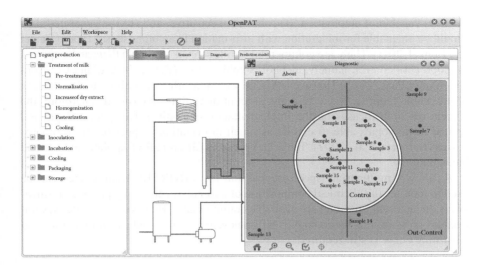

FIGURE 11.12 "OpenPAT" interface.

Multivariate models reduce the process to the study of models of each unit, being able to create a robust model of the process. Exploring the records of each unit allows creating a multivariate model that translates the operation in different operating conditions. It is then possible to predict how such process behaves under various process conditions and to estimate how the production line will react to perturbations.

11.4.3.2 Multivariate Console

After obtaining and testing process models of each unit, their implementation can be done in a multivariate console.

The "OpenPAT," shown in Figure 11.12, is an example of a multivariate console, which, after model implementation, can predict how the process behaves under different process conditions and can estimate the production line reaction to disturbances and to effects of, for example, changing raw materials, among other process variables.

11.5 PAT BIOTECHNOLOGY APPLICATIONS

Bioprocess monitoring is, generally, performed by frequent at-line sampling gathered from fermenters and taken to the laboratory for key compounds' quantification. However, this type of analysis is time consuming, laborious, costly, and not suitable for fast determinations during, for example, fermentation stages. The modern bioprocesses industry, for example, food and beverage, needs both fast and reliable process quality control methods and techniques to provide real-time information to assure the quality and consistency of the final product. An ideal bioprocess monitoring technology is usually considered by including the following characteristics: (a) rapid, (b) nondestructive, (c) multianalyte information, (d) operable in (near) real time, (e) capable of automation, (f) robust, (g) sensitive, and (h) the generated data should be amenable for integration with information from other sensor types (Beutel and Henkel, 2011). There are commercial online sensors available that can be applied to bioprocess control, namely, alcohol, soluble solids (Brix), extract, and CO_2 sensors (AntonPaar, 2012). However, this information offers a shallow understanding of the whole bioprocess due to a low amount of analytes measured. Therefore, the current research focuses on the implementation of noninvasive spectroscopic and optical technologies able to extract as much information as possible from bioprocesses (Berrueta et al., 2007). Table 11.3 presents different PAT applications of food biotechnology.

As shown, multivariate spectroscopic analyzers (NIR, IR, Raman, etc.) can be used to monitor the chemical composition of a food biotechnology process noninvasively, in near real time, and are available commercially. However, even after several decades of work, this technology is still a novelty for most industrial processes and a learning curve is still required.

11.6   FUTURE PERSPECTIVES OF PAT

When a bioprocess is understood, appropriate control strategies can be employed that will allow a reduction in process variability and improve benefits. To reach the point of predictive control, relevant process measurements must be made and their values must be used to make decisions on bioprocesses. The development of new PAT tools, including sensors and data analysis strategies, continues. The developments tend to increase from the need to solve a particular problem and arise from collaborative innovation between analytical chemists (in the industry and in academia) and scientists and process engineers. Automation and control of manufacturing processes from material management to packaging, inventory control, and product distribution are all part of the food chain supply, and new PAT solutions must arise.

Several spectroscopic techniques and their applications to the food area have been discussed and reviewed. Different technologies will appear in the next few years, but there are important questions such as can it be smaller and faster? How does one get the data in the format needed? Are calibrations necessary? And does the instrument require a computer at all?

As large amounts of data will be generated by modern PAT tools, the opportunity to interface analyzers to controllers, analyze and store the data, archive data, and communicate them to a central system must be addressed. These data can be subsequently used to understand the bioprocess and act on it, making adjustments to improve it almost automatically.

Spectroscopic and chromatographic technologies will be improved and new technologies such as nanotechnologies and lab-on-a-chip will be integrated into intelligent systems making the remote control of bioprocesses a reality. One important feature for the application of PAT is the usage of web browsers so that technicians and engineers can instantaneously access the global process information. Consequently, relevant process information must be perceptible for engineers or technicians, to facilitate their communication.

TABLE 11.3
Publications on Food Biotechnology PAT

Process/ Product	Aim	Reference
Ethanol production	Monitor ethanol concentration of a fermentation by using NIR spectroscopy as a noninvasive technique.	Cavinato et al. (1990)
Red wine fermentation	Development of a tool to monitor wine fermentations, to avoid sample preparation and extraction procedures. By combining NIR spectroscopy with chemometrics, it was possible to classify wine fermentations in terms of grape variety, yeast strain, and temperature.	Cozzolino et al. (2006)
Yogurt fermentation	Fusion of sensors to monitor several variables in yogurt fermentation. Using electronic nose, NIR spectroscopy, and other probes (pH, O_2, and temperature). ANN chemometrics technique revealed that sensor fusion was important to understand yogurt fermentation process.	Cimander et al. (2002)
White wines oxidation	Monitor the oxidation of white wines to assess its spoilage. Here, a cyclic voltammetry technique was used as the sensor. Cyclic voltammetry and multivariate analysis revealed an important tool for monitoring oxidative reactions in white wines.	Martins et al. (2008)
Beer fermentation	Manage beer fermentation by UV–VIS–SWNIR spectroscopy. The spectral signal was preprocessed by multivariate statistics revealing good predictions in several variables of the brewing process.	Silva et al. (2009)
Grape maturation	Monitor grape maturation by a portable device called VinePAT based on VIS–SWNIR spectroscopy. Mapping of the vineyard in terms of grape maturation and phenolic profiles.	Martins et al. (2009)
Meat spoilage	Meat quality parameters (color, pH, drip loss, and fat) were calibrated by NIR spectroscopy. Reasonably good correlation predictions were obtained for fat, pH, and color revealing a good technique for fast quality management.	Savenije et al. (2006)
Horticultural products	NIR spectroscopy to evaluate quality parameters of fruits and vegetables (texture, Brix, dry matter, solid content, and acidity). The spectroscopy revealed to be an important noninvasive technique suitable for horticultural products.	Nicolai et al. (2007)
Cheese contamination	UV–VIS–SWNIR spectroscopy to identify yeasts and fungi contamination during cheese storage. UV–VIS–SWNIR spectroscopy is a viable technology for the identification of microorganism contamination in the cheese surface and may provide a tool for managing food quality and safety for cheese manufacturers and distributors.	Silva et al. (2007)

REFERENCES

Abdi, H. 2003. Partial least squares regression (PLS-regression). In: M. Lewis-Beck, A. Bryman and T. Futing (eds.), *Encyclopedia for Research Methods for the Social Sciences* (pp. 792–795). Thousand Oaks (CA): Sage.

Andreas, S. 1997. SAP-R/3 in process industries: Expectations, experiences and outlooks. *ISA Transactions, 36*(3), 161–166.

AntonPaar, 2012. Anton Paar products. http://www.anton-paar.com/Products/2_USA_en (Access date: 2012-04-20).

Ariyama, K., Horita, H., and Yasui, A. 2004. Application of inorganic element ratios to chemometrics for determination of the geographic origin of Welsh onions. *Journal of Agricultural and Food Chemistry, 52*(19), 5803–5809.

Arteaga, F. and Ferrer-Riquelme, A. J. 2009. Missing data. In: Editors-in-chief: D. B. Stephen, T. Romà and W. Beata (eds.), *Comprehensive Chemometrics*, (pp. 285–314). Oxford: Elsevier.

Arvanitoyannis, I. S., Chalhoub, C., Gotsiou, P., Lydakis-Simantiris, N., and Kefalas, P. 2005a. Novel quality control methods in conjunction with chemometrics (multivariate analysis) for detecting honey authenticity. *Critical Reviews in Food Science and Nutrition, 45*(3), 193–203.

Arvanitoyannis, I. S., Katsota, M. N., Psarra, E. P., Soufleros, E. H., and Kallithraka, S. 1999. Application of quality control methods for assessing wine authenticity: Use of multivariate analysis (chemometrics). *Trends in Food Science & Technology, 10*(10), 321–336.

Arvanitoyannis, I. S., Tsitsika, E. V., and Panagiotaki, P. 2005b. Implementation of quality control methods (physico-chemical, microbiological and sensory) in conjunction with multivariate analysis towards fish authenticity. *International Journal of Food Science & Technology, 40*(3), 237–263.

Arvanitoyannis, I. S. and van Houwelingen-Koukaliaroglou, M. 2003. Implementation of chemometrics for quality control and authentication of meat and meat products. *Critical Reviews in Food Science and Nutrition, 43*(2), 173–218.

Bakeev, K. A. 2010. *Process Analytical Technology: Spectroscopic Tools and Implementation Strategies for the Chemical and Pharmaceutical Industries*, Second Edition, Chichester (UK): John Wiley & Sons, Ltd.

Berrueta, L. A., Alonso-Salces, R. M., and Héberger, K. 2007. Supervised pattern recognition in food analysis. *Journal of Chromatography A, 1158*(1–2), 196–214.

Beutel, S. and Henkel, S. 2011. *In situ* sensor techniques in modern bioprocess monitoring. *Applied Microbiology and Biotechnology, 91*(6), 1493–1505.

Bluma, A., Höpfner, T., Lindner, P., Rehbock, C., Beutel, S., Riechers, D. et al. 2010. *In-situ* imaging sensors for bioprocess monitoring: State of the art. *Analytical and Bioanalytical Chemistry, 398*(6), 2429–2438.

Brereton, R. G. 2003. *Chemometrics: Data Analysis for the Laboratory and Chemical Plant*, Chichester (UK): John Wiley & Sons, Ltd.

Brosilow, C. and Joseph, B. 2002. *Techniques of Model-Based Control*, Upper Saddle River (NJ): Prentice-Hall.

Bylesjö, M., Cloarec, O., and Rantalainen, M. 2009. Normalization and closure. In: Editors-in-chief: D. B. Stephen, T. Romà and W. Beata (eds.), *Comprehensive Chemometrics*, (pp. 109–127). Oxford: Elsevier.

Cardoso, D. R., Andrade-Sobrinho, L. G., Leite-Neto, A. F., Reche, R. V., Isique, W. D., Ferreira, M. M. C. et al. 2004. Comparison between cachaça and rum using pattern recognition methods. *Journal of Agricultural and Food Chemistry, 52*(11), 3429–3433.

Casale, M., Sáiz Abajo, M. J., Gonzáez Sáiz, J. M., Pizarro, C., and Forina, M. 2006. Study of the aging and oxidation processes of vinegar samples from different origins during storage by near-infrared spectroscopy. *Analytica Chimica Acta, 557*(1), 360–366.

Cavinato, A. G., Mayes, D. M., Ge, Z., and Callis, J. B. 1990. Noninvasive method for monitoring ethanol in fermentation processes using fiber-optic near-infrared spectroscopy. *Analytical Chemistry,* 62(18), 1977–1982.

Chen, Z., Lovett, D., and Morris, J. 2011. Process analytical technologies and real time process control a review of some spectroscopic issues and challenges. *Journal of Process Control,* 21(10), 1467–1482.

Cimander, C., Carlsson, M., and Mandenius, C. F. 2002. Sensor fusion for on-line monitoring of yoghurt fermentation. *Journal of Biotechnology,* 99(3), 237–248.

Ciosek, P., Sobanski, T., Augustyniak, E., and Wróblewski, W. 2006. ISE-based sensor array system for classification of foodstuffs. *Measurement Science and Technology,* 17, 6.

Cozzolino, D., Cynkar, W. U., Shah, N., and Smith, P. 2011. Multivariate data analysis applied to spectroscopy: Potential application to juice and fruit quality. *Food Research International,* 44(7), 1888–1896.

Cozzolino, D., Kwiatkowski, M., Dambergs, R., Cynkar, W. U., Janik, L. J., Skouroumounis, G. et al. 2008a. Analysis of elements in wine using near infrared spectroscopy and partial least squares regression. *Talanta,* 74(4), 711–716.

Cozzolino, D., Kwiatkowski, M., Parker, M., Cynkar, W., Dambergs, R., Gishen, M. et al. 2004. Prediction of phenolic compounds in red wine fermentations by visible and near infrared spectroscopy. *Analytica Chimica Acta, 513*(1), 73–80.

Cozzolino, D., Parker, M., Dambergs, R. G., Herderich, M., and Gishen, M. 2006. Chemometrics and visible-near infrared spectroscopic monitoring of red wine fermentation in a pilot scale. *Biotechnology and Bioengineering,* 95(6), 1101–1107.

Cozzolino, D., Smyth, H. E., Cynkar, W., Janik, L., Dambergs, R. G., and Gishen, M. 2008b. Use of direct headspace–mass spectrometry coupled with chemometrics to predict aroma properties in Australian Riesling wine. *Analytica Chimica Acta, 621*(1), 2–7.

da Costa, R. S., Santos, S. R. B., Almeida, L. F., Nascimento, E. C. L., Pontes, M. J. C., Lima, R. A. C. et al. 2004. A novel strategy to verification of adulteration in alcoholic beverages based on Schlieren effect measurements and chemometric techniques. *Microchemical Journal,* 78(1), 27–33.

Daszykowski, M., Kaczmarek, K., Vander Heyden, Y., and Walczak, B. 2007. Robust statistics in data analysis—A review: Basic concepts. *Chemometrics and Intelligent Laboratory Systems,* 85(2), 203–219.

Delwiche, S. R. and Gaines, C. S. 2005. Wavelength selection for monochromatic and bichromatic sorting of *Fusarium*-damaged wheat. *American Society of Agricultural Engineers,* 21(4), 681–688.

Dodd, T. H., Hale, S. A., and Blanchard, S. M. 2004. Electronic nose analysis of *Tilapia* storage. *Transactions-American Society of Agricultural Engineers,* 47(1), 135–140.

Esbensen, K. H., Guyot, D., Westad, F., and Houmøller, L. P. 2002. *Multivariate Data Analysis—In Practice: An Introduction to Multivariate Data Analysis and Experimental Design.* Multivariate Data Analysis, Aalborg University, Esbjerg, Denmark.

FDA. 2004. PAT guidance for industry a framework for innovative pharmaceutical development, manufacturing and quality assurance. http://www.fda.gov/downloads/Drugs/.../Guidances/ucm070305.pdf (Access date: 19-03-2012).

Ganguly, J. and Vogel, G. 2006. Process analytical technology (PAT) and scalable automation for bioprocess control and monitoring—A case study. *Pharmaceutical Engineering,* 26(1), 8.

Geladi, P. 2003. Chemometrics in spectroscopy. Part 1. Classical chemometrics. *Spectrochimica Acta Part B: Atomic Spectroscopy,* 58(5), 767–782.

Geladi, P. and Kowalski, B. R. 1986. Partial least-squares regression: A tutorial. *Analytica Chimica Acta, 185,* 1–17.

Gibbon, G. A. 1996. A brief history of LIMS. *Laboratory Automation & Information Management,* 32(1), 1–5.

Gnoth, S., Jenzsch, M., Simutis, R., and Lübbert, A. 2007. Process analytical technology (PAT): Batch-to-batch reproducibility of fermentation processes by robust process operational design and control. *Journal of Biotechnology, 132*(2), 180–186.

Gurden, S. P., Martin, E. B., and Morris, A. J. 1998. The introduction of process chemometrics into an industrial pilot plant laboratory. *Chemometrics and Intelligent Laboratory Systems, 44*(1–2), 319–330.

Henrikson, F. W. 1995. Chapter 3—On-line quality control: Advances in sensor technology. In: G. G. Anilkumar (ed.), *Food Processing*, (pp. 37–57). Amsterdam: Elsevier Science B.V.

Hernández-Caraballo, E. A., Avila-Gomez, R. M., Capote, T., Rivas, F., and Pérez, A. G. 2003. Classification of Venezuelan spirituous beverages by means of discriminant analysis and artificial neural networks based on their Zn, Cu and Fe concentrations. *Talanta, 60*(6), 1259–1267.

Hinz, D. 2006. Process analytical technologies in the pharmaceutical industry: The FDA's PAT initiative. *Analytical and Bioanalytical Chemistry, 384*(5), 1036–1042.

Jackson, J. E. 1991. *A User's Guide to Principal Components*, New York: Wiley.

Jung, Y. M. 2004. Principal component analysis based two-dimensional correlation spectroscopy for noise filtering effect. *Vibrational Spectroscopy, 36*(2), 267–270.

Jurado, J. M., Alcázar, A., Pablos, F., Martín, M. J., and González, A. G. 2005. Classification of aniseed drinks by means of cluster, linear discriminant analysis and soft independent modelling of class analogy based on their Zn, B, Fe, Mg, Ca, Na and Si content. *Talanta, 66*(5), 1350–1354.

Kizil, R. and Irudayaraj, J. 2006. Discrimination of irradiated starch gels using FT–Raman spectroscopy and chemometrics. *Journal of Agricultural and Food Chemistry, 54*(1), 13–18.

Koch, K. H. 1999. *Process Analytical Chemistry: Control, Optimization, Quality, Economy*, Berlin, Germany: Springer.

Kokkinofta, R. I. and Theocharis, C. R. 2005. Chemometric characterization of the Cypriot spirit "Zivania". *Journal of Agricultural and Food Chemistry, 53*(13), 5067–5073.

Korenovska, M. and Suhaj, M. 2005. Identification of some Slovakian and European wines origin by the use of factor analysis of elemental data. *European Food Research and Technology, 221*(3), 550–558.

Krzanowski, W. J. 2000. *Principles of Multivariate Analysis: A User's Perspective*: USA: Oxford University Press .

Lopes, J. A., Costa, P. F., Alves, T. P., and Menezes, J. C. 2004. Chemometrics in bioprocess engineering: Process analytical technology (PAT) applications. *Chemometrics and Intelligent Laboratory Systems, 74*(2), 269–275.

MacGregor, J. F. and Kourti, T. 1995. Statistical process control of multivariate processes. *Control Engineering Practice, 3*(3), 403–414.

Martins, R., Lopes, V. V., and Ferreira, A. C. S. 2009. WinePAT: Facing the complexity of grape quality management and delivering a high-throughput device. In: *Alabe 2009—A Inovação no Sector Vitivinícola*, (pp. 19). Portugal.

Martins, R. C., Oliveira, R., Bento, F., Geraldo, D., Lopes, V. V., Guedes de Pinho, P. et al. 2008. Oxidation management of white wines using cyclic voltammetry and multivariate process monitoring. *Journal of Agricultural and Food Chemistry, 56*(24), 12092–12098.

Menezes, J. C. 2011. 3.43—Process analytical technology in bioprocess development and manufacturing. In: Editor-in-chief: M. Y. Murray (ed.), *Comprehensive Biotechnology* (2nd ed.), (pp. 501–509). Burlington: Academic Press.

Miller, J. N. and Miller, J. C. 2005. *Statistics and Chemometrics for Analytical Chemistry*, New York: Prentice-Hall.

Nicolai, B. M., Beullens, K., Bobelyn, E., Peirs, A., Saeys, W., Theron, K. I. et al. 2007. Nondestructive measurement of fruit and vegetable quality by means of NIR spectroscopy: A review. *Postharvest Biology and Technology, 46*(2), 99–118.

O'Farrell, M., Lewis, E., Flanagan, C., Lyons, W. B., and Jackman, N. 2005. Combining principal component analysis with an artificial neural network to perform online quality assessment of food as it cooks in a large-scale industrial oven. *Sensors and Actuators B: Chemical, 107*(1), 104–112.

Ordaz-Ortiz, J. J., Devaux, M. F., and Saulnier, L. 2005. Classification of wheat varieties based on structural features of arabinoxylans as revealed by endoxylanase treatment of flour and grain. *Journal of Agricultural and Food Chemistry, 53*(21), 8349–8356.

Palmer, M., Pons, G. X., and Linde, M. 2004. Discriminating between geographical groups of a Mediterranean commercial clam (*Chamelea gallina Veneridae*) by shape analysis. *Fisheries Research, 67*(1), 93–98.

Pons, M. A. 1992. Bioprocess monitoring and control. *Biotechnology Techniques, 6*(3), 188–288.

Pontes, M. J. C., Santos, S. R. B., Araújo, M. C. U., Almeida, L. F., Lima, R. A. C., Gaiao, E. N. et al. 2006. Classification of distilled alcoholic beverages and verification of adulteration by near infrared spectrometry. *Food Research International, 39*(2), 182–189.

Ritthiruangdej, P., Kasemsumran, S., Suwonsichon, T., Haruthaithanasan, V., Thanapase, W., and Ozaki, Y. 2005. Determination of total nitrogen content, pH, density, refractive index, and Brix in Thai fish sauces and their classification by near-infrared spectroscopy with searching combination moving window partial least squares. *Analyst, 130*(10), 1439–1445.

Rosas, J. G., Blanco, M., González, J. M., and Alcalá, M. 2012. Real-time determination of critical quality attributes using near-infrared spectroscopy: A contribution for process analytical technology (PAT). *Talanta, 97*, 163–170.

Rosegrant, M. W., Paisner, M. S., Meijer, S., and Witcover, J. 2001. *Global Food Projections to 2020: Emerging Trends and Alternative Futures*, International Food Policy Research Institute, New York.

Savenije, B., Geesink, G. H., Van Der Palen, J. G. P., and Hemke, G. 2006. Prediction of pork quality using visible/near-infrared reflectance spectroscopy. *Meat Science, 73*(1), 181–184.

Shinskey, F. G. 1994. *Feedback Controllers for the Process Industries*, New York: McGraw-Hill.

Siebert, K. J. 2001. Chemometrics in brewing—A review. *Journal of the American Society of Brewing Chemists, 59*(4), 147–156.

Siebert, K. J. and Lynn, P. Y. 2005. Comparison of methods for measuring protein in beer. *Journal of the American Society of Brewing Chemists, 63*(4), 163–170.

Sikorska, E., Górecki, T., Khmelinskii, I. V., Sikorski, M., and De Keukeleire, D. 2006. Monitoring beer during storage by fluorescence spectroscopy. *Food Chemistry, 96*(4), 632–639.

Silva, J. S., Jesus, P., Martins, R. C., Vicente, A., and Teixeira, J. A. 2007. Feasibility of yeast and fungi monitoring during cheese maturation using fibre optic sensor. In: *Congresso Nacional MICRO'07-BIOTEC'07*, (pp. 77). Lisbon.

Silva, R. G., Silva, J. S., Vicente, A. A., Teixeira, J. A., and Martins, R. C. 2009. *In-situ*, real-time bioreactor monitoring by fiber optics sensors. *BioSignals International Conference on Bio-Inspired Systems and Signal Processing*. Porto, pp. 14–17, January 2009.

Singh, R., Gernaey, K. V., and Gani, R. 2009. Model-based computer-aided framework for design of process monitoring and analysis systems. *Computers and Chemical Engineering, 33*(1), 22–42.

Skibsted, E. and Engelsen, S. B. 2010. Spectroscopy for process analytical technology (PAT). In: Editor-in-chief: L. John (ed.), *Encyclopedia of Spectroscopy and Spectrometry* (2nd ed.), (pp. 2651–2661). Oxford: Academic Press.

Sohn, M., Himmelsbach, D. S., Kays, S. E., Archibald, D. D., and Barton, F. E. 2005. NIR–FT/Raman spectroscopy for nutritional classification of cereal foods. *Cereal Chemistry,* 82(6), 660–665.

Sola-Larranaga, C. and Navarro-Blasco, I. 2006. Preliminary chemometric study of minerals and trace elements in Spanish infant formulae. *Analytica Chimica Acta,* 555(2), 354–363.

Svenstrup, G., Heimdal, H., and Nørgaard, L. 2005. Rapid instrumental methods and chemometrics for the determination of pre-crystallization in chocolate. *International Journal of Food Science & Technology,* 40(9), 953–962.

Tewari, J., Dixit, V., and Malik, K. 2010. On-line monitoring of residual solvent during the pharmaceutical drying process using non-contact infrared sensor: A process analytical technology (PAT) approach. *Sensors and Actuators B: Chemical,* 144(1), 104–111.

Tony Liu, D. and William Xu, X. 2001. A review of web-based product data management systems. *Computers in Industry,* 44(3), 251–262.

Trevisan, M. G. and Poppi, R. J. 2006. Química analítica de processos. *Quimica Nova,* 29(5), 1065.

Tzouros, N. E. and Arvanitoyannis, I. S. 2001. Agricultural produces: Synopsis of employed quality control methods for the authentication of foods and application of chemometrics for the classification of foods according to their variety or geographical origin. *Critical Reviews in Food Science and Nutrition,* 41(4), 287–319.

Vojinović, V., Cabral, J. M. S., and Fonseca, L. P. 2006. Real-time bioprocess monitoring: Part I: *In situ* sensors. *Sensors and Actuators B: Chemical,* 114(2), 1083–1091.

Wang, N., Zhang, N., and Wang, M. 2006. Wireless sensors in agriculture and food industry—Recent development and future perspective. *Computers and Electronics in Agriculture,* 50(1), 1–14.

Wold, S., Esbensen, K., and Geladi, P. 1987. Principal component analysis. *Chemometrics and Intelligent Laboratory Systems,* 2(1–3), 37–52.

Wold, S., Sjöström, M., and Eriksson, L. 2001. PLS-regression: A basic tool of chemometrics. *Chemometrics and Intelligent Laboratory Systems,* 58(2), 109–130.

Worldometers. 2012. World population clock. http://www.worldometers.info/world-population/(Access date: 2012-04-20).

WRAP. 2010. A review of waste arisings in the supply of food and drink to UK households. http://www.wrap.org.uk/sites/files/wrap/Waste%20arisings%20in%20the%20supply%20of%20food%20and%20drink%20toUK%20households,%20Nov%202011.pdf (Access date: 20-04-2012).

Wu, H., White, M., and Khan, M. A. 2011. Quality-by-design (QbD): An integrated process analytical technology (PAT) approach for a dynamic pharmaceutical co-precipitation process characterization and process design space development. *International Journal of Pharmaceutics,* 405(1–2), 63–78.

Yu, W., Hertel, T. W., Preckel, P. V., and Eales, J. S. 2004. Projecting world food demand using alternative demand systems. *Economic Modelling,* 21(1), 99–129.

Zeaiter, M., Roger, J. M., and Bellon-Maurel, V. 2005. Robustness of models developed by multivariate calibration. Part II: The influence of pre-processing methods. *TrAC Trends in Analytical Chemistry,* 24(5), 437–445.

Section III

Adding Value to Food Processing By-Products—The Role of Biotechnology

12 Dairy

Carla Oliveira, Lucília Domingues,
Lígia Rodrigues, and José A. Teixeira

CONTENTS

12.1 GENERAL INTRODUCTION

Whey is the principal by-product of cheese manufacture that has been traditionally considered a waste (Fuda et al. 2004). Its management has often involved the use of the most economical disposal methods, including discharge into water streams and onto fields, or simple processing into low-value commodity powders (Smithers et al. 1996). However, the increasing restrictions and environmental concerns have encouraged the development of novel valorization strategies, since cheese whey retains the majority of milk nutrients, namely lactose, proteins, and other valuable components present in minor concentrations. The composition of cheese whey and the huge amounts produced—world whey production is over than 186 million tons per year (Affertsholt 2009)—makes this residue an important raw material for several applications. Cheese whey began to be recognized as a valuable resource and gained increasing interest for its potential use in functional food, nutraceuticals, pharmaceuticals, and cosmetics (Marshall 2004; Alhaj et al. 2007; Saxena et al. 2009). Exploring the full potential of cheese whey requires the development of advanced purification technologies, in particular for protein separation, as well as in the case of lactose, processes that allow its transformation in value-added compounds.

12.2 WHEY PROTEINS PURIFICATION

Whey is a dilute solution composed of lactose, a variety of proteins, minerals, vitamins, and fat. Whey proteins, constituting about 0.7% (w/v) of the total weight of whey, include β-lactoglobulin (BLG), α-lactalbumin (ALA), bovine serum albumin (BSA), and immunoglobulins (IgG) (Jovanovic et al. 2007). ALA and BLG together comprise about 70% (w/w) of the total whey proteins and are responsible for the hydration, gelation, emulsifying, and foaming properties of whey. Besides these, whey contains numerous minor proteins, such as lactoferrin (LF), lactoperoxidase (LP), osteopontin (OPN), lysozyme (LZ), protein components of the MFGM (milk fat globule membrane), proteose–peptone components (PP), lactophorin, β-casein fragments and glycomacropeptide (GMP) (only found in sweet whey), among others (Strange et al. 1992; Gonzalez-Martinez et al. 2002; Perez et al. 2006; El-Sayed and Chase 2011). Functional properties such as solubility, foaming, water sorption, viscosity, emulsification, and gelation make whey proteins an indispensable ingredient in food industry (Albreht and Vovk 2012).

Health benefits of the individual whey proteins that have been reported in the literature include the transport of retinol, palmitate, fatty acids, vitamin D, and cholesterol by BLG (Pérez and Calvo 1995); the induction of apoptosis in tumor cells by ALA and LF (Permyakov and Berliner 2000; Rodrigues et al. 2009); the prevention of cancer by BSA (Madureira et al. 2007); the host defense against organisms requiring iron by LF (Madureira et al. 2007; Rodrigues and Teixeira 2009); the antimicrobial and antiviral activity by LP (Rodrigues and Teixeira 2009); and the antiviral activity against HIV by IgG (Heeboll-Nielsen et al. 2004). Dairy proteins commonly available today are typically concentrates of caseins or whey proteins, being the isolated proteins less common. The isolated protein could be used by individuals with special nutritional needs to tailor their diet, thus improving health (Etzel 2004).

A very large volume of whey is produced worldwide each year, thus representing an excellent source of the above-mentioned proteins. Nevertheless, these proteins are present in whey in dilute concentrations requiring the development of concentration, fractionation, and purification techniques in order to turn them commercially attractive. Whey can be used as animal feedstock, for lactose production or for the production of whey powders and individual whey proteins. Usually, whey protein products are available in three major forms: concentrates (WPC), isolates (WPI), and hydrolysates (WPH) (Huffman and Harper 1999). However, the lack of consistency in the gross composition and functionality of whey powders has limited their acceptance by the food industry (Morr and Ha 1993). Also, the commercial WPCs can develop a stale off-flavor due to the presence of lipid and protein impurities (Morr and Ha 1991; Zydney 1998). Additionally, the unique nutritional, therapeutic, and functional characteristics of the individual whey proteins are largely unrealized in these whey products due to interactions between components and degradation during processing (El-Sayed and Chase 2011). For these reasons, much effort has been aimed at recovering the individual (purified) whey proteins with well-characterized functional and biological properties. Protein functions have been related to their native structure, which depends on pH, temperature, pressure, and solvent effects (Wit 1998). Changes in native structure affect functional properties, so there has been a renewed interest in developing efficient separation and purification processes that prevent denaturation and loss of biological activity (Lozano et al. 2008). Furthermore, there is also an increasing need to produce these individual whey proteins on a large scale without damaging protein structure or their functional properties (Zydney 1998).

The dairy industry has conducted many efforts to develop efficient separation technologies that enable the production of new products. However, some of the early techniques proved to be impractical and can only be applied for lab-scale purposes (Marshall 2004; Liang et al. 2006). Examples of these are salting out, precipitation which is based on selective solubility in presence of solvents, and acid/heat separation which utilizes the differences in thermal stability in acidic conditions. A technical process should be simple, rapid, and nondenaturing and must guarantee a high yield and product quality. None of the above-cited procedures fulfill these prerequisites for successful scaling-up. Several processes have been proposed for commercial-scale production of whey protein fractions. These processes fall into two main categories: (a) chromatography using selective adsorption or selective elution and (b) membrane separation. Membrane separation processes are volume-dependent separation methods, wherein the equipment capacity and cost of manufacture is proportional to the volume of solution processed and not to the mass of the product. For dilute protein solutions such as whey, large volumes of liquid must be processed to recover a fixed mass of protein. Chromatographic processes are less volume dependent because adsorbent capacity depends mostly on the mass of protein recovered, not the volume of liquid processed (Doultani et al. 2004), as well as the concentration of the adsorbed species. Other methods were introduced recently including aqueous two-phase separation (ATPS) and magnetic fishing. The feasibility of these methods for scale-up processing is still under investigation.

12.2.1 CHROMATOGRAPHIC PROCESSES

Within the last decade there has been increasing interest in liquid chromatographic processes because of the growing biotechnology industry and the special needs of the pharmaceutical and chemical industries (Smithers et al. 2008). Chromatographic methods include ion-exchange chromatography in which polyvalent, charged proteins bind to oppositely charged ion exchangers; affinity chromatography which exploits the specific interactions that can occur between proteins and immobilized ligands; and a more general class normally referred to simply as adsorption chromatography where the mechanism of adsorption is often uncertain. Examples of the latter class include the use of hydroxyapatite where adsorption is due to hydrophilic interactions, and other noncharged adsorbents where hydrophobic interactions are implicated in the adsorption process.

Several examples have been reported on chromatographic processes involving selective adsorption and selective elution of whey proteins. Using affinity chromatography, Gurgel and coworkers (2000) immobilized an hexapeptide (WHWRKR) for the adsorption of ALA, while Gambero et al. (1997) used a chemically modified silica with b-diketoamine groups to recover the same protein from bovine milk whey. For BLG adsorption, Wang and Swaisgood (1993) and Noppe et al. (1998)

immobilized retinal on Celite, and Vyas and collaborators (2002) immobilized trans-retinal on calcium biosilicate. The authors reported 80%, 95%, and 95% BLG recovery using a packed-bed column, a stirred tank or a fluidized-bed column, respectively. The fluidized-bed column was found to be the most suitable for further scale-up. Other examples include the BLG adsorption kinetics studies on immuno-chromatographic support (Puerta et al. 2002); the use of thiophilic chromatography on a T-gel to purify IgG from sweet cheese whey (Konecny et al. 1994); and the use of heparin affinity chromatography to extract minor proteins, namely LF and growth factors (Ounis et al. 2008).

Using hydrophobic chromatography, Chaplin (1986) separated bovine whey proteins and caseins on a fast protein liquid chromatography column, Phenyl-Superose; Yoshida (1989) isolated whey LF in a Butyl Toyopearl 650M column; Machold et al. (2002) studied the selectivity and separation efficiency of whey proteins onto different HIC sorbents from various manufacturer's; Santos et al. (2011) isolated BLG (45.2%) with 99.6% purity from a whey protein concentrate (WPC80) using the HiPrep Octyl Sepharose FF column; Hahn et al. (2003) compared different hydrophobic interaction media regarding dynamic binding capacity, recovery and mass transfer properties; and Conrado et al. (2005) described the operating parameters related to the use of Streamline Phenyl for the recovery of ALA from cow milk whey in an expanded-bed adsorption mode of operation using a theoretical model. ALA at a purity of 79% was obtained using this method. Moreover, Sousa et al. (2008) studied the separation of different forms of PP3 by hydrophobic interaction chromatography with a dual salt system, in four different adsorbents (butyl-, octyl-, phenyl-, and epoxy-sepharose).

Ion-exchange chromatography has been extensively used for whey protein purification purposes, either using pure systems of model whey proteins or whey protein systems. Studies have been conducted on the two-component, competitive adsorption model protein systems involving the two major whey proteins (ALA and BLG), as well as other standard proteins from whey. The separation of cytochrome c, a1-acid glycoprotein, ovalbumin and BLG on polyethyleneimine (PEI)-silica gel column using a linear gradient from 0.025 to 0.50 M potassium phosphate at pH 6.8 was reported by Flashner et al. (1983). Wahlgren and coworkers (1993) studied the adsorption of a mixture of BLG, LF, and LZ on hydrophilic silica surfaces. BLG (negative charge) was adsorbed in small amounts, while LF and LZ (positive charge) were adsorbed in higher amounts. The adsorption mechanism was found to be either by cooperative adsorption leading to higher amounts of adsorbed proteins or by competitive adsorption depending on the solution conditions, the protein, and the surface. Moreover, Outinen et al. (1996) evaluated the adsorption characteristics of ALA and BLG from whey on 11 strongly basic anion-exchange adsorbents by performing frontal analysis. Significant fractionation of the two proteins was obtained with Diaion HPA and 75–80% recovery of ALA was obtained in optimized batch adsorption experiments. Kim and Kuga (2002) studied the separation of the two BLG variants (A and B) using a cellulose-based anion-exchanger bearing water-soluble polycations. Also, Stojadinovic and coworkers (2012) used a diethylaminoethyl Sephadex column (with potential for scaling up) to purify the two BLG isoforms from defatted whey. Albreht and Vovk (2012) proposed the separation and isolation of major whey proteins (ALA and BLG) from a WPI using short monolithic columns (CIMac DEAE). Results suggest clear advantages of the developed methods, as well as the possibility of scaling-up. For all the above-mentioned examples, the protein separation was based on selective elution following adsorption rather than selective adsorption.

Few competitive adsorption studies have been carried out on ion-exchange membranes. Examples include the work reported by Weinbrenner and Etzel (1994) that studied the competitive adsorption of ALA and BSA to a sulfopropyl ion-exchange membrane. It was found that competitive adsorption caused displacement of bound BSA monomer by the more strongly bound BSA dimer, thus showing that even apparently single protein systems may display multicomponent competitive behavior. In the two protein experiment, ALA was competitively displaced by the more strongly bound BSA monomer and dimer, indicating that the binding strength was in the order: BSA dimer > BSA monomer > ALA. El-Sayed and Chase (2009) studied the single- and two-component cation-exchange adsorption of ALA and BLG onto SP sepharose FF and found a strong evidence for competitive adsorption as ALA displaced and eluted all of the BLG from the column in a pure form.

Several attempts have been made to separate and purify the different components of whey by means of either column or membrane ion-exchange techniques. Lucas et al. (1998) extracted ALA selectively from acid casein whey protein concentrate using anion-exchange membrane chromatography. Ulber and coworkers (2001) used a cation exchange membrane to obtain a protein stream of LF and LP from whey. Bhattacharjee et al. (2006) were able to separate ALA and BLG from whey using a strong anion exchange membrane. Goodall and coworkers (2008) designed rennet whey breakthrough curves to show that ALA and BSA were displaced from the strong and weak anion-exchange membranes by BLG. Membrane chromatography can overcome some of the limitations of packed-bed column chromatography but preparation of adsorptive membranes usually involves complex and harsh chemical modification. Saufi and Fee (2011) proposed the use of mixed matrix membranes for the extraction of the major proteins from bovine whey. A single membrane was synthesized by incorporating 42.5 wt% Lewatit MP500 anionic resin and 7.5 wt% SP Sepharose cationic resin into an ethylene vinyl alcohol base polymer casting solution. The proposed method was found to be suitable for the recovery of individual whey proteins (ALA, BLG, and LF) on a larger scale.

Many studies have been reported involving the use of column chromatography with anion exchangers, cation exchangers, or a combination of both to separate whey proteins. Santos et al. (2012) used the anionic-exchange chromatography column Mono Q 5%50 GL to study the separation of ALA, BLG, and BSA; both from a mixed pure system and from whey protein concentrate (WPC80). The method described by the authors is based on the use of an ionic column (Mono Q) and a salt gradient elution by increasing the ionic strength of the elution buffer (Tris–HCl 20 mM plus 0–1 M NaCl). The proposed method was found to be suitable to fractionate the major whey proteins from the WPC80 in different fractions, namely one fraction containing all the ALA and immunoglobulins; another fraction containing all the BSA; and two distinct fractions each containing a different variant of BLG. A 60.5% (w/w) recovery of the two main BLG variants was obtained. Carrere and coworkers (1996) studied the recovery of ALA and BLG from sweet whey protein using a fluidized ion-exchange chromatographic process. Vogt and Freitag (1997) evaluated the suitability of anion-exchange and hydroxyapatite displacement chromatography for the processing of dairy whey. Lan et al. (2002) used a liquid–solid circulating fluidized-bed ion-exchange extraction system for continuous protein recovery from cheese whey.

The use of cation-exchange chromatography of whey proteins has been less reported. Hahn and collaborators (1998) developed a fractionation scheme for IgG, LF and LP based on the cation-exchangers S-HyperD F, S Sepharose FF, Fractogel EMD-S 650 (S), and Macro-Prep High S. Andersson and Mattiasson (2006) evaluated the simulated moving-bed technology (SMB) to chromatographically separate LP and LF from WPC using the cation-exchanger Streamline-SP. SMB technology presented several advantages as compared to conventional batch chromatography regarding raised productivity, product concentration, reduced buffer consumption, as well as more efficient use of raw material. Lozano and coworkers (2008) used cation-exchange chromatography only as a final purification step after isolating BLG from bovine whey by means of precipitation. Doultani et al. (2004) developed a cation-exchange column chromatography process that used selective elution to fractionate proteins from whey. ALA was obtained with a recovery of more than 90%, while almost all of the BLG was recovered but with a low purity as it was mixed with substantial amounts of BSA and IgG.

As previously mentioned there are few reports on the use of cation-exchange adsorption of BLG and ALA. Moreover, all the column-chromatography whey purification studies have been based on selective elution rather than selective adsorption. Recently, a novel separation process based on selective adsorption was developed for the purification of ALA and BLG from WPC mixtures using cation-exchange chromatography (El-Sayed and Chase 2010). Almost all of the BLG in the feed was recovered, with 78% being recovered at 95% purity and a further 20% at 86% purity. In addition, 67% of ALA was recovered, 48% at 54% purity, and 19% at 60% purity.

Finally, a number of studies have used a combination of cation- and anion exchange steps to purify these proteins. Gerberding and Byers (1998) described a preparative ion-exchange chromatographic process for the separation and recovery of the four major proteins ALA, BLG, BSA, and IgG, as well as lactose from sweet dairy whey. In that study, it was found that an anion-exchange step was most effective in separating BLG from the whey feed mixture of pH 5.8 while a cation-exchange step was used to further recover the IgG. The obtained yields for ALA, BLG, IgG, and BSA were 36%, 94%, 6%, and 21%, respectively. The low recovery and incomplete separation observed made the process impractical for commercial application. Furthermore, Ye et al. (2000) used cation-exchange chromatography to recover LP and LF by applying salt-gradient elution. Then, the effluent pH was adjusted and it was passed through an anion-exchange column to recover ALA and BLG. Next, the pH of the buffer irrigating the column was adjusted and BLG was eluted.

12.2.2 MEMBRANE PROCESSES

Membrane separation methods can be used as an alternative to the chromatographic ones for the separation of whey proteins. Membrane techniques can be grouped into three main types, namely conventional filtration, ion-exchange membrane chromatography (discussed above), and electro-separation. The conventional filtration methods include the ultrafiltration and diafiltration modes. Cheang and Zydney (2004) studied the use of a two-stage tangential flow filtration system for the purification of ALA and BLG from a whey protein isolate. Using 100 and 30 kDa membranes in series, ALA was obtained at 90% yield. Almecija et al. (2007) explored the potential of membrane ultrafiltration for the fractionation of clarified whey. Using a 300 kDa tubular ceramic membrane in continuous diafiltration mode, the highest permeate yields for ALA and BLG were 56% and 33%, respectively, while BSA, IgG, and LF were mostly retained. Moreover, Muller and coworkers (1999) studied the separation of ALA using different ultrafiltration modes of operation. Using continuous concentration up to a high volume reduction ratio or combined continuous concentration-diafiltration yielded 90% of ALA.

Regarding the electro-separation methods, these include electro-dialysis and electro-acidification. The latter is a recent technology that provides the combined effect of demineralization and acidification by using the properties of bipolar membranes to dissociate water molecules at their interfaces followed by cation-exchange membranes to demineralize by migration of low molecular weight ionic species. Bazinet and collaborators (2004) demonstrated the feasibility of bipolar membrane electro-acidification for whey protein separation. This technology allowed the separation of 98% pure BLG with a 44% recovery yield. On the other hand, Ndiaye et al. (2010) evaluated the feasibility of separating LF from whey using electro-dialysis with an ultrafiltration membrane system of 500 kDa. The highest migration rate for LF was obtained at pH 3 with a migration yield of 15%.

Furthermore, the use of combined methods for the separation of whey proteins has been reported, namely chromatography/membrane separation, chemical treatment/membrane separation, or magnetic fishing/ion-exchange or affinity adsorption. Xu and collaborators (2000) separated bovine IgG and GMP from whey, by selective adsorption of GMP on a polystyrene anion-exchanger IRA93 at pH 4.7, followed by IgG concentration as a result of the selective removal of ALA, BLG, and BSA using the same anion exchanger at pH 7.0 followed by a 100 kDa Amicon YM membrane. Konrad et al. (2000) isolated BLG from whey through a peptic treatment of whey and membrane UF. BLG was produced with a recovery of 67% and a purity of 94%. Also, Konrad and Kleinschmidt (2008) isolated ALA from sweet whey by means of UF followed by a tryptic treatment of permeate, and then a final UF/DF step of the hydrolysate. Up to 15% overall recovery of ALA was achieved with a purity of 90–95%.

12.2.3 OTHER PURIFICATION PROCESSES

One of the recent techniques for whey protein purification is magnetic fishing (Safarik and Safarikova, 2004). It involves the use of magnetic particles bearing an immobilized affinity ligand, a hydrophobic ligand, or ion-exchange groups, as well as magnetic biopolymer particles that have affinity to the desired component. The target protein binds to these particles forming a complex which is easily and rapidly removed from the sample with an appropriate magnetic separator. After washing out the contaminants, the isolated target compound can be eluted. Heeboll-Nielsen and collaborators (2004) reported the use of super-magnetic ion exchangers to fractionate whey proteins. Crude bovine whey was allowed to contact with a super-magnetic cation exchanger to adsorb LF and LP and then the supernatant was contacted with a super-magnetic anion-exchanger to adsorb BLG. The adsorbed proteins were then salt eluted. LF was also recovered from crude whey by adsorption of the protein to magnetic micro cation-exchanger particles followed by elution involving high-gradient magnetic separation (Meyer et al. 2007). The same protein was isolated from acid whey by magnetic affinity separation, employing super-magnetic polyglycidyl methacrylate particles derivatized with heparin. This technology proved to be fast, and has a potential for scaling-up (Chen et al. 2007).

Aqueous two-phase separation (ATPS) has been widely used for the separation of biological material due to its high water content which provides mild extraction conditions (Perumalsamy and Murugesan 2012). It was first used for whey protein purification by Chen (1992) and Coimbra et al. (1994). A polymeric-saline two-phase system composed of polyethylene glycol (PEG) and potassium phosphate was used for the isolation and separation of BLG and ALA from whey. The partition behavior of ALA and BLG on PEG/$(NH_4)_2SO_4$ system was also studied (Rodrigues et al. 2001). Mass recovery yields of 96.7% for ALA in the upper phase and 83.8% for BLG in the lower phase was obtained, in 18% (w/w) PEG 900/14% (w/w) $(NH_4)_2SO_4$ system, at pH 7. This system was found to be an economical alternative for the recovery and separation of the two whey proteins. An efficient and inexpensive separation of both proteins from whey could be achieved, by using 16% (w/w) PEG 900/15% (w/w) $(NH_4)_2SO_4$, at pH 7.5. The technique was further developed by Jara and Pilosof (2011) that used a two-phase system composed of whey protein concentrate and polysaccharide hydroxypropyl methylcellulose at pH 6.5 to study the relative partitioning of ALA and BLG. Furthermore, Rodrigues and coworkers (2003) used ATPS to recover PP3 from cheese whey. Recovery was optimal using a 16% (w/w) Reppal PES 100—24% (w/w) PEG 600 system, at pH 7, giving a mass recovery yield of 99% and a purity of 83% for PP3 in the upper phase. Nandini and Rastogi (2011) proposed an integrated approach for the purification of LP from milk whey by coupling ATPS with ultrasound-assisted ultrafiltration. LP was obtained in the bottom phase, while contaminant proteins partitioned to the top phase. PEG 6000-potassium phosphate system was found to be suitable for the maximum activity recovery of LP 150.70% leading to 2.31-fold purity.

Size-exclusion chromatography is generally used as a final purification step following ion exchange and affinity methods. Preliminary studies conducted by El-Sayed and Chase (2009) using size-exclusion chromatography as a polishing step to increase purity of ALA obtained from a two-stage ion-exchange chromatography process, showed that purity could be increased from 60% to over 90%. In addition, Rojas et al. (2004) used size-exclusion chromatography to purify cheese whey ALA and BLG produced, respectively, in the polymeric and saline phases of ATP systems. A high degree of protein purification of 99.7% for the saline phase and 99.6% for the polymeric phase was obtained.

In summary, different techniques have been used for the purification of individual whey proteins. These include conventional chromatography and membrane techniques, in addition to other techniques that have been recently introduced for whey protein purification such as electro-separation, magnetic fishing, and ATPS. Column chromatography and membrane separation remain the most commonly used techniques for whey protein fractionation. Ion-exchange chromatography methods produce higher yields but lower purity as compared with other chromatography methods such as affinity, hydrophobic, or hydroxyapatite. As for membrane techniques, pressure-driven separation methods produce reasonable yields, but low-purity percentages. Nevertheless, higher purity levels can be achieved when these techniques are coupled with other techniques such as ion-exchange chromatography and chemical treatment. Alternatively, the purity can increase when an additional driving force is added such as the potential difference/bipolarity effect in electro-acidification. Magnetic fishing, electro-separation, and ATPS proved to be the more promising techniques for whey protein fractionation particularly when coupled with other traditional techniques. However, their suitability for scale-up is yet to be evaluated. The technical and economic feasibility of these coupled systems will probably be one of the major future challenges facing the whey protein fractionation industry.

12.3 ALCOHOLIC FERMENTATION OF LACTOSE

The lactose content of cheese whey (ca. 5% w/v) and the presence of other essential nutrients for microbial growth make this dairy by-product a potential feedstock for the production of valuable compounds through fermentation processes. Bioproducts obtained from alcoholic fermentation of whey lactose include bioethanol and whey-based beverages.

12.3.1 KEFIR-LIKE WHEY BEVERAGES

The production of a functional whey beverage simulating kefir obtained upon whey fermentation by kefir grains could be an interesting alternative for cheese whey utilization. Cheese whey fermentation by kefir microorganisms could decrease the high lactose content in cheese whey, producing mainly lactic acid and other metabolites such as aroma compounds contributing to the flavor and the texture and increasing carbohydrate solubility and sweetness of the end product.

Kefir is a refreshing, naturally carbonated fermented dairy beverage with a slightly acidic taste, yeasty flavor, and creamy consistency (Powell et al. 2007). Kefir is traditionally made by inoculating fresh milk with kefir grains and differs from other fermented milk in its unique starter. Kefir grains are small (0.3–3.5 cm in diameter), irregularly shaped, yellow-white granules mostly composed by proteins and polysaccharides and enclose a complex microflora. It contains lactic acid bacteria (LAB) (including *Lactobacillus*, *Lactococcus*, *Leuconostoc*, and *Streptococcus* spp), acetic acid bacteria, and a mixture of yeasts. This microflora is coupled together with casein and complex sugars by a matrix of polysaccharides denominated kefiran (Güzel-Seydim et al. 2005), which is a water-soluble branched glucogalactan. LAB and yeasts exist in kefir grains in a complex symbiotic relationship and are responsible for lactic acid and alcoholic fermentation, respectively. The mixed yeast culture of kefir grains usually contains lactose-fermenting yeasts (*Kluyveromyces lactis, Kluyveromyces marxianus*, and *Torula kefir*), as well as nonlactose-fermenting yeasts (*Saccharomyces cerevisiae*) (Farnworth 2005). Since lactose-fermenting kefir yeasts have the potential to consume cheese whey lactose, kefir grains have been used for beverage production using this lactose-rich waste.

First, cheese whey utilization by kefir grains has been studied for potable alcohol production (Athanasiadis et al. 2002) indicating the ability of this biocatalyst to produce high yields in alcoholic fermentations. Then, the production of kefir-like whey beverages has been investigated using as substrate a cheese whey–milk mixture (Paraskevopoulou et al. 2003), whole cheese whey and deproteinized cheese whey (Athanasiadis et al. 2004; Magalhães et al. 2011a,b). Kefir grains were able to utilize lactose from both cheese whey and deproteinized cheese whey and produce similar amounts of ethanol, lactic acid, acetic acid, and aromatic compounds (higher alcohols and ethyl esters, mostly) to those obtained during milk fermentation (Magalhães et al. 2011a,b). Nevertheless, more time was needed for total lactose consumption when using cheese whey and deproteinized cheese whey than when using milk (Magalhães et al. 2011b). The organoleptic character of the produced whey-based kefir beverages was well accepted by the tasters (Athanasiadis et al. 2004; Magalhães et al. 2011b).

The characterization of kefir grains and beverages obtained from milk and whole/ deproteinized cheese whey fermentation showed a consistent grain structure and microbiota, including probiotic bacteria, which confers probiotic label to kefir beverages (Magalhães et al. 2010). In addition, cheese whey fermented with kefir grains was found to have strong inhibitory properties against pathogenic bacteria (Londero et al. 2011). Altogether, these data encourage the production of kefir-like whey beverages, adding at the same time an important value to cheese whey. Nevertheless, one important aspect that should be taken into consideration is the effect of kefir grains on the proteolysis of milk/whey proteins (Ferreira et al. 2010).

Recently, some reports have been conducted to optimize fermentation of cheese whey by kefir grains, namely in what concerns the experimental parameters influencing lactose uptake rate (Golfinopoulos et al. 2011) and the preservation/utilization of kefir grains for obtaining higher fermentation efficiency (Papapostolou et al. 2008; Londero et al. 2012).

Other perspectives for kefir grains application in whey valorization strategies have also been presented, such as on the production of water-soluble polysaccharide (Rimada and Abraham 2001) on single cell protein production (Koutinas et al. 2005) and more recently, on starter culture production from whey for use in cheese ripening (Koutinas et al. 2009). Moreover, beverage development from whey fermentation with immobilized kefir-yeast has been explored (Athanasiadis et al. 2004; Kourkoutas et al. 2002b). Continuous fermentation of modified whey (containing 1% raisin extract and molasses) using immobilized kefir-yeast on delignified cellulosic material resulted in a quality product (Kourkoutas et al. 2002b). Therefore, the process was proposed for the production of potable alcohol or a novel low-alcoholic content drink (Kourkoutas et al. 2002b).

In summary, the production of beverages with high nutritional and therapeutic value from cheese whey using kefir grains has a promising applicability. From the industrial point of view, besides being an eco-friendly technology, due to the use of a very polluting industrial waste, it is also cost friendly, as the cost of raw materials (whey) are negligible and the use of kefir grains, an easily precipitating granular biomass, leads to the avoidance of centrifugal separators that have a high-energy demand and require high industrial investment (Koutinas et al. 2007).

12.3.2 ALCOHOLIC BEVERAGES

The production of an alcoholic beverage by bioconversion of whey is also an alternative of great interest for reuse of this industrial by-product. Various research efforts have been done on this subject in the last few decades. However, most of the reports on alcohol production from whey are based on the addition of fruit juices, such as mango, banana, pineapple, guava, and strawberries, to whey (Kourkoutas et al. 2002b).

Alcoholic beverages produced from whey fermentation include whey beer and wines (Kosikowski and Wzorek 1977; Kosikowski 1979), low alcohol content drinks (Parrondo et al. 2000a,b; Kourkoutas et al. 2002a) and distilled drinks (Dragone et al. 2009).

Lactose-fermenting yeasts such as *Kluyveromyces fragilis* (Parrondo et al. 2000a,b) and *K. marxianus* (Kourkoutas et al. 2002a) have been employed as biocatalysts for obtaining low alcohol content drinks from whey, in both batch and continuous fermentations. *K. marxianus* was immobilized on delignified cellulosic material and used for high-temperature batch fermentation of whey (Kourkoutas et al. 2002a). The advantage of high-temperature fermentation is that it can give a unique character to the product and thus acceptance by the consumers. Indeed, a product with a good quality and a distinctive aroma potential was obtained (Kourkoutas et al. 2002a).

Recently, an alcoholic beverage (35.4% v/v ethanol) of acceptable organoleptic characteristics was produced by distillation of the fermented broth obtained by large-scale continuous whey fermentation with *K. marxianus* (Dragone et al. 2009). Forty volatile compounds were identified in the distilled product, and most of them were similar to those reported for other alcoholic beverages, although with different concentration values. Higher alcohols (mainly isoamyl alcohol, isobutanol, and 1-propanol) and ethyl esters (mainly ethyl acetate) were the most dominant compounds present, contributing thus to the greatest proportion of the total aroma. Some short- and long-chain fatty acid esters that contribute to fruity and flowery aroma were also present, and the volatile compounds that can be harmful to health (methanol, acetaldehyde, and ethyl acetate) were found at low levels.

Furthermore, a "tea fungus" was used to produce whey beverages with low ethanol content from fermentation of three types of cheese whey (fresh sweet, fresh acid, and reconstituted sweet) (Belloso-Morales and Hernandez-Sanchez 2003). "Tea fungus" is a symbiotic culture of acetic acid bacteria and yeasts used in the production of a sour beverage called "kombucha" by fermentation of black tea and sugar. The whey-based beverages obtained with that culture were not sparkling and strongly sour and salty (Belloso-Morales and Hernandez-Sanchez 2003).

Overall, the alcoholic beverages produced from whey fermentation present characteristics and quality comparable to other alcoholic drinks and thus have commercial potential; however, additional research efforts still have to be done to achieve industrial implementation.

12.3.3 BIOETHANOL

Whey permeate is the lactose-rich stream obtained after separation of whey proteins. The permeate remains a major pollutant since it retains the lactose, which represents more than 70% of the total whey solids and is largely responsible for the whey polluting load. Therefore, it creates disposal problems, in terms of volumes produced and polluting load, almost equal to the disposal of raw whey (Guimarães et al. 2010). One of the alternatives proposed for the treatment and valorization of whey or permeate is the production of ethanol by fermentation (Guimarães et al. 2010; Panesar and Kennedy 2012). For that, whey or permeate is usually concentrated (e.g., by ultrafiltration and/or reverse osmosis process) to attain higher lactose content and thus higher ethanol titer at the end of fermentation.

The number of microorganisms that can use lactose as a source of carbon and energy is limited, in spite of including bacteria, yeasts, and filamentous fungi. Although the yeasts that assimilate lactose aerobically are widespread, those that ferment lactose are rather rare (Guimarães et al. 2010). There are numerous reports about the alcoholic fermentation of lactose/whey, particularly using yeasts, namely *K. marxianus*, *K. fragilis*, *Candida pseudotropicalis* (synonym of *C. kefyr*), and recombinant *S. cerevisiae* strains. In fact, these *Kluyveromyces* and *Candida* species are currently synonyms (Guimarães et al. 2010). The less used strain is *C. pseudotropicalis*; nevertheless, interesting yields of ethanol have been obtained with this yeast (e.g., Ghaly and El-Taweel 1995a, 1997a,b; Szczodrak et al. 1997). For example, a maximum ethanol titer of 41 g L^{-1} was produced from deproteinized whey (10% lactose) in shake flasks, with a corresponding productivity of 0.85 g L h^{-1} (Szczodrak et al. 1997).

The yeasts *K. marxianus* and *K. fragilis* are by far the most used microorganisms in the production of ethanol from whey. Nevertheless, some inhibitory effects in the fermentation of high concentrated lactose/whey media have been reported (above 100–150 g L^{-1} lactose, or above 200 g L^{-1} in some cases), such as osmotic sensitivity, low ethanol tolerance, and inhibition by high salt concentration. This leads to low fermentations with high residual sugar. Some specific experimental conditions have been found to minimize such effects, namely in whey fermentation: the implementation of fed-batch fermentation systems may circumvent lactose and salts inhibition (Ozmihci and Kargi 2007c) and the nutritional supplementation of the medium may alleviate ethanol inhibition (Janssens et al. 1983). Other concerns are the limitation in the whey nutrient content and the oxygen availability (e.g., Vienne and von Stockar 1985; Zafar et al. 2005). Janssens et al. (1983) reported a strong improvement in the fermentation ability of *K. fragilis* when concentrated whey (200 g L^{-1} lactose), already supplemented with 0.5% bacto-peptone, was further supplemented with ergosterol and linoleic acid in Tween 80. Other supplements tested in whey include ammonium sulfate, urea, or peptone (Mahmoud and Kosikowski 1982), yeast extract, vitamins and minerals (Vienne and von Stockar 1985), and NH_4Cl and KH_2PO_4 salts (Kargi and Ozmihci 2006). In what concerns the oxygen effect, low oxygen levels favored fermentative metabolism of whey permeate (lactose concentration range tested: 1–240 g L^{-1}) by *K. marxianus*, leading to higher ethanol volumetric productivities in hypoxic, followed by anoxic and then aerobic conditions (Silveira et al. 2005). In addition, improvements in lactose fermentation by using

Kluyveromyces yeasts in co-cultures with other microorganisms, particularly with *S. cerevisiae*, have been reported (Guo et al. 2010).

A significant advantage of some *K. marxianus* strains is their natural ability to grow and ferment at elevated temperatures (>40°C), which enables cost savings in ethanol production bioprocesses, mainly due to reduced cooling cost (Fonseca et al. 2008). Furthermore, the solubility of lactose solutions increases with the temperature increase, which is relevant for fermentation of high concentrated lactose/whey media. Thermotolerant *K. marxianus* strains have been reported to be capable of growing aerobically at 52°C on lactose and whey permeate (Banat and Marchant 1995) and one of these strains was shown to produce ethanol from whey at 45°C (Kourkoutas et al. 2002a). Nevertheless, ethanol evaporation is a concern when operating at high temperatures.

The production of ethanol from whey by *Kluyveromyces* spp has been mainly conducted in batch and continuous systems but also in feed-batch systems (Table 12.1).

Up to 81 g L^{-1} of maximum ethanol titer has been reported for nonsupplemented fermentations (20% lactose; shake flasks) with *Kluyveromyces* spp (Dragone et al. 2011; Table 12.1). The implementation of an experimental factorial design showed that only the initial lactose concentration, among all the variables tested (fermentation temperature, initial lactose concentration, and inoculum concentration), had a significant effect on the ethanol production by *K. fragilis* (Dragone et al. 2011). Moreover, initial lactose concentrations up to 20% favored the bioconversion to ethanol, but higher lactose concentration values (between 20 and 25%) drastically affected this reaction (Dragone et al. 2011), as reported in other works.

Higher productivity (17.2 g L^{-1} h^{-1}) was obtained when operating in continuous mode in a tubular bioreactor with *K. fragilis* cells immobilized in charcoal pellets and using 15% of initial lactose concentration (Gianetto et al. 1986; Table 12.1). Indeed, feed-batch fermentations had a positive effect on ethanol production, has above indicated (Ferrari et al. 1994; Ozmihci and Kargi 2007c; Table 12.1).

Finally, it is noteworthy the kinetic models that have been developed to describe fermentation of lactose/whey by *Kluyveromyces* spp, which are essential for their optimization and industrial implementation (e.g., Longhi et al. 2004; Zafar et al. 2005; Ozmihci and Kargi 2007a).

S. cerevisiae is the microorganism more widely used in industrial fermentation processes and its exploitation for lactose/whey fermentation has attracted much attention. Nevertheless, this yeast is not capable of metabolizing lactose and thus different strategies were conducted to achieve that end.

The initial strategies involved the use of pre-hydrolyzed lactose solutions (solutions of glucose and galactose). However, this implies a lactose hydrolysis step (e.g., using an enzyme as β-galactosidase) and usually results in long diauxic fermentations because *S. cerevisiae* consumes glucose preferentially to galactose. Nevertheless, some authors have proposed for whey fermentation the co-immobilization of the β-galactosidase enzyme with *S. cerevisiae* cells (Hahn-Hägerdal 1985; Roukas and Lazarides 1991; Lewandowska and Kujawski 2007; Staniszewski et al. 2009). *S. cerevisiae* co-immobilized with β-galactosidase in calcium alginate yielded four times more ethanol titer (52 g L^{-1}) and productivity (4.5 g L^{-1} h^{-1}) in continuous concentrated whey fermentation (15% lactose), comparing to immobilized *K. fragilis*

TABLE 12.1
Selected Studies on Ethanol Production from Whey by *Kluyveromyces fragilis* and *K. marxianus*

Organism	Lactose Content (%)	Operation Type	Ethanol Productivity (g L⁻¹)	Maximum Ethanol Titer (g L⁻¹)	Reference
K. fragilis	24	Batch[a]	0.2	80	Gawel and Kosikowski (1978)
	24	Batch	0.6	72	Mahmoud and Kosikowski (1982)
	15[b]	Batch[f]	2.0	71	Janssens et al. (1983)
	20[c]		1.4	86	
	10[b]	Continuous	7.1	47	Janssens et al. (1984)
	15	Continuous	1.1	13	Hahn-Hägerdal (1985)*
	15	Continuous	17.2	18	Gianetto et al. (1986)*
	10	Fed-batch	3.3	64	Ferrari et al. (1994)
	5.5	Continuous	14.5	20	Kleine et al. (1995)*
	20	Batch[e]	1.8	81	Dragone et al. (2011)
K. marxianus	6.5[d]	Batch	0.52	26	Rosenberg et al. (1995)
	10[d]	Batch	3.1	43	Grba et al. (2002)
		Fed-batch	4.9	59	
	17	Batch[f]	1.0–1.5	76–80	Silveira et al. (2005)
	15	Batch	0.4	80	Kargi and Ozmihci (2006)
	10	Continuous	0.74	32	Ozmihci and Kargi (2007a)
	10–12.5	Continuous	0.54	29	Ozmihci and Kargi (2007b)
	12.5	Fed-batch	5.3	63	Ozmihci and Kargi (2007c)
	7.5	Batch[e]	0.55	40	Ozmihci and Kargi (2007d)
	5	Continuous	0.4	20	Ozmihci and Kargi (2008)*
	5	Continuous	0.36	16	Jędrzejewska and Kozak (2011)
K. marxianus and *S. cerevisiae*	10	Batch[e]	0.88	42	Guo et al. (2010)*

Source: Adapted from Guimarães PMR, Teixeira JA, Domingues L 2010. *Biotechnol Adv* 28(3):375–384.

* Immobilized cells.
[a] 3 L bottles.
[b] Whey supplemented with 0.5% peptone.
[c] Whey supplemented with 0.5% peptone, ergosterol, linoleic acid, and Tween 80.
[d] Whey supplemented with yeast extract and salts.
[e] Shake flasks.
[f] 1 L flasks (the remaining works were performed in bioreactors).

cells (Hahn-Hägerdal 1985; Table 12.1). More recently, a similar biocatalyst has been applied in a hybrid system coupling whey fermentation with a pervaporation module for ethanol separation (Staniszewski et al. 2009). In addition, enhanced ethanol productivity (1.0 g L^{-1} h^{-1}) was obtained using permeabilized *K. marxianus* cells as the source of β-galactosidase for simultaneous hydrolysis of lactose in concentrated whey-based medium (6.5% lactose) and fermentation by *S. cerevisiae* (batch operation), as compared to direct fermentation using *K. marxianus* (Rosenberg et al. 1995; Table 12.1).

Other strategies consisted of the construction of lactose-consuming (Lac$^+$) *S. cerevisiae* strains by using recombinant DNA technology. This has been attempted by several ways such as, protoplast fusion, expression of heterologous β-galactosidases (secreted to the medium or released by means other than secretion), or simultaneous expression of heterologous lactose permease and β-galactosidase. The generation of hybrids of *S. cerevisiae* and *Kluyveromyces* spp by protoplasts fusion was one of the first approaches to create Lac$^+$ *S. cerevisiae* strains. Hybrids of *S. cerevisiae* and *K. lactis* were able to ferment lactose in sweet and salted whey (Tahoun et al. 1999, 2002). However, the genetic stability of the hybrids strains was a concern (Farahnak et al. 1986).

Lac$^+$ *S. cerevisiae* strains have been constructed by the transfer of lactose metabolization genes mainly from *E. coli* (*lacZ*), *K. lactis* (*lac4*), and *A. niger* (*lacA*) (reviewed in Domingues et al. 2010). In some cases, the main goal was the production of heterologous β-galactosidase (reviewed in Oliveira et al. 2011). Nevertheless, the production of ethanol was still evaluated, namely when using cheese whey as culture medium. As a relevant example, a flocculent *S. cerevisiae* strain secreting high levels of *A. niger* β-galactosidase produced ethanol from lactose/whey with close to theoretical yields in batch and in high cell density continuous fermentations with complete lactose utilization (Domingues et al. 2002, 2005). This recombinant system could be of interest for the dairy industry as the recombinant β-galactosidase was secreted from flocculent *S. cerevisiae* cells at the same time as ethanol was produced while using whey permeate as substrate. Moreover, with a simple ultrafiltration step the produced enzyme could be easily applied for lactose bioconversion.

Several approaches based on the release of intracellular recombinant enzymes (e.g., from *E. coli* and *K. lactis*) were designed to obtain Lac$^+$ strains. For instance, *E. coli* β-galactosidase was released in the culture medium by recombinant *S. cerevisiae* with the overexpression of the transcriptional activator *GAL4*, which induced partial lysis of the mother cells (Porro et al. 1992; Compagno et al. 1995). The amount of the released β-galactosidase was enough to support growth in lactose/whey-based culture medium. Ethanol production was observed in stationary phase with interesting yields (73–84% of the theoretical conversion yield, Porro et al. 1992) but low productivities (up to 1 g L^{-1} h^{-1}) (Porro et al. 1992; Compagno et al. 1995). Interestingly, diauxic growth was not observed (Porro et al. 1992). Nevertheless, cell lysis has a negative impact on downstream processing, which represents a disadvantage over the secretion approach.

Other alternative approach for obtaining Lac$^+$ strains involves the simultaneous expression in *S. cerevisiae* of *K. lactis lac12* and *lac4* genes (which encode lactose permease, a cell membrane protein responsible for the intracellular uptake of lactose,

and intracellular β-galactosidase, respectively). The first *S. cerevisiae* Lac⁺ strains
obtained by transfer of the *K. lactis lac* genes grew slowly in lactose (Sreekrishna and
Dickson 1985) or presented low ethanol production (Rubio-Texeira et al. 1998). A flocculent *S. cerevisiae* Lac⁺ strain, obtained by transfer of the same genes (Domingues
et al. 1999b), showed improved performance in lactose medium and ethanol conversion yield close to the theoretical value (Domingues et al. 1999a). This phenotype was
acquired after an adaptation period, where the strain was maintained in periodically-
refreshed liquid lactose medium. When operating in continuous high cell density
system (6 L air-lift bioreactor), using cheese whey permeate as substrate, an ethanol
productivity near 10 g L⁻¹ h⁻¹ was obtained (Domingues et al. 2001). The use of 2-fold
concentrated cheese whey permeate was also considered, resulting in a fermentation
product with 5% (w/v) ethanol (Domingues et al. 2001). However, it was not possible to
operate continuously using the high cell density bioreactor with concentrated cheese
whey due to a deflocculating effect attributed to the salts concentration (Domingues
et al. 2001). Later, this recombinant flocculent strain was subjected to a long-term evolutionary engineering process that improved its lactose fermentation capacity and also
its flocculation ability (Guimarães et al. 2008a,b). In batch fermentations, the evolved
strain produced a maximum of 8% (v/v) ethanol from mineral medium with 150 g L⁻¹
lactose, with a productivity of 1.5–2.0 g L⁻¹ h⁻¹ (Guimarães et al. 2008c). It was also
capable of fermenting concentrated cheese whey (150 g L⁻¹ lactose) producing 7%
(v/v) ethanol (Guimarães et al. 2008a). However, in this case, the fermentation was
much slower and consequently the productivity was low (0.46 g L⁻¹ L⁻¹) (Guimarães
et al. 2008a). The supplementation of whey with 10 g L⁻¹ of corn steep liquor significantly enhanced fermentation, resulting in the production of 7.4% (v/v) ethanol from
150 g L⁻¹ initial lactose in shake-flask fermentations, with a corresponding productivity of 1.2 g L⁻¹ h⁻¹ (Silva et al. 2010). The flocculation capacity of this yeast strain
enabled stable operation of a repeated-batch process in a 5.5 L air-lift bioreactor, with
simple biomass recycling by sedimentation of the yeast flocs. During five consecutive
batches, the average ethanol productivity was 0.65 g L⁻¹ h⁻¹ and ethanol accumulated
up to 8% (v/v) with lactose-to-ethanol conversion yields over 80% of theoretical. The
stability and robustness of this strain was demonstrated by the high viability (>97%)
and plasmid retention (>84%) during operation.

Strains having a highly flocculent phenotype, such as the above-mentioned
strains, are particularly interesting for application in high cell density fermentations,
which usually result in enhanced process productivity. These systems have the par-
ticularity of minimizing the risk of contamination due to the high biomass concentration attained (Domingues et al. 2000). In addition, these strains present important
advantages for industrial application namely, easy and inexpensive cell recycling
for repeated-batch operation and simplification of the downstream processing, since
flocculated cells may be easily separated from the fermentation broth by sedimentation at the end of fermentation, as well as improved ethanol tolerance and cell viability (Zhao and Bai 2009; Silva et al. 2010).

In terms of effective industrial application, there are already established processes to produce ethanol from whey in some countries, namely in Ireland, New
Zealand, United States, and Denmark (Guimarães et al. 2010).

12.4 FERMENTATION AND BIOTRANSFORMATION OF LACTOSE FROM WHEY INTO VALUABLE PREBIOTICS

Industrial biotechnology is considered the potential basis of a future knowledge-based bioeconomy toward a more sustainable society (Carrez 2006). Process-integrated biotechnologies, especially the use of biocatalysts, are expected to guide the implementation of greener production processes in a number of industries (Tundo et al. 2000; Jiménez-González et al. 2011; Watson 2012). The unique capabilities of natural catalysts (microorganisms and enzymes) regarding their high specificity and selectivity under mild pH, temperature, and pressure conditions can be exploited to develop sustainable processes which can compete with the classical chemical routes available or even attain the discovery of novel routes for the synthesis of new products by means of synthetic biology approaches (Rodrigues and Kluskens 2011).

Dairy industries produce worldwide 186 million tons of whey per year as a residue, mainly composed of lactose (up to 70% of the total dry material). As previously mentioned the disposal of whey or any other process streams containing high amounts of lactose is becoming increasingly difficult due to more restrictive legislations and rising costs of waste treatment. Lactose is a low value-added sugar and new technologies for its transformation into more value-added products with application in food and pharmaceutical industries could dramatically increase its market value (Ganzle et al. 2008). Only a small amount of the lactose is used nowadays as a feedstock for the chemical, enzymatic, and microbiological conversion into higher market value lactose derivatives such as lactulose, lacitol, lactobionic acid, lacto-sucrose, and galactooligosaccharides (GOS) (Ganzle et al. 2008; Guimarães et al. 2010). These compounds are currently used as prebiotic or low-caloric food ingredients, as chelating agent in calcium supplements, in the treatment of constipation and chronic hepatic diseases, and in the cold storage of transplant organs (Ganzle et al. 2008; Torres et al. 2010).

12.4.1 LACTOSE DERIVATIVES

Lactulose is a synthetic sugar that cannot be degraded by animal or human digestive enzymes. It reaches the colon unchanged, where it is metabolized by intestinal bacteria with production of short-chain fatty acids (Nilsson and Nyman 2005). Its specific bifidogenic effect is less pronounced when compared with fructooligosaccharides (FOS) or GOS (Bouhnik et al. 2004). Currently, the main applications of lactulose are as a drug for the treatment of constipation and hepatic encephalopathy and as a prebiotic food ingredient. The lactulose market is expected to increase in a near future due to aging of the population in industrialized countries, increasing unhealthy styles of living, growth of populations in lesser developed countries, as well as to the promising research on the (co)therapeutic use of lactulose in numerous potential applications (Panesar and Kumari, 2011). Lactulose is currently produced by chemical isomerization of lactose but the main drawback is the presence of difficult to remove colored by-products. Enzymatic processes using β-galactosidases (Lee et al. 2004) and β-glycosidases (Mayer et al. 2010) as biocatalysts lead to a reduced number of by-products, arising as promising alternatives for lactulose production.

Lactitol is produced by chemical hydrogenation of lactose; its relative sweetness as compared to sucrose is 0.3–0.4. Lactitol hydrolysis to galactose and sorbitol in the human intestine is strongly dependent on the intestinal microflora; being most of the lactitol metabolized to short-chain fatty acids by the colonic microflora while a part of the galactose is resorbed (Dills 1989). Lactitol is a strong laxative and its metabolism by humans is insulin independent. Polyols such as mannitol, xylitol, or arabitol, lactitol are applied as noncaloric sweeteners in calorie-reduced and diabetic foods. Moreover, lactitol is used as an alternative to lactulose in the treatment of hepatic encephalopathy (Als-Nielsen et al. 2004).

Commercial lactobionic acid is produced by chemical oxidation of lactose (Gerling 1998). Alternatively, it can be produced using a glucose–fructose oxidoreductase from *Zymomonas mobilis* as described by Satory and coworkers (1997). Lactobionic acid is a strong chelator of calcium and therefore is used in calcium supplements in pharmaceuticas. Furthermore, these chelating properties enable lactobionic acid application as ion sequestrant in detergents. Additionally, lactobionic acid is a key component of solutions used to cold store transplant organs, since it prevents hypothermically induced cell swelling and reduces oxidative injury during storage (Southard and Belzer 1995).

Lactosucrose is a trisaccharide derived from transfructosylation of lactose and is commercially produced in Japan. This lactose derivative is not resorbed in the upper intestine and therefore is available for hydrolysis and metabolism by the colonic microflora, that may be considered a prebiotic although not as well documented as is the case for FOS or GOS. Lactosucrose has a bifidogenic effect and its consumption was reported to decrease fecal pH and to inhibit growth of colonic clostridia (Ogata et al. 1993). Transfructosylation of lactose is carried out by bacterial or fungal fructosyltransferases using sucrose or raffinose as fructosyl donor. Fructosyltransferases catalyze the hydrolysis of sucrose, as well as the transfructosylation to acceptor carbohydrates. Fructosyltransferases catalyze the formation of high molecular weight fructan polymers and the formation of oligosaccharides in addition to sucrose hydrolysis. Depending on the type of polymer formed, fructosyltransferases are referred to as levansucrases or inulosucrases (Tieking and Ganzle 2005; van Hijum et al. 2006). The formation of lactosucrose by levansucrase was first described using an enzyme from *Rhanella aquatilis* (Ohtsuka et al. 1992). Because the spectrum of acceptor carbohydrates is essentially comparable in all bacterial levansucrases characterized so far (Cote and Ahlgren, 1993; Tieking et al. 2005), it can be assumed that the synthesis of lactosucrose from sucrose and lactose is a general property of these bacterial enzymes. Levansucrase activity is frequently found in food fermenting LAB, particularly *Lactobacillus reuteri*, *Lactobacillus pontis*, and *Lactobacillus acidophilus*. These organisms have been successfully used to generate high amounts of oligosaccharides in food fermentations (Tieking et al. 2005).

12.4.2 Prebiotics

Prebiotics are nondigestible food ingredients that stimulate the growth and/or activity of beneficial bacteria in the colon (Gibson and Roberfroid 1995; De Preter et al. 2011). These compounds are not digested nor adsorbed in the small intestine; therefore, they are classified as dietary fiber contributing to a diet recognized as beneficial to human health (Englyst and Englyst, 2005). The end products of the metabolism of prebiotics by intestinal bacteria are lactate and short-chain fatty acids which contribute to the beneficial effects of prebiotics (Topping and Clifton, 2001). Moreover, prebiotics are generally used to stimulate growth and metabolism of bifidobacteria and their consumption results in an increase in both the occurrence and number of bifidobacteria isolated from fecal material (Cummings et al. 2001). The establishment of metabolically active bifidobacteria is considered to be beneficial for the host.

Beneficial effects that have been reported include the regulation of bowel habit, stabilization of the gut mucosal barrier and the prevention of diarrhea, and increased mineral absorption (Cummings et al. 2001; Hopkins and Macfarlane, 2003). More than 130 different human milk oligosaccharides have been identified. Most compounds carry lactose at the reducing end and are substituted with N-acetylglucosamine, L-fucose, and/or N-acetylneuraminic acid (sialic acid). Additionally, possible functions of these compounds are the stimulation of the immune system, antimicrobial, and/or anti-inflammatory protection, and to act as soluble receptors for pathogens (Kunz et al. 2000).

Typically, prebiotics are carbohydrates with functional properties that vary depending on the chain length and type of linkage between the monomer units. Lactose is the main substrate for the enzymatic production of a novel kind of prebiotics the so-called GOS (Neri et al. 2009; Husain 2010; Torres et al. 2010). The yield and type of GOS produced strongly depends on the enzymatic source, lactose concentration in the feeding stock, use of free or immobilized enzyme, reaction medium, and operational conditions. Therefore, there is a broad margin for the enzymatic production of new lactose derivatives with potential applications.

The formation of oligosaccharides from monosaccharides by the action of mineral acids (the so-called chemical synthesis) is well known. Also, the adequate conditions for oligosaccharide production during acidic hydrolysis of lactose and the resulting oligosaccharide structures formed have been reported (Huh et al. 1990, 1991). During this chemical process a complex mixture of disaccharides and trisaccharides is formed, with a variety of linkages with α- and β-anomeric configurations, and anhydro-sugars. Probably due to the lack of product specificity and extreme conditions used for the acidic hydrolysis of lactose, this process is not used on a large scale for the production of GOS. The preferred mode for GOS synthesis is by enzymatic catalysis from lactose using glycosyltransferases (EC 2.4) or glycoside hydrolases (EC 3.2.1) (De Roode et al. 2003). Although highly region-selective, stereo-selective, and efficient, these enzymes are not used for industrial GOS production due to their unavailability, prohibitive prices of commercial enzyme preparations, and the need of specific sugar nucleotides as substrates. Currently, GOS are industrially produced using the catalytic activity of glycoside hydrolases. These enzymes are more readily available than glycosyltransferases but are generally less stereo-selective (Tzortzis and Vulevic 2009). Converting lactose into GOS by β-galactosidases is a kinetically controlled reaction, by means of the competition between hydrolysis and transgalactosylation. Specifically, during this conversion, the thermodynamically favored hydrolysis of lactose, which generates D-galactose and D-glucose, competes with the transferase activity that generates a complex mixture of various galactose-based di- and oligosaccharides of different structures. Hence, knowledge of the reaction time course (or lactose conversion) is required to determine the point of maximum yield of the desired product.

Glycoside hydrolases with β-galactosidase activity occur in a variety of microorganisms from the super-kingdoms Archaea, Bacteria, and Eukaryota. Some of these enzymes have been expressed in host organisms, and/or purified by a combination of several conventional techniques, such as salting-out fractionation, ion exchange, gel filtration, hydroxyapatite, and hydrophobic interaction chromatography (Nakayama and Amachi 1999). A variety of glycoside hydrolases with β-galactosidase activity are commercially available for use in food processing (Panesar et al. 2006), and some of them are already used in the industrial production of GOS. Nevertheless, there is continuous interest in finding microorganisms with well-fitted properties for industrial uses and able to produce specific GOS mixtures with better yields.

GOS are commercially produced from lactose using fungal β-galactosidases. The β-galactosidases hydrolyze lactose to glucose and galactose and alternatively catalyze the transgalactosylation of lactose to produce GOS (Torres et al. 2010). Depending on the source of the enzymes, the oligosaccharides have predominantly β(1–4) and/or β(1–6) linkages. For example, the β-galactosidase of *Kluyveromyces lactis* produces predominantly β-(1–6) oligosaccharides (6′-galactosyl-lactose and β-D-Gal(1–6)D-Gluc), the β-galactosidase of *Sterigmatomyces elviae* produces predominantly 4′-galactosyl-lactose whereas *Bacillus circulans* β-galactosidase forms β-(1–2), β-(1–3), β-(1–4), or β-(1–6) linkages to produce a large variety of oligosaccharides (Onishi et al. 1995; Yanahira et al. 1995). Glucose, galactose, mannose, fructose, maltodextrins, N-acetylneuraminic acid, glucuronic acid, and a number of aromatic compounds have been shown to act as galactose acceptor for β-galactosidases, providing a virtually unlimited variety of oligosaccharides (Lee et al. 2004; Miyasato and Ajisaka 2004; Bridiau et al. 2006). Both 4′-galactosyl lactose and 6′-galactosyl lactose are considered to have prebiotic properties (Torres et al. 2010). However, different oligosaccharide preparations may vary with respect to technological benefits such as flavor-enhancing properties, sweetness, hygroscopicity, and solubility.

Generally, the yield of GOS synthesis from lactose using glycoside hydrolases can be increased using high lactose concentrations; decreasing the water thermodynamic activity; continuously removing the final product and/or inhibitors from the reaction medium; and modifying the enzyme. Data gleaned from the literature show that the transgalactosylation reaction of β-galactosidases is favored at high lactose concentrations and typical oligosaccharide yields range from 10% to 40% (GOS/initial lactose). For instance, the effect of lactose concentration on lactose turnover by a β-galactosidase from *Lactobacillus delbrueckii* spp *bulgaricus* has been reported (Boon et al. 2000). Moreover, since lactose solubility is relatively low at room temperature (Roos 2009), high temperatures are generally desired. These high incubation temperatures strongly favor oligosaccharides formation over lactose hydrolysis (Vasiljevic and Jelen 2003). Glycoside hydrolases from *Sulfolobus solfataricus* (Park et al. 2008), *Pyrococcus furiosus* (Hansson et al. 2001), *Thermus caldophilus* (Choi et al. 2003), *Thermotoga maritima* (Ji et al. 2005) are some examples of enzymes that can be used at temperatures around 80°C. An increase in reaction temperature significantly increases GOS yield when glycoside hydrolases from *Aspergillus aculeatus* (Cardelle-Cobas et al. 2008), *S. solfataricus*, or *P. furiosus* (Hansson et al. 2001) were used. Other approaches to optimize oligosaccharide synthesis by β-galactosidases include screening for enzymes with high preference for transgalactosylation (Boon et al. 2000; Vasiljevic and Jelen 2003; Cheng et al. 2006), enzyme immobilization to achieve an improved enzyme stability at high temperatures (Albayrak and Yang 2002), and use of high hydrostatic pressure to improve the yield of oligosaccharides (Ganzle et al. 2008). Moreover, protein engineering was successfully used to obtain truncated or modified β-galactosidases with high preference for production of oligosaccharides (Jorgensen et al. 2001).

Furthermore, LAB produces β-galactosidases offer substantial potential for the production of GOS. These bacteria are known to be good producers of extracellular β-galactosidases that enable GOS production from lactose (Hung et al. 2001). Also, LAB have been widely used in food fermentations and exhibit rapid anaerobic growth on low-cost raw materials (e.g., whey). Therefore, GOS may be produced from crude cellular extracts without costly downstream processing (Vasiljevic and Jelen 2003). Moreover, GOS may be produced *in situ* during food fermentations or by using whey to produce food-grade GOS preparations.

In summary, lactose is nowadays the most underused dairy component. The huge amount of available lactose resulting from cheese manufacture, together with its low market value is driving the development of novel approaches for its conversion into valuable derivatives. Some of the lactose derivatives discussed above are just a few examples of the new trends in surplus lactose management. Additional major opportunities may arise in using lactose for various fermentation applications. Lactose used to be one of the cheapest fermentation carbohydrates on the market for some time. It is not inconceivable that lactose, in its crudest form as whey or UF permeate, could become an economical source of biogas or bioethanol. Finally, lactose derivatives possess a number of health benefits and constitute an interesting alternative for lactose-intolerance individuals. Therefore, their production from lactose would also benefit the industrial dairy processors in opening up new attractive markets for dried whey and other dairy products containing lactose.

REFERENCES

Affertsholt T, 3A Business Consulting 2009, *International Whey Market Overview, ADPI/ABI Annual Conference*, Chicago, IL, April 26–28, 2009.

Albayrak N, Yang ST. 2002. Production of galacto-oligosaccharides from lactose by *Aspergillus oryzae* β-galactosidase immobilized on cotton cloth. *Biotechnol Bioeng* 77:8–19.

Albreht A, Vovk I. 2012. Applicability of analytical and preparative monolithic columns to the separation and isolation of major whey proteins. *J Chromatogr A* 1227:210–218.

Alhaj OA, Kanekanian AD, Peters AC. 2007. Investigation on whey proteins profile of commercially available milk-based probiotics health drinks using fast protein liquid chromatography (FPLC). *Br Food J* 109(6):469–480.

Almecija MC, Ibanez R, Guadix A, Guadix EM. 2007. Effect of pH on the fractionation of whey proteins with a ceramic ultrafiltration membrane. *J Membr Sci* 288:28–35.

Als-Nielsen B, Gluud LL, Gluud C. 2004. Non-absorbable disaccharides for hepatic encephalopathy: Systematic review of randomized trials. *Br Med J* 328:1046–1050.

Andersson J, Mattiasson B. 2006. Simulated moving bed technology with a simplified approach for protein purification separation of lactoperoxidase and lactoferrin from whey protein concentrate. *J Chrom A* 1107:88–95.

Athanasiadis I, Boskou D, Kanellaki M, Kiosseoglou V, Koutinas AA. 2002. Whey liquid waste of the dairy industry as raw material for potable alcohol production by kefir granules. *J Agric Food Chem* 50:7231–7234.

Athanasiadis I, Paraskevopoulou A, Blekas G, Kiosseoglou V. 2004. Development of a novel whey beverage by fermentation with kefir granules. Effect of various treatments. *Biotechnol Prog* 20:1091–1095.

Banat IM, Marchant R. 1995. Characterization and potential industrial applications of five novel, thermotolerant, fermentative, yeast strains. *World J Microbiol Biotechnol* 11: 304–306.

Bazinet L, Ippersiel D, Mahdavi B. 2004. Fractionation of whey proteins by bipolar membrane electro-acidification. *Innovat Food Sci Emerg Technol* 5:17–25.

Belloso-Morales G, Hernandez-Sanchez H. 2003. Manufacture of a beverage from cheese whey using a "tea fungus" fermentation. *Rev Latinoam Microbiol* 45:5–11.

Bhattacharjee S, Bhattacharjee C, Datta S. 2006. Studies on the fractionation of b-lactoglobulin from casein whey using ultrafiltration and ion-exchange membrane chromatography. *J Membr Sci* 275:141–150.

Boon MA, Janssen AEM, van't Riet K. 2000. Effect of temperature and enzyme origin on the enzymatic synthesis of oligosaccharides. *Enz Microb Technol* 26:271–281.

Bouhnik Y, Raskine L, Simoneau G, Vicaut E, Neut C, Flourie B, Brouns F, Bornet FR. 2004. The capacity of nondigestible carbohydrates to stimulate fecal *Bifidobacteria* in healthy humans: A double-blind, randomized, placebo-controlled, parallel group, dose-response relation study. *Am J Clin Nutr* 80:1658–1664.

Bridiau N, Taboubi S, Marzouki M, Legoy MD, Maugard T. 2006. β-Galactosidase catalyzed selective galactosylation of aromatic compounds. *Biotechnol Prog* 22:326–330.

Cardelle-Cobas A, Villamiel M, Olano A, Corzo N. 2008. Study of galacto-oligosaccharide formation from lactose using pectinex ultra SP-L. *J Sci Food Agric* 88(6):954–961.

Carrere H, Bascoul A, Floquet P, Wilhelm AM, Delmas H. 1996. Whey proteins extraction by fluidized ion exchange chromatography: Simplified modelling and economical optimization. *Chem Eng J Biochem Eng J* 64(3):307–317.

Carrez D. 2006. *Industrial or White Biotechnology: A Policy Agenda for Europe*. Brussels: European Association for Bioindustries.

Chaplin LC. 1986. Hydrophobic interaction fast protein liquid chromatography of milk proteins. *J Chrom A* 363(2):329–335.

Cheang B, Zydney A. 2004. A two stage ultrafiltration process for fractionation of whey protein isolate. *J Membr Sci* 231:156–167.

Chen JP. 1992. Partitioning and separation of alpha-lactalbumin and beta-lactoglobulin PEG/potassium phosphate two phase systems. *J Bioimaging* 73:360–365.

Chen L, Guo C, Guan Y, Liu H. 2007. Isolation of lactoferrin from acid whey by magnetic affinity separation. *Sep Purif Technol* 56:168–174.

Cheng C-C, Yu M-C, Cheng T-C, Sheu D-C, Duan K-J, Tai W-L. 2006. Production of high-content galacto-oligosaccharide by enzyme catalysis and fermentation with *Klyveromyces marxianus*. *Biotechnol Lett* 28(11): 793–797.

Choi JJ, Oh EJ, Lee YJ, Suh DS, Lee JH, Lee SW, Shin HT, Kwon ST. 2003. Enhanced expression of the gene for beta-glycosidase of *Thermus caldophilus* GK24 and synthesis of galacto-oligosaccharides by the enzyme. *Biotechnol Appl Biochem* 38:131–136.

Coimbra JR, Thommes J, Kula MR. 1994. Continuous separation of whey proteins with aqueous two-phase system in a Graesser contactor. *J Chrom A* 668:85–94.

Compagno C, Porro D, Smeraldi C, Ranzi BM. 1995. Fermentation of whey and starch by transformed *Saccharomyces cerevisiae* cells. *Appl Microbiol Biotechnol* 43:822–825.

Conrado LS, Veredas V, Nobrega ES, Santana CC. 2005. Concentration of α-lactalbumin from cow milk whey through expanded bed adsorption using a hydrophobic resin. *Braz J Chem Eng* 22(4):501–509.

Cote L, Ahlgren J. 1993. Metabolism in microorganisms. Part 1. Levan and levansucrase. In: *Science and Technology of Fructans* (pp. 141–161). Boca Raton, FL: CRC Press.

Cummings JH, Macfarlane GT, Englyst HN. 2001. Prebiotic digestion and fermentation. *Am J Clin Nutr* 73:415S–420S.

De Preter V, Hamer HM, Windey K, Verbeke K. 2011. The impact of pre- and/or probiotics on human colonic metabolism: Does it affect human health? *Mol Nutr Food Res* 55(1): 46–57.

De Roode BM, Franssen ACR, van der Padt A, Boom RM. 2003. Perspectives for the industrial enzymatic production of glycosides. *Biotechnol Prog* 19(5):1391–1402.

Dills WL. 1989. Sugar alcohols as bulk sweeteners. *Ann Rev Nutr* 9:161–186.

Domingues L, Dantas MM, Lima N, Teixeira JA. 1999a. Continuous ethanol fermentation of lactose by a recombinant flocculating *Saccharomyces cerevisiae* strain. *Biotechnol Bioeng* 64:692–697.

Domingues L, Teixeira JA, Lima N. 1999b. Construction of a flocculent *Saccharomyces cerevisiae* fermenting lactose. *Appl Microbiol Biotechnol* 51:621–626.

Domingues L, Guimarães PMR, Oliveira C. 2010. Metabolic engineering of yeast strains for lactose/whey metabolisation. *Bioeng Bugs* 1:164–171.

Domingues L, Lima N, Teixeira JA. 2001. Alcohol production from cheese whey permeate using genetically modified flocculent yeast cells. *Biotechnol Bioeng* 72:507–514.

Domingues L, Lima N, Teixeira JA. 2005. *Aspergillus niger* β-galactosidase production by yeast in a continuous high cell density reactor. *Process Biochem* 40:1151–1154.

Domingues L, Lima N, Teixeira JA. 2000. Contamination of a high-cell-density continuous bioreactor. *Biotechnol Bioeng* 68:584–587.

Domingues L, Teixeira JA, Penttila M, Lima N. 2002. Construction of a flocculent *Saccharomyces cerevisiae* strain secreting high levels of *Aspergillus niger* β-galactosidase. *Appl Microbiol Biotechnol* 58:645–650.

Doultani S, Turhan KN, Etzel MR. 2004. Fractionation of proteins from whey using cation exchange chromatography. *Process Biochem* 39:1737–1743.

Dragone G, Mussatto SI, Almeida e Silva JB, Teixeira JA. 2011. Optimal fermentation conditions for maximizing the ethanol production by *Kluyveromyces fragilis* from cheese whey powder. *Biomass Bioenerg* 35:1977–1982.

Dragone G, Mussatto SI, Oliveira JM, Teixeira JA. 2009. Characterisation of volatile compounds in an alcoholic beverage produced by whey fermentation. *Food Chem* 112:929–935.

El-Sayed MH, Chase HA. 2009. Single and two-component cation-exchange adsorption of the two pure major whey proteins. *J Chrom A* 1216:8705–8711.

El-Sayed MH, Chase HA. 2010. Purification of the two major proteins from whey concentrate using a cation-exchange selective adsorption process. *J Biotechnol Prog* 26(1):192–199.

El-Sayed MMH, Chase HA. 2011. Trends in whey protein fractionation. *Biotechnol Lett* 33:1501–1511.

Englyst KN, Englyst HN. 2005. Carbohydrate bioavailability. *Br J Nutr* 84:1–11.

Etzel MR. 2004. Manufacture and use of dairy protein fractions. *J Nutr* 134(4): 996S–1002S.

Farahnak F, Seki T, Ryu DD, Ogrydziak D. 1986. Construction of lactose-assimilating and high ethanol producing yeasts by protoplast fusion. *Appl Environ Microbiol* 51:362–367.

Farnworth ER. 2005. Kefirda complex probiotic. *Food Sci Technol Bull: Funct Foods* 2:1–17.

Ferrari MD, Loperena L, Varela H. 1994. Ethanol production from concentrated whey permeate using a fed-batch culture of *Kluyveromyces fragilis*. *Biotechnol Lett* 16:205–210.

Ferreira IM, Pinho O, Monteiro D, Faria S, Cruz S, Perreira A et al. 2010. Short communication: Effect of kefir grains on proteolysis of major milk proteins. *J Dairy Sci* 93:27–31.

Flashner M, Ramsden H, Crane LJ. 1983. Separation of proteins by high-performance anion-exchange chromatography. *J Anal Biochem* 135(2):340–344.

Fonseca GG, Heinzle E, Wittmann C, Gombert AK. 2008. The yeast *Kluyveromyces marxianus* and its biotechnological potential. *Appl Microbiol Biotechnol* 79:339–354.

Fuda E, Jauregi P, Pyle DL. 2004. Recovery of lactoferrin and lactoperoxidase from sweet whey using colloidal gas aphrons (CGAs) generated from an anionic surfactant, AOT. *Biotechnol Prog* 20:514–525.

Gambero A, Kubota LT, Gushikem Y, Airoldi C, Granjeiro JM, Taga EM, Alcantara EFC. 1997. Use of chemically modified silica with β-diketoamine groups for separation of α-lactalbumin from bovine milk whey by affinity chromatography. *J Colloid Interface Sci* 185(2):313–316.

Ganzle MG, Haase G, Jelen P. 2008. Lactose: Crystallization, hydrolysis and value-added derivatives. *Int Dairy J* 18:685–694.

Gawel J, Kosikowski FV. 1978. Improving alcohol fermentation in concentrated ultrafiltration permeates of cottage cheese whey. *J Food Sci* 43:1717–1719.

Gerberding SJ, Byers CH. 1998. Preparative ion-exchange chromatography of proteins from dairy whey. *J Chromatogr A* 808:141–151.

Gerling KG. 1998. Large-scale production of lactobionic acid—Use and new applications. *Int Dairy Fed* 9804:251–261.

Ghaly AE, El-Taweel AA. 1995. Effect of micro-aeration on the growth of *Candida pseudotropicalis* and production of ethanol during batch fermentation of cheese whey. *Bioresour Technol* 52:203–217.

Ghaly AE, El-Taweel AA. 1997a. Continuous ethanol production from cheese whey fermentation by *Candida pseudotropicalis*. *Energy Sour A Recover Util Environ* 19:1043–1063.

Ghaly AE, El-Taweel AA. 1997b. Kinetic modelling of continuous production of ethanol from cheese whey. *Biomass Bioenergy* 12:461–472.

Gianetto A, Berruti F, Glick BR, Kempton AG. 1986. The production of ethanol from lactose in a tubular reactor by immobilized cells of *Kluyveromyces fragilis*. *Appl Microbiol Biotechnol* 24:277–281.

Gibson GR, Roberfroid MB. 1995. Dietary modulation of the human colonic microbiota: Introducing the concept of prebiotics. *J Nutr* 125:1401–1412.

Golfinopoulos A, Kopsahelis N, Tsaousi K, Koutinas AA, Soupioni M. 2011. Research perspectives and role of lactose uptake rate revealed by its study using 14C-labelled lactose in whey fermentation. *Bioresour Technol* 102:4204–4209.

Gonzalez-Martinez C, Becerra M, Cafer A, Albors A, Carot J, Chiralt A. 2002. Influence of substituting milk powder for whey powder on yogurt quality. *Trends Food Sci Technol* 13:334–340.

Goodall S, Grandison AS, Jauregi PJ, Price J. 2008. Selective separation of the major whey proteins using ion exchange membranes. *J Dairy Sci* 91:1–10.

Grba S, Stehlik-Tomas V, Stanzer D, Vahcic N, Skrlin A. 2002. Selection of yeast strain *Kluyveromyces marxianus* for alcohol and biomass production on whey. *Chem Biochem Eng Q* 16:13–16.

Guimarães PMR, François J, Parrou JL, Teixeira JA, Domingues L. 2008a. Adaptive evolution of a lactose-consuming *Saccharomyces cerevisiae* recombinant. *Appl Environ Microbiol* 74:1748–1756.

Guimarães PMR, Le Berre V, Sokol S, François J, Teixeira JA, Domingues L. 2008b. Comparative transcriptome analysis between original and evolved recombinant lactose consuming *Saccharomyces cerevisiae* strains. *Biotechnol J* 3:1591–1597.

Guimarães PMR, Teixeira JA, Domingues L. 2008c. Fermentation of high concentrations of lactose to ethanol by engineered flocculent *Saccharomyces cerevisiae*. *Biotechnol Lett* 30:1953–1958.

Guimarães PMR, Teixeira JA, Domingues L. 2010. Fermentation of lactose to bio-ethanol by yeasts as part of integrated solutions for the valorisation of cheese whey. *Biotechnol Adv* 28(3):375–384.

Guo X, Zhou J, Xiao D. 2010. Improved ethanol production by mixed immobilized cells of *Kluyveromyces marxianus* and *Saccharomyces cerevisiae* from cheese whey powder solution fermentation. *Appl Biochem Biotechnol* 160:532–538.

Gurgel PV, Carbonell RG, Swaisgood HE. 2000. Fractionation of whey proteins with a hexapeptide ligand affinity resin. *Bioseparation* 9:385–392.

Güzel-Seydim Z, Wyffels JT, Seydim AC, Greene AK. 2005. Turkish kefir and kefir grains: Microbial enumeration and electron microscopic observation. *Int J Dairy Technol* 58:25–29.

Hahn-Hägerdal B. 1985. Comparison between immobilized *Kluyveromyces fragilis* and *Saccharomyces cerevisiae* coimmobilized with β-galactosidase, with respect to continuous ethanol production from concentrated whey permeate. *Biotechnol Bioeng* 27:914–916.

Hahn R, Deinhofer K, Machold C, Jungbauer A. 2003. Hydrophobic interaction chromatography of proteins: II. Binding capacity, recovery and mass transfer properties. *J Chromatogr B* 790(1–2):99–114.

Hahn RP, Schulz M, Schaupp C, Jungbauer A. 1998. Bovine whey fractionation based on cation-exchange chromatography. *J Chromatogr A* 795:277–287.

Hansson T, Kaper T, van der Oost J, de Vos WM, Adlercreutz P. 2001. Improved oligosaccharide synthesis by protein engineering of beta-glucosidase CelB from hyperthermophilic *Pyrococcus furiosus*. *Biotechnol Bioeng* 73(3):203–210.

Heeboll-Nielsen A, Justesen SFL, Thomas ORT. 2004. Fractionation of whey proteins with high capacity superparamagnetic ion-exchangers. *J Biotechnol* 113(1–3): 247–262.

Hopkins MJ, Macfarlane GT. 2003. Non digestible oligosaccharides enhance bacterial colonization resistance against *Clostridium difficile in vitro*. *Appl Environ Microbiol* 69:1920–1927.

Huffman LM, Harper WJ. 1999. Maximizing the value of milk through separation technologies. *J Dairy Sci* 82:2238–2244.

Huh KT, Toba T, Adachi S. 1990. Oligosaccharide formation during the hydrolysis of lactose with hydrochloric-acid and cation-exchange resin. *Food Chem* 38(4):305–314.

Huh KT, Toba T, Adachi S. 1991. Oligosaccharide structures formed during acid-hydrolysis of lactose. *Food Chem* 39(1):39–49.

Hung M-N, Xia Z, Hu N-T, Lee BH. 2001. Molecular and biochemical analysis of two β-galactosidases from *Bifidobacterium infantis* HL96. *Appl Environ Microbiol* 67:4256–4263.

Husain Q. 2010. Beta-galactosidases and their potential applications: A review. *Crit Rev Biotechnol* 30(1):41–62.

Janssens JH, Bernard A, Bailey RB. 1984. Ethanol from whey—Continuous fermentation with cell recycle. *Biotechnol Bioeng* 26:1–5.

Janssens JH, Burris N, Woodward A, Bailey RB. 1983. Lipid-enhanced ethanol production by *Kluyveromyces fragilis*. *Appl Environ Microbiol* 45:598–602.

Jara F, Pilosof AMR. 2011. Partitioning of alpha-lactalbumin and beta-lactoglobulin in whey protein concentrate/hydroxypropyl methyl cellulose aqueous two-phase systems. *Food Hydrocoll* 25(3):374–380.

Jędrzejewska M, Kozak K. 2011. Ethanol production from whey permeate in a continuous anaerobic bioreactor by *Kluyveromyces marxianus*. *Environ Technol* 32:37–42.

Ji ES, Park NH, Oh DK. 2005. Galacto-oligosaccharide production by a thermostable recombinant beta-galactosidase from *Thermotoga maritima*. *World J Microbiol Biotechnol* 21(5):759–764.

Jiménez-González C, Poechlauer P, Broxterman QB, Yang B-S, Ende D, Baird J et al. 2011. Key green engineering research areas for sustainable manufacturing: A perspective from pharmaceutical and fine chemicals manufacturers. *Org Process Res Dev* 15(4): 900–911.

Jorgensen F, Hansen OC, Sougaard P. 2001. High-efficiency synthesis of oligosaccharides with a truncated β-galactosidase from *Bifidobacterium bifidum*. *Appl Microbiol Biotechnol* 57:647–652.

Jovanovic S, Barac M, Macej O, Vucic T, Lacnjevac C. 2007. SDS–PAGE analysis of soluble proteins in reconstituted milk exposed to different heat treatments. *Sensors* 7:371–383.

Kargi F, Ozmihci S. 2006. Utilization of cheese whey powder (CWP) for ethanol fermentations: Effects of operating parameters. *Enzyme Microb Technol* 38:711–718.

Kim Y, Kuga S. 2002. Ion-exchange separation of proteins by polyallylamine-grafted cellulose gel. *J Chromatogr A* 955:191–196.

Kleine R, Achenbach S, Thoss S. 1995. Whey disposal by deproteinization and fermentation. *Acta Biotechnol* 15:139–148.

Konecny P, Brown RJ, Scouten WH. 1994. Chromatographic purification of immunoglobulin G from bovine milk whey. *J Chromatogr A* 673(1):45–53.

Konrad G, Kleinschmidt T. 2008. A new method for isolation of native alpha-lactalbumin from sweet whey. *Int Dairy J* 18:47–54.

Konrad G, Lieske B, Faber W. 2000. A large-scale isolation of native beta-lactoglobulin: Characterization of physicochemical properties and comparison with other methods. *Int Dairy J* 10:713–721.

Kosikowski FV. 1979. Whey utilization and whey products. *J Dairy Sci* 62:1149–1160.

Kosikowski FV, Wzorek W. 1977. Whey wine from concentrates of reconstituted acid whey powder. *J Dairy Sci* 60:1982–1986.

Kourkoutas Y, Dimitropoulou S, Kanellaki M, Marchant R, Nigam P, Banat IM et al. 2002a. High-temperature alcoholic fermentation of whey using *Kluyveromyces marxianus* IMB3 yeast immobilized on delignified cellulosic material. *Bioresour Technol* 82:177–181.

Kourkoutas Y, Psarianos C, Koutinas AA, Kanellaki M, Banat IM, Marchant R. 2002b. Continuous whey fermentation using kefir yeast immobilized on delignified cellulosic material. *J Agric Food Chem* 50:2543–2547.

Koutinas AA, Athanasiadis I, Bekatorou A, Iconomopoulou M, Blekas G. 2005. Kefir yeast technology: Scale-up in SCP production using milk whey. *Biotechnol Bioeng* 89:788–796.

Koutinas AA, Athanasiadis I, Bekatorou A, Psarianos C, Kanellaki M, Agouridis N, Blekas G. 2007. Kefir–yeast technology: Industrial scale-up of alcoholic fermentation of whey, promoted by raisin extracts, using kefir–yeast granular biomass. *Enzyme Microbiol Technol* 41:576–582.

Koutinas AA, Papapostolou H, Dimitrellou D, Kopsahelis N, Katechaki E, Bekatorou A, Bosnea LA. 2009. Whey valorisation: A complete and novel technology development for dairy industry starter culture production. *Bioresour Technol* 100:3734–3739.

Kunz C, Rudloff S, Baier W, Klein N, Strobel S. 2000. Oligosaccharides in human milk: Structural, functional and metabolic aspects. *Annual Rev Nutr* 20:699–722.

Lan Q, Bassi A, Zhu JX, Margaritis A. 2002. Continuous protein recovery from whey using liquid–solid circulating fluidized bed ion-exchange extraction. *J Biotech Bioeng* 78(2):157–163.

Lee YJ, Kim CS, Oh DK. 2004. Lactulose production by beta-galactosidase in permeabilized cells of *Kluyveromyces lactis*. *Appl Microbiol Biotechnol* 64:787–793.

Lewandowska M, Kujawski W. 2007. Ethanol production from lactose in a fermentation/pervaporation system. *J Food Eng* 79:430–437.

Liang M, Chen VYT, Chen H, Chen W. 2006. A simple and direct isolation of whey components from raw milk by gel filtration chromatography and structural characterization by Fourier transform Raman spectroscopy. *Talanta* 69:1269–1277.

Londero A, Hamet MF, De Antoni GL, Garrote GL, Abraham AG. 2012. Kefir grains as a starter for whey fermentation at different temperatures: Chemical and microbiological characterisation. *J Dairy Res* 79:262–271.

Londero A, Quinta R, Abraham AG, Sereno R, De Antoni G, Garrote GL. 2011. Inhibitory activity of cheese whey fermented with kefir grains. *J Food Prot* 74:94–100.

Longhi LGS, Luvizetto DJ, Ferreira LS, Rech R, Ayub MAZ, Secchi AR. 2004. A growth kinetic model of *Kluyveromyces marxianus* cultures on cheese whey as substrate. *J Ind Microbiol Biotechnol* 31:35–40.

Lozano JM, Giraldo GI, Romero CM. 2008. An improved method for isolation of β-lactoglobulin. *Int Dairy J* 18:55–63.

Lucas D, Rabiller-Baudry M, Millesime L, Chaufer B, Daufin G. 1998. Extraction of alpha-lactalbumin from whey protein concentrate with modified inorganic membranes. *J Membr Sci* 148:1–12.

Machold C, Deinhofer K, Hahn R, Jungbauer A. 2002. Hydrophobic interaction chromatography of proteins. I. Comparison of selectivity. *J Chromatogr A* 972(1):3–19.

Madureira AR, Pereira CI, Gomes AMP, Pintado ME, Malcata FX. 2007. Bovine whey proteins—Overview on their main biological properties. *Food Res Int* 40(10):1197–1211.

Magalhães KT, Dias DR, de Melo Pereira GV, Oliveira JM, Domingues L, Teixeira JA et al. 2011a. Chemical composition and sensory analysis of cheese whey-based beverages using kefir grains as starter culture. *Int J Food Sci Tech* 46:871–878.

Magalhães KT, Dragone G, de Melo Pereira GV, Oliveira JM, Domingues L, Teixeira JA et al. 2011b. Comparative study of the biochemical changes and volatile compound formations during the production of novel whey-based kefir beverages and traditional milk kefir. *Food Chem* 126:249–253.

Magalhães KT, Pereira MA, Nicolau A, Dragone G, Domingues L, Teixeira JA et al. 2010. Production of fermented cheese whey-based beverage using kefir grains as starter culture: Evaluation of morphological and microbial variations. *Bioresour Technol* 101:8843–8850.

Mahmoud MM, Kosikowski FV. 1982. Alcohol and single cell protein production by *Kluyveromyces* in concentrated whey permeates with reduced ash. *J Dairy Sci* 65:2082–2087.

Marshall K. 2004. Therapeutic applications of whey proteins. *Altern Med Rev* 9(2):136–156.

Mayer J, Kranz B, Fischer L. 2010. Continuous production of lactulose by immobilized thermostable beta-glycosidase from *Pyrococcus furiosus*. *J Biotechnol* 145(4):387–393.

Meyer A, Berensmeier S, Franzreb M. 2007. Direct capture of lactoferrin from whey using magnetic micro-ion exchangers in combination with high-gradient magnetic separation. *React Funct Polym* 67:1577–1588.

Miyasato M, Ajisaka K. 2004. Regioselectivity in β-galactosidase-catalyzed transglycosylation for the enzymatic assembly of D-galactosyl-D-mannose. *Biosci Biotechnol Biochem* 68:2086–2090.

Morr CV, Ha EYW. 1991. Off-flavours of whey protein concentrates: A literature review. *Int Dairy J* 1(1):1–11.

Morr CV, Ha EYW. 1993. Whey protein concentrates and isolates: Processing and functional properties. *Crit Rev Food Sci Nutr* 33:431–476.

Muller A, Daufin G, Chaufer B. 1999. Ultrafiltration modes of operation for the separation of alpha-lactalbumin from acid casein whey. *J Membr Sci* 153:9–21.

Nakayama T, Amachi T. 1999. Beta-galactosidase, enzymology. In: Flickinger MC, Drew SW, eds. *Encyclopedia of Bioprocess Technology: Fermentation, Biocatalysis, and Bioseparation* (pp. 1291–1305). New York, NY: John Wiley & Sons, Inc.

Nandini KE, Rastogi NK. 2011. Integrated downstream processing of lactoperoxidase from milk whey involving aqueous two-phase extraction and ultrasound-assisted ultrafiltration. *Appl Biochem Biotechnol* 163:173–185.

Ndiaye N, Pouliot Y, Saucier L, Beaulieu L, Bazinet L. 2010. Electroseparation of bovine lactoferrin from model and whey solutions. *Sep Purif Technol* 74:93–99.

Neri DFM, Balcão VM, Costa RS, Rocha ICAP, Ferreira EMFC, Torres DPM, Rodrigues LR, Carvalho Jr. LB, Texeira JA. 2009. Galacto-oligosaccharides production during lactose hydrolysis by free *Aspergillus oryzae* beta-galactosidase and immobilized on magnetic polysiloxane–polyvinyl alcohol. *Food Chem* 115:92–99.

Nilsson U, Nyman M. 2005. Short-chain fatty acid formation in the hindgut of rats fed oligosaccharides varying in monomeric composition, degree of polymerisation and solubility. *Br J Nutr* 94:705–713.

Noppe W, Haezebrouck P, Hanssens I, De Cuyper M. 1998. A simplified purification procedure of alpha-lactalbumin from milk using Ca_2 dependent adsorption in hydrophobic expanded bed chromatography. *Bioseparation* 8:153–158.

Ogata Y, Fujita K, Ishigami H, Hara K, Terada A, Hara H, Fujimori I, Misuoka T. 1993. Effect of a small amount of 4G-beta-D-galactosylsucrose (lactosucrose) on fecal flora and fecal properties. *J Jap Soc Nutr Food Sci* 46:317–323.

Ohtsuka K, Hino S, Fukushima T, Ozawa O, Kanematsu T, Uchida T. 1992. Characterization of levansucrase from *Rahnella aquatilis* JCM-1683. *Biosci Biotechnol Biochem* 56:1373–1377.

Oliveira C, Guimaraes PM, Domingues L. 2011. Recombinant microbial systems for improved β-galactosidase production and biotechnological applications. *Biotechnol Adv* 29:600–609.

Onishi N, Yamashiro A, Yokozeki K. 1995. Production of galacto-oligosaccharide from lactose by *Sterigmatomyces elviae* CBS8119. *Appl Environ Microbiol* 61:4022–4025.

Ounis WB, Gauthier SF, Turgeon SL, Roufik S, Pouliot Y. 2008. Separation of minor protein components from whey protein isolates by heparin affinity chromatography. *Int Dairy J* 18:1043–1050.

Outinen M, Tossavainen O, Syvaoja EL. 1996. Chromatographic fractionation of alpha lactalbumin and beta lactoglobulin with polystyrenic strongly basic anion exchange resins. *Lebensmittel-Wissenschaft Technol* 29(4):340–343.

Ozmihci S, Kargi F. 2007a. Continuous ethanol fermentation of cheese whey powder solution: Effects of hydraulic residence time. *Bioprocess Biosyst Eng* 30:79–86.

Ozmihci S, Kargi F. 2007b. Effects of feed sugar concentration on continuous ethanol fermentation of cheese whey powder solution (CWP). *Enzyme Microbiol Technol* 41:876–880.

Ozmihci S, Kargi F. 2007c. Ethanol fermentation of cheese whey powder solution by repeated fed-batch operation. *Enzyme Microbiol Technol* 41:169–174.

Ozmihci S, Kargi F. 2007d. Kinetics of batch ethanol fermentation of cheese-whey powder (CWP) solution as function of substrate and yeast concentrations. *Bioresour Technol* 98:2978–2984.

Ozmihci S, Kargi F. 2008. Ethanol production from cheese whey powder solution in a packed column bioreactor at different hydraulic residence times. *Biochem Eng J* 42:180–185.

Panesar PS, Kennedy JF. 2012. Biotechnological approaches for the value addition of whey. *Crit Rev Biotechnol* 32(4):327–348.

Panesar PS, Kumari S. 2011. Lactulose: Production, purification and potential applications. *Biotechnol Adv* 29(6): 940–948.

Panesar PS, Panesar R, Singh RS, Kennedy JF, Kumar H. 2006. Microbial production, immobilization and applications of beta-D-galactosidase. *J Chem Technol Biotechnol* 81(4):530–543.

Papapostolou H, Bosnea LA, Koutinas AA, Kanellaki M. 2008. Fermentation efficiency of thermally dried kefir. *Bioresour Technol* 99:6949–6956.

Paraskevopoulou A, Athanasiadis A, Blekas IG, Koutinas AA, Kanellaki M, Kiosseoglou V. 2003. Influence of polysaccharide addition on stability of a cheese whey kefir–milk mixture. *Food Hydrocolloids* 17:615–620.

Parrondo J, Garcia LA, Diaz M. 2000a. Production of an alcoholic beverage by fermentation of whey permeate with *Kluyveromyces fragilis* I: Primary metabolism. *J Inst Brewing, London* 106(6):367–375.

Parrondo J, Garcia LA, Diaz M. 2000b. Production of an alcoholic beverage by fermentation of whey permeate with *Kluyveromyces fragilis* II: Aroma composition. *J Inst Brewing, London* 106(6):377–382.

Park HY, Kim HJ, Lee JK, Kim D, Oh DK. 2008. Galactooligosaccharide production by a thermostable beta-galactosidase from *Sulfolobus solfataricus*. *World J Microbiol Biotechnol* 24(8):1553–1558.

Perez MD, Calvo M. 1995. Interaction of β-lactoglobulin with retinol and fatty acids and its role as a possible biological function for this protein: A review. *J Dairy Sci* 78(5):978–988.

Perez OE, Wargon V, Pilosof AMR. 2006. Gelation and structural characteristics of incompatible whey proteins/hydroxypropyl methylcellulose mixtures. *Food Hydrocol* 207:966–974.

Permyakov EA, Berliner LJ. 2000. Alpha-lactalbumin: Structure and function. *FEBS Lett* 473(3): 269–274.

Perumalsamy M, Murugesan T. 2012. Extraction of cheese whey proteins (α-lactalbumin and β-lactoglobulin) from dairy effluents using environmentally benign aqueous biphasic system. *Int J Chem Environ Eng* 3(1):50–54.

Porro D, Martegani E, Ranzi BM, Alberghina L. 1992. Lactose/whey utilization and ethanol production by transformed *Saccharomyces cerevisiae* cells. *Biotechnol Bioeng* 39:799–805.

Powell JE, Witthuhn RC, Todorov SD, Dicks LMT. 2007. Characterization of bacteriocin ST8KF produced by a kefir isolate *Lactobacillus plantarum* ST8KF. *Int Dairy J* 17(3):190–198.

Puerta A, Jaulmes A, De Frutos M, Diez-Masa JC, Vidal-Madjar C. 2002. Adsorption kinetics of beta-lactoglobulin on a polyclonal immunochromatographic support. *J Chromatogr A* 953(1–2):17–30.

Rimada PS, Abraham AG. 2001. Polysaccharide production by kefir grains during whey fermentation. *J Dairy Res* 68:653–661.

Rodrigues LR, Kluskens L. 2011. Synthetic biology & bioinformatics: Prospects in the cancer arena. In: HS Lopes, LM Cruz (eds.), *Computational Biology and Applied Bioinformatics*, Vol. 8, pp. 159–186. Rijeka, Croatia: InTech.

Rodrigues LR, Teixeira JA, Schmitt F, Paulsson M, Lindmark M H. 2009. Lactoferrin and cancer disease prevention. *Crit Rev Food Sci Nutr* 49(3):203–217.

Rodrigues LR, Teixeira JA. 2009. Potential applications of whey proteins in the medical field. In: J Coimbra, J Teixeira (eds.), *Engineering Aspects of Milk and Dairy Products*, Vol. 10, pp. 221–252. Boca Raton, FL: CRC Press.

Rodrigues LR, Venâncio A, Teixeira JA. 2001. Partitioning and separation of alpha-lactalbumin and beta-lactoglobulin in polyethylene glycol/ammonium sulphate aqueous two-phase systems. *Biotechnol Lett* 23:1893–1897.

Rodrigues LR, Venâncio A, Teixeira JA. 2003. Recovery of the proteose peptone component 3 from cheese whey in Reppal PES 100/polyethylene glycol aqueous two-phase systems. *Biotechnol Lett* 25:651–655.

Rojas EG, Coimbra JS, Minim LA, Zuniga AD, Saraiva SH, Minim VP. 2004. Size-exclusion chromatography applied to the purification of whey proteins from the polymeric and saline phases of aqueous two-phase systems. *Proc Biochem* 39:1751–1759.

Roos YH. 2009. Solid and liquid states of lactose. In: McSweeney PLH, Fox PF, eds. *Lactose, Water, Salts and Minor Constituents*. 3rd ed. (pp. 17–33). New York, NY: Springer.

Rosenberg M, Tomaska M, Kanuch J, Sturdik E. 1995. Improved ethanol production from whey with *Saccharomyces cerevisiae* using permeabilized cells of *Kluyveromyces marxianus*. *Acta Biotechnol* 15:387–390.

Roukas T, Lazarides HN. 1991. Ethanol production from deproteinized whey by β-galactosidase coimmobilized cells of *Saccharomyces cerevisiae*. *J Ind Microbiol* 7:15–18.

Rubio-Texeira M, Castrillo JI, Adam AC, Ugalde UO, Polaina J. 1998. Highly efficient assimilation of lactose by a metabolically engineered strain of *Saccharomyces cerevisiae*. *Yeast* 14:827–837.

Safarik I, Safarikova M. 2004. Magnetic techniques for the isolation and purification of proteins and peptides. *BioMagnetic Res Technol* 2:7.

Santos MJ, Teixeira JA, Rodrigues LR. 2011. Fractionation and recovery of whey proteins by hydrophobic interaction chromatography. *J Chromatogr B* 879:475–479.

Santos MJ, Teixeira JA, Rodrigues LR. 2012. Fractionation of the major whey proteins and isolation of β-lactoglobulin variants by anion exchange chromatography. *Sep Purif Technol* 90:133–139.

Satory M, Fuhrlinger M, Haltrich D, Kulbe KD, Pittner F, Nidetzky B. 1997. Continuous enzymatic production of lactobionic acid using glucose–fructose oxidoreductase in an ultrafiltration membrane reactior. *Biotechnol Lett* 19:1205–1208.

Saufi SM, Fee CJ. 2011. Simultaneous anion and cation exchange chromatography of whey proteins using a customizable mixed matrix membrane. *J Chromatogr A* 1218:9003–9009.

Saxena A, Tripathi BP, Kumar M, Shahi VK. 2009. Recent progress in protein separation by membrane technology. *Adv Colloid Interface Sci* 145(1–2):1–22.

Silva AC, Guimaraes PM, Teixeira JA, Domingues L. 2010. Fermentation of deproteinized cheese whey powder solutions to ethanol by engineered *Saccharomyces cerevisiae*: Effect of supplementation with corn steep liquor and repeated-batch operation with biomass recycling by flocculation. *J Ind Microbiol Biotechnol* 37:973–982.

Silveira WB, Passos F, Mantovani HC, Passos FML. 2005. Ethanol production from cheese whey permeate by *Kluyveromyces marxianus* UFV-3: A flux analysis of oxido-reductive metabolism as a function of lactose concentration and oxygen levels. *Enzyme Microbiol Technol* 36:930–936.

Smithers GW. 2008. Whey and whey proteins—From "gutter-to-gold." *Int Dairy J* 18: 695–704.

Smithers GW, Ballard FJ, Copeland AD, Silva KJ, Dionysius DA, Francis GL, Godard C, Griece PA, McIntosh GH, Mitchell IR, Pearce RJ, Regester GO. 1996. New opportunities from the isolation and utilization of whey proteins *J Dairy Sci* 79(8):1454–1459.

Sousa A, Passarinha LA, Rodrigues LR, Teixeira JA, Mendonça A, Queiroz JA. 2008. Separation of different forms of proteose peptone 3 by hydrophobic interaction chromatography with a dual salt system. *Biomed Chromatogr* 22:447–449.

Southard JH, Belzer FO. 1995. Organ preservation. *Ann Rev Med* 46:235–247.

Sreekrishna K, Dickson RC. 1985. Construction of strains of *Saccharomyces cerevisiae* that grow on lactose. *Proc Natl Acad Sci USA* 82:7909–7913.

Staniszewski M, Kujawski W, Lewandowska M. 2009. Semi-continuous ethanol production in bioreactor from whey with co-immobilized enzyme and yeast cells followed by pervaporative recovery of product—Kinetic model predictions considering glucose repression. *J Food Eng* 91:240–249.

Stojadinovic M, Burazer L, Ercili-Cura D, Sancho A, Buchert J, Velckovic TC, Stanic-Vucinic D. 2012. One-step method for isolation and purification of native β-lactoglobulin from bovine whey. *J Sci Food Agric* 92:1432–1440.

Strange ED, Malin EL, Van Hekken DL, Basch J. 1992. Chromatographic and electrophoretic methods used for analysis of milk proteins. *J Chrom* 624(1–2):81–102.

Szczodrak J, Szewczuk D, Rogalski J, Fiedurek J. 1997. Selection of yeast strain and fermentation conditions for high-yield ethanol production from lactose and concentrated whey. *Acta Biotechnol* 17:51–61.

Tahoun MK, El-Nemr TM, Shata OH. 2002. A recombinant *Saccharomyces cerevisiae* strain for efficient conversion of lactose in salted and unsalted cheese whey into ethanol. *Nahrung* 46:321–326.

Tahoun MK, El-Nemr TM, Shata OH. 1999. Ethanol from lactose in salted cheese whey by recombinant *Saccharomyces cerevisiae*. *Z Lebensm-Unters-Forsch A Eur Food Res Technol* 208:60–64.

Tieking M, Ganzle MG. 2005. Exopolysaccharides from cereal-associated *Lactobacilli*. *Trends Food Sci Technol* 16:79–84.

Tieking M, Kuhnl W, Ganzle MG. 2005. Evidence for formation of heterooligosaccharides by *Lactobacillus sanfranciscensis* during growth in wheat sourdough. *J Agric Food Chem* 53:2456–2461.

Topping DL, Clifton PM. 2001. Short-chain fatty acids and human colonic function: Roles of resistant starch and nonstarch polysaccharides. *Physiol Rev* 81:1031–1064.

Torres D, Gonçalves MPF, Teixeira JA, Rodrigues LR. 2010. Galacto-oligosaccharides: Production, properties, applications, and significance as prebiotics. *Compr Rev Food Sci Food Saf* 9(5):438–454.

Tundo P, Anastas P, Black D, Breen J, Collins T, Memoli S, Miyamoto J, Polyakoff M, Tumas W. 2000. Synthetic pathways and processes in green chemistry. Introductory overview. *Pure Appl Chem* 72(7): 1207–1228.

Tzortzis G, Vulevic J. 2009. Galacto-oligosaccharide prebiotics. In: Charalampopoulos D, Rastall RA, eds. *Prebiotics and Probiotics Science and Technology* (pp. 207–244). New York, NY: Springer.

Ulber R, Plate K, Demmer T, Buchholz H, Scheper T. 2001. Downstream processing of bovine lactoferrin from sweet whey. *Acta Biotechnol* 21:27–34.

Van Hijum SAFT, Kralj S, Ozimek LK, Kijkhuizen L, van Geel Schutten IGH. 2006. Structure–function relationships of glucansucrase and fructansucrase enzymes from lactic acid bacteria. *Microbiol Mol Biol Rev* 70:157–176.

Vasiljevic T, Jelen P. 2003. Oligosaccharide production and proteolysis during lactose hydrolysis using crude cellular extracts from lactic acid bacteria. *Lait* 83:453–467.

Vienne P, von Stockar U. 1985. An investigation of ethanol inhibition and other limitations occurring during the fermentation of concentrated whey permeate by *Kluyveromyces fragilis*. *Biotechnol Lett* 7:521–526.

Vogt S, Freitag R. 1997. Comparison of anion exchange and hydroxyapatite displacement chromatography for the isolation of whey proteins. *J Chromatogr* 760:125–137.

Vyas HK, Izco JM, Jimenez-Flores R. 2002. Scale-up of native beta-lactoglobulin affinity separation process. *J Dairy Sci* 85(7):1639–1645.

Wahlgren MC, Arnebrant T, Paulson MA. 1993. Adsorption from solutions of β-lactoglobulin mixed with lactoferrin or lysozyme onto silica and methylated silica surfaces. *J Colloid Interface Sci* 158:46–53.

Wang Q, Swaisgood HE. 1993. Characteristics of beta-lactoglobulin binding to the all-trans-retinal moiety covalently immobilized on Celite™. *J Dairy Sci* 76:1895–1901.

Watson WJW. 2012. How do the fine chemical, pharmaceutical, and related industries approach green chemistry and sustainability? *Green Chem* 14:251–259.

Weinbrenner WF, Etzel MR. 1994. Competitive adsorption of α-lactalbumin and bovine serum albumin to a sulfopropyl ion-exchange membrane. *J Chromatogr A* 662(2): 414–419.

Wit JN. 1998. Nutritional and functional characteristics of whey proteins in food products. *J Dairy Sci* 81:597–608.

Xu Y, Sleigh R, Hourigan J, Johnson R. 2000. Separation of bovine immunoglobulin G and glycomacropeptide from dairy whey. *Proc Biochem* 36(5):393–399.

Yanahira S, Kobayashi T, Suguri T, Nakakoshi M, Miura S, Ishikawa H, Nakajima I. 1995. Formation of oligosaccharides from lactose by *Bacillus circulans* β-galactosidase. *Biosci Biotechnol Biochem* 59:1021–1026.

Ye X, Yoshida S, Ng TB. 2000. Isolation of lactoperoxidase, lactoferrin, α-lactalbumin, β-lactoglobulin B and β-lactoglobulin A from bovine rennet whey using ion exchange chromatography. *Int J Biochem Cell Biol* 32(11–12):1143–1150.

Yoshida S. 1989. Preparation of lactoferrin by hydrophobic interaction chromatography from milk acid whey. *J Dairy Sci* 72(6):1446–1450.

Zafar S, Owais M, Salleemuddin M, Husain S. 2005. Batch kinetics and modelling of ethanolic fermentation of whey. *Int J Food Sci Technol* 40:597–604.

Zhao XQ, Bai FW. 2009. Yeast flocculation: New story in fuel ethanol production. *Biotechnol Adv* 27:849–856.

Zydney AL. 1998. Protein separations using membrane filtration: New opportunities for whey fractionation. *Int Dairy J* 8:243–250.

13 Recovery and Biotechnological Production of High-Value-Added Products from Fruit and Vegetable Residues

Elisa Alonso González, Ana Torrado Agrasar,
María Luisa Rúa Rodríguez, Lorenzo Pastrana Castro,
and Nelson Pérez Guerra

CONTENTS

13.1 INTRODUCTION

According to the recent Food and Agricultural Organization (FAO) data (FAO Database (2012) at http://faostat.fao.org), one-third of the food produced around the world for human consumption is lost. This means that about 1.3 billion tonnes are wasted per year along the food supply chain, from the initial agricultural production down to the final household consumption (Gustavsson et al. 2011). The commodities of agricultural origin represent the main part of these losses due to mechanical damage and/or spillage during harvesting, post-harvest handling and storage, distribution, and consumption. Moreover, the industrial processing of crops is a source of losses as well as wastes and by-products during the washing, peeling, slicing, and boiling processes as well as other transformations.

The part of the initial production of fruits and vegetables lost or wasted at different stages of the food supply chain is higher than that of cereals, oilseeds, and tubers. The wastes of these last commodities reach around 50% around the world (Gustavsson et al. 2011). Although fruits are not considered to have the same status as cereals from a food safety point of view, the added value associated with the commerce of fruits is higher than that of cereals, particularly in some regional communities.

The fruit and vegetable industry (producers, manufacturers, and retailers) is an important sector in economic and employment terms in many countries. Therefore, fruit and vegetable losses have a great economic and environmental impact as they represent a waste of the resources used in production (land, water, energy, and inputs) and a decrease in the economic value of the food produced. Moreover, owing to the large volume of these wastes and their high organic matter content, these materials constitute a high pollution load, which makes necessary the application of expensive treatments before they are discharged into the environment. It should be noted that this problem is more acute in low-income countries than in developed countries due to two factors: the higher postharvest losses in production and distribution in low-income countries, and tropical and subtropical fruit processing has a smaller edible portion and higher ratio of by-products compared to the processing of fruits from the temperate zone (Schieber et al. 2003).

The consumption of fruits and their derivatives has increased over the recent years in the Western countries due to their health benefits, such as a reduced risk of coronary heart disease and stroke, as well as certain types of cancer (Pearson et al. 2012; Ćetković et al. 2012). Consequently, the production of wastes from the industrial fruit transformation has also increased. Since the composition of these residues and by-products is normally very similar to the original raw material, recovery, bioconversion, and the use of the valuable constituents from these wastes is an interesting alternative for valorizing these materials. Nevertheless, some specific aspects need to be highlighted depending on the origin of the wastes. Fruit and vegetable industry wastes are often made up of fruit parts (seeds, peels) that are enriched in certain valuable components (e.g., oils, polyphenols), which makes it easier to reintegrate them into the productive cycle of wastes as new raw materials. The wastes from postharvest losses due to socioeconomic factors that make the fruit/vegetable unacceptable to consumers, such as small size or bad appearance (Kader 2005), have practically the same composition as commercial fruits, and therefore, they could be valorized without applying different constraints to those of commercial fruits. Nevertheless, in some cases, owing to biological or external physical causes of deterioration such as enzymatic changes, microbial contamination, or insect damage, the composition of wastes from postharvest losses (Dalmadi et al. 2006; Tortoe et al. 2007) is modified, thus reducing their safety and the possibilities of obtaining some profit from them.

It should be noted that the profitability of fruit and vegetable waste by-products is also influenced by factors other than their chemical composition, such as economic factors (purchasing expenditure, transport costs), operational factors (management of by-products and wastes, drying costs), and also image and marketing aspects for application in cosmetics and in nutraceuticals. In this sense, several constraints for valorizing these by-products can be highlighted. A low degree of standardization of wastes due to the heterogeneity of the batches and lack of specifications for cultivars, provenance, storage time, and the conditions of processed fruits and vegetables often make a waste valorization treatment difficult. When the valorization process is not located at the fruit and vegetable processing plants, transporting the by-products and the drying process can be expensive (Peschel et al. 2006).

Comparing the efforts made to develop the biorefinery concept, it should be noted that, in most cases, fruit and vegetable wastes have a wider diversity of valuable compounds than other lignocellulosic feedstock, which makes valorizing them more interesting. Thus, the integral exploitation of fruits and vegetables could have economic and environment benefits, leading to a greater diversity of products for use in the food, energy, fine chemical, and pharmaceutical industries (Ayala-Zavala et al. 2010). In this context, three main alternatives can be envisaged to take advantage of fruit and vegetable wastes: composting for use in agriculture, use as microbial substrates in fermentative processes to obtain a variety of value-added bioproducts, including biofuels, and recovery of natural ingredients such as nutritive compounds and functional or bioactive molecules by means of biotechnological (enzymatic and fermentative), physical, and chemical procedures (Murthy and Naidu 2012).

This chapter summarizes some recent developments in the valorization of the most important fruit and vegetable wastes, focusing first on the recovery of some of their valuable biomolecules, namely, antioxidants and dietary fiber, and second on the bioconversion of their fermentable compounds to produce high-value metabolites. The wastes/by-products, biomolecules and metabolites, and processes discussed here were selected according to their economic relevance and impact as well as their suitability for a potential application.

13.2 RECOVERY OF VALUABLE BIOMOLECULES FROM FRUIT AND VEGETABLE WASTES AND BY-PRODUCTS

13.2.1 ANTIOXIDANTS

The most abundant by-products from industrial fruit and vegetable processing plants are peel and seeds, which have been reported to contain high amounts of phyto-chemical compounds with antioxidant properties (Ayala-Zavala et al. 2010). Many of these phytochemicals are polyphenols whose antioxidant activity depends on many factors, such as their molecular weight and chemical structure (Zhang 1999), the presence of monomeric and polymeric forms (Hagerman et al. 1998), the extraction mode, including the type and polarity of the extracting solvent, and the assay method and the substrate used (Heinonen et al. 1998).

These phenolic antioxidant compounds from fruit by-products have several advantages: they can be used in food, pharmaceutical, and cosmetic products as substitutes for "chemical preservatives" or as active ingredients because they are considered to be completely safe, unlike synthetic antioxidants (Peschel et al. 2006; Murthy and Naidu 2012). In the food industry, the antioxidant compounds derived from wastes can be used to prevent lipid peroxidation by scavenging oxygen free radicals. They can also be used to increase the stability and shelf life of food products (Makris et al. 2007). In health products (pharmaceutical and cosmetic products), they are considered as nonsynthetic ingredients or natural supplements, which is a property that is required in nutraceutical or cosmeceutical products (Peschel et al. 2006).

Polyphenols from fruit wastes have also been shown to be chemopreventive agents against cancer (Torres et al. 2002) and diseases related to inflammatory processes (Aviram and Fuhrman 2002), inhibitors of platelet aggregation (Hertog et al. 1995), and also antidiabetics (Al-Awwadi et al. 2004).

In addition, for many fruits, the phenolic content of the raw waste-derived extracts is higher than in the original fruits and vegetables (Peschel et al. 2006). In spite of this and the vast quantities of plant residues produced by the food processing industry, only antioxidant extracts derived from grape seed and olive waste have been developed successfully in Europe (Alonso et al. 2002; Amro 2002). Nevertheless, based on the many scientific studies focused on obtaining antioxidants from agro- and food wastes, this list could be easily extended.

13.2.1.1 Grape

According to FAO data, grapes are the world's largest fruit crop. The main grape producers are Italy, France, and Spain (FAO Database (2012) at http://faostat.fao.org). The whole grape pomace that remains after pressing in the wine industry is approximately 13% by weight of the processed grapes (Torres et al. 2002), resulting in more than six million tonnes of waste generated worldwide per year (Rosales-Soto et al. 2012). Grape pomace, consisting of skins, seeds, and stems, is a rich source of antioxidant polyphenols, such as proanthocyanidins, catechins, and glycosylated flavonols, among others (Ayala-Zavala et al. 2010). Anthocyanins and flavonoids, in particular, have been found to be among the major phenolic compounds responsible for the beneficial effects of consuming red wine (Soleas et al. 2002).

Grape seed extract is a commercial product obtained from separated, extracted, and purified grape seeds. This extract contains high amounts of total polyphenols (around 10% on a dry weight basis) and is particularly rich in catechins and their isomers and polymers (Makris et al. 2007). Another related product is grape seed flour (GSF), which can be used as a functional ingredient in food. A recent paper described how GSF from Merlot and Cabernet Sauvignon was included in certain bakery products (cereal bars, pancakes, and noodles), obtaining high antioxidant activity and, at the same time, good consumer acceptability (Rosales-Soto 2012). The antioxidant activity of flavanol compounds is very stable even if they are submitted to high temperatures. Dried grape seeds (obtained after color extraction and alcohol distillation of the wine pomace) still have large flavanol concentrations and significant antioxidant activity (González-Paramás et al. 2004).

The grape pomace cell wall is a complex mixture containing 15% insoluble proanthocyanidins, lignin, and structural proteins and phenols, and the two last components are cross-linked to the lignin–carbohydrate framework (Chamorro et al. 2012). It must be noted that although there are some potent individual antioxidant polyphenolic compounds in grape pomace, there is not always a correlation between the specific polyphenols identified in the grape pomace and the antioxidant activity measured. This suggests that there could be a synergic action between the polyphenols, indicating that the antioxidant activity is the result of the total polyphenolic content but not with particular compounds (Alonso et al. 2002).

Distilled grape pomace pressing liquors are a distillery effluent containing phenolic compounds that, due to polymerization resulting from the thermal treatment, become more active than those in grape pomace (Pinelo et al. 2005). Despite polymerization, there are still appreciable amounts of gallic, cinnamic (p-coumaric), and gentisic acids in these liquors, as well as other potent antioxidant molecules, including catechins (mainly, catechin and epicatechin), flavonols (quercetin, kaempferol, and myricetin), and benzoic acids (Pinelo et al. 2005).

Different procedures have been described for recovering polyphenols from grape pomace. Although solid–liquid extraction (SLE) with solvents (methanol and ethanol) is the most conventional method, new techniques have been developed in recent years. Pulsed electric fields (PEF), ultrasound, and high hydrostatic pressure have been proposed for improving the extraction of polyphenols (Corrales et al. 2008; Puértolas et al. 2010). When supercritical extraction (SFE) by running carbon dioxide coupled with ethanol as a modifier was used, the phenolic concentrations of the extracts were twice those obtained by conventional SLE, which employs 96% ethanol and water. Moreover, different compounds were obtained with the different methods. Basically, proanthocyanidins were present in SLE extracts, whereas SFE extracts contained a mixture of gallic acid, catechin, and epicatechin (Pinelo et al. 2007). High-voltage electrical discharges (HVED) have electrical and mechanical effects on the product, which are caused by shock waves. This technique consists of submerging two electrodes in an aqueous solution to introduce energy through a plasma channel formed by a high-current/high-voltage electrical discharge. Recently, HVED was applied successfully for intensifying the extraction of polyphenols from grape pomace and has been shown to be more efficient than PEF for extracting polyphenols from grape skins (Boussetta et al. 2011). Charcoal and polymeric resins have been used for recovering and purifying phenolic compounds from distilled grape pomace pressing liquors, giving better results than the use of resins (Scordino et al. 2004).

The release of polyphenols from grape pomace mash can be improved using enzymatic treatments. Since phenolics are linked to cell wall polysaccharides, it has been proposed that cell-wall-hydrolyzing enzymes, such as pectinase, cellulase, hemicellulase, and glucanase, release cell wall complex polyphenols (Maier et al. 2008). The extraction of phenolic compounds from grape pomace was reported to be enhanced by using pectinase and tannase combined (Chamorro et al. 2012).

Particular consideration should be given to *trans*-resveratrol (3,5,4′-trihydroxystilbene). This compound has gained significant worldwide attention and commercial interest due to its potential beneficial properties for human health and the possibility of gaining economic benefits from grape pomace. It has been reported that it is able to inhibit or retard cardiovascular disease and cancer in animals (Baur et al. 2006). Resveratrol can be obtained more efficiently and more cheaply by SFE (Casas et al. 2010).

13.2.1.2 Apple

Apples and their derived products are one of the main fruits consumed all over the world and they are a major polyphenol source in the Western countries (Bellion et al. 2010). The main industrially processed apple products are apple juice, jelly, and cider. These productions generate a by-product, the apple pomace, which represents about 20–35% of the total fruit production (Bhushan et al. 2008). This waste is a heterogeneous mixture rich in polyphenol compounds consisting of peel, core, seed, calyx, stem, and soft tissue (Adil et al. 2007). Several approaches have been developed to valorize apple pomace, the most important of which are for recovering pectin (Schieber et al. 2003) and natural antioxidants (Foo and Lu 1999) as well as using it as a fermentation substrate to produce different value-added bioproducts (Ajila et al. 2011).

The polyphenolic content depends on the apple varieties and the parts of the fruit. Thus, apple peels have a higher polyphenolic content than the flesh. Stress conditions have also been reported to cause the accumulation of antioxidants in apple peel (Wolfe et al. 2003). The main polyphenols present in apples are monomeric and oligomeric flavonols, dihydrochalcones, anthocyanidins, and in lower proportions, catechins and procyanidins. The other important individual compounds present in apples are chlorogenic acid, phloretin glucosides, and quercetin glucosides (Wijngaard et al. 2009).

It has been reported that the antioxidant activity of apple polyphenols is higher than that of vitamin C and could be more effective in preventing diseases than dietary supplements. It has been found that apple extracts inhibit the growth of colon and liver cancer cells *in vitro* in a dose-dependent manner (Eberhardt et al. 2000). Similarly, Barth et al. (2005) used a rat model in colon carcinogenesis studies and found significantly increased effectiveness of cloudy apple juice containing polyphenols against colon cancer induced by dimethylhydrazine. Nevertheless, the antioxidant activity of apple polyphenols is not always responsible for the health effects of consuming apple pomace. Polyphenol-rich extracts from apple pomace and apple peel effectively diminish deoxyribonucleic acid (DNA) oxidation damage that can be attributed to the induction of cellular defense rather than to the radical scavenging activity of polyphenols/procyanidins (Bellion et al. 2010). Apple pomace also has other healthy properties. Ethanolic and methanolic extracts show antiviral activity and are able to inhibit both HSV-1 and HSV-2 replication in Vero cells without a cytotoxic effect (Suárez et al. 2010).

An important constraint when apple pomace is used as a source of polyphenols is that these compounds are present in a bound form with carbohydrates, such as glycosides, in nature. Their antioxidant activity in this form is low, therefore reducing their health functionality when they are ingested into the body via food or nutraceuticals (Vattem and Shetty 2003). Proanthocyanidins are the main polyphenols that cannot be extracted from apples. These polymers are chains of catechin, epicatechin, and their gallic acid esters, and procyanidins are the most common family. It has been reported that proanthocyanidins account for a major fraction of the total flavonoids ingested in the Western diet. In addition, these compounds showed a greater inhibitory effect against the survival of HeLa, HepG2, and HT-29 human cancer cells (Tow et al. 2011).

Therefore, different strategies have been proposed for enhancing the recovery of polyphenols from apple pomace, including improving the extraction process and enzymatic and fermentative pretreatments to release free phenolics from the mash. The conventional extraction of polyphenols is generally performed by maceration with mixtures of organic solvents such as aqueous methanol, acetone, and ethanol. Although water is not better than organic mixtures, it has also been studied as an extractant for recovering apple pomace phenolics due to the environmental and health advantages, accessibility, and price (Çam and Aaby 2010). In any case, the maceration technique has a very low cost efficiency and is highly time consuming. For this reason, other extraction techniques have been proposed, such as pressurized liquid extraction, microwave-assisted extraction, and ultrasound-assisted extraction, and have been found to increase the extraction yield by 20% (Virot et al. 2010). Pressurized liquid extraction of apple pomace increased the antioxidant activity by 2.4 times in comparison to the traditional SLE; however, unwanted compounds such as hydroxymethylfurfural were formed (Wijngaard and Brunton 2009). Ultrasound-assisted extraction yielded a total phenolic content that was 30% higher than the content obtained by conventional extraction. The extracts obtained by ultrasound showed higher antioxidant activity, confirming that the main polyphenols were not degraded under the applied conditions (Pingret et al. 2012). Ultrasonication and microwave-assisted extraction methods have also been reported to successfully extract polyphenolics from fermented apple pomace (Ajila et al. 2011).

Enzymes have been used to improve the recovery of phenolic compounds from apple pomace. Liquefaction with pectolytic enzymes increases the polyphenolic content, especially for dihydrochalcones and quercetin glycosides (Will et al. 2000). It has been found that the polyphenol contents in cloudy apple juices significantly increase after pectinex yield mash, pectinex smash XXL, and pectinex XXL maceration. Moreover, the radical scavenging activity of cloudy apple juices that were treated with pectinase was higher than the untreated reference samples (Oszmiański et al. 2009). In addition, enzymatic-treated pomace can be used in the production of puree-enriched cloudy apple juices to improve the polyphenolic contents and juice yields. This possibility was successfully assayed and cloudy apple juices with a high content of extractable and nonextractable procyanidins were obtained (Oszmiański et al. 2011).

13.2.1.3 Tomato

Tomatoes are one of the vegetables most consumed worldwide because they can be eaten either fresh or as industrially processed products, such as canned tomatoes, sun-dried tomatoes, juices, ketchup, pastes, purees, salads, sauces, and soups (Ćetković et al. 2012). These industrial transformations produce tomato seeds and peels as wastes, which can reach 40% of the weight of the processed tomatoes (Strati and Oreopoulou 2011a,b)

Several works suggest that, as they are greatly consumed in different presentations, tomatoes provide a significant proportion of the total antioxidants in the diet, mainly in the form of carotenes and phenolic compounds (Martínez-Valverde et al. 2002; Benakmoum et al. 2008).

The skin in industrial tomato wastes is richer than the pulp and seed fractions and contains higher levels of the total phenolics, total flavonoids, ascorbic acid, and carotenes. The antioxidant activity is also higher (Toor and Savage 2005). For example, it has been reported that the carotenoid content of dry tomato by-products is nearly 6 times higher in peel than in seed by-product (Knoblich et al. 2005).

Carotenes, due to the conjugated double bonds in their structure, are responsible for the natural yellow, orange, and red colors of fruits and vegetables, and have antioxidant properties (Strati and Oreopoulou 2011a,b). Carotenoids have health-promoting functions; they are a source of provitamin A and they enhance the immune system and reduce the risk of cancer and cardiovascular disease (Fraser and Bramley 2004).

Lycopene is the most important carotene present in tomato skin and its presence varies significantly with ripening and the variety of tomato. It is associated with the content of insoluble solids (Sharma and Le Maguer 1996). Several health claims are associated with consuming lycopene. Lycopene is protective against some types of cancer as it quenches singlet oxygen and can act as a scavenger of peroxyl radicals. At the cellular level, lycopene is associated with the induction of cell–cell communication and growth control (Martínez-Valverde et al. 2002). Other epidemiological studies suggest that lycopene inhibits the formation of oxidized products of low-density lipoprotein (LDL) cholesterol, and consequently, consuming them could prevent cardiovascular disease (Argawal and Rao 1998; Weisburger 1998; Arab and Steek 2000).

The conventional methods for obtaining lycopene from tomato wastes are SLE with organic solvents such as hexane, ethanol, acetone, methanol, tetrahydrofuran, benzene, or petroleum ether (Lin and Chen 2003). Several variables have been studied for improving lycopene extraction, including temperature, time, the number of extraction steps in the process, the solvent mixture to waste ratio, the particle size of the dried tomato waste, and the type of pure organic solvents and their mixtures (Strati and Oreopoulou 2011a,b). The main advantages of these procedures, from an industrial point of view, are their simplicity and low cost. Recently, supercritical fluids have been proposed as a way of avoiding lycopene degradation by light, oxygen, and high temperatures. The main inconvenience of this alternative is that high pressures are necessary for obtaining reasonable extraction yields due to the relatively low solubility of carotenes compared to their solubility in organic solvents (Mattea et al. 2009).

Several interesting applications of lycopene from tomato wastes have been proposed. For example, enriching low-quality edible oils (such as refined olive oils) with carotenoids and lycopene were found to improve their quality and avoid rancidity, and they have been used to elaborate new functional foods (Benakmoum et al. 2008). Similarly, the intake of lycopene-rich foods cooked with monounsaturated fat acts as a prevention against coronary heart disease (Ahuja et al. 2003).

13.2.1.4 Coffee

Every day, millions of people drink coffee, which explains why this fruit is the most important food commodity worldwide (Lashermes et al. 2008). Postharvest losses reach around 15–20% of the coffee fruit production, and therefore, it is not used to obtain commercialized green coffee and is discarded during processing (Esquivel and Jiménez 2012).

Various by-products, such as coffee pulp, cherry husk, parchment husk, silver skin, and spent waste, are obtained in the industrial processing of the coffee fruit to remove the shell and mucilaginous part from the coffee cherries (Murthy and Naidu 2012). Coffee berries are industrially processed to obtain the internationally traded green coffee and several by-products are obtained depending on the processing method followed. The waste obtained in the dry process is composed of skin, pulp, mucilage, and parchment, all together in a single fraction (coffee husks) (Prata and Oliveira 2007). In contrast, in wet processing, two main fractions are obtained: the skin and pulp are recovered in one fraction (43.2% w/w of the whole fruit), and mucilage and soluble sugars are recovered in a second fraction when fermentation is not used (11.8% w/w). In addition, the parchment reaches 6.1% w/w (Bressani 1978).

Flavanols, hydroxycinnamic acids, flavonols, and anthocyanidins are the four major classes of polyphenols contained in coffee husks, skin, and pulp (Esquivel and Jiménez 2012). In fresh coffee pulp, chlorogenic acid represents 42.2% of the total of the identified phenolic compounds. The other phenolic compounds found are epicatechin, 3,4-dicaffeoylquinic acid, 3,5-dicaffeoylquinic acid, 4,5-dicaffeoylquinic acid, catechin, rutin, protocatechuic acid, and ferulic acid (Ramírez-Martínez 1988). The anthocyanin cyanidin-3-rutinoside has been described in both fresh coffee husks (Prata and Oliveira 2007) and peels and pulp derived from wet-processed fruits (Esquivel and Jiménez 2012). In addition, proanthocyanidins are also present in the fresh coffee pulp and their concentration is greater in yellow coffee varieties than in red coffee varieties (Esquivel and Jiménez 2012).

Almost 50% of the world's coffee production is processed for soluble coffee. This process consists of extracting and concentrating water-soluble compounds and yields spent coffee as a by-product. This by-product has been reported to be a source of bioactive compounds, such as caffeine, trigonelline, and chlorogenic acids (Ramalakshmi et al. 2009). The extracts from spent coffee (from the production of instant—soluble—coffee) have been found to show strong radical-scavenging, antioxidant activity. It has also been suggested that the antitumor activity is limited to the anti-inflammatory and antiallergic action of the above-mentioned extracts (Ramalakshmi et al. 2009).

SLE using a mixture of isopropanal and water as solvents was used to recover phenolic compounds from the coffee by-products (coffee pulp, husk, silver skin, and spent coffee). The yields were improved when the by-products were pretreated with viscozyme and 70% of the antioxidant activity remained (Murthy and Naidu 2012).

13.2.1.5 Olive

Generally, the industrial process to obtain olive oil leads to 35 kg of olive cake and 440 L of olive mill wastewater (OMW) for 100 kg of processed olives. This huge quantity of olive by-products is a major environmental problem in the Mediterranean countries due to the high-polluting power of OMW that is associated with its high chemical and biochemical oxygen demand, and total solids content (Ghanbari et al. 2012).

Olive cake is a source of a great variety of phenolic compounds, the most important of which are hydroxytyrosol, oleuropein, tyrosol, caffeic acid, p-coumaric acid, vanillic acid, verbascoside, elenolic acid, catechol, and rutin. Hesperidin and quercetin are only detected in defatted olive cake, but not in full-fat olive cake, after extraction using alkaline hydrolysis. These compounds have antioxidant properties. In addition, it has been reported that the above-mentioned polyphenols have anticarcinogenic, anti-inflammatory, antimicrobial, antihypertensive, antidyslipidemic, cardiotonic, laxative, and antiplatelet properties. Consequently, the extract of the olive cake has potential uses in the pharmaceutical and nutraceutical industries (Ghanbari et al. 2012).

OMW constitutes a rich source of hydroxytyrosol, which is the major compound, and also tyrosol, caffeic acid, p-coumaric acid, homovanillic acid, protocatechuic acid, 3,4-dihydroxymandelic acid, vanillic acid, and ferulic acid, among others. In addition, 98% of the olive content of biophenol remains in the OMW extract. This last polyphenol has free radical-scavenging and metal-chelating properties and its antioxidant activity *in vitro* has been reported to be higher than vitamins E and C on a molar basis (Rice-Evans et al. 1997). The other biological activities of polyphenols from OMW include an antioxidant effect on intestinal human epithelial cells, anti-inflammatory activity through the inhibition of 5-lipoxygenase, biocide activity (including antiviral, molluscicidal, antibacterial, and antifungal activities), as well as cardioprotective, antiatherogenic, and antitumor activities (Ghanbari et al. 2012).

Finally, it must be pointed out that the by-product resulting from the filtration process of extra virgin olive oil contains high amounts of phenolic compounds such as hydroxytyrosol, tyrosol, decarboxymethyl oleuropein aglycone, and luteolin (Lozano-Sánchez et al. 2011). Besides its antioxidative activities, oleuropein has antiplatelet aggregation properties and it has been suggested that it may prevent thrombotic complications associated with platelet hyperaggregability (Zbidi et al. 2009).

13.2.2 DIETARY FIBER

Dietary fiber (DF) is a heterogeneous and highly complex mixture of compounds, mainly made up of carbohydrates of vegetable origin that are resistant to digestion and absorption in the small intestine. This concept includes oligosaccharides and polysaccharides (cellulose, hemicelluloses, pectins, gums, resistant starch, and inulin) that may be associated with lignin and other noncarbohydrate components (e.g., polyphenols, waxes, saponins, cutin, phytanes, and resistant protein) (Elleuch et al. 2011).

Although DF is usually classified in a simplistic way as soluble or insoluble, other functional properties could be more relevant for classifying it, such as viscosity, gel-forming capabilities, or the rate at which it is fermented by the gut flora. Viscous fiber, formerly called water-soluble fiber, includes pectin, gums, and mucilage, whereas nonviscous fiber (water-insoluble fiber) includes cellulose, hemicellulose, and lignin (Riccioni et al. 2012).

Thus, DF has both health and technological functions. The health benefits associated with an increased intake of DF include an increase in fecal bulk and it also stimulates colonic fermentation, reduces postprandial blood glucose (reduces insulin responses), and reduces preprandial cholesterol levels (Champ et al. 2003). It has also been reported that it reduces the risk of coronary heart disease, diabetes, obesity, and some forms of cancer (Mann and Cummings 2009).

In addition, DF has interesting food technological properties because it increases the water-holding capacity, oil holding capacity, emulsification and/or gel formation. DF is used in bakery and dairy products, jams, meats, and soups to improve the rheological properties of the foods, avoid synergesis, stabilize emulsions, and extend shelf life (Riccioni et al. 2012).

Fruit and vegetable by-products and wastes are the main industrial source of DF, particularly viscous fiber, as they have a high DF content and the consumer perceives them as natural ingredients. Pectin is the most valuable soluble fiber obtained from the by-products of the fruit juice industry, namely, from apple pomace and citrus peels (Zykwinska et al. 2008). Although pectin basically consists of D-galacturonic acid units (usually methylated) joined in chains by means of α-(1-4) glycosidic linkage, its fine chemical structure is still not well understood because other sugars can be present and the molecular size and degree of esterification varies according to the source and the conditions applied during isolation (Srivastava and Malviya 2011).

Apple pomace contains 10–15% pectin, in dry weight. This pectin has superior gelling properties to citrus pectin, but because the phenolic constituents that are coextracted with pectin are oxidized, it is brown in color, which limits its incorporation into light-colored foods.

The worldwide citrus production is more than 88 million tonnes and one-third of this is mainly processed to obtain juice, which leads to waste that accounts for 50% of the original whole fruit mass. Compared with apple pomace, citrus peel has a higher content of soluble DF (Marín et al. 2007). Recovering pectins from fruit by-products requires a compromise because increasing the concentration and molecular weight of the pectins extracted leads to high-viscosity solutions, but decreases the efficiency of extraction and separation of solids (Srivastava and Malviya 2011).

The conventional water-based method for recovering pectins from fruit by-products involves extracting the pectin in acidic conditions (pH up to 2 with mineral or organic acids) at relatively moderate temperatures (below 70°C) and then precipitation using ethanol or isopropyl alcohol. Since this procedure generates large amounts of effluents and undesired degradation of pectins, alternative enzymatic methods have been developed. These new alternatives are based on using cellulases and proteases to destroy cellulose/xyloglucan and the protein complex to release pectins from the cell walls (Zykwinska et al. 2008).

13.3 USE OF FRUITS WITH LOW COMMERCIAL VALUE AND AGRO-INDUSTRIAL AND AGRICULTURAL RESIDUES AS FERMENTATION SUBSTRATES FOR PRODUCING VALUE-ADDED PRODUCTS

Agriculture and the agro-based industry produce large amounts of fruits with a low commercial value, wastes, residues, and by-products. The disposal of these residues is not only a serious environmental pollution problem but also a loss of valuable biomass and nutrients, and consequently, a great economic loss (Laufenberg et al. 2003; Zhong-Tao et al. 2009). Composting is an interesting procedure for transforming these materials into safe soil amendments by means of the organic matter stabilization that results from the aerobic development of successive populations of endogenous microorganisms. This process makes it possible to transform the carbohydrates, proteins, lipids, and lignin present in the wastes into CO_2, biomass, heat, and a humus-like end-product (Tuomela et al. 2000). The increase in the temperature during the thermophilic phase of composting due to the microbial metabolism destroys the pathogenic species and sanitizes the composted material.

Nevertheless, the availability, high carbohydrate content, and high degree of biodegradability of fruits and agro-industrial wastes also make them excellent candidates for being used as cheap substrates for the biotechnological production of high-value-added products, including ethanol, L-glutamic acid, ergot alkaloids, penicillin, fruity aroma, lactic and citric acid, pigments, enzymes (Pandey et al. 2000; Pérez and Guerra 2009; Guerra et al. 2009), feed/fodder (Laufenberg et al. 2003), or alcoholic beverages (Alonso et al. 2010, 2011), among others. This approach would not only contribute to preventing the environmental pollution associated with the disposal of these wastes but could also be beneficial for farmers by increasing their incomes, and for manufacturers and consumers, by decreasing fuel and food costs (Alonso et al. 2010; Rose et al. 2010).

In the following sections, some examples are given of the potential applications of different agricultural and agro-industrial residues and fruits of low commercial value as fermentation substrates for producing different value-added products of industrial interest.

13.3.1 PRODUCTION OF ETHANOL AND DISTILLED ALCOHOLIC BEVERAGES FROM VEGETABLE BY-PRODUCTS

13.3.1.1 Ethanol

The production of ethanol from different agricultural residues is one of the most promising alternatives as it is economical and efficient and avoids the environmental problems associated with waste disposal. For example, apple pomace (McIntosh apple variety at 85% initial moisture), the main residue of the apple cider and juice processing industries, has been used as a substrate for producing ethanol by solid-state fermentation (SSF) with *Saccharomyces cerevisiae* ATCC 24702 (Ngadi and Correia 1992). The maximum ethanol concentration obtained was 20.79 (g ethanol/g dry weight of apple pomace) after 40 h of fermentation.

More recently, saccharification and fermentation processes have been combined to produce ethanol from different agro-industrial residues. Thus, mixtures of steam-pretreated wheat straw (SPWS) and presaccharified wheat meal (PWM) were converted into ethanol in simultaneous saccharification and fermentation processes with an ordinary baker's yeast of *S. cerevisiae* (Erdei et al. 2010). The fermentation of a mixture of SPWS containing 2.5% water-insoluble solids (WIS) and PWM containing 2.5% WIS resulted in an ethanol concentration of 56.5 g/L.

Similarly, simultaneous saccharification and fermentation of raw cassava flour at 150 or 250 g/L with raw cassava starch-degrading enzyme (30 U/g flour) and *Penicillium* sp. GXU20 gave ethanol yields of 53.3 g/L (after 48 h of incubation) and 75.6 g/L (after 60 and 72 h of fermentation), respectively (Lin et al. 2011).

Another study examined the use of cotton stalk residue, an important by-product of cotton, for ethanol production using thermotolerant *Pichia kudriavzevii* HOP-1 (Kaur et al. 2012). Before fermentation, cotton stalks were shredded and baled after the cotton was picked, and the residues were air dried to reduce their moisture content, milled, and screened to a particle size of 1 mm. The cotton stalk powder was alkali treated with NaOH at 1–4% (w/v), autoclave-sterilized, and treated with an ozone concentration of 45 mg/L at a rate of 0.37 L/min for 150 min. Then, the 4% alkali-treated and ozone-treated cotton stalks were hydrolyzed by using 20 filter paper cellulase units/gram dried substrate (FPU/gds), 45 IU/gds β-glucosidase, and 15 IU/gds pectinase (50°C, 120 rpm for 72 h). After enzymatic hydrolysis during 48 h, the 4% alkali-treated samples contained 42.29 g/L glucose and 6.82 g/L xylose, whereas the ozone-treated biomass had 24.13 g/L of glucose and 8.3 g/L xylose. Separate hydrolysis and SSF with *P. kudriavzevii* HOP-1 gave ethanol concentrations of 19.82 g/L from alkali-treated samples and 10.96 g/L from ozone-treated samples. However, simultaneous saccharification and fermentation of the alkali-treated cotton stalks after 12-h prehydrolysis gave an ethanol concentration of 19.48 g/L (Kaur et al. 2012).

Oberoi et al. (2012) recently studied the ethanol production by *P. kudriavzevii* HOP-1 in alkali-treated rice straw using simultaneous saccharification (20 FPU/gds cellulase, 50 IU/gds β-glucosidase, and 15 IU/gds pectinase) and SSF of the alkali-treated substrate. The ethanol concentration obtained after these two simultaneous processes were carried out during 24 h was 24.25 g/L.

Hydrothermally pretreated peel from Kinnow mandarin (*Citrus reticulata*) has also been assayed as a substrate for ethanol production via simultaneous saccharification (with a crude filtrate produced by an *Aspergillus oryzae* strain) and SSF with *P. kudriavzevii* HOP-1. In this case, the ethanol concentration obtained was 33.87 g/L after 12 h of incubation (Sandhu et al. 2012).

Sharma et al. (2007) attempted to optimize the fermentation parameters for ethanol production from Kinnow waste and banana peels by using simultaneous saccharification and SSF. However, in this case, the simultaneous processes were carried out using the cellulase enzyme and a coculture of *S. cerevisiae* G and *Pachysolen tannophilus* MTCC 1077, and the substrates were mixed at a ratio of 4:6 (Kinnow waste:banana peels). According to these researchers, the optimal ethanol production (26.84 g/L) was obtained at a temperature of 30°C, an inoculum size of *S. cerevisiae* G of 6% (v/v) and *P. tannophilus* MTCC 1077 of 4% (v/v), and an incubation period of 48 h with agitation during the first 24 h.

The potential of residual wood (*Eucalyptus grandis*) chips from the cellulose industry for producing ethanol has also been studied (Luci et al. 2011). This residue was first subjected to acid treatment with diluted sulfuric acid for xylose production, and then the hemicellulosic hydrolysate (50 g/L xylose) was fermented with *Pichia stipitis* CBS5774. The remaining solid fraction generated after pretreatment was subjected to saccharification and fermentation processes using a strain of *S. cerevisiae*. Enzymatic hydrolysis was carried out with a commercial cellulase preparation containing FPasic, β-glucosidase, and CMCasic activities of 100 FPU/U, 125 U/mL, and 5500 U/mL, respectively. The final ethanol concentration obtained after the fermentation of hemicellulosic hydrolysate by *P. stipitis* was 15.3 g/L, and after the simultaneous saccharification and fermentation processes by commercial enzymes and *S. cerevisiae*, it was 28.7 g/L.

An innovative approach for recovering ethanol from solid- state fermented apple pomace and the concomitant production of animal feed has also been proposed by Joshi and Devrajan (2008). In this study, two types of solid-state fermented apple pomace samples obtained with *S. cerevisiae* or with *Candida utilis* and *Kloeckera* spp. as sequential interactive cocultures were used to recover ethanol by applying four different methods: hydraulic pressing, direct distillation, steam distillation, and vacuum distillation. The authors then evaluated the physicochemical characteristics (moisture, titratable acidity, reducing and total sugars, crude and soluble proteins) of the remaining solid materials obtained after ethanol separation with the different recovery methods. The steam distillation method provided the highest separation efficiency and the lowest nutritional loss in the remaining solid material. These results and those obtained by other researchers (Joshi et al. 2000) suggest that the solid residue obtained after ethanol separation could be used as an additive in animal feed.

The common dates and date wastes of the *Deglet-Nour* variety have a very low commercial value and represent between 16% and 19% of the total date production in Algeria (Acourene and Ammouche 2011). In a recent study, the potential of these wastes as substrates for ethanol production with *S. cerevisiae* SDB under submerged fermentation was investigated and optimized in a syrup previously obtained by heating the date wastes. The optimum ethanol concentration (136 g/L) was obtained at a fermentation period of 72 h, inoculum content of 4% (w/v), sugar concentration of 180 g/L, and ammonium phosphate concentration of 1 g/L (Acourene and Ammouche 2011).

Another study evaluated producing ethanol from a mixture of coffee pulp and mucilage of the Colombian variety *Coffea arabica* L., by using a commercial dry baker's yeast *S. cerevisiae* (Levapan, Bogotá, Colombia) and panela (dehydrated and solidified cane juice). The juice obtained from the filtered pulp and mucilage was mixed and then this wort was subjected to an acid hydrolysis treatment with heating to increase the sugar content. The ethanol yield obtained after 40 h of fermentation was 25.44 kg of ethanol per cubic meter of juice pulp and mucilage, resulting from the 64.40 kg/m^3 of the total sugars (Navia et al. 2011).

Cocoa pod is another agricultural waste that has been evaluated for ethanol production due to its high carbohydrate content (43.9–45.2%) and the high availability of this waste in cocoa-growing countries (Samah et al. 2011). For this purpose, ripe cocoa pods were subjected to different thermal and acid treatments to select the best method for hydrolyzing cellulose to glucose. The highest glucose content in the hydrolysates (307 g/L) was obtained with 1.0 M HCl at 75°C during 4 h of incubation. The fermentation of this hydrolysate with *S. cerevisiae* at 30°C gave an ethanol concentration of 173 g/L after approximately 26 h of fermentation.

All these promising results are encouraging for the further scale-up production of ethanol from the above-mentioned residues (Kaur et al. 2012).

13.3.1.2 Distilled Alcoholic Beverages

The mountain regions of Galicia, an autonomous community located in northwest Spain, produce a large amount of berries (e.g., black mulberry, black currant, red raspberry, and arbutus berry) that are not commercialized.

Black mulberries (*Morus nigra* L.) and black currants (*Ribes nigrum* L.) are good sources of vitamins, antioxidants, nonvolatile organic acids, and phenolic acids that contribute to the quality of taste and aroma (Soufleros et al. 2004; Varming et al. 2004). In addition, these fruits have been reported to have natural therapeutic qualities (Darias-Martín et al. 2003; Alonso et al., 2010). Arbutus berries (*Arbutus unedo* L.) are a good source of antioxidants, such as flavonoids, vitamins C and E, and carotenoids (Pallauf et al. 2008), and have been used to treat arterial hypertension (Cavaco et al. 2007) or as antiseptics, diuretics, and laxatives in traditional folk medicine (Ayaz et al. 2000; Pabuçcuoglu et al. 2003). Similarly, raspberries (*Rubus idaeus* L.) have been reported to contain high levels of ellagic acid (Juranic et al. 2005), which has shown to have antiviral activity (Corthout et al. 1991) and anticancer potential (Stoner and Morse 1997). However, the commercialization on a large scale of these fruits as fresh products is unviable mainly because their commercial value is practically null, they have a short shelf life since they ripen quickly, and it is difficult to harvest large amounts of them due to their geographic dispersion (Darias-Martín et al. 2003; Alonso et al. 2010, 2011).

The inhabitants of the Canary Islands (Tenerife, La Gomera, La Palma, El Hierro, and Lanzarote) collect the black mulberry and prepare homemade juices and alcoholic beverages (from black mulberry syrup) for medicinal purposes, to control type II diabetes mellitus and inflammations of the throat, tongue, and mouth (Darias-Martín et al. 2003).

Although an aromatic distillate from the arbutus berry (*Aguardente de medronho*) has traditionally been produced in the Algarve region (Portugal) on a small scale (Alarcão et al. 2001), the fermentation process is artisanally carried out under uncontrolled conditions by the wild microbiota of the fruits during 4–5 weeks, depending on the weather conditions. In these conditions, the fruits are overfermented or not completely fermented leading to the appearance of some organoleptic defects, such as acidity, lack of flavor, or even off-flavors, in the fermented products

(Cavaco et al. 2007). As a consequence, the final quality of the alcoholic beverages obtained is highly variable (Aloys and Angeline 2009).

The successful production of distilled alcoholic beverages using a reproducible fermentation procedure has recently been proposed as an alternative for the revalorization of berries with the additional advantage of a potential increase in the income of Galician farmers (Alonso et al. 2010, 2011). Four different alcoholic beverages with their own distinctive quality characteristics were produced by SSF of four berries (black mulberry, black currant, red raspberry, and arbutus berry) with *S. cerevisiae* IFI83, a high ethanol-producing yeast strain, and the further distillation of the fermented fruits (Alonso et al. 2010, 2011). In the four distillates, the mean methanol concentrations were much lower than the maximum acceptable levels fixed by the European Council (Regulation EC 110/2008) for fruit spirits. In addition, the mean levels of ethanol and volatile substances in these distillates were higher than the minimum limits (38.5 and 200 g/hL absolute alcohol) fixed by the aforementioned regulation for fruit distillates (Alonso et al. 2010, 2011).

13.3.2 PRODUCTION OF ENZYMES AND HIGH-VALUE-ADDED PRODUCTS

13.3.2.1 Enzymes

13.3.2.1.1 Amylases

Different studies have shown that a great number of agro-industrial residues and forages can potentially be used as low- cost carbon sources for enzyme production with different microorganisms, mainly using SSF.

The production of α-amylase under SSF by *Bacillus cereus* MTCC 1305 was investigated using wheat bran and rice flake manufacturing waste as substrates (Anto et al. 2006). The highest enzyme production (122 U/gds) was obtained with wheat bran by using an inoculum size of 10% (v/w), a substrate:moisture ratio of 1:1, a pH of 5, a temperature of 55°C, and by supplementing the dry substrate with glucose at a concentration of 0.04 g/gds.

The production of the same enzyme by submerged fermentation with *C. guilliermondii* CGL-A10 was optimized in the syrup obtained from date wastes. The highest amylase production (2304.19 μmol/L/min) was obtained after 72 h of incubation at 30°C, with an initial pH of 6.0 and concentrations of potassium phosphate and urea of 6.0 g/L and 5.0 g/L, respectively (Acourene and Ammouche 2011).

13.3.2.1.2 Cellulases and Hemicellulases

Rice straw and wheat straw were found to be the best carbon sources for the maximum production of xylanase (900.2 and 656.6 U/gds), endoglucanase (32.9 and 30.8 U/gds), β-glucosidase (7.48 and 6.78 U/gds), and FPase (2.44 and 1.37 U/gds) by *Myceliophthora* sp. IMI 387099 (Badhan et al. 2007). Sugarcane bagasse and corncob supported a relatively high production of xylanase (620.1 and 411.6 U/gds), but the production of the other three enzymes on these substrates was relatively low. However, wheat bran led to the lowest xylanase activity (128.9 U/gds), although the production of endoglucanase (26.6 U/gds) and β-glucosidase (5.49 U/gds) was comparable to that observed on rice straw and wheat straw (Badhan et al. 2007).

Further optimization of enzyme production on rice straw showed that adding CH_3COONH_3, KH_2PO_4, and $(NH_4)_2SO_4$ (1.3%) to the fermentation medium at optimal concentrations improved the production of xylanase (2366 U/gds), endoglucanase (42 U/gds), β-glucosidase (9 U/gds), and FPase (10 U/gds) using the same microbial strain (Badhan et al. 2007).

The use of mixtures of wheat bran and pretreated sugarcane bagasse increased the productions of endoglucanase (282.4 U/gdm), β-glucosidase (58.9 U/gdm), and xylanase (10.0 U/gdm) with *Penicillium echinulatum* 9A02S1, in comparison with the productions obtained in the controls, in which the two carbon sources were assayed separately (Camassola and Dillon 2007).

Da Silva et al. (2005) assayed the production of avicelase, xylanase, and CMCase by *Thermoascus aurantiacus* Miehe using different agricultural residues such as wheat bran, sugarcane bagasse, orange bagasse, corncob, green grass, dried grass, sawdust, and corn straw as the substrate source. In this study, wheat bran was the best carbon source for avicelase production (1.5 U/mL). However, the highest CMCase and xylanase productions were obtained with corncob (60.0 and 107.0 U/mL), dried grass (60.0 and 99.0 U/mL), green grass (59.0 and 102.0 U/mL), and corn straw (59.0 and 97.0 U/mL).

Other researchers (Alves-Prado et al. 2010) have studied xylanase and CMCase production with *Neosartorya spinosa* P2D19 in SSF with different types of substrates: wheat bran, corncob, cassava bran, wheat bran plus sawdust, corn straw, and sugarcane bagasse. The highest xylanase and carboxymethyl cellulase activities were obtained after 72 h of incubation with wheat bran (15.1 and 3.6 U/mL) and corncob (8.5 and 0.2 U/mL). However, corn straw was not a good carbon source for producing the two enzymes.

In another study, Facchini et al. (2011) assayed sugarcane bagasse, wheat bran, soy bran, crushed corncob, crushed corn, green grass, hay ensilage, *Eucalyptus* sawdust, orange peel, rice straw, and corn straw for producing β-1,4-endocellulase and endoxylanase, two fibrolytic enzymes that could be used as ruminant feed additives. Although the results obtained showed the suitability of these residues as low-cost carbon sources for producing the two enzymes by *Aspergillus japonicus* C03, soy bran and wheat bran were found to be the best substrates for xylanase (224.62 U/g) and cellulase (19.38 U/g) production, respectively. The other substrates assayed (corn straw, *Eucalyptus* sawdust, and orange peel) were poor substrates for xylanase and cellulase production. In other experiments, the same researchers observed that when soy bran was added to crushed corncob (1:3 w/w) or wheat bran was added to sugarcane bagasse (3:1 w/w), xylanase and cellulase activities increased by 8.5% and 1.4%, respectively, compared with the productions obtained on soy bran and wheat bran.

Apple pomace is a waste produced in large quantities as a by-product of juice extraction from apples. A small part (25–30%) of this waste is currently used for animal feed or as fertilizer, fuel, or industrial materials, but most of this residue is discarded as an industrial waste (Zhong-Tao et al. 2009). Therefore, different studies have investigated using this residue for different bioproductions with potential industrial application. For example, Sun et al. (2010) investigated the feasibility of using apple pomace as a substrate for cellulase production by *Trichoderma* sp. GIM 3.0010 under SSF. According to the results obtained by these researchers, the optimum cellulase production (4.2 U/gds) was obtained at 32°C for an initial moisture level of 70% and inoculum size of 2×10^8 spores per flask. Supplementing apple pomace with lactose (2%, w/w) and corn-steep solid (1%, w/w) was found to enhance enzyme production (7.6 U/gds).

13.3.2.1.3   Pectinases

Zheng and Shetty (2000) studied the production of polygalacturonase by *Lentinus edodes* CY-35 through SSF using three fruit processing wastes as substrates. The enzyme activities obtained after 40 days of fermentation on strawberry, apple, and cranberry pomaces were 29.4, 20.1, and 14.0 U per gram of dried pomace, respectively. Interestingly, adding polygalacturonic acid to media prepared with apple and cranberry pomaces led to an increase in polygalacturonase production, but adding the inducer compound to strawberry pomace medium did not affect the enzyme production significantly (Zheng and Shetty 2000).

More recently, Kiran et al. (2010) carried out a study to optimize endo-polygalacturonase production by overproducing mutants of *A. niger* GHRM5 in SSF on apple pomace. The maximum enzyme activity obtained was found to be 4.27 U under the optimal medium composition, which was (g/100 g): urea, 0.38; $(NH_4)_2SO_4$ 1.26, K_2HPO_4 0.65, $MgSO_4$ 0.02, $FeSO_4$ 0.03, apple pomace 11.80, sugarcane bagasse 23.10, and 70 mL distilled water. Glucose (13.76 g/L) was added to maintain the final concentration of the fermentation medium.

Pectinases production by *A. niger* P-6021 in slurry-state fermentation by using wheat bran mixed with the peel of *Citrus changshan-huyou* (a natural hybrid of pomelo *Citrus grandis* L. Osbeck and orange *Citrus sinensis* L. Osbeck), has also been reported (Zhong and Cen 2005). The optimal composition (w/w) of the culture medium was wheat bran: 6%, *Citrus changshan-huyou* peel: 4%, $(NH_4)SO_4$: 2%, KCl: 0.8%, MgSO4: 0.25%, and K_2HPO_4: 0.25%. In the optimized culture medium, the maximal enzyme activities obtained were 42 U/L in the case of polymethyl galacturonase, 6.7 U/L in the case of polymethyl galacturonate esterase, and 4.3 U/L in the case of polymethyl galacturonate lyase, after 3 days of fermentation at 180 rpm and 30°C.

Maller et al. (2011) studied the production of a thermostable polygalacturonase by a strain of *A. niveus* isolated from mango (*Mangifera indica*) in both liquid and solid media supplemented with different agro-industrial wastes. In submerged fermentation, the best carbon sources for enzyme production were orange peel (*C. sinensis*), passion fruit peel (*Passiflora edulis*), and lemon peel (*C. latifolia*), but rice straw (*Oryza sativa*) was not a good carbon source. However, in SSF, lemon peel (*C. latifolia*) and passion fruit peel (*P. edulis*) were the best inducers for polygalacturonase production.

13.3.2.1.4   Multienzyme Biofeed

Zhong-Tao et al. (2009) studied using apple pomace for multienzyme (proteinase, cellulase, and pectinase) biofeed production by *A. niger* strains M2 and M3 in SSF. The other objectives of this study were to biodegrade the antinutritional factors, such as the pectin and tannins in apple pomace, and nutritionally enrich the fermented substrate. A mixture of apple pomace and cottonseed powder (1:1, w/w), supplemented with 1% (w/w) $(NH_4)_2SO_4$ and 0.1% (w/w) KH_2PO_4 was found to be the best substrate. The highest activities of pectinase (21,168 U/g), proteinase (3585 U/g), and cellulase (1208 U/g) were obtained at an inoculum level of 0.4% (w/w) of well-mixed spores of the strains M2 and M3 (2:1, w/w), at a temperature of 30°C and 48 h of incubation. In these conditions, the biodegradation rates of pectin and tannins reached 99.0% and 66.1%, respectively.

13.3.2.2 Citric Acid

Pineapple peel, a by-product of pineapple juice extraction, is a potential substrate for producing citric acid by *A. niger* ACM 4992 in SSF (Tran et al. 1998). The highest level of citric acid (19.4 g/100 g of dry fermented pineapple waste) was produced at a 65% (w/w) initial moisture content, 3% (v/w) methanol, at a temperature of 30°C, an initial pH of 3.4, a particle size of 2 mm, and 5 ppm Fe^{2+}.

Citric acid production has also been investigated on apple pomace with *A. niger* BC1 in a multilayer packed-bed solid-state bioreactor (Shojaosadati and Babaeipour 2002). The optimum citric acid production (124 g/kg dry apple pomace) was obtained in the following conditions: aeration rate, 0.8 L/min; bed height, 10 cm; particle size, 0.60–2.33 mm; and moisture content, 78% (w/w).

Prado et al. (2005) assayed the production of citric acid by *A. niger* LPB 21 using cassava bagasse (CB) as the substrate, in SSF on a semipilot scale. The experiments were carried out in a horizontal drum bioreactor and in a tray-type bioreactor (with three bed thicknesses: 2, 4, and 6 cm). In this study, gelatinization at different starch percentages in CB was carried out to make the starch structure more susceptible to being consumed by the citric acid-producing strain. According to the results obtained, the highest production of the organic acid (26.9 g/100 g of dry CB) was obtained in a horizontal drum bioreactor using 100% gelatinized CB after 144 h of fermentation. However, SSF using 80% gelatinized CB in a tray-type bioreactor with a bed thickness of 4 cm gave a citric acid production (26.3 g/100 g of dry CB) comparable to that obtained in the horizontal drum bioreactor.

Acourene and Ammouche (2011) recently reported using a syrup from date wastes for optimizing citric acid production with *A. niger* ANSS-B5 under submerged fermentation. The highest citric acid production (98.42 g/L) was obtained at a temperature of 30°C, initial sugar concentration of 150 g/L, methanol concentration of 3%, initial pH of 3.5, and concentrations of ammonium nitrate and potassium phosphate of 2.5 g/L.

13.3.2.3 Biogas

Since pineapple peel is rich in cellulose, hemicellulose, and other carbohydrates, this waste could have two important industrial applications. On the one hand, its ensilage generates methane that can be used as a biogas. On the other hand, the digested slurry obtained after the anaerobic fermentation of the peels could be used as an additive in feeds for monogastric animals, poultry, and fish (Rani and Nand 2004). For example, ensilaging of pineapple peel converted 55% of the carbohydrates into volatile fatty acids and the initial biological oxygen demand of the waste was reduced by 91% at the end of the treatment. The digestion of ensilaged pineapple peel during 6 months produced a biogas yield of 0.67 m^3/kg volatile solids (VS) added with a methane content of 65%. However, when fresh and dried pineapple peels were digested, the yields decreased to 0.55 and 0.41 m^3/kg of VS added and a methane content of 51% and 41%, respectively (Rani and Nand 2004).

Another study showed that this waste has a high potential for industrial biogas production (Paepatung et al. 2009) compared to other agricultural residues. In fact, the maximum specific methane production rate with pineapple peel (36.77 mL CH_4/d) was higher than the rate with cassava pulp, empty fruit bunches, water hyacinth, and rice straw.

13.3.2.4 Animal Feed

The solid residue (peels, seeds, and pulp) produced in the industrial production of juices, flavors, and concentrates from apples has also been assayed for producing a nutritionally enriched substrate with increased digestibility to be used as the ruminant feed (Villas-Bôas et al. 2003). The study involved the SSF of apple pomace either individually or sequentially with *C. utilis* CCT 3469 and *Pleurotus ostreatus* CCT 2603. The fermentation with *C. utilis* alone after 6 days of fermentation provided better results than the treatment with *P. ostreatus* after 30 days. Thus, the fermentation with *C. utilis* led to an increase in the crude protein level (from 4.1% to 8.3% w/w), digestibility (from 58.5% to 63.3% w/w), and in mineral content (from 0.07% to 0.09% w/w for phosphorus, 0.10% to 0.12% w/w for calcium, and 0.43% to 0.53% w/w for potassium). The sequential fermentation by *C. utilis* (6 days) and *P. ostreatus* (30 days) provided an increase in both the crude protein level (from 4.1% to 10.5%, w/w) and in the mineral content (w/w) (from 0.07% to 0.13% for phosphorus, from 0.10% to 0.13% for calcium, and from 0.43% to 0.67% for potassium) after 60 days of fermentation. However, digestibility decreased by 10.2% and the total digestible nutrients decreased by 4.5%, which makes fermented apple pomace a less appropriate substrate for feeding ruminants. According to these results, the authors (Villas-Bôas et al. 2003) propose that the treatment with *C. utilis* alone is the most efficient alternative for converting apple pomace into a more nutritive substrate for ruminant feed.

Another study by Joshi et al. (2000) showed the potential of using fermented apple pomace as a component of animal feed. In this case, fermentation of this residue with five yeast strains (*S. cerevisiae*, *C. utilis*, *Torula utilis*, *Schizosaccharomyces pombe*, and *Kloeckera* spp.) led to an increase in crude protein, crude fat, ascorbic acid, and minerals. Adding the dried fermented apple pomace to a standard broiler feed (1:1 ratio) led to a regular increase (from 270.1 to 537.9 g) in the mean body weight of animals during 8 weeks of treatment.

13.3.2.5 Biosurfactants

Rodrigues et al. (2006) studied the production of biosurfactants by probiotic bacteria (*L. lactis* 53 and *S. thermophilus* A) on synthetic and waste media and reported the feasibility and even better results obtained using yeast extract- and peptone-supplemented sugarcane molasses in comparison to a conventional medium (MRS) for lactic acid bacteria. Ghribi et al. (2011) evaluated the feasibility of using orange peels and soy beans as substrates for biosurfactant production by *B. subtilis* SPB1. The optimal biosurfactant production (4.45 g/L) was found when the fermentation medium was composed of orange peels (15.5 g/L), soy beans (10 g/L), and diluted seawater (30%). This biosurfactant concentration was 2 times higher than that obtained in a defined medium containing basal salts (KH_2PO_4, K_2HPO_4, KCl, $MgSO_4$, $FeSO_4$, and

$CaCl_2$), urea, a trace element solution containing $ZnSO_4$, $MnSO_4$, $CuSO_4$, and NaBr, and glucose (Ghribi and Ellouze-Chaabouni 2011).

There are some reports describing the potential of different lignocellulosic wastes for the production of biosurfactants by lactic acid bacteria. Moldes et al. (2007) assayed the acid hydrolysate of the hemicellulosic fraction of barley bran, trimming vine shoots, corncobs, and *Eucalyptus globulus* chips as substrates for *Lactobacillus pentosus*, getting the best biosurfactant productions with trimming vine shoots, and Rodríguez et al. (2010) also described the use of the enzymatic hydrolysate of the cellulosic fraction of this waste for biosurfactant production by *L. lactis*.

13.3.2.6 Riboflavin

Pujari and Chandra (2000a) used agro-industrial by-products to optimize a culture medium for riboflavin production by *Eremothecium ashbyii* NRRL 1363 wild type. The enhanced vitamin production (1.21 mg/mL) was obtained at the optimal concentrations of molasses (50.0 g/L, glucose equivalents), sesame seed cake (50.0 g/L), yeast extract (2.0 g/L), KH_2PO_4 (2.0 g/L), $MgSO_4 \cdot 7H_2O$ (0.12 g/L), and NaCl (1.13 g/L). However, the optimal medium composition for enhanced riboflavin production (1.77 mg/mL) by an ultraviolet (UV) mutant of *E. ashbyii* UV-18-57 was different (Pujari and Chandra 2000b). In this case, the optimal concentrations for the medium components were: molasses (30.85 g/L), sesame seed cake (39.80 g/L), KH_2PO_4 (1.48 g/L), and $MgSO_4 \cdot 7H_2O$ (0.07 g/L).

13.3.2.7 Gibberellic Acid

The production of this hormone in SSF by *Gibberella fujikuroi* NRRL 2278 was assayed in culture media prepared with different concentrations of cassava (*Manihot esculenta*) flour obtained from fresh cassava roots (Tomasini et al. 1997). The production medium contained glycerol, 2%; glucose, 1%; lactose, 2%; KH_2PO_4, 0.05%; $(NH_4)_2SO_4$, 0.01%; and $MgSO_4$, 0.05%. The pH was adjusted to 4.5. This medium was supplemented with cassava (50% initial moisture content) to obtain a final concentration of 400 g/L. For SSFs, two fermentation media were used in which the concentration of solids was multiplied by two (medium 2×) and four (medium 4×). Although the strain NRRL 2278 produced high amounts of biomass in the 4× medium, the maximum concentration of gibberellic acid (240 mg/kg of the initial dry solid medium) reached in both media after 36 h of fermentation was very similar.

Sugarcane bagasse, one of the largest cellulosic agro-industrial by-products (Pandey and Soccol 1998), was used as an inert support for gibberellic acid production by SSF with *G. fujikuroi* NRRL 2278 (Tomasini et al. 1997). In this case, sugarcane bagasse pith (70% initial moisture content) was impregnated with the production medium (glycerol, 2%; glucose, 1%; lactose, 2%; KH_2PO_4, 0.05%; $(NH_4)_2SO_4$, 0.01%; and $MgSO_4$, 0.05%). SSF on this residue showed excellent growth but gibberellin was retained by the bagasse during the extraction process.

The other applications of this important agro-industrial residue (as a carbon source or as an inert support in both submerged and solid-state fermentations) for different biotechnological processes with different microbial strains are summarized in Tables 13.1 through 13.3 (Pandey and Soccol 1998; Pandey et al. 2000; Pérez and Guerra 2009; Guerra et al. 2009).

TABLE 13.1

Some Examples of Using Sugarcane Bagasse or Bagasse Pith as the Carbon Source in Submerged Fermentation Processes for Different Biotechnological Productions

Production	Substrates	Microbial Culture	Main Results	References
Single cell protein and cellulase	Untreated (washed with distilled water) and alkali-treated (at room temperature and at 100°C, washed with water at pH 2) sugarcane bagasse	*Trichoderma reesei* QM 9414	The highest cellulase and biomass yields were obtained in both untreated bagasse (16 IU and 0.57 g biomass per gram of cellulose consumed) and alkali-treated bagasse at room temperature (19 IU and 0.78 g biomass per gram of cellulose consumed)	Aiello et al. (1996)
Single cell protein and cellulase	Modified Czapek medium containing different amounts of sugarcane bagasse previously treated with 4% NaOH, autoclaved for 30 min, washed to extract the alkali, dried at 60°C and ground to a 40 mesh size	*Aspergillus terreus* strain	Optimum biomass protein production (21–28% (w/w) and protein recovery (11–14.5 g/100 g bagasse) were obtained for an alkali-treated bagasse substrate concentration of 1.5% (w/v), a pH of 4.5, a temperature of 35°C, 1:5 (v/v) culture broth, a 4% (v/v) inoculum and continuous agitation during seven days. Optimum cellulase activities were in the range of 0.85–1.2 U/mL and 0.08–0.11 U/mL with the substrates carboxymethyl cellulose and filter paper, respectively	El-Nawwi and El-Kader (1996)
Single cell protein	Sugarcane bagasse piths with previous dry or wet pretreatment with NaOH	*Cellulomonas flavigena* CDBB-B-532 and *Xanthomonas* sp. ATCC-31920	The maximum digestibility of the bagasse (78%) and biomass concentration (4.5 g/L) were respectively obtained by using the dry method	Rodríguez-Vázquez et al. (1992)
	Sugarcane bagasse pith	*Cellulomonas* sp. and *Bacillus subtilis*	The yield obtained was 0.17 g protein/g bagasse pith	Molina et al. (1983)

Product	Substrate	Microorganism	Description	Reference
	Sugarcane bagasse	*Cellulomonas flavigena* and *Xanthomonas* sp.	In continuous culture both microorganisms co-existed between 30 and 40°C, at pH 5.8–7.0 and D = 0.04–0.11 h^{-1}	Ponce and De la Torre (1992)
	Sugarcane bagasse pith	*Cellulomonas* sp. and *Pseudomonas* sp.	Fed-batch cultivation of the mixed culture on this substrate led to a high biomass production (19.4 g/L)	Rodríguez and Gallardo (1993)
	Sugarcane bagasse piths pretreated with NaOH, Ca(OH)$_2$, NH$_4$OH, and H$_2$O$_2$	*Cellulomonas flavigena* CDBB-B-532 and *Xanthomonas* sp. ATCC-31920	Maximum biomass concentrations (2.34 and 2.15 g/L) were obtained in materials pretreated with NaOH and Ca(OH)$_2$	Rodríguez-Vázquez and Díaz-Cervantes (1994)
	Untreated and treated (delignified with sodium chlorite) sugarcane bagasse	*Pleurotus ostreatus* NRRL-2366	The amounts of protein obtained were 19.4 and 15.3 g in the untreated and chemically pretreated bagasse, respectively. The protein-rich fermentation product has a suitable amino acid composition for use as an animal feed	El-Sayed et al. (1994)
Cellulases	Liquid hydrolysate from sugarcane bagasse pretreated with hot water (varying from 0 to 80% hydrolysate) and mixed with 10 g/L of sorbitol (carbon source) or the solid residue from the same pretreatment operation supplemented with varying cellulose concentrations (5, 10, 15 and 25 g/L)	*Trichoderma reesei* Rut C30	Amounts of liquid hydrolysate higher than 40% decreased the growth rate of the producing strain. Maximum cellulose production (0.6 FPU/mL) was obtained with pretreated bagasse solids supplemented with 25 g/L cellulose	Bigelow and Wyman (2002)
Cellulases and hemicellulases	L-arabinose, D-xylose, oat spelt xylan, sugarcane bagasse, or corn straw as carbon source	*Acremonium zeae* EA0802 and *Acremonium* sp. EA0810	*Acremonium* sp. EA0810 produced the highest exoglucanase, endoglucanase, and xylanase activities when the sugarcane bagasse was used as a carbon source	De Almeida et al. (2011)

continued

TABLE 13.1 (continued)

Some Examples of Using Sugarcane Bagasse or Bagasse Pith as the Carbon Source in Submerged Fermentation Processes for Different Biotechnological Productions

Production	Substrates	Microbial Culture	Main Results	References
Protein enriched feed	Untreated and chemically pretreated (delignified using sodium chlorite) sugarcane bagasse submerged in a liquid medium containing nitrogen and phosphorous sources, different mineral salts and yeast extract	*Pleurotus ostreatus* NRRL-2366	After 14 days of fermentation, an increment of 22.6% and 18.0% of crude protein content was obtained for the untreated and chemically pretreated sugarcane bagasse, respectively	El-Sayed et al. (1994)
Cellulases and ligninases	Sun-dried sugarcane bagasse treated with H_2SO_4, H_2O_2 or with H_2O_2 plus NaOH (20 g) in vinasse (60 mL)	*Pleurotus sajor-caju* CCB020, *Pleurotus ostreatus*, *Pleurotus ostreatoroseus* CCB440, and *Trichoderma reesei*	Treating the bagasse with 2% H_2O_2 + 1.5% NaOH + autoclave favoured enzyme production for all cultures. The *P. ostreatus* strain produced the highest levels of laccase (325.23 IU/L), manganese-peroxidase (27.69 IU/L) activities on the 21st day of cultivation, and peroxidase activity (19.84 IU/L) after 15 days of fermentation. *T. reesei* produced the highest levels of exoglucanase (18.35 IU/L) and endoglucanase (5.88 IU/L) activities after 18 and 24 days of incubation, respectively	Aguiar et al. (2010)

Note: FPU, filter paper units and IU, international units.

TABLE 13.2

Some Examples of Using Sugarcane Bagasse or Bagasse Pith as the Carbon Source in Solid-State Fermentation Processes for Different Biotechnological Productions

Production	Substrates	Microbial Culture	Main Results	References
Protein enriched feed	Sugarcane bagasse washed, dried, and milled to 35 and 80 mesh particle size powders	*Polyporus* sp. strains BH₁ and BW₁	After 8 days of fermentation with both strains, the highest digestibility of the substrate and protein content was obtained in the SSF medium by using an inoculum/dry substrate ratio of 1.0:1.5 (v/v), pH of 5.5, 75% moisture content and a temperature of 28°C	Nigam (1990)
Inulinase	Sugarcane bagasse supplemented with soybean bran and corn steep liquor	*Kluyveromyces marxianus* NRRL Y-7571	Maximum inulinase activity (250 U/g of dry bagasse) was achieved when the sugarcane bagasse with a particle size in the range of 9/32 mesh was supplemented with 20% (w/w) corn steep liquor, 5% (w/w) soybean bran, and inoculated with 1×10^{10} cells/mL	Mazutti et al. (2007)
Phenol oxidase	Sugarcane bagasse	*Flammulina velutipes* and *Trametes versicolor*	Phenol oxidase was only found with *F. velutipes*	Pal et al. (1995)
Xylanase	Alkali treated sugarcane bagasse supplemented with different nitrogen sources: ((NH₄)₂SO₄), urea, or soymeal	A mixed culture of *Trichoderma reesei* LM-UC4E1 with *Aspergillus niger* ATCC 10864 or with *Aspergillus phoenicis* QM 329	The highest xylanase productivity (5500–5900 IU/L·h) was obtained when the alkali-treated bagasse was supplemented with soymeal	Gutiérrez-Correa and Tengerdy (1998)
Acetyl esterase	Wheat straw, sugarcane bagasse, rice husk, and rice straw	*Melanocarpus albomyces* IIS-68	The highest acetyl esterase activity (21 U/gs) was obtained on untreated bagasse	Jain (1995)
Xylanases and cellulases	Sugarcane bagasse	*Neosartorya spinosa* P2D19	Relatively low xylanase (0.5 U/mL) and cellulase (0.05 U/mL) activities were obtained after 72 h of fermentation	Alves-Prado et al. (2010)

continued

TABLE 13.2 (continued)

Some Examples of Using Sugarcane Bagasse or Bagasse Pith as the Carbon Source in Solid-State Fermentation Processes for Different Biotechnological Productions

Production	Substrates	Microbial Culture	Main Results	References
Xylanases, cellulases, and avicelase	Sugarcane bagasse (2 g) mixed with sterilised distilled water (3 mL)	*Aspergillus japonicus* C03	The mean xylanase and cellulase activities obtained after 5 days of fermentation were 101.77 and 13.25 U/g, respectively	Facchini et al. (2011)
	Sugarcane bagasse (10 g) mixed with sterilised distilled water (10 mL)	*Thermoascus aurantiacus* Miehe	The mean avicelase, xylanase, and cellulase activities obtained after 4 days of fermentation were 0.6, 3.0, and 11.0 U/mL, respectively	Da Silva et al. (2005)
Pectinases	Sugarcane bagasse, orange bagasse, and wheat bran	*Thermoascus aurantiacus* 179-5	The maximum pectin lyase production (40 180 U/gdss) was obtained in a medium containing 10% sugar cane bagasse and 90% orange bagasse after 8 days of fermentation. The maximum polygalacturonase activity (43 U/gdss) was obtained after 4 and 6 days of incubation, when wheat bran or orange bagasse were respectively used as substrates. However, the maximum production of this enzyme (37 U/gdss) was obtained in a medium containing 30% sugar cane bagasse and 70% wheat bran after 4 days of fermentation	Martins et al. (2002)
Ethanol	Sugarcane bagasse after steam pretreatment at different temperatures (160, 170, 180, and 190° C) with 1% (w/w) phosphoric acid	*Escherichia coli* strain MM160	The highest titer (30 g/L ethanol) and yield (0.21 g ethanol/g bagasse dry weight) were obtained after incubation for 122 h using 14% dry weight slurries of pretreated bagasse (180°C)	Geddes et al. (2011)
Pigments	Wet sugarcane bagasse containing PGY medium	*Monascus purpurea* strain IAM 8081	Fungus produced red and yellow pigments in wet bagasse containing PGY medium with corn oil	Chiu and Chan (1992)

Note: IU, international units; gs, grams of dry solid substrate; and PGY medium containing peptone, glucose, and yeast extract.

TABLE 13.3
Some Examples of Using Sugarcane Bagasse or Bagasse Pith as an Inert Support in Solid-State Fermentation Processes for Different Biotechnological Productions

Production	Support and Culture Media	Microbial Culture	Main Results	References
Penicillin	Sugarcane bagasse (10–30 mesh) impregnated with the inoculated (2×10^6 spores/mL) liquid production medium, which contained glucose and phenylacetic acid	*Penicillium chrysogenum* Wis. 54-1255	Increased penicillin production (1.68 μg/gdm) was obtained by using a large particle size (14 mm) support (sugar cane bagasse). Cultures with closer packing densities (0.35) produced 20% more penicillin. Agitation did not produce a negative effect on penicillin production if moisture loss during the operation was restituted	Barrios-González et al. (1993)
L-glutamic acid	Sugar cane bagasse impregnated with a medium containing glucose, urea, mineral salts, and vitamins	*Brevibacterium* sp.	The yields (80 mg glutamic acid/gds) were highest when bagasse of mixed particle sizes was moistened to a 85–90% moisture level with the medium containing 10% glucose	Nampoothiri and Pandey (1996)
Ergot alkaloids	Washed sugarcane pith bagasse impregnated with inoculated liquid medium	*Claviceps purpurea* 1029c	After 120 h of fermentation, the highest ergot alkaloids levels (505.5 μg/g IDM) were obtained in the washed sugarcane pith bagasse impregnated with inoculated liquid medium containing sucrose (200 g/L), urea (1.73 g/L), oxalate-NH_3 (9.6 g/L), KH_2PO_4 (1.0 g/L), $MgSO_4 \cdot 7H_2O$ (0.625 g/L), and tryptophan (1.0 g/L) with an agitation rate of 4 L/h per column.	Trejo et al. (1993)
L(+)-lactic acid	Sugarcane bagasse (10 g), $CaCO_3$ (10 g), and 80 mL of the fermentation medium containing glucose	*Rhizopus oyzae* NRRL 395	The maximum lactic acid production (137 g/L) and yield (76%) were obtained at an inoculation rate of 2×10^6 spores/g glucose, a glucose concentration of 180 g/L and with an aeration rate of 20 mL/min	Soccol et al. (1994)

continued

TABLE 13.3 (continued)
Some Examples of Using Sugarcane Bagasse or Bagasse Pith as an Inert Support in Solid-State Fermentation Processes for Different Biotechnological Productions

Production	Support and Culture Media	Microbial Culture	Main Results	References
Citric acid	Sugar cane bagasse (10 g) with inoculated (10^5–10^6 spores/mL) liquid medium (75 mL) containing sucrose (14%) or Indian Cane molasses, and supplemented or not with 3% methanol.	Aspergillus niger strain 3/1	The highest average yields of citric acid under solid state conditions were obtained in the presence of methanol in sucrose medium (62.8%) and in molasses medium (56.4%)	Lakshminarayana et al. (1975)
Amylases	Sugarcane bagasse impregnated with the production medium containing varying concentrations of starch	Aspergillus niger strain UO-01	The optimum amylase levels (457.82 EU/gds) were obtained for a particle size of bagasse in the range of 6–8 mm, an incubation temperature of 30.2°C, a pH value of 6.0, with a bagasse moisture content of 75.3%, inoculum concentration of 1×10^7 spores/gds and concentrations of starch, yeast extract and KH_2PO_4 of 70.5, 11.59, and 9.83 mg/gds, respectively	Pérez and Guerra (2009)
	Sugarcane bagasse impregnated with the production medium containing varying concentrations of starch	Aspergillus oryzae strain FQB-01	The optimum conditions for high amylase production (539 EU/gd) under solid state fermentation were a bagasse particle size in the range of 5–10 mm, incubation temperature of 32.5°C, pH of 5.9, bagasse moisture content of 75%, starch concentration of 70.5 mg/gds, and inoculum size of 1.4×10^7 spores/gds.	Guerra et al. (2009)
Laccase	Sugarcane bagasse impregnated with the culture medium and with ethanol as a laccase inducer	Pycnoporus cinnabarinus ss3	The highest laccase activity (80 U/gds) in the ethanol-treated bagasse was obtained after 28 days of fermentation	Meza et al. (2006)

Product	Organism	Substrate/Medium	Observations	Reference
Pectinases	*Aspergillus niger* CH4	Sugar cane bagasse pith impregnated (70%) with the nutritive medium containing pectin as inducer and sucrose as carbon source and a mineral solution containing urea (2.4 g/L), $(NH_4)_2SO_4$ (9.8 g/L), KH_2PO_4 (5.0 g/L), $FeSO_4 \cdot 7H_2O$ (1.0 mg/L), $ZnSO_4 \cdot 7H_2O$ (0.8 mg/L), $MgSO_4 \cdot 7H_2O$ (4.0 mg/L), and $CuSO_4 \cdot 5H_2O$ (1.0 mg/L)	Maximum production of pectinases (1750 U/g WS) was obtained after 45 h of fermentation in sugar cane bagasse with a pectin/sucrose ratio of 1/2	Trejo et al. (1991)
Fruity aroma	*Ceratocystis fimbriata* Ellis & Halst. CBS 374-83	Wheat bran and cassava bagasse were supplemented with an oligoelement solution with the following composition: $Fe(NO_3)_3$ (723.8 mg/L), $ZnSO_4 \cdot 7H_2O$ (439.8 mg/L), $MnSO_4 \cdot 7H_2O$ (203.0 mg/L). Sugar cane bagasse was supplemented with a synthetic medium containing glucose, (200.0 g/L), urea (7.6 g/L), $(NH_4)_2SO_4$ (18.0 g/L), KH_2PO_4 (4.0 g/L), $Ca(NO_3)_2$ (4.0 g/L), $MgSO_4 \cdot 7H_2O$ (3.0 g/L), oligoelement solution (8.0 mL/L), and leucine (167.0 mmol/L)	Sugar cane bagasse complemented with a synthetic medium containing glucose (200 g/L) gave a fruity aroma, while the addition of leucine to the same medium gave a strong banana aroma	Christen et al. (1997)

Note: gds, grams of dry support; gdm, grams of dry medium; IDM, initial dry matter; WS, wet sample.

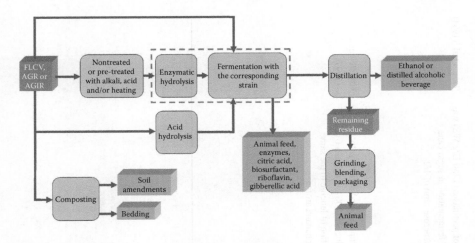

FIGURE 13.1 Scheme of the processes to produce different value-added products from fruit with low commercial value (FLCV), agricultural residues (AGR), and agro-industrial residues (AGIR).

Figure 13.1 shows a general scheme of the processes commonly used to revalorize agricultural and agro-industrial residues and fruits of low commercial value by producing different useful products of higher value.

REFERENCES

Acourene, S. and A. Ammouche. 2011. Optimization of ethanol, citric acid, and α-amylase production from date wastes by strains of *Saccharomyces cerevisiae, Aspergillus niger* and *Candida guilliermondii. J Ind Microbiol Biotechnol* 39:759–766.

Adil, I. H., Cetin, H. I., Yener, M. E., and A. Bayindirli. 2007. Subcritical (carbon dioxide plus ethanol) extraction of polyphenols from apple and peach pomaces, and determination of the antioxidant activities of the extracts. *J Supercrit Fluids* 43:55–63.

Aguiar, M. M., Romanholo, L. F., and R. T. Rosim. 2010. Use of vinasse and sugarcane bagasse for the production of enzymes by lignocellulolytic fungi. *Braz Arch Biol Technol* 53:1245–1254.

Ahuja, K. D. K., Ashton, E. L., and M. J. Ball. 2003. Effects of a high monounsaturated fat, tomato-rich diet on serum levels of lycopene. *Eur J Clin Nutr* 57:832–841.

Aiello, C., Ferrer, A., and A. Ledesma. 1996. Effect of alkaline treatments at various temperatures on cellulase and biomass production using submerged sugarcane bagasse fermentation with *Trichoderma reesei* QM 9414. *Bioresour Technol* 57:13–18.

Ajila, C. M., Brar, S. K., Verma, M., Tyagi, R. D., and J. R. Valéro. 2011. Solid-state fermentation of apple pomace using *Phanerocheate chrysosporium*—Liberation and extraction of phenolic antioxidants. *Food Chem* 126:1071–1080.

Al-Awwadi, N., Bornet, A., Azay, J., Araiz, C., Delbosc, S., Cristol, J-P., Linck, N., Cros, G., and P-L. Teissedre. 2004. Red wine polyphenols alone or in association with ethanol prevent hypertension, cardiac hypertrophy, and production of reactive oxygen species in the insulin-resistant fructose-fed rat. *J Agric Food Chem* 52:1008–1016.

Alarcão-E-Silva, M. L. C. M. M., Leitão, A. E. B., Azinheira, H. G., and M. C. A. Leitão. 2001. The arbutus berry: Studies on its color and chemical characteristics at two mature stages. *J Food Compost Anal* 14:27–35.

Alonso, A. M., Guillén, D. A., Barroso, C. G., Puertas, B., and A. García. 2002. Determination of antioxidant activity of wine byproducts and its correlation with polyphenolic content. *J Agric Food Chem* 50:5832–5836.

Alonso, E., Torrado, A., Pastrana, L., Orriols, I., and N. P. Guerra. 2010. Production and characterization of distilled alcoholic beverages obtained by solid-state fermentation of black mulberry (*Morus nigra* L.) and black currant (*Ribes nigrum* L.). *J Agric Food Chem* 58:2529–2535.

Alonso, E., Torrado, A., Pastrana, L., Orriols, I., and N. P. Guerra. 2011. Solid-state fermentation of red raspberry (*Rubus ideaus* L.) and arbutus berry (*Arbutus unedo* L.) and characterization of their distillates. *Food Res Int* 44:1419–1426.

Aloys, N. and N. Angeline. 2009. Traditional fermented foods and beverages in Burundi. *Food Res Int* 42:588–594.

Alves-Prado, H. F., Pavezzi, F. C., Leite, R. S., de Oliveira, V. M., Sette, L. D., and R. Dasilva. 2010. Screening and production study of microbial xylanase producers from Brazilian Cerrado. *Appl Biochem Biotechnol* 161:333–346.

Amro, B., Aburjai, T., and S. Al-Khalil. 2002. Antioxidative and radical scavenging effects of olive cake extract. *Fitoterapia* 73:456–461.

Anto, H., Trivedi, U., and K. Patel. 2006. Alpha amylase production by *Bacillus cereus* MTCC 1305 using solid-state fermentation. *Food Technol Biotechnol* 44:241–245.

Arab, L., and S. Steek. 2000. Lycopene and cardiovascular disease. *Am J Clin Nutr* 71:1691–1695.

Argawal, S. and A.V. Rao. 1998. Tomato lycopene and low density lipoprotein oxidation: A human dietary intervention study. *Lipids* 33:981–984.

Aviram, M. and B. Fuhrman. 2002. Wine flavonoids protect against LDL oxidation and atherosclerosis. *Ann NY Acad Sci* 957:146–161.

Ayala-Zavala, J. F., Rosas-Domínguez, C., Vega-Vega, V., and G. A. González-Aguilar. 2010. Antioxidant enrichment and antimicrobial protection of fresh-cut fruits using their own byproducts: Looking for integral exploitation. *J Food Sci* 75:175–181.

Ayaz, F. A., Kucukislamoglu, M., and M. Reunanen. 2000. Sugar, non volatile and phenolic acid composition of strawberry tree (*Arbutus unedo* L. var. *ellipsoidea*) fruits. *J Food Compost Anal* 13:171–177.

Badhan, A. K., Chadha, B. S., Kaur, J., Saini, H.S., and M. K. Bhat. 2007. Production of multiple xylanolytic and cellulolytic enzymes by thermophilic fungus *Myceliophthora* sp. IMI 387099. *Bioresour Technol* 98:504–510.

Barrios-González, J., González, H., and A. Mejía. 1993. Effect of particle size, packing density and agitation on penicillin production in solid state fermentation. *Biotechnol Adv* 11:525–537.

Barth, S. W., Fähndrich, C., Bub, A., Dietrich, H., Watzl, B., Will, F., Briviba, K., and G. Rechkemmer. 2005. Cloudy apple juice decreases DNA damage, hyperproliferation and aberrant crypt foci development in the distal colon of DMH-initiated rats. *Carcinogenesis* 26:1414–1421.

Baur, J. A., Pearson, K. J., Price, N. L., Jamieson, H. A., Lerin, C., Kalra, A., Prabhu, V. V. et al. 2006. Resveratrol improves health and survival of mice on a high-calorie diet. *Nature* 444:337–342.

Bellion, P., Digles, J., Will, F., Dietrich, H., Baum, M., Eisenbrand, G., and C. Janzowski. 2010. Polyphenolic apple extracts: Effects of raw material and production method on antioxidant effectiveness and reduction of DNA damage in Caco-2 cells. *J Agric Food Chem* 58:6636–6642.

Benakmoum, A., Abbeddou, S., Ammouche, A., Kefalas, P. and D. Gerasopoulos. 2008. Valorisation of low quality edible oil with tomato peel waste. *Food Chem* 110:684–690.

Bhushan, S., Kalia, K., Sharma, M., Singh, B., and P. S. Ahuja. 2008. Processing of apple pomace for bioactive molecules. *Crit Rev Biotechnol* 28:285–296.

Bigelow, M. and E. W. Wyman. 2002. Cellulase production on bagasse pretreated with hot water. *Appl Biochem Biotechnol* 98–100:921–934.

Boussetta, N., Vorobiev, E., Deloison, V., Pochez, F., Falcimaigne-Cordin, A., and J.L. Lanoisellé. 2011. Valorisation of grape pomace by the extraction of phenolic antioxidants: Application of high voltage electrical discharges. *Food Chem* 128:364–370.

Bressani, R. 1978. The by-products of coffee berries. In *Coffee Pulp, Composition, Technology, and Utilization*, ed. J. E. Braham and R. Bressani, 5–10. Canada: INCAP.

Çam, M. and K. Aaby. 2010. Optimization of extraction of apple pomace phenolics with water by response surface methodology. *J Agric Food Chem* 58:9103–9111.

Camassola, M. and A. J. P. Dillon. 2007. Production of cellulases and hemicellulases by *Penicillium echinulatum* grown on pretreated sugar cane bagasse and wheat bran in solid-state fermentation. *J Appl Microbiol* 103:2196–2204.

Casas, L., Mantell, C., Rodríguez, M., Martínez de la Ossa, E., Roldán, A., De Ory, I., Caro, I., and A. Blandino. 2010. Extraction of resveratrol from the pomace of *Palomino fino* grapes by supercritical carbon dioxide. *J Food Eng* 96:304–308.

Cavaco, T., Longuinho, C., Quintas, C., and I. Saraiva. 2007. Chemical and microbial changes during the natural fermentation of strawberry tree (*Arbutus unedo* L.) fruits. *J Food Biochem* 31:715–725.

Ćetković, G., Savatović, S., Čanadanović-Brunet, J. Djilasa, S., Vulića, J., Mandićb, A., and D. Ćetojević-Siminc. 2012. Valorisation of phenolic composition, antioxidant and cell growth activities of tomato waste. *Food Chem* 133:938–945.

Chamorro, S., Viveros, A., Alvarez, I., Vega, E., and A. Brenes. 2012. Changes in polyphenol and polysaccharide content of grape seed extract and grape pomace after enzymatic treatment. *Food Chem* 133:308–314.

Champ, M., Langkilde, A. M., Brouns, F., Kettlitz, B., Collet, Y., and B. Le. 2003. Advances in dietary fibre characterisation. 1. Definition of dietary fibre, physiological relevance, health benefits and analytical aspects. *Nutr Res Rev* 16:71–82.

Chiu, S. W., and S. M. Chan. 1992. Production of pigments by *Monascus purpurea* using sugar cane bagasse in roller bottle cultures. *World J Microbiol Biotechnol* 8:68–70.

Christen, P., Meza, J. C., and S. Revah. 1997. Fruity aroma production in solid state fermentation by *Ceratocystis fimbriata*: Influence of the substrate type and the presence of precursors. *Mycol Res* 101:911–919.

Corrales, M., Toepfl, S., Butz, P., Knorr, D., and B. Tauscher. 2008. Extraction of anthocyanins from grape by-products assisted by ultrasonics, high hydrostatic pressure or pulsed electric fields: A comparison. *Innov Food Sci Emerg Technol* 9:85–91.

Corthout, J., Peiters, L. A., Claeys, M., Vanden Berghe, D. A., and A. J. Vleitinck. 1991. Antiviral ellagitannins from *Spondia mombin*. *Phytochemistry* 30:1129–1130.

Da Silva, R., Lago, E. S., Merheb, C. W., Macchione, M. M., Park, Y. K., and E. Gomes. 2005. Production of xylanase and CMCase on solid state fermentation in different residues by *Thermoascus aurantiacus* Miehe. *Braz J Microbiol* 36:235–241.

Dalmadi, I., Rapeanu, G., Smout, C., and M. Hendrick. 2006. Characterization and inactivation by thermal and pressure processing of strawberry (*Fragaria ananassa*) polyphenol oxidase: A kinetic study. *J Food Biochem* 30:56–76.

Darias-Martín, J., Lobo-Rodrigo, G., Hernández-Cordero, J., Díaz-Díaz, E., and C. Díaz-Romero. 2003. Alcoholic beverages obtained from black mulberry. *Food Technol Biotechnol* 41:173–176.

de Almeida, M. N., Guimarães, V. M., Bischoff, K. M., Falkoski, D. L., Pereira, O. L., Gonçalves, D. S., and S. T. de Rezende. 2011. Cellulases and hemicellulases from endophytic *Acremonium* species and its application on sugarcane bagasse hydrolysis. *Appl Biochem Biotechnol* 165:594–610.

El-Sayed, S. A., Zaki, M. T., and A.W.A. El-Khair. 1994. Bioconversion of sugarcane bagasse into a protein-rich product by white rot fungus. *Resour Conserv Recycl* 12:195–200.

Eberhardt, M. V., Lee, C. Y., and R. H. Liu. 2000. Nutrition: Antioxidant activity of fresh apples. *Nature* 405:903–904.

El-Nawwi, S. A. and A. A. El-Kader. 1996. Production of single-cell protein and cellulase from sugarcane bagasse: Effect of culture factors. *Biomass Bioenergy* 11:361–364.

Elleuch, M., Bedigian, D., Roiseux, O., Besbes, S., Blecker, C., and H. Attia. 2011. Dietary fibre and fibre-rich by-products of food processing: Characterisation, technological functionality and commercial applications: A review. *Food Chem* 124:411–421.

Erdei, B., Barta, Z., Sipos, B., Réczey, K., Galbe, M., and G. Zacchi. 2010. Ethanol production from mixtures of wheat straw and wheat meal. *Biotechnol Biofuels* 3, art. no. 16:1–9. http://www.ncbi.nlm.nih.gov/pmc/articles/PMC2912878/pdf/1754-6834-3-16.pdf.

Esquivel, P. and V. M. Jiménez. 2012. Functional properties of coffee and coffee by-products. *Food Res Int* 46:488–495.

Facchini, F. D., Vici, A. C., Reis, V. R., Jorge, J. A., Terenzi, H. F., Reis, R. A., and M. L. Polizeli. 2011. Production of fibrolytic enzymes by *Aspergillus japonicus* C03 using agro-industrial residues with potential application as additives in animal feed. *Bioprocess Biosyst Eng* 34:347–355.

FAO Database. http://faostat.fao.org (accessed March 20, 2012).

Foo, L. Y. and Y. Lu. 1999. Isolation and identification of procyanidins in apple pomace. *Food Chem* 64:511–518.

Fraser, P. and P. Bramley. 2004. The biosynthesis and nutritional uses of carotenoids. *Prog Lipid Res* 43:228–265.

Geddes, C. C., Mullinnix, M. T., Nieves, I. U., Peterson, J. J., Hoffman, R. W., York, S. W., Yomano, L. P., Miller, E. N., Shanmugam, K. T., and L. O. Ingram. 2011. Simplified process for ethanol production from sugarcane bagasse using hydrolysate-resistant *Escherichia coli* strain MM160. *Bioresour Technol* 102:2702–2711.

Ghanbari, R., Anwar, F., Alkharfy, K. M., Gilani, A. H., and N. Saari. 2012. Valuable nutrients and functional bioactives in different parts of olive (*Olea europaea* L.). A review. *Int J Mol Sci* 13:3291–3340.

Ghribi, D. and S. Ellouze-Chaabouni. 2011. Enhancement of *Bacillus subtilis* lipopeptide biosurfactants production through optimization of medium composition and adequate control of aeration. *Biotechnol Res Int* 2011:1–6.

Ghribi, D., Mnif, I., Boukedi, H., Kammoun, R., and S. Ellouze-Chaabouni. 2011. Statistical optimization of low-cost medium for economical production of *Bacillus subtilis* biosurfactant, a biocontrol agent for the olive moth *Prays oleae*. *Afr J Microbiol Res* 5:4927–4936.

González-Paramás, A. M., Esteban-Ruano, S., Santos-Buelga, C., Pascual-Teresa, S., and J. C. Rivas-Gonzalo. 2004. Flavanol content and antioxidant activity in winery products. *J Agric Food Chem* 52:234–238.

Guerra, N. P., Pastrana, L., and R. Pérez. 2009. Amylase production by *Aspergillus niger* in solid state fermentations. In *Biochemical Engineering*, ed. F. E. Dumont, and J. A. Sacco, 185–205. New York: Nova Science Publishers, Inc.

Gustavsson, J., Cederberg, C., Sonesson, U., van Otterdijk, R., and A. Meybeck. 2011. Global food losses and food waste. Food and Agriculture Organization of the United Nations, Rome. http://www.fao.org/fileadmin/user_upload/ags/publications/GFL_web.pdf (accessed March 10, 2012).

Gutiérrez-Correa, M. and R. P. Tengerdy. 1998. Xylanase production by fungal mixed culture solid state fermentation on sugar cane bagasse. *Biotechnol Lett* 20:45–47.

Hagerman, A. E., Riedl, K. M., Jones, G. A., Sovik, K. N., Ritchard, N. T., Hartzfeld, P. W., and T. L. Riechel. 1998. High molecular weight plant polyphenolics (tannins) as biological antioxidants. *J Agric Food Chem* 46:1887–1892.

Heinonen, M., Meyer, A. S., and E. N. Frankel. 1998. Antioxidant activity of berry phenolics on human low-density-lipoprotein and liposome oxidation. *J Agric Food Chem* 46:107–112.

Hertog, M. G., Kromhout, D., Aravanis, C., Blackburn, H., Buzina, R., Fidanza, F., Giampaoli, S. 1995. Flavonoid intake and long-term risk of coronary heart disease and cancer in the seven countries study. *Arch Int Med* 155:381–386.

Jain, A. 1995. Production of xylanase by thermophilic *Melanocarpus albomyces* IIS-68. *Process Biochem* 30:705–709.

Joshi, V. K. and A. Devrajan. 2008. Ethanol recovery from solid state fermented apple pomace and evaluation of physico-chemical characteristics of the residue. *Nat Prod Rad* 7:127–132.

Joshi, V. K., Gupta, K., Devrajan, A., Lal, B. B., and S. P. Arya. 2000. Production and evaluation of fermented apple pomace in the feed of broilers. *J Food Sci Technol* 37:609–612.

Juranic, Z., Zizak, Z., Tasic, S., Petrovic, S., Nidzovic, S., Leposavic, A., and T. Stanojkovic. 2005. Antiproliferative action of water extracts of seeds or pulp of five different raspberry cultivars. *Food Chem* 93:39–45.

Kader, A. A. 2005. Increasing food availability by reducing postharvest losses of fresh produce. *Acta Hortic* 682:2169–2175.

Kaur, U., Oberoi, H. S., Bhargav, V. K., Shivappa, R. S., and S. S. Dhaliwal. 2012. Ethanol production from alkali- and ozone-treated cotton stalks using thermotolerant *Pichia kudriavzevii* HOP-1. *Ind Crop Prod* 37:219–226.

Kiran, R. R. S., Konduri, R., Rao, G. H., and G. M. Madhu. 2010. Statistical optimization of endo-polygalacturonase production by overproducing mutants of *Aspergillus niger* in solid-state fermentation. *J Biochem Technol* 2:154–157.

Knoblich, M., Anderson, B., and D. Latshaw. 2005. Analyses of tomato peel and seed byproducts and their use as a source of carotenoids. *J Sci Food Agric* 85:1166–1170.

Lakshminarayana, K., Chaudhary, K., Ethiraj, S., and P. Tauro. 1975. A solid state fermentation method for citric acid production using sugar cane bagasse. *Biotechnol Bioeng* 7:291–293.

Lashermes, P., Andrade, A. C., and H. Etienne. 2008. Genomics of coffee, one of the world's largest traded commodities. In *Genomics of Tropical Crop Plants*, ed. P. H. Moore and R. Ming, 203–225. New York: Springer.

Laufenberg, G., Kunz, B., and M. Nystroem. 2003. Transformation of vegetable waste into value added products: (a) the upgrading concept; (b) practical implementations. *Bioresour Technol* 87:167–198.

Lin, C. H. and B. H. Chen. 2003. Determination of carotenoids in tomato juice by liquid chromatography. *J Chromatogr A* 1012:103–109.

Lin, H. J., Xian, L., Zhang, Q. J., Luo, X. M., Xu, Q. S., Yang, Q., Duan, C. J., Liu, J. L., Tang, J. L., and J. X. Feng. 2011. Production of raw cassava starch-degrading enzyme by *Penicillium* and its use in conversion of raw cassava flour to ethanol. *J Ind Microbiol Biotechnol* 38:733–742.

Lozano-Sánchez, J., Giambanelli, E., Quirantes-Piné, R., Cerretani, L., Bendini, A., Segura-Carretero, A., and A. Fernández-Gutiérrez. 2011. Wastes generated during the storage of extra virgin olive oil as a natural source of phenolic compounds. *J Agric Food Chem* 59:11491–11500.

Luci, N., Silva, C., Betancur, G. J. V., Vasquez, M. P., Gomes, E. B., and N. Pereira. 2011. Ethanol production from residual wood chips of cellulose industry: Acid pretreatment investigation, hemicellulosic hydrolysate fermentation, and remaining solid fraction fermentation by SSF process. *Appl Biochem Biotechnol* 163:928–936.

Maier, T., Göppert, A., Kammerer, D. R., Schieber, A., and R. Carle. 2008. Optimisation of a process for enzyme assisted pigment extraction from grape (*Vitis vinifera* L.) pomace. *Eur Food Res Technol* 227:267–275.

Makris, D. P., Boskou, G., and N. K. Andrikopoulos. 2007. Polyphenolic content and *in vitro* antioxidant characteristics of wine industry and other agri-food solid waste extracts. *J Food Compost Anal* 20:125–132.

Maller, A., Damásio, A. R. L., da Silva, T. M., Jorge, J. A., Terenzi, H. F., and M. L. T. M. Polizeli. 2011. Biotechnological potential of agro-industrial wastes as a carbon source to thermostable polygalacturonase production in *Aspergillus niveus*. *Enzyme Res* 289206: 1–6.

Mann, J. I. and J. H. Cummings. 2009. Possible implications for health of the different definitions of dietary fibre. *Nutr Metab Cardiovasc Dis* 19:226–229.

Marín, F. R., Soler-Rivas, C., Benavente-García, O., Castillo, J., and J. A. Pérez-Alvarez. 2007. By-products from different citrus processes as a source of customized functional fibres. *Food Chem* 100:736–741.

Martínez-Valverde, I., Periago, M. J., Provan, G., and A. Chesson. 2002. Phenolic compounds, lycopene and antioxidant activity in commercial varieties of tomato (*Lycopersicon esculentum*). *J Sci Food Agric* 82:323–330.

Martins, E. S., Silva, D., da Silva, R., and E. Gomes. 2002. Solid state production of thermostable pectinases from thermophilic *Thermoascus aurantiacus*. *Process Biochem* 37: 949–954.

Mattea, F., Martín, A. and M. J. Cocero. 2009. Carotenoid processing with supercritical fluids. *J Food Eng* 93:255–265.

Mazutti, M., Ceni, G., Di Luccio, M., and H. Treichel. 2007. Production of inulinase by solid-state fermentation: Effect of process parameters on production and preliminary characterization of enzyme preparations. *Bioprocess Biosyst Eng* 30:297–304.

Meza, J. C., Sigoillot, J. C., Lomascolo, A., Navarro, D., and R. Auria. 2006. New process for fungal delignification of sugar-cane bagasse and simultaneous production of laccase in a vapor phase bioreactor. *J Agric Food Chem* 54:3852–3858.

Moldes, A. B., Torrado, A. M., Barral, M. T., and J. M. Domínguez. 2007. Evaluation of biosurfactant production from various agricultural residues by *Lactobacillus pentosus*. *J Agric Food Chem* 55:4481–4486.

Molina, O.E., Galvez, N.I.P., and D. A. Callieri. 1983. Bacterial protein production from sugar cane bagasse pith. *Acta Cient Venez* 34:59–64.

Murthy, P. S. and M. M. Naidu. 2012. Recovery of phenolic antioxidants and functional compounds from coffee industry by-products. *Food Bioprocess Technol* 5:897–903.

Nampoothiri, K. M. and A. Pandey. 1996. Solid state fermentation for L-glutamic acid production using *Brevibacterium* sp. *Biotechnol Lett* 18:199–204.

Navia, D. P., Velasco, R. J., and J. L. Hoyos. 2011. Production and evaluation of ethanol from coffee processing by-products. *Vitae* 18:287–294.

Ngadi, M. O. and L. R. Correia. 1992. Kinetics of solid-state ethanol fermentation from apple pomace. *J Food Eng* 17:97–116.

Nigam, P. 1990. Investigation of some factors important for solid state fermentation of sugar cane bagasse for animal feed production. *Enzyme Microb Technol* 12:808–811.

Oberoi, H. S., Babbar, N., Sandhu, S. K., Dhaliwal, S. S., Kaur, U., Chadha, B. S., and V. K. Bhargav. 2012. Ethanol production from alkali-treated rice straw via simultaneous saccharification and fermentation using newly isolated thermotolerant *Pichia kudriavzevii* HOP-1. *J Ind Microbiol Biotechnol* 39:557–566.

Oszmiański, J., Wojdylo, A., and J. Kolniak. 2009. Effect of enzymatic mash treatment and storage on phenolic composition, antioxidant activity, and turbidity of cloudy apple juices. *J Agric Food Chem* 57:7078–7085.

Oszmiański, J., Wojdylo, A., and J. Kolniak. 2011. Effect of pectinase treatment on extraction of antioxidant phenols from pomace, for the production of puree-enriched cloudy apple juices. *Food Chem* 127:623–631.

Pabuçcuoglu, A., Kivçak, B., Bas, M., and T. Mert. 2003. Antioxidant activity of *Arbutus unedo* leaves. *Fitoterapia* 74:597–599.

Paepatung, N., Nopharatana, A., and W. Songkasiri. 2009. Bio-methane potential of biological solid materials and agricultural wastes. *As J Energy Environ* 10:19–27.

Pal, M., Calvo, A. M., Terron, M. C., and A. E. González. 1995. Solid state fermentation of sugar cane bagasse with *Flammulina velutipes* and *Trametes versicolor*. *World J Microbiol Biotechnol* 11:541–545.

Pallauf, K., Rivas-Gonzalo, J. C., del Castillo, M. D., Cano, M. P., and S. de Pascual-Teresa. 2008. Characterization of the antioxidant composition of strawberry tree (*Arbutus unedo* L.) fruits. *J Food Compos Anal* 21:273–281.

Pandey, A. and C. R. Soccol. 1998. Bioconversion of biomass: A case study of ligno-cellulosics bioconversions in solid state fermentation. *Braz Arch Biol Technol* 41:379–390.

Pandey, A., Soccol, C. R., Nigam, P., and V. T. Soccol. 2000. Biotechnological potential of agro-industrial residues. I: Sugarcane bagasse. *Bioresour Technol* 74:69–80.

Pearson, N., Ball, K., and D. Crawford. 2012. Parental influences on adolescent fruit consumption: The role of adolescent self-efficacy. *Health Educ Res* 27:14–23.

Pérez, R. and N. P. Guerra. 2009. Optimization of amylase production by *Aspergillus oryzae* in submerged and solid-state fermentation using sugarcane bagasse as solid support material. *World J Microbiol Biotechnol* 25:1929–1939.

Peschel, W., Sánchez-Rabaneda, F., Diekmann, W., Plescher, A., Gartzía, I., Jiménez, D., Rosa Lamuela-Raventós, R., Buxaderas, S., and C. Codina. 2006. An industrial approach in the search of natural antioxidants from vegetable and fruit wastes. *Food Chem* 97:137–150.

Pinelo, M., Rubilar, M., Sineiro, J., and M. J. Núñez. 2005. A thermal treatment to increase the antioxidant capacity of natural phenols: Catechin, resveratrol and grape extract cases. *Eur Food Res Technol* 221:284–290.

Pinelo, M., Ruiz-Rodríguez, A., Sineiro, J., Señoráns, F. J., Reglero, G., and M. J. Núñez. 2007. Supercritical fluid and solid–liquid extraction of phenolic antioxidants from grape pomace: A comparative study. *Eur Food Res Technol* 226:199–205.

Pingret, D., Fabiano-Tixier, A. S., Le Bourvellec, C., Renard, C. M. G. C., and F. Chemat. 2012. Lab and pilot-scale ultrasound-assisted water extraction of polyphenols from apple pomace. *J Food Eng* 111:73–81.

Ponce, N. T. and M. De la Torre. 1992. Interactions in a mixed culture composed of *Cellulomonas flavigena* and *Xanthomonas* sp. growing in continuous culture on sugarcane bagasse. *Appl Microbiol Biotechnol* 40:531–534.

Prado, F. C., Vandenberghe, L. P. S., Woiciechowski, A. L., Rodrígues-León, J. A., and C. R. Soccol. 2005. Citric acid production by solid-state fermentation on a semi-pilot scale using different percentages of treated cassava bagasse. *Braz J Chem Eng* 22:547–555.

Prata, E. R. B. A. and L. S. Oliveira. 2007. Fresh coffee husks as potential sources of anthocyanins. *LWT-Food Sci Technol* 40:1555–1560.

Puértolas, E., López, N., Condón, S., Álvarez, I., and J. Raso. 2010. Potential applications of PEF to improve red wine quality. *Trends Food Sci Technol* 21:247–255.

Pujari, V. and T. S. Chandra. 2000a. Statistical optimization of medium components for improved synthesis of riboflavin by *Eremothecium ashbyii*. *Bioprocess Biosyst Eng* 23:303–307.

Pujari, V. and T. S. Chandra. 2000b. Statistical optimization of medium components for enhanced riboflavin production by a UV-mutant of *Eremothecium ashbyii*. *Process Biochem* 36:31–37.

Ramalakshmi, K., Rao, L. J. M., Takano-Ishikawa, Y., and M. Goto. 2009. Bioactives of low grade green coffee and spent coffee in different *in vitro* model systems. *Food Chem* 115:79–85.

Ramírez-Martínez, J. R. 1988. Phenolic compounds in coffee pulp: Quantitative determination by HPLC. *J Sci Food Agric* 43:135–144.

Rani, D. S. and K. Nand. 2004. Ensilage of pineapple processing waste for methane generation. *Waste Manage* 24:523–528.

Rice-Evans, C., Miller, N. J., and G. Paganga. 1997. Antioxidant properties of phenolic compounds. *Trends Plant Sci* 2:152–159.

Riccioni, G., Sblendorio, V., Gemello, E., de Bello, B., Scotti, L., Cusenza, S., and N. D'Orazio. 2012. Dietary fibers and cardiometabolic diseases. *Int J Mol Sci* 13:1524–1540.

Rodrigues, L. R., Teixeira, J.A., and R. Oliveira. 2006. Low-cost fermentative medium for biosurfactant production by probiotic bacteria. *Biochem Eng J* 32:135–142.

Rodríguez, H., and R. Gallardo. 1993. Single cell protein from bagasse pith by a mixed bacterial culture. *Acta Biotechnol* 13:141–149.

Rodríguez, N., Torrado, A., Cortés, S., and J. M. Domínguez. 2010. Use of waste materials for *Lactococcus lactis* development. *J Sci Food Agric* 90:1726–1734.

Rodríguez-Vazquez, R. and D. Diaz-Cervantes. 1994. Effect of chemical solutions sprayed on sugarcane bagasse pith to produce single cell protein: Physical and chemical analysis of pith. *Bioresour Technol* 47:159–164.

Rodríguez-Vazquez, R., Villanueva-Ventura, G., and E. Rios-Leal. 1992. Sugarcane bagasse pith dry pretreatment for single cell protein production. *Bioresour Technol* 39:17–22.

Rosales-Soto, M. U., Brown, K., and C. F. Ross. 2012. Antioxidant activity and consumer acceptance of grape seed flour-containing food products. *Int J Food Sci Technol* 47:592–602.

Rose, D. J., Inglett, G. E., and S. X. Liu. 2010. Utilisation of corn (*Zea mays*) bran and corn fiber in the production of food components. *J Sci Food Agric* 90:915–924.

Samah, O. A., Sias, S., Hua, Y. G., and N. N. Hussin. 2011. Production of ethanol from cocoa pod hydrolysate. *ITB J Sci* 43:87–94.

Sandhu, S. K., Oberoi, H. S., Dhaliwal, S. S., Babbar, N., Kaur, U., Nanda, D., and D. Kumar. 2012. Ethanol production from Kinnow mandarin (*Citrus reticulata*) peels via simultaneous saccharification and fermentation using crude enzyme produced by *Aspergillus oryzae* and the thermotolerant *Pichia kudriavzevii* strain. *Ann Microbiol* 62:655–666.

Schieber, A., Hilt, P., Streker, P., Endress, H.U., Rentschler, C., and R. Carle. 2003. A new process for the combined recovery of pectin and phenolic compounds from apple pomace. *Innov Food Sci Emerg Technol* 4:99–107.

Scordino, M., Di Mauro, A., Passerini, A., and E. Maccarone. 2004. Adsorption of flavonoids on resins: Cyanidin 3-glucoside. *J Agric Food Chem* 52:1965–1972.

Sharma, N., Kalra, K. L., Oberoi, H. S., and S. Bansal. 2007. Optimization of fermentation parameters for production of ethanol from Kinnow waste and banana peels by simultaneous saccharification and fermentation. *Indian J Microbiol* 47:310–316.

Sharma, S. K., and M. Le Maguer. 1996. Lycopene in tomatoes and tomato pulp fractions. *Ital J Food Sci* 2:107–113.

Shojaosadati, S. A. and V. Babaeipour. 2002. Citric acid production from apple pomace in multi-layer packed bed solid-state bioreactor. *Process Biochem* 37:909–914.

Soccol, C. R., Marin, B., Raimbault, M., and J. M. Lebeault. 1994. Potential of solid state fermentation for production of L(+)-lactic acid by *Rhizopus oyzae*. *Appl Microbiol Biotechnol* 41:286–290.

Soleas, G. J., Grass, L., Josephy, P. D., Goldberg, D. M., and E. P. Diamandis. 2002. A comparison of the anticarcinogenic properties of four red wine polyphenols. *Clin Biochem* 35:119–124.

Soufleros, E. H., Mygdalia, A. S., and P. Natskoulis. 2004. Characterization and safety evaluation of the traditional Greek fruit distillate "Mouro" by flavor compounds and mineral analysis. *Food Chem* 86:625–636.

Srivastava, P. and R. Malviya. 2011. Sources of pectin, extraction and its applications in pharmaceutical industry—An overview. *Indian J Nat Prod Resour* 2:10–18.

Stoner, G. D. and M. A. Morse. 1997. Isocyanates and plant polyphenols as inhibitors of lung and esophageal cancer. *Cancer Lett* 114:113–119.

Strati, I. F. and V. Oreopoulou. 2011a. Effect of extraction parameters on the carotenoid recovery from tomato waste. *Int J Food Sci Technol* 46:23–29.

Strati, I. F. and V. Oreopoulou. 2011b. Process optimisation for recovery of carotenoids from tomato waste. *Food Chem* 129:747–752.

Suárez, B., Álvarez, A. L., Diñeiro-García, Y., del Barrio, G., Picinelli-Lobo, A., and F. Parra. 2010. Phenolic profiles, antioxidant activity and *in vitro* antiviral properties of apple pomace. *Food Chem* 120:339–342.

Sun, H., Ge, X., Hao, Z., and M. Peng. 2010. Cellulase production by *Trichoderma* sp. on apple pomace under solid state fermentation. *Afr J Biotechnol* 9:163–166.

Tomasini, A., Fajardo, C., and J. Barrios-González. 1997. Gibberellic acid production using different solid-state fermentation systems. *World J Microbiol Biotechnol* 13:203–206.

Toor, R. K. and G. P. Savage. 2005. Antioxidant activity in different fractions of tomatoes. *Food Res Int* 38:487–494.

Torres, J. L., Varela, B., García, M. T., Carilla, J., Matito, C., Centelles, J. J., Cascante, M., Sort, X., and R. Bobet. 2002. Valorization of grape (*Vitis vinifera*) byproducts. Antioxidant and biological properties of polyphenolic fractions differing in procyanidin composition and flavonol content. *J Agric Food Chem* 2002:7548–7555.

Tortoe, C., Orchard, J., and A. Beezer. 2007. Prevention of enzymatic browning of apple cylinders using different solutions. *Int J Food Sci Technol* 42:1475–1481.

Tow, W. W., Premier, R., Jing, H., and S. Ajlouni. 2011. Antioxidant and antiproliferation effects of extractable and nonextractable polyphenols isolated from apple waste using different extraction methods. *J Food Sci* 76:163–172.

Tran, C. T., Sly, L. I., and D. A. Mitchell. 1998. Selection of a strain of *Aspergillus* for the production of citric acid from pineapple waste in solid-state fermentation. *World J Microbiol Biotechnol* 14:399–404.

Trejo, M. R., Lonsane, B. K., Raimbault, M., and S. Roussos. 1993. Spectra of ergot alkaloids produced by *Claviceps purpurea* 1029c in solid-state fermentation system: Influence of the composition of liquid medium used for impregnating sugar-cane pith bagasse. *Process Biochem* 28:23–27.

Trejo, M. R., Oriol, E., López-Canales, A., Roussos, S., Viniegra, G., and M. Raimbault. 1991. Pectic enzymes production by *Aspergillus niger* in solid state fermentation. *Micol Neotrop Apl* 4:49–62.

Tuomela, M., Vikman, M., Hatakka, A., and M. Itävaara. 2000. Biodegradation of lignin in a compost environment: A review. *Bioresour Technol* 72:169–183.

Varming, C., Andersen, M. L., and L. Poll. 2004. Influence of thermal treatment on black currant (*Ribes nigrum* L.) juice aroma. *J Agric Food Chem* 52:7628–7636.

Vattem, D.A. and K. Shetty. 2003. Enrichment of phenolic antioxidants by solid-state bioprocessing. *Agro Food Ind Hi-Tech* 14:54–57.

Villas-Bôas, S. G., Espósito, E., and M. M. de Mendonça. 2003. Bioconversion of apple pomace into a nutritionally enriched substrate by *Candida utilis* and *Pleurotus ostreatus*. *World J Microbiol Biotechnol* 19:461–467.

Virot, M., Tomao, V., Le Bourvellec, C., Renard, C. M. C. G., and F. Chemat. 2010. Towards the industrial production of antioxidants from food processing by-products with ultrasound-assisted extraction. *Ultrason Sonochem* 17:1066–1074.

Weisburger, J. H. 1998. Evaluation of the evidence on the role of tomato products in disease prevention. *Proc Soc Exp Biol Med* 218:140–143.

Wijngaard, H. H. and N. Brunton. 2009. The optimization of extraction of antioxidants from apple pomace by pressurized liquids. *J Agric Food Chem* 57:10625–10631.

Wijngaard, H. H., Rossle, C., and N. Brunton. 2009. A survey of Irish fruit and vegetable waste and by-products as a source of polyphenolic antioxidants. *Food Chem* 116:202–207.

Will, F., Bauckhage, K., and H. Dietrich. 2000. Apple pomace liquefaction with pectinases and cellulases: Analytical data of the corresponding juices. *Eur Food Res Technol* 211:291–297.

Wolfe, K., Wu, X., and R. H. Liu. 2003. Antioxidant activity of apple peels. *J Agric Food Chem* 51:609–614.

Zbidi, H., Salido, S., Altarejos, J., Pérez-Bonilla, M., Bartegi, A., and J. A. Rosado. 2009. Olive tree wood phenolic compounds with human platelet antiaggregant properties. *Blood Cell Mol Dis* 42:279–285.

Zhang, H. Y. 1999. Theoretical methods used in elucidating activity differences of phenolic antioxidants. *J Am Oil Chem Soc* 76:745–748.

Zheng, Z. X. and K. Shetty. 2000. Solid state production of polygalacturonase by *Lentinus edodes* using fruit processing wastes. *Process Biochem* 35:825–830.

Zhong, W. and P. Cen. 2005. Pectinase production by *Aspergillus niger* P-6021 on citrus *Changshan-huyou* peel in slurry-state fermentation. *Chinese J Chem Eng* 13:510–515.

Zhong-Tao, S., Lin-Mao, T., Cheng, L., and D. Jin-Hua. 2009. Bioconversion of apple pomace into a multienzyme bio-feed by two mixed strains of *Aspergillus niger* in solid state fermentation. *Electron J Biotechnol* 12:1–13.

Zykwinska, A., Boiffard, M. H., Kontkanen, H., Buchert, J., Thibault, J. F., and E. Bonnin. 2008. Extraction of green labeled pectins and pectic oligosaccharides from plant byproducts. *J Agric Food Chem* 56:8926–8935.

14 Valorization and Integral Use of Seafood By-Products

M.E. López-Caballero, B. Giménez,
M.C. Gómez-Guillén, and P. Montero

CONTENTS

14.1 INTRODUCTION

Fish products are a source of nutrients of high biological value. However, waste from their production and processing represents an important economic value and an environmental problem difficult to manage. Biotechnological processes are currently applied for example to improve the production of active compounds. However, sometimes the biotechnology applied to these products is lagging behind compared to other commodities, because scientific research and efforts have been much focused to preserve and find out some uses of by-products, in an attempt to valorize them while minimizing economic losses. In this sense, this chapter reviews from a general perspective the integral utilization of fishery products.

14.2 MUSCLE PROTEIN

Food and Agriculture Organization of the United Nations (FAO) have reported that seafood consumption has undergone major changes over the past four decades, considering that globally the increase has been continuous. Values of seafood consumption have not been indicated in this chapter, since the consumption is highly variable depending on diverse factors ranging from economic conditions to consumption habits. The estimated seafood consumption per capita ranges from less than 1 kg in a country to more than 100 kg in another. Differences are also evident within countries, as consumption is usually significantly higher in coastal areas.

Currently, capture fisheries still provide the majority of the total food fish supply, but there is a significant growth of aquaculture (FAO, 2010). Marketable species are very diverse depending on the kind of species (pelagic or demersal, cold water or warm water, lean or fat), their size, seasonal variations, and so forth. This whole set of extrinsic and intrinsic characteristics determine the properties of the raw material and its suitability to be processed, and therefore the generation of by-products to a greater or lesser extent.

Despite the great variety of existing seafood species, only some of them are consumed worldwide. Some of the marine fish species most consumed are hake, cod, haddock, anchovies, herring, whiting, tuna, bonito, mackerel, and swordfish. Squid, cuttlefish, shrimps, and prawns are outstanding among shellfish species. Regarding fresh water species, carp, catfish, and other cyprinids, tilapia, and freshwater crustaceans can be considered species of significant consumption.

The optimal way in which seafood should be marketed is very different depending on the species. Fish is a poikilothermic animal and therefore it is highly perishable in most cases. Once caught, evisceration after fish death is a highly desirable step, but this practice may not always be profitable depending on the fish size and other factors; obviously it is not the same to eviscerate a tuna or a herring. The marketing method should also be taken into account: whole, filleted, ready to eat, and so on. In each of these cases, the waste products derived from processing is different and this will be discussed in this chapter. Some similarity occurs with certain shellfish species, such as cephalopods and crustaceans, which are usually sold whole, but they are also marketed peeled. Viscera, skin, and shells are the main residues and make up about 30–40% of the total weight. They are mostly accumulated in the processing industry as waste.

14.2.1 MINCE, PROTEIN CONCENTRATED AND RESTRUCTURED PRODUCTS

Muscle proteins are the most valuable part of both fish and shellfish and they are usually consumed directly as fillets, slices, and so on. However, some muscle tissue is thrown together with thorns and bones after processing, manually or by machine, and this constitutes the so-called trimmings. Trimmings may involve approximately 20% of muscle protein depending on the species and the processing, for example, hake or sole. This protein may be collected by a mechanical deboning machine, in which the flesh is pressed through perforations (1–3 mm diameter) in a rotating drum, and later restructured. Restructured seafood products are products made from minced and/or chopped seafood muscle, with or without other ingredients.

Other sources of myofibrillar protein may be undervalued/underutilized species. These species are not usually captured or, when captured as a by-catch, they are usually thrown into the sea. The reasons for lack of commercial value can be very different. Sometimes it may occur that the volume of catch is not large enough to be marketed whole or in fillets. At other times fish may contain many thorns and direct consumption is not suitable for certain populations such as children, sick people, and so on. Fish also may have an unpleasant odor or color. In all these cases, the mince will have to be obtained if muscle protein is intended to be used as material for human consumption. This requires passing the bones after filleting or the whole degutted fish through a deboning machine, to get the mince separately from the rest of the compounds such as bones and skin. When there is no need to remove undesirable compounds, the mince thus obtained can be considered a finished product. Cryoprotectants and antioxidants are usually added to minced fish; the latter are needed if the mince comes from a fatty or semi-fatty species. The mince yield is very high and there is only a partial removal of lipids, therefore it retains all the fish muscle's nutritional and healthy properties (Gómez-Guillén et al., 1996). The main destination of this mince is the production of fish fingers, nugget, burgers, fillet, and so on. However, sometimes, as in the case of blue whiting, fish muscle texture is excessively soft and the mince does not have enough consistency to give shape to the restructured product. In this case, the mince is reformulated with other ingredients such as salts, fibers, proteins, and starches (Pérez-Mateos and Montero, 2000).

In some cases it has been possible to relate the poor texture shown by a fish with a specific problem, such as Pacific whiting that shows a high endogenous autolytic activity (Yongsawatdigul et al., 1997). In this case the removal of the enzymes by washing is of great importance as described below.

Regarding cephalopods, there are a large number of species distributed throughout the world. They have a high metabolic rate and therefore a high proteolytic activity, a high content of muscle protein and a low fat content, are white colored, and a rubbery texture that is highly appreciated in some societies. The texture of the mantle may be variable, hard as in the case of *Illex argentinus* or soft as in *Dosidicus gigas*. Depending on the size (may be a few millimeters to one meter) cephalopods may be sold whole round, whole cleaned, tube-shaped, rings, and the like. Some cephalopods like these species are considered as by-products because of the texture of the mantle. Cephalopods have a poor gelling capacity due to their high proteolytic activity (Gómez-Guillén et al., 1996, 1997; Ayensa et al., 1999, 2002). As a consequence, the preparation of restructured products from cephalopods involves obtaining of a protein concentrate with gelling capacity.

Strategies for restructured product development are mainly two: (a) Formulation with ingredients that enhance the gelling properties and confer the adequate texture and binding properties (Gómez-Guillén et al., 1997, 2002a). This method will be especially suitable if it is not necessary to remove undesirable compounds. Furthermore, this strategy gives a high yield; (b) washing and preparing surimi, to remove proteases and improve their ability to gel. In this case the yield is lower but the gelling capacity is improved (Careche et al., 2004). This method is especially suitable in some cases in which there are other problems associated. For instance, it is the case of large individuals of giant squid (*Dosidicus gigas*), where high physiological concentrations of nonprotein nitrogen compounds are accumulated such as ammonia (Márquez-Ríos et al., 2007), together with the presence of a bitter taste associated with peptides and free amino acids (Sánchez-Brambilla et al., 2004) that have to be removed.

14.2.1.1 Washed Mince

The incorporation of one or more washing steps to remove diverse undesirable compounds that could have some species or trimmings is a common stage in the processing before restructuring. The yield drops considerably with the washing process. One wash step may be enough to eliminate strong fishy tastes or smells. It should be taken into account that the more extensive this process, the lower the yield and the higher the by-product or waste generation. Nutrients beneficial to health, for instance fish lipids, are also lost during washing.

However, a repeated number of consecutive washes is made more often, for which the traditional Japanese term "*surimi*" is used. The number of washings is usually three. The pH values are adjusted to near neutrality in all the washings, and some salts may be added in the last washing to favor water elimination. In this process the objective is to obtain myofibrillar protein as pure as possible and separated from other components such as sarcoplasmic proteins, lipids, and so on, as well as from undesirable compounds such as metals, ammonium, or enzymes. After washing, a stage of refining is carried out. Finally, once pressed, the washed and refined muscle is added with cryoprotectants that confer stability to protein during the frozen storage.

Most of the surimi is produced from white fish because the yield is higher. However, surimi can also be produced from fatty fish, although the yield is lower because of the abundance of blood, fat, and pigments. The more washing steps and further refining, the lower the yield.

At the end of the 1990s, other surimi processing methods appeared, based on the solubilization of fish proteins in acidic or alkaline conditions to improve both the processing yield and the protein stability (Hultin and Kelleher, 1999, 2000). The lower the pH, the more the myosin that was denatured. During acidification, the elastic modulus of acid-treated myosin was found to decrease with HCl to a higher extent than with other acids used. The viscous modulus of myosin treated at pH 1.5 was significantly higher than that of myosin treated at pH 2.5 (Raghavan and Kristinsson, 2007). The unfolding/refolding treatment increased the elastic modulus of thermally treated myosin, especially when pH 11 was used. Furthermore, KOH induced greater denaturation and higher gelling ability (higher values of the elastic modulus, G') than NaOH. Thus, the myosin molecule is affected by the type of acid/alkali used and the pH attained and the effect is dependent on the species (Raghavan and Kristinsson, 2008a). As an advantage, the sarcoplasmic protein recovery in the acid/alkali process is higher than in the traditional method, and therefore the wash water is lower in solids. All contaminating materials with different density from that of proteins can be removed by gravity, for example, through centrifugation. Furthermore, the lipids can also be efficiently removed.

Protein functionality obtained by acid/alkali process has also been studied. The acid and alkaline treatment improves the functional properties such as emulsification and gelation of myosin with respect to native myosin. However, there is no improvement in the gel strength of myofibrillar proteins obtained by acid or alkaline process, although there is improvement in their emulsifying properties (Kristinsson and Hultin, 2003). This may be attributed to a partial unfolding of myosin, which may improve the functional properties of cod myosin and myofibrillar proteins.

This process may be used in other seafood muscle processing, adapting the process to the specific needs of each species. One of the main technological problems in the case of cephalopods is the high prevalence of proteases which interferes with gelation and impairs gelification (Gómez-Guillén et al., 1997; Ayensa et al., 1999, 2002). In some species, as *Dosidicus gigas* and especially in large individuals, there are additional problems described above as the high quantities of ammonia accumulated in the muscle (Márquez-Ríos et al., 2007), as well as a bitter taste associated with peptides and free amino acids (Sánchez-Brambilla et al., 2004) have to be removed. This is the typical case where the washing process is needed to obtain a protein concentrate, eliminating unpleasant compounds contained in the muscle, which on the other hand are very low molecular weight and easily removable. A method for producing a functional protein concentrate from cephalopod muscle is based on the solubilization of the mantle at very low ionic strength and neutral pH (0.16 M NaCl and 0.1% NaHCO$_3$) with further precipitation (pH 4.7–4.9) (Careche et al., 2004). The resulting gels had a pleasant taste and odor unlike the raw material. The addition of 250 ppm of EDTA inhibits the proteolysis and enhances the gel capacity.

14.2.1.2 Washing Waters

In all cases, the loss of sarcoplasmic proteins during washing involves a considerable loss of yield which is not always necessary. In this sense, some strategies have been developed to recover protein (sarcoplasmic and myofibrillar) from the wash water. Niki et al. (1985) adjusted the pH to 10 of the first wash water to recover the protein by centrifugation. Afterward, the pH value was adjusted to 5, and the water was heated to 80°C for protein coagulation. The yield of protein recovery was about 20% of the surimi products. Subsequently, Huang et al. (1997), instead of using ionic strength to solubilize and precipitate proteins, directly applied ohmic heating to coagulate proteins from the water used during the washing process of Pacific whiting mince. A 33% of the protein was recovered by this method.

Besides concentrating the protein, increasing productivity and reducing the amount of organic matter discharged into the environment, it is interesting to maintain the functional properties of proteins. Therefore, other strategies used consisted of the application of microfiltration followed by ultrafiltration to concentrate soluble proteins in the wash water of cephalothorax, krill muscle, and minced fish (Huidobro et al., 1998; Montero and Gómez-Guillén, 1998; Stine et al., 2012). Organic substances and microorganisms are removed simultaneously. Most of the muscle proteins thus extracted were less than 67 kDa, while there was a large amount of 100–150 kDa proteins from the cephalothorax. Ultrafiltration maintains the functional properties of proteins, which could be added back in the restructured fish products or used as food ingredients.

In addition to the recovery of protein material, pigments are of great value in the washing waters from crustacean. Astaxanthin is a ketocarotenoid oxidized from β-carotene that represents between 74% and 98% of the total pigments in crustaceans shell (Shahidi et al., 1998). Ferraro et al. (2010) review the properties of this pink pigment: natural origin, no toxicity, high versatility and hydro- and liposoluble, it is a precursor of vitamin A and antioxidant, making it ideal for medical applications and food technology. Part of the compounds recovered from the washing water can be used as a basis for soup concentrates, prepared meals, and so on.

14.2.1.3 Restructured Products

Obviously not all muscles have the same texture or the same sensory acceptance; some of them are too soft or too hard. In these cases, once the minced fish and/ or surimi has been obtained, the stage of formulating the restructured products is important, since different ingredients are added to muscle protein to define texture and other desirable properties in this stage. It is necessary to assess whether there are problems with odors, flavors, or contaminants which could devalue the raw material and therefore, proceed to their elimination by washing as described above, at the expense of decreased yield and a higher cost; or on the contrary, it is only a matter of inadequate or poor texture and binding properties which can be solved with the addition of other compounds that also helps to increase the yield.

The traditional restructured products have been barely modified. It is the case of fish fingers or other shapes of children food, often covered with breadcrumbs. However, after the marketing of surimi in the United States and other Western countries, the gelation of muscle has been commonly used as a way to develop new products. These were initially analogous products to those addressed in the respective countries, and afterwards to any kind of product that imagination may offer. During this process several ingredients or additives can be added. These ingredients can be categorized as functional technological or functional nutraceuticals, and often are included in both categories.

In this sense there are many possibilities of ingredients. The most frequently used are nonmuscle proteins both from animals and plants (egg white, bovine albumin, whey, casein, isolated soy protein, gluten, etc.), starches, fibers (alginate, carrageenan, carboxymethylcellulose, gums, agar and other insoluble fibers), salts, enzymes. Among enzymes, TGase is widely used for its ability to establish covalent bonds at low temperature (Montero and Pérez-Mateos, 2001; Montero et al., 2005, 2010; Park, 2005).

The choice of one or another ingredient or combination of a selection of them, will depend on the raw material (species, seasonality, product, and processing previously performed) and the desirable characteristics in the final product.

Fish "sawdust" is powdered muscle produced by sawing blocks of frozen hake. It is an underutilized material with low odor, white color, and very poor functional properties. Therefore, it is difficult to use in reconstituted products without the addition of ingredients to enhance these properties. To enhance the gel-forming capacity of sawdust from hake muscle (*Merluccius australis*), different formulations were studied. It was observed that it was possible to obtain acceptable and even quite good gelling properties with some of the formulations tested (Borderías et al., 1996).

The effect of the addition of nonmuscle proteins in minced sardine gels is highly dependent on the mince quality. In general, addition of ingredients improves the properties of the restructured product when the mince used has a poor gelling ability. It has been observed that in high functional quality sardine mince, any protein addition interfered with gelation of muscle proteins. In low-quality mince, however, egg white or soy protein considerably improved the gel strength. Casein reduced all the rheological parameters studied. Regarding salt concentration, high-quality mince did not gel with low salt (1.5%), whereas low-quality mince gelled with both 1.5% and 2.5% salt concentrations. Gels made with 1.5% NaCl exhibited lower gel strength, elasticity, and cohesiveness than gels with the higher salt concentration (Gómez-Guillén et al., 1996).

From a technological point of view, the introduction of fibers improves water binding, thickening, emulsion capacity, and gelling properties of products made with minced muscle, especially where the raw material used is of poor functional quality. In gels made from surimi, the addition of dietary fibers does not always improve gel properties, especially in high-quality surimi. In addition the antioxidant capacity of some dietary fibers, attributed to their chelant action on metal ions, is particularly convenient in the fatty fish restructured products. From a physiological point of view, the addition of dietary fiber to a functional product like fish would further complement its healthy characteristics, could add beneficial effects such as reducing cholesterolemia, modifying the glucemic response, reducing nutrient availability, prebiotic capacity, and so on (Borderías et al., 2005).

Another example of the dual role that may exert some ingredients may be the addition of certain extracts of plants that act as antioxidants during processing and preservation of products, or remain as potential functional antioxidants. The antioxidant activity of quercetin and rosemary extracts were studied in minced Atlantic mackerel after homogenization to form a batter and gelation induced by both heat (conventional and microwave) and high pressure. It was deduced from FRAP (reducing capacity) and DPPH (antiradical scavenging capacity) results that both extracts showed antioxidant capacity after processing. The rosemary extract was more effective at protecting lipids from oxidation; whereas protein oxidation was prevented by both antioxidants, although quercetin was the most efficient antioxidant in those batches subjected to thermal treatment for gel formation (conventional and microwave) (Montero et al., 2005).

Impurities, odors, pigments, proteins, and lipids are removed during washing. However, the presence of ω-3 polyunsaturated fatty acids (PUFAs) in food (intrinsic or through fortification with these oils) is increasingly demanded. For this reason, there are already some works that incorporate polyunsaturated oils in formulations of surimi gel, trying not to devalue their properties. These authors evaluated the gelling ability of surimi from Alaska pollack fortified with various concentrations of ω-3 PUFA. This study showed that the nutritional value and gelation of surimi seafood can be enhanced without altering texture properties by addition of ω-3 PUFAs-rich oils (Pietrowski et al., 2012).

14.3 COLLAGEN/GELATIN

14.3.1 GELATIN EXTRACTION

The waste from fish processing after filleting can account for as much as 75% of the total catch weight. About 30% of such waste consists of skins and bones with high collagen content. The collagen crosslinking degree is highly variable and depends on the collagen type, tissue, animal species, age, and so forth. Since gelatin is derived from denatured collagen, its properties and gelling abilities (involving a partial renaturation of denatured collagen molecules) are going to depend on all these parameters (Ledward, 1986).

Interstitial collagen molecules are composed of three α-chains intertwined in the so-called collagen triple helix, mainly stabilized by intra- and inter-chain hydrogen bonding. Its particular structure is the result of an almost continuous repeating of the Gly–X–Y– sequence, where X is mostly proline and Y is mostly hydroxyproline. The very short N- and C-terminal regions, called telopeptides, do not form triple helical structures; they are largely made up of lysine and hydroxylysine (Hyl) residues, and their aldehyde derivatives, involving intra- and inter-molecular covalent cross-links (Asghar and Henrickson, 1982). The basic unit of collagen fibrils is constituted by four to eight collagen molecules in cross-section, stabilized and reinforced by covalent bonds. Thus, the typical strong, rigid nature of skins, tendons, and bones is due to the basic structure formed by many of these cross-linked collagen fibrils.

To convert insoluble native collagen into gelatin a pretreatment is required to break down the noncovalent bonds and disorganize the protein structure, allowing swelling and cleavage of intra- and intermolecular bonds to solubilize the collagen (Stainsby, 1987). Subsequent heat treatment cleaves the hydrogen and covalent bonds to destabilize the triple helix, resulting in helix-to-coil transition and conversion into soluble gelatin (Djabourov et al., 1993). The degree of conversion of collagen into gelatin is related to the severity of both the pretreatment and the extraction process, depending on the pH, temperature, and extraction time (Johnston-Banks, 1990). Two types of gelatin are obtainable, depending on the pretreatment. These are known commercially as type-A gelatin, obtained in acid pretreatment conditions, and type-B gelatin, obtained in alkaline pretreatment conditions.

A mild acid pretreatment is normally used for type-A gelatin extraction from fish skins, since their collagenous material is characterized by a low degree of intra and interchain covalent cross-linking (Montero et al., 1990; Norland, 1990). Extraction processes have been modified to improve rheological properties or extraction yields, for instance by using different organic acids for pretreatment of the skins (Gómez-Guillén and Montero, 2001; Giménez et al., 2005a; Songchotikunpan et al., 2008), different salts for washing the skins (Giménez et al., 2005b), high-pressure treatment (Gómez-Guillén et al., 2005; Kim et al., 2010) or pepsin-aided digestion (Nalinanon et al., 2008; Giménez et al., 2009a). In addition, collagen from fish skins is more susceptible to deteriorate when compared to mammal sources. After degutting and filleting of fish, skins are often kept together with the rest of the by-products, being subject to rapid enzymatic and microbial damage. This problem has led to the need of studying methods for the stabilization of these skins. Freezing flounder skins was reported to decrease gel strength and reduce the subsequent renaturation ability of the corresponding gelatin (Fernández-Díaz et al., 2003). In contrast, drying Dover sole skins with glycerol, ethanol, or dry salt, slightly lowered the viscoelastic properties and the gelling and melting points, without any appreciable changes over the course of storage at room temperature for 160 days (Giménez et al., 2005c). Similarly, Bower et al. (2010) found that the dehydration process using different dessicants does not harm the functional properties of gelatin extracted from Alaska pollack skins.

Rheological properties of gelatins are largely dependent on the molecular properties, especially with respect to two main factors: (i) the molecular weight distribution, which results mainly from processing conditions and (ii) the amino acid composition, which is species specific (Gómez-Guillén et al., 2002b, 2011). Marine gelatins have long been known to have worse rheological properties than mammalian gelatins, particularly in the case of gelatins from cold-water fish species, such as cod, salmon, or Alaska pollack (Leuenberger, 1991; Gudmundsson and Hafsteinsson, 1997; Haug et al., 2004; Zhou et al., 2006). This has mainly been attributed to the lower number of Pro + Hyp-rich collagen regions that are most likely involved in the formation of nucleation zones conducive to the formation of triple helical structures (Ledward, 1986). Nevertheless, recent studies have indicated that warm-water fish gelatins (from yellowfin tuna, tilapia, grass carp, or catfish), while not superior to mammalian gelatins, might afford similar thermostability and rheological properties, depending on the species and processing conditions (Choi and Regenstein, 2000; Jamilah and Harvinder, 2002; Cho et al., 2005; Avena-Bustillos et al., 2006; Zhou et al., 2006; Kasankala et al., 2007; Yang et al., 2007; Ninan et al., 2010).

14.3.2 SOURCES

The huge number of species having very different intrinsic characteristics, increased the interest of the scientific community in optimizing the extracting conditions as well as characterizing the yields, and physico-chemical and functional properties of the resulting gelatins, obtained mainly from skin and bone residues (Gómez-Guillén et al., 2009; Karim and Bhat, 2009). For example, in the last few years, gelatins with reasonably good gelling properties have been extracted from the skins and bones of bigeye snapper (Benjakul et al., 2009; Binsi et al., 2009), cuttlefish (Aewsiri et al., 2009), greater lizardfish (Taheri et al., 2009), grouper (Rahman and Al-Mahruoqi, 2009), hoki (Mohtar et al., 2010) or giant catfish (Jongjareonrak et al., 2010). The skins of processed or semi-processed fish products, such as salted and marinated herring or cold-smoked salmon, have also been proposed as an usable source of gelatin (Kolodziejska et al., 2008).

Scales may account for around 5% of the material contained in fish collagenous waste, but unlike skins, they are very rich in Ca phosphate compounds; therefore, removal of Ca with acids or chelators should be a critical pretreatment in order to obtain the final yield, purity, and gel strength of the gelatin (Wang and Regenstein, 2009; Zhang et al., 2011). Collagen or gelatin extracted from fish scales has been reported for sea bream and red tilapia (Ikoma et al., 2003), black drum and sheepshead (Ogawa et al., 2004), sardine (Nomura et al., 1996; Harada et al., 2007), grass carp (Li et al., 2008), deep-sea redfish (Wang et al., 2008), Asian silver carp (Wang and Regenstein, 2009) or lizardfish (Wangtueai and Noomhorm, 2009).

Besides filleting, other industrial processes also generate considerable amounts of collagen-rich wastes. For example, the solid waste from surimi processing, which may range from 50% to 70% of the original raw material, could also be the initial material for obtaining collagen or gelatin from under-utilized fish resources (Morrissey et al., 2005; Norziah et al., 2009). Within the same manufacturing process, crude collagen from refiner discharge of Pacific whiting surimi processing was found to present higher functional properties (emulsifying activity, cooking stability, water and oil absorption capacity) than other by-products such as skins and frames (Kim and Park, 2004, 2005). Precooked fins constitute another important waste product from canned tuna processing which has been proposed as a promising source for high-performance gelatin extraction (Aewsiri et al., 2008). As for fish scales, an exhaustive demineralization step is needed, since tuna fin showed an ash content as high as 40%.

14.3.3 FUNCTIONAL PROPERTIES

Gelatin is a protein compound which in itself may be considered a highly digestible dietary food ideal as a complement in certain types of diet. For technological purposes it can be used as an ingredient to improve the elasticity, consistency, and stability of foods; moreover, its use for encapsulation and film formation makes it of interest to the pharmaceutical and photographic industries. Apart from basic physico-chemical properties, such as composition parameters, solubility, transparency, color, odor, and taste, the main attributes that best define the overall commercial quality of gelatin are gel strength and thermal stability. For standardizing purposes, measurement of gel strength is determined using the so-called Bloom test, which consists of performing a well-defined protocol at a given gelatin concentration (6.67%), temperature (10°C), and maturation time (17 h), thus allowing gel strength to be expressed in the normalized "bloom value" (Wainewright, 1977).

Besides their basic hydration properties, such as swelling and solubility, the main food, photographic, cosmetic, and pharmaceutical applications of gelatin are based on properties that can be divided into two groups: (i) properties associated with their gelling behavior, that is, gel formation, texturizing, thickening, and water-binding capacity, and (ii) properties related to their surface behavior, which include emulsion and foam formation and stabilization, adhesion and cohesion, protective colloid function, and film-forming capacity.

Gel formation, viscosity, and texture are closely related properties determined mainly by the structure, molecular size, and temperature of the system (Djabourov et al., 1993). The most widespread single use of gelatin in food products is in water gel desserts, due to its unique melt-in-the-mouth property. Other hydrocolloids also have thermo-reversible characteristics, but they generally melt at higher temperatures. By increasing gelatin concentrations or by using gelatin mixtures (of cold and warm water fish), desserts made from fish skin gelatin were found to be more similar to desserts made from high bloom pork skin gelatin (Zhou and Regenstein, 2007). Moreover, new textures and appearance, offering ample versatility in product development, can be provided by introducing a gas phase into gelatin-gel-based products, such as fruit jellies or marshmallows (Zúñiga and Aguilera, 2009).

Gelatin or collagen chains in solution may be covalently cross-linked to form matrices capable of swelling in the presence of aqueous solutions, forming what are commonly known as gelatin hydrogels. Hydrogels, which are characterized by their hydrophilicity and insolubility in water, have the ability to swell to an equilibrium volume while preserving their shape. The super-swelling properties, degradation rate and capacity for controlled release of drugs of gelatin hybrid hydrogels produced by mixing with synthetic polymers were exhaustively reviewed by Zohuriaan-Mehr et al. (2009). To increase the gel strength, setting time, and thermostability of fish gelatins, different polysaccharides, such as k-carrageenan and/or gellan (Haug et al., 2004; Pranoto et al., 2007) or hydroxypropylmethylcellulose (Chen et al., 2009), as well as enzymatic crosslinkers, such as transglutaminase (Kolodziejska et al., 2004) have also been used.

Gelatin and soluble collagen exhibit suitable emulsifying and foaming properties based on the presence of charged groups in the protein side chains containing either hydrophilic or hydrophobic amino acids. Both hydrophobic and hydrophilic parts tend to migrate toward surfaces, hence reducing the surface tension of aqueous systems and forming the required identically charged film around the components of the dispersed phase, which can be additionally strengthened by gel formation (Schrieber and Gareis, 2007). Gelatins with good emulsifying properties have been obtained from cod bones (Kim et al., 1996), tuna fin (Aewsiri et al., 2008), bigeye snapper (Binsi et al., 2009) or gray triggerfish skins (Jellouli et al., 2011). Similarly, the foam capacity of gelatin from farmed giant catfish was found to be higher than that from calf skin, the difference possibly being due to the higher content of hydrophobic amino acid residues and almost 4 times greater viscosity in the former (Jongjareonrak et al., 2010).

Gelatin has been extensively studied on account of its film-forming ability and its usefulness as an outer film to protect food from drying and exposure to light and oxygen (Arvanitoyannis, 2002). The effect on film properties of fish gelatin attributes was reviewed recently by Gómez-Guillén et al. (2009). The molecular weight distribution and amino acid composition, which are the main factors influencing the physical and structural properties of gelatin, are also believed to play a key role in the mechanical and barrier properties of the resulting films. Weaker and more deformable films are normally obtained when low-molecular weight fragments predominate in a given gelatin preparation (Muyonga et al., 2004a; Carvalho et al., 2008). Films made from warm-water fish species, such as Nile perch or channel catfish, have exhibited mechanical and water resistance comparable to those of films made from mammalian gelatin (Muyonga et al., 2004b; Zhang et al., 2007). In contrast, gelatins from cold water fish gave films with extremely high water solubility (Pérez-Mateos et al., 2009).

The use of complex systems and natural cross-linkers has also been reported as a means of reinforcing microcapsules made of gelatin (Strauss and Gibson, 2004;

Yeo et al., 2005). Microcapsules are of special interest as they can entrap functional components in a carrier and provide protection against oxidation or degradation during storage. Moreover, encapsulation can be used to control the release of functional components from the food product or the bioactive packaging when ingested in the body (Chiu et al., 2007; Bao et al., 2009). Microcapsules based on gelatin have also been developed to protect and improve the survival of lactic acid bacteria and bifidobacteria (Lian et al., 2002; Weissbrodt and Kunz, 2007; Li et al., 2009).

14.4   OTHER COLLAGENOUS MATERIAL

Seafood by-products also serve as a source for obtaining high-value compounds in biomedicine and cosmetics. Cartilage is formed by a matrix of collagen associated with proteoglycans, macromolecules with an axial protein to which the glycosaminoglycans chondrotin sulfate and keratan sulfate are covalently linked (Murado et al., 2010a). Chondroitin sulfate is a sugar backbone of repeated disaccharide units [D-glucuronic acid (GlcUA)β1-3N-acetyl-D-galatosamine (GalNAc)], an ubiquitous component of the cell surface and extracellular matrix that participates in diverse biological processes such as growth factor signaling and the development of the nervous system (Sugahara et al., 2003). The chondroitin sulfate is isolated from ray fish cartilage, and its structure and neurite outgrowth promoting activity was investigated in relation to the potential application to nerve regeneration (Hashiguchi et al., 2011). Murado et al. (2010a) optimized the different stages to obtain chondrotin sulfate from Thornback skate ray (*Raja clavata*); the combination of enzymatic and chemical hydrolysis, selective precipitation together with membranes technology led to the formulation of a quick and highly efficient process with low consumption of reagents and high purity for the chondroitin sulfate.

Hyaluronic acid, composed of alternate disaccharide units of [β(1−4)-D-glucuronic acid-β(1−3)-N-acetyl-D-glucosamine] linkages, is the only nonsulfate glycosaminoglycan that is widely distributed in connective, epithelial, and neural tissues such as skin, cartilage, and the vitreous humor (Oh et al., 2008; Li et al., 2012; Xu et al., 2012), all of them are part of the seafood wastes. Thus, the fish eyeball (vitreous humor) is a source for obtaining hyaluronic acid of high purity (99.5%), useful for clinical and cosmetic application (Murado et al., 2010b). Because of its natural origin, the hyaluronic acid is biocompatible, nonimmunogenic and noninflammatory. This macromolecule has complex biological functions ranging from matrix organization, construction of the extracellular matrix, cell adhesion and migration, angiogenesis and morphogenesis, wound healing, and inflammatory responses to cancer metastasis (Oh et al., 2008; Li et al., 2012; Xu et al., 2012). For these reasons, the hyaluronic acid is a starting material for the construction of hydrogels with desired morphology, stiffness, and bioactivity, as a promising material for tissue repair and regeneration, drug delivery, cell encapsulation, among others (Li et al., 2012; Xu et al., 2012). Hydrogels made of other natural biopolymers (chitosan, gelatin, chondrotin sulfate, etc.) have also been reported (Li et al., 2012).

In addition to collagen, which is about 30%, fish bone consists of 60–70% of inorganic substances, mainly composed of calcium and hydroxyapatite. Fish bone material from processing of large fish is a useful calcium source, essential element for humans. Calcium must be converted into an edible form by different methods before it can be incorporated into calcium-fortified food (Kim and Mendis, 2006). In this connection, piglets getting salmon bone treated with enzymes had significantly higher calcium absorption than piglets getting boiled fish bone or calcium carbonate (Malde et al., 2010). These authors stated that fish bone can be a useful and well-absorbed calcium source, suitable in food, feed, or as a supplement.

Regarding hydroxyapatite, it has been introduced as a bone graft in medical and dental applications because of their similar chemical composition. Hydroxyapatite does not break under physiological conditions, it is thermodynamically stable at physiological pH, and takes part actively in bone bonding (Kim and Mendis, 2006).

14.5 CHITOSAN

Chitosan is a family of polymers derived from chitin by N-deacetylation in variable percentage (represented by the degree of deacetylation), which identifies the content of free amino groups—NH_2 of a block copolymer consisting of 2-amino-2-deoxy-D-glucose (glucosamine) and 2-acetamido-2-deoxy-D-glucose (D-glucosacetoamide) (Kim and Mendis, 2006).

14.5.1 EXTRACTION PROCESSES

Chitosan is obtained by alkaline deacetylation of chitin, a natural polysaccharide and the second most-abundant renewable natural polymer after cellulose. Chitin is a major constituent of crustacean shells, exoskeletons of insect and cell walls of certain fungi, which confers resistance and stability (Kumar et al., 2004; Rinaudo, 2006) The α-chitin is the most abundant and stable form in crustaceans (Zhang et al., 2005). Because of the resistance of some acetyl groups in *trans* position between C_2 and C_3, the deacetylation must be carried out with a long and severe alkaline treatment, which also requires temperatures above 100°C (Kumar, 2000). The deacetylation is affected by the alkali concentration, particle size, and density of chitin. These last two factors affect the penetration of the alkali in the amorphous and crystalline polymer, so that chitosans are generally achieved with 75–80% deacetylation. The degradation of the polymer chain also occurs during the process of deacetylation. To reduce oxidation during the deacetylation, reducing agents are employed, which also increase the viscosity of the acid solutions of chitosan (Aranaz et al., 2010). The degree of deacetylation is one of the most important chemical properties of chitosan, and determines its potential application, toxicity, solubility, free amino group content, and enables differentiation from the chitin (Guinesi and Cavalheiro, 2006).

Regarding the extraction process, Kurita (2006) reported that crustacean shells are treated with hydrochloric acid to remove metals and salts; then decalcified shells are exposed to an alkaline media to hydrolyze proteins and pigments until α-chitin is obtained. The demineralization and deproteinization steps have been carried out in different ways, for example, via enzymatic fermentation, using enzymes (proteases from *Bacillus subtilis*, Sini et al., 2007) or microorganisms (*Pseudomonas malto-philia*, Qin and Agboh, 1997). It is possible to recover proteins, carotenoids, and so on, throughout the different steps of the process, which increases the viability of the chitosan extraction process, so far expensive in reagents and resources.

In contrast to chitin, the presence of free amino groups in chitosan allows dissolution of the biopolymer in acid medium (pH $\leq$ 6.5), which confers excellent properties of biocompatibility, biodegradability, and affinity with other polymers. The chitosan is degraded by enzymatic hydrolysis and some authors relate it to the degree of crystallinity, which is controlled primarily by the degree of deacetylation (Muzzarelli, 1997; Suh and Matthew, 2000). Solubilization occurs by protonation of $-NH_2$ of the C_2 unit D-glucosamine in acid medium, so that the polysaccharide is converted into a soluble polyelectrolyte (Rinaudo, 2006). The solubility depends on several factors: deacetylation degree, molecular weight, acid concentration, ionic strength, and concentration of the biopolymer (Santos et al., 2004). Chitosan is positively charged in solution due to protonation of the original amino groups ($-NH_3^+$). This polymer is indeed an alkaline or neutral polysaccharide, on the contrary to other commercial polysaccharides such as cellulose, agar–agar, pectin, and so forth, which are neutral or acidic in nature (Sing and Ray, 2000).

The high viscosity and low solubility of chitosan at neutral pH, caused largely by the high molecular weight of this polymer, limits its use as dietary fiber *in vivo*. However, the modification of chitin and chitosan led to polymers with low viscosity and low short-chain length, making them easily soluble in aqueous solutions, and thus readily absorbable *in vivo* (Jeon et al., 2000).

14.5.2 FUNCTIONAL PROPERTIES

Chitosan is a biopolymer with multiple applications based on its functional properties (Ferraro et al., 2010). From a technological point of view, it is of great interest regarding the antimicrobial power, filmogenic ability, texturizing, and binding properties (Hardinge-Lyme, 2001) and antioxidant behavior (Kamil and Jeon, 2002), which contribute to extend the shelf-life of foods (Dallan and da Luz Moreira, 2007; Kumar et al., 2004). Other applications are its use in water purification, immobilization of enzymes, and nutraceuticals' encapsulation (Shahidi and Abuzaytoun, 2005). Chitosan is also of great importance in reducing intestinal absorption of lipids and due to its behavior as dietary fiber, acting as a regulator of intestinal motility (Shahidi et al., 1999). Furthermore, it is a hypocholesterolemic agent, and therefore it may be used as a dietary supplement (Maezaki et al., 1993; Gallaher et al., 2002). Chitosan has also an effect on coagulation of blood, restoration of injury, bone regeneration, immune activity, and so on (Kumar, 2000; Dallan and da Luz Moreira, 2007).

Concerning the antimicrobial properties, chitosan possesses a broad spectrum of activity. It was observed that chitosan acts fast on fungi and bacteria and the activity against pathogenic organisms is comparable with some antibiotics used in clinical practice (Chen et al., 2002; Goy et al., 2009). Antimicrobial activity of chitosan depends on several factors such as: type of chitosan (deacetylation degree, molecular weight), pH, temperature, reaction time, presence of other components that can interact with the biopolymer, type of microorganism, and age of the microbial cell (Devlieghere et al., 2004). The mechanism of action has not been fully elucidated, although many theories have been proposed; the most accepted is the change in cell

permeability due to interactions between the polycationic chitosan and electronega-
tive charges on the cell surface. This interaction allows the removal of intracellular
electrolytes and protein constituents (Papineau et al., 1991; Sudarshan et al., 1992).
Other mechanisms mentioned in the literature suggest that microbial interactions
with DNA inhibit protein synthesis (Hadwiger et al., 1986; Sudarshan et al., 1992)
and metal chelation or absorption of essential nutrients for the cell (Wang et al.,
2005).

The antimicrobial activity is directly related to the protonation and to the number
of amino groups linked to C_2 of the polymeric structure of the chitosan that can produce significant electrostatic interactions. A high content of amino groups increases
the antimicrobial activity, which corresponds to a greater degree of deacetylation
and favors the inhibitory effect against microorganisms. In the solid state, chitosan
interacts at a local level, surface to surface, rather than at a broader and deep level
that occurs in the liquid state (López-Caballero et al., 2005; Gómez-Estaca et al.,
2011). The antibacterial activity of chitosan is closely correlated with the characteristics of the cell surface. Gram-negative bacteria have an outer membrane containing
lipopolysaccharides causing a hydrophilic surface. The lipid component contains
inside anionic groups (phosphate, carboxyl) that contribute to the stability against
electrostatic interactions with divalent cations (Helander et al., 2001). The negatively
charged cell surface in Gram-negative bacteria is greater than that of Gram-positive
bacteria, allowing the inhibitory effect of chitosan to be more effective in Gram-
negative bacteria because its absorption is greater (Chung et al., 2004).

14.6   VISCERA-DERIVED PRODUCTS

14.6.1   SOURCE OF FUNCTIONAL COMPOUNDS

Internal organs are a significant percentage, both in volume and cost, in the seafood
processing industry. For example, in the tuna canning process, only about one-third
of the whole fish is used, and the industry generates as much as 70% solid wastes
from original fish material (Herpandi et al., 2011). Most of the seafood by-products
are used to obtain products such as fish oil, fish meal, fertilizers, pet food, and fish
silage (Kim and Mendis, 2006), although in fact, these residues may be an important
source of high-value biological material of high value that could be used as functional ingredients.

The fat content of fish (depending mainly on the species, geographic, season, etc.)
varies from 2% to 30%. The fish oils are mostly composed of two types of fatty acids
(FFAA) (eicosapentenoic acid, EPA, and docosahexanoic acid, DHA), PUFA, classified as ω-3 fatty acids, found mainly in marine animals (Kim and Mendis, 2006).
Bio-oils are produced by methods such as pyrolysis, fermentation, hydrolysis, physi-
cal/chemical extraction and conversion processes (Dermirbas et al., 2006), being the
methods and conditions of extraction vital for quality of oils.

The hylsa fish (*Hilsa ilisa*) viscera was used for the production of PUFA linoleic,
EPA, and DHA (Patil and Nag, 2011). The use of the viscera can be commercialized
for the production of PUFA concentrates (via acetone treatment followed by the urea
inclusion compound-based fractionation), which has potential value for large-scale

production. In tuna, only the head, meat, and bones but not the viscera are used in oil production (Herpandi et al., 2011).

The fish oil intake is related to promotion of human health against diseases. Kim and Mendis (2006) reviewed the clinical effects and reported that there is an inverse relationship between the amount of ω-3 fatty acid level present in the blood and the ocurrence of heart diseases, showing an antiatherogenic and antithrombotic effect. The intake of fish oils also exerts beneficial effects against diabetes mellitus, exhibited anti-inflammatory action and increases survival of people with autoimmune diseases. Furthermore, ω-3 fatty acids are also associated with the brain development, vision, and functions of the reproductive system. The PUFA concentrates that are devoid of more saturated fatty acids are better for human consumption than fish oils themselves, as they allow the daily intake of total lipid to be kept as low as possible (Herpandi et al., 2011).

14.6.2 SOURCE OF ENZYMES

Viscera also represent a source of enzymes that may be used to hydrolyze fish by-products. Thus, Khantaphant et al. (2011) studied the effects of pretreatments (washing and/or membrane separation and vice versa) on antioxidative activities of protein hydrolysates obtained after treatment with proteases from pyloric ceca, from brownstripe red snapper (*Lutjanus vitta*). Functional and antioxidant activity of protein hydrolysates from ornate threadfin bream (*Nemipterus hexodon*) muscle using skipjack tune pepsin were studied (Nalinanon et al., 2011). Several enzymes are commercially extracted from marine fish viscera, including pepsin, trypsin, chymotrypsin, and collagenases (Byun et al., 2003). In addition, the viscera can be used as the substrate from which, after an enzymatic treatment, a protein material with potential biological activity (antioxidant, antihypertensive, etc.) may be obtained. In this regard, Swapna et al. (2011) reported that enzymatic hydrolysis with four commercial proteases of viscera from carp (*Catla catla*) and rohu (*Labeo rohita*) can be effective to recover biomolecules such as lipids, protein hydrolysates, and collagen, thus reducing the organic load and diminishing the pollution problem.

Sometimes commercial enzymes and enzymes derived from viscera are used simultaneously. In this regard, Bougatef et al. (2010) used enzymes (Alcalase®, enzyme from *Aspergillus clavatus* ES1, alkaline protease from *B. licheniformis* and crude enzyme extract from the viscera of sardine) to hydrolyze heads and viscera from sardinelle to obtain antioxidative peptides. Khantaphant et al. (2011) studied the biological properties of protein hydrolyases from brownstripe red snapper obtained with pyloric ceca and commercial proteases (Alcalase and Flavourzyme).

14.6.3 VISCERA AND MICROORGANISMS

The fermentation is a biological method in which the microorganisms (LAB) are used to generate acid for preservation of waste or for recovery of biomolecules with functionality (Amit et al., 2010). Fermentation with native LAB (*Enterococcus faecium, Enterococcus faecalis,* and *Pediococcus acidilactici*) of viscera from catla and rohu can be an effective method for lipids and protein recovery. The process resulted in a >90% recovery of oil present in the material against no recovery in unfermented viscera, and resulted in a >50% degree of hydrolysis of proteins. Lipids recovered were rich in PUFA, which have several health benefits, being a better alternative to fish oil. The fatty acid profile of lipids was not affected by the fermentation process. The antioxidant and antimicrobial properties (e.g., against *E. coli*) of the fermented viscera indicated the possibility to use them as food ingredients in animal feed formulations (Amit et al., 2011).

Viscera are also used as a base substrate to obtain active molecules from microbial action. Thus, LAB (*Lactobacillus plantarum, Lactobacillus buchneri, Lactobacillus casei* ssp *casei, Lactococcus lactis* ssp *lactis, Leuconostoc mesenteroides* ssp *Mesenteroides*, and *Pediococcus acidilactici*) were able to ferment the waste material (viscera from swordfish, thornback ray, and shark) and thus produced organic acids, mainly lactic and acetic acid to preserve and generate ingredients for animal feed (Vázquez et al., 2011).

Beyond its role in the fermentation of fishery by-products, the importance of lactic acid bacteria (LAB) is questionable since in natural storage under aerobic conditions, vacuum packaging and modified atmospheres, lactic acid bacteria are not of much concern. The role of LAB in marine products depends on the fish species, treatment and storage conditions, bacterial species and strains, and interaction among them (Leroi, 2010). However, LAB may become important in relation to the lightly preserved fish, where several situations can take place: LAB may have no particular effect, LAB may be responsible for spoilage or even LAB may exert a bioprotective effect against undesirable bacteria during fish storage. The lightly preserved fish, despite containing viable LAB, are not considered as probiotic for humans because they are not consumed in sufficient quantity. The seafood technology has a long way to go, since fish lack the presence of certain elements (carbohydrates or sugars/salts as those present in milk, meat, etc.), in their composition that favor the acidification by lactic acid bacteria that allows the natural preservation of the product (Leroi, 2010).

14.7 HYDROLYSATES

Seafood processing discards by-products that have long been recognized as wastes and account for approximately three-quarters of the total weight of the catch. Owing to this large quantity of seafood processing wastes annually generated, a great deal of attention has been paid to obtain valuable components such as fish oil, collagen and gelatin, enzymes, minerals, chitin, or protein hydrolysates from these by-products.

Enzymatic hydrolysis is one of the methods used for the recovery of valuable components from the protein-rich by-products of sea food. Several proteolytic enzymes from microbes, plants, and animals have been employed for the hydrolysis of seafood processing by-products (Simpson et al., 1998). This technology has been extensively used to improve functional properties of proteins and recently has also been used as an efficient way to recover peptides with interesting biological activities.

14.7.1 FUNCTIONAL PROPERTIES

Peptides produced by enzymatic proteolysis have smaller molecular size and less secondary structure than the parent proteins. Hence, they are expected to have increased solubility and improved foaming, gelling, and emulsifying capacity (Chobert et al., 1988). Numerous studies have demonstrated that selective enzymatic hydrolysis of fish proteins allows improving their functional properties, including solubility, water holding, oil holding, emulsifying, and foaming characteristics. Some of these studies have been carried out on shark protein (Diniz and Martin, 1996), salmon protein (Kristinsson and Rasco, 2000a; Gbogouri et al., 2004; Sathivel et al., 2005), herring (Liceaga-Gesualdo and Li-Chan, 1999; Sathivel et al., 2003), capelin (Shahidi et al., 1995), sardine (Quaglia and Orban, 1987, 1990), cod backbones (Slizyte et al., 2009), round scad muscle (Thiansilakul et al., 2007), Pacific whiting (*Merluccius productus*) muscle (Pacheco-Aguilar et al., 2008), grass carp (*Ctenopharyngodon idella*) skin (Wasswa et al., 2007). Furthermore, fish protein hydrolysates (FPHs) produced under controlled conditions yield high nutritional value and reduced bitterness (Liceaga-Gesualdo and Li-Chan, 1999; Kristinsson and Rasco, 2000a,b), and they can contribute to water holding, texture, gelling, whipping, and emulsification properties when added to food (Kristinsson, 2007).

The functional properties of FPHs and therefore their uses as food ingredients are directly affected by the characteristics of the hydrolysates, mainly molecular weight (Alder-Nissen, 1986) and hydrophobicity (Turgeon et al., 1992). Protease specificity affects size, amount, and amino acid sequences of the resultant peptides and therefore the functional properties of the hydrolysates (Gauthier et al., 1993).

In general, FPHs have an excellent solubility over a wide range of pH. This is a substantially useful characteristic for many food applications and plays an important role in other functional properties such as emulsifying and foaming properties (Gbogouri et al., 2004; Kristinsson and Rasco, 2000b). FPH have shown good foaming and emulsion properties (Giménez et al., 2009b; Thiansilakul et al., 2007; Slizyte et al., 2009) and therefore they can be used as ingredients in emulsion and foam-based products, aiding in the formation and stabilization of both emulsions and foams. The molecular size of FPH and hence the degree of hydrolysis play an important role in surface activity. Extensive hydrolysis has been reported to result in a higher solubility but also in a drastic loss of emulsifying and foaming properties of hydrolysates (Mahmoud, 1994; Jeon et al., 1999; Kong et al., 2007; Souissi et al., 2007). Thus, a lower foaming and/or emulsifying capacity was found as the degree of hydrolysis increased for sardine muscle, salmon muscle, or heads and viscera of sardinella (Quaglia and Orban, 1990; Kristinsson and Rasco, 2000a; Souissi et al., 2007). Small peptides easily diffuse to and absorb at the interface, but they are less efficient in lowering the interface tension than larger peptides obtained in limited hydrolysis (Kristinsson and Rasco, 2000b; Gbogouri et al., 2004; Kong et al., 2007).

14.7.2 BIOACTIVE PROPERTIES

Dietary proteins are a source of biologically active peptides, which are inactive within the sequence of the parent protein but can be liberated during gastrointestinal digestion, food processing or fermentation. Once they are released, bioactive peptides can affect numerous physiological functions of the organism.

There is a growing body of scientific evidence that shows that many protein hydrolysates and peptides derived from fish, molluscs, and crustacean underutilized species and seafood processing by-products may promote human health and aid the prevention of chronic diseases (Kim et al., 2008; Kim and Wijesekara, 2010). Numerous biological activities have been reported for these protein hydrolysates and peptides, such as antioxidant, ACE inhibitory, antihypertensive, calcium-binding, antimicrobial, anticancer, anticoagulant, or hormonal activities (Fahmi et al., 2004; Rajapakse et al., 2005a; Jung and Kim, 2007; Liu et al., 2008; Martínez-Álvarez et al., 2008; Giménez et al., 2009b; Alemán et al., 2011a).

Marine bioactive peptides may be produced by solvent extraction, enzymatic hydrolysis, or microbial fermentation. However, especially in food and pharmaceutical industries, the enzymatic hydrolysis is the method most widely used because of lack of residual organic solvents or toxic chemicals in the resulting hydrolysates (Kim and Wijesekara, 2010). A number of commercial proteases have been used for the production of these hydrolysates and peptides from seafood wastes, including trypsin, chymotrypsin, pepsin, Alcalase, Flavourzyme, Neutrase, Properase E, Pronase, collagenase, bromelain, and papain (Kim et al., 2001a; Mendis et al., 2005a; Lin and Li, 2006; Yang et al., 2008; Foh et al., 2010). Besides commercial proteases, the use of enzymatic extracts from fish viscera has also been reported to obtain bioactive hydrolysates from seafood wastes of different species (Jung et al., 2005; Bougatef et al., 2008; Phanturat et al., 2010). Protease specificity affects size, amount, composition of free amino acid and peptides and their amino acid sequence, which in turn influences the biological activity of the hydrolysates (Chen et al., 1995; Jeon et al., 1999; Wu et al., 2003).

The average molecular weight of protein hydrolysates is one of the most important factors that determine their biological properties (Jeon et al., 1999; Park et al., 2001). Peptide fractions from protein hydrolysates may show different effectiveness for a given biological activity. An ultrafiltration membrane system could be a useful and industrially advantageous method for obtaining peptide fractions with a desired molecular size and a higher bioactivity, depending on the composition of the starting hydrolysate and the activity studied (Jeon et al., 1999; Korhonen and Pihlanto, 2003; Cinq-Mars and Li-Chan, 2007; Picot et al., 2010). This system has been successfully applied in the fractionation and functional characterization of silver carp by-product protein hydrolysates (Zhong et al., 2011), gelatin hydrolysates from squid or cobia skins (Lin and Li, 2006; Yang et al., 2008), or tuna liver hydrolysates (Ahn et al., 2010); and also as a first step in the isolation and further purification of bioactive peptides from different seafood by-products (Kim et al., 2001a; Mendis et al., 2005a; He et al., 2006; Zhao et al., 2007; Alemán et al., 2011b).

14.7.2.1 Antioxidant Activity

Since the antioxidant effect of peptides was first reported (Marcuse, 1960), numerous studies have been conducted to investigate antioxidant properties of hydrolysates and peptides from plant or animal sources such as rice bran, sunflower protein, milk casein, egg-yolk protein, and so forth (Suetsuna et al., 2000; Sakanaka and Tachibana, 2006; Megías et al., 2008; Revilla et al., 2009).

Recently, several studies have demonstrated that hydrolysates and peptides derived from protein fish by-products such as heads, viscera, frames, or collagenous materials act as potential antioxidants. Antioxidant hydrolysates have been widely obtained from cutoffs resulting from mechanically deboned fish or from muscle protein of undervalued species such as grass carp, giant squid, or Pacific hake (*Merluccius productus*) (Rajapakse et al., 2005b; Ren et al., 2008; Samarayanaka and Li-Chan, 2008). One potential problem in the preparation of protein hydrolysates from muscle sources is the presence of unsaturated lipids and pro-oxidants such as heme proteins, especially in dark muscle. In these cases, protein isolates prepared from muscle by alkali solubilization can be used as the substrate for enzyme hydrolysis (Raghavan and Kristinsson, 2008b; Dekkers et al., 2011). Fish frames, which include heads, bones, and tails, are abundant by-products from fish filleting processing plants with a considerable high protein concentration. Enzymatic hydrolysis of fish frames from different species such as hoki, cod, or yellowfin sole has given hydrolysates with a noticeable antioxidant activity (Jeon et al., 1999; Jun et al., 2004; Kim et al., 2007). Collagenous materials (skins, fins, scales, backbones) are also an important source of antioxidant hydrolysates and peptides. Gelatin hydrolysates from different species such as hoki, Alaska Pollack, cobia, squid, or tuna (Kim et al., 2001a; Je et al., 2007; Nam et al., 2008; Yang et al., 2008; Giménez et al., 2009b) have shown antioxidant activities even higher than those derived from muscle sources, as a consequence of a higher content of hydrophobic amino acids which could increase their solubility in lipids and therefore enhance their antioxidative activity (Kim et al., 2001a). Furthermore, the high percentage of Gly and Pro has also been associated with the higher antioxidant activity of gelatin peptides when compared with peptides from other sources (Rajapakse et al., 2005b). Protein hydrolysates obtained from residues of shellfish processing industry have been also an important source of peptides with antioxidant activity, as those obtained from *Acetes chinensis* (Cao et al., 2009) or from a mixture of *Penaeus braziliensis* and *Penaeus subtilis* (Guerard et al., 2007).

The exact mechanism underlying the antioxidant activity of peptides has not been fully understood, yet various studies have shown that they are inhibitors of lipid peroxidation, scavengers of free radicals and chelators of transition metal ions. Antioxidative properties of the peptides are related to their amino acid composition, structure, and hydrophobicity. Dávalos et al. (2004) working on the activity of individual amino acids reported that Trp, Tyr, Met showed the highest antioxidant activity, followed by Cys, His, and Phe. The rest of the amino acids did not show any antioxidant activity. However, many peptides have been described to have antioxidant capacity without containing any of the above-mentioned proton-donating amino acid residues in their sequences. Thus, Kim et al. (2001a) isolated two peptides composed of 13 and 16 amino acid residues, respectively from Alaska pollack skin, both of which contained a Gly residue at the C-terminus and the repeating motif Gly–Pro–Hyp. The peptide Asn–Gly–Pro–Leu–Gln–Ala–Gly–Gln–Pro–Gly–Glu–Arg was purified from squid skin gelatin with a valuable free radical quenching capacity (Mendis et al., 2005a).

In addition to the presence of proper amino acids, their correct positioning in peptide sequence plays an important role in antioxidant activity. Peptide conformation has been also claimed to influence antioxidant capacity, showing both synergistic and antagonistic effects, as far as the antioxidant activity of free amino acids is concerned (Hernández-Ledesma et al., 2005). As mentioned above, the specificity of protease used for hydrolysis may determine the size and the sequence of the peptides, and as a result the antioxidant activity. Thus, alkaline proteases (e.g., Alcalase) showed higher activities than the acid or neutral proteases (e.g., Flavourzyme or Neutrase), which is consistent with the data obtained by Foh et al. (2010) in Nile tilapia muscle hydrolysates and Dong et al. (2008) in Silver carp muscle hydrolysates. Furthermore, it has been found that antioxidant activity of Alcalase gelatin-derived hydrolysates was higher than that of the other enzyme hydrolysates such as those obtained by collagenase, pepsin, trypsin, chymotrypsin, papain, or Neutrase (Qian et al., 2008; Alemán et al., 2011c). However, the highest antioxidant activity was obtained in the peptic hydrolysate of yellowfin sole frame protein (degree of hydrolysis of 22%) and not in the hydrolysates obtained with Alcalase, α-chymotrypsin, papain, Pronase E, Neutrase, and trypsin (Jun et al., 2004).

Moreover, antioxidant activity is strongly related to molecular weight of peptides. According to Gómez-Guillén et al. (2010), although antioxidant activity was found in all the peptide fractions from squid skin hydrolysate, this was higher in the fractions with lower molecular weight. Similarly, the peptide fraction with the lowest molecular weight (<3.5 kDa) from sardinella viscera protein hydrolysate was found to exhibit the highest radical scavenging activity (Barkia et al., 2010).

The antioxidant activity of hydrolysates and peptides derived from seafood by-products has been determined by various *in vitro* methods such as those based on the scavenging capacity of different radicals (DPPH, ABTS, carbon-centered, peroxyl, superoxide, hydroxyl), reducing power, chelating ability, inhibition of the peroxidation of linoleic acid or inhibition of the formation of thiobarbituric acid substances (Mendis et al., 2005b; Klompong et al., 2007; Theodore et al., 2008; Bougatef et al., 2010; Alemán et al., 2011c). In addition, it has been reported that hydrolysates and peptides derived from seafood by-products may protect living cells against free radical-mediated oxidative damage. Therefore, scavenging of free radical species is an important mechanism by which antioxidant peptides enhance cell viability against oxidation-induced cell death. Kim et al. (2001a) reported that a peptide isolated from Alaska pollack skin gelatin was able to protect rat liver cells from oxidant injury induced by *tert*-butyl hydroperoxide (t-BHP). In the same manner, purified peptides from squid muscle and tuna dark muscle significantly enhanced the viability of human embryonic lung fibroblasts when cellular oxidative stress was induced by t-BHP (Rajapakse et al., 2005b; Je et al., 2008). Moreover, results from one study revealed that peptides isolated from hoki skin gelatin were capable of enhancing the expression of antioxidative enzymes such as glutathione peroxidase, catalase, and superoxide dismutase in human hepatoma cells (Mendis et al., 2005b). Some studies have been conducted to confirm the *in vivo* antioxidant activity of collagen and gelatin peptides, and some convincing data have been obtained for animal models. Recently, some collagen and gelatin peptides from marine sources were reported to act protectively against ultraviolet radiation-induced damage on mice skin (Hou et al., 2009; Zhuang et al., 2009). Furthermore, loach muscle peptides have been reported to increase the antioxidative enzyme activities of glutathione peroxidase, catalase, and superoxide dismutase in mice subjected to exhaustive swimming test (You et al., 2011a).

14.7.2.2 Antihypertensive/ACE Inhibitory Activity

Antihypertensive peptides are peptide molecules which when ingested may lower blood pressure within the body through inhibition of vasoactive enzymes such as angiotensin converting enzyme (ACE). ACE plays an important role in the regulation of blood pressure by means of the renin–angiotensin system, and inhibition of this enzyme is considered to be a useful therapeutic approach in the treatment of hypertension (Odentti et al., 1982; Chen et al., 2007).

Since the discovery of ACE inhibitory peptides in snake venom, many studies have been attempted in the synthesis of ACE inhibitors, such as Captopril, Enalapril, Alacepril, and Lisinopril, which are currently used extensively in the treatment of hypertension and heart failure in humans (Odentti, 1977; Patchett et al., 1980). However, synthetic ACE inhibitors are believed to have certain side effects (Atkinson and Robertson, 1979). Therefore, over the last 10 years many researchers worldwide have paid considerable attention to find ACE inhibitors from natural sources such as food proteins, less potent than the synthetic ones but without known side effects. Although the major natural source of ACE inhibitory peptides identified to date is milk, these peptides have been isolated from many other animal and plant protein sources, such as blood proteins (Mito et al., 1996), ovalbumin (Miguel et al., 2004), maize (Miyoshi et al., 1991), chickpea (Yust et al., 2003), soy (Wu and Ding, 2001) or muscle proteins from pig, cattle, fish, and chicken (Fujita and Yoshikawa, 1999; Fujita et al., 2000; Katayama et al., 2003; Jang and Lee, 2005; Ahhmed and Muguruma, 2010).

In general, seafood protein by-products have also demonstrated to be a good source of antihypertensive peptides by enzymatic digestion, although they have not been so extensively studied as other sources. Protein hydrolysates with ACE inhibitory activity have been prepared from heads and viscera of sardinelle with a crude enzyme extract of sardine (Bougatef et al., 2008). Muscle protein hydrolysates from cod and salmon were found to exhibit higher ACE inhibitory activity than other by-products such as skins, frames, or heads after *in vitro* gastrointestinal digestion (Dragnes et al., 2009). The peptic hydrolysate of Bigeye tuna (*Thunnus obesus*) dark muscle was reported to contain the ACE inhibitory peptide Trp–Pro–Glu–Ala–Ala–Glu–Leu–Met–Met–Glu–Val–Asp–Pro ($IC_{50} = 21.6$ µM; Qian et al., 2007). Frames from different fish species have also rendered ACE inhibitory hydrolysates and peptides by enzymatic hydrolysis. Je et al. (2004) isolated an effective inhibitory peptide from Alaska pollack frame protein hydrolysate (Phe–Gly–Ala–Ser–Thr–Arg–Gly–Ala; $IC_{50} = 14.7$ µM), whereas the sequence Met–Ile–Phe–Pro–Gly–Ala–Gly–Gly–Pro–Glu–Leu ($IC_{50} = 22.3$ µM) was purified from a yellowfin sole frame protein hydrolysate (Jung et al., 2006). Potent ACE inhibitory hydrolysates and peptides have been obtained from collagenous materials such as fish skins (Byun and Kim, 2001; Park et al., 2009; Nagai et al., 2006), fish cartilage (Nagai et al., 2006), scales (Fahmi et al., 2004), and squid tunics (Alemán et al., 2011b). Some of the peptide sequences identified from collagenous hydrolysates were Gly–Pro–Leu ($IC_{50} = 2.6$ µM), purified from Alaska Pollack skin gelatin hydrolysate (Byun and Kim, 2001); or Val–Ile–Tyr and Val–Tyr ($IC_{50} = 7.5$ and 16 µM, respectively), the most potent peptides sequences isolated from sea bream scale hydrolysate. In addition, ACE-inhibitory peptides have also been prepared from undervalued shellfish species and processing by-products. For example, effective inhibitory peptides have been isolated from fresh water clam (*Corbicula fluminea*) hydrolysates (Val–Lys–Pro, Val–Lys–Lys) (Tsai et al., 2006), or shrimp (*Acetes chinensis*) hydrolysates. Several potent peptide sequences have been isolated from *A. chinensis* in different studies: Leu–His–Pro (IC_{50} = 1.6 µM; Cao et al., 2010), Ile–Phe–Val–Pro–Ala–Phe, Phe–Cys–Val–Leu–Arg–Pro (IC_{50} = 3.4 and 12.3 µM, respectively; He et al., 2006).

Although the relationship between the structure and the activity of ACE inhibitory peptides has not yet been established, these peptides have certain common features. Most of them are relatively short sequences with low molecular mass, since the active site of ACE cannot accommodate large peptide molecules. Binding to ACE is strongly influenced by the C-terminal tripeptide sequence, which may interact with subsites at the active site of the enzyme (Odentti and Cushman, 1982). ACE prefers substrates or inhibitors that contain hydrophobic amino acid residues (aromatic or branched side chains) at each of the three C-terminal positions (Cheung et al., 1980; Murray and FitzGerald, 2007). The presence of Arg or Lys on the C-terminal position has also been reported to contribute substantially to the inhibitory activity (Cheung et al., 1980; Ariyoshi, 1993; Meisel, 2003). For example, the ACE inhibitory activity described for collagen and gelatin hydrolysates and peptides may be related to the high content of hydrophobic amino acids, as well as to the high content of Pro. This amino acid seems to be one of the most effective ones in increasing the ACE inhibitory activity and it has been described in many of the naturally occurring ACE peptide inhibitors (Gómez-Ruiz et al., 2004; Quirós et al., 2007; Pihlanto et al., 2008; Contreras et al., 2009), especially in those derived from collagenous sources (Byun and Kim, 2001; Kim et al., 2001b; Ichimura et al., 2009; Shimizu et al., 2010; Alemán et al., 2011b).

The *in vivo* effects of antihypertensive peptides are usually tested in spontaneously hypertensive rats (SHR), which constitute an accepted model for human essential hypertension (Fitz-Gerald et al., 2004). Many of the ACE inhibitory peptides isolated from seafood protein by-products have been already tested *in vivo*, and a significant antihypertensive effect has been reported (Fahmi et al., 2004; Jung et al., 2006; Tsai et al., 2006; Qian et al., 2007; He et al., 2008; Cao et al., 2010). A significant decrease in blood pressure has been achieved following a single oral administration to SHR (10 mg/kg) of the peptide Gly–Asp–Leu–Gly–Lys–Thr–Thr–Thr–Val–Ser–Asn–Trp–Ser–Pro–Pro–Lys–Trp–Lys–Asp–Thr–Pro, purified from tuna frame protein hydrolysate (Lee et al., 2010). In another study, systolic blood pressure of SHR was reduced in a similar way to that of Captopril when the ACE inhibitory peptide Met–Ile–Phe–Pro–Gly–Ala–Gly–Gly–Pro–Glu–Leu, isolated from yellowfin sole frame protein hydrolysate, was orally administered at 10 mg/kg (Jung et al., 2006).

14.7.2.3 Antimicrobial Activity

The relationship between peptide characteristics and antimicrobial activity has not yet been clearly stated. Several factors, such as amino acid composition, sequence, molecular weight, and type of bacteria need to be taken into account (Di Bernardini et al., 2011). As a common feature, the main properties of antimicrobial peptides affecting their activity are as follows: they present 50 amino acids (or less), they have a positive charge due to an excess of basic (mainly lysine and/or arginine) over acidic residues, they contain about 50% hydrophobic amino acids and molecular weight below 10 kDa. They often fold into three-dimensional amphipathic structures stabilized by cysteine disulfide bridges (linear peptides lacking cysteines tend to fold, in contact with the membranes in a variety of amphiphatic helixes, pleated-sheets, loops, or extended structures), in which hydrophilic positively charged domains are well defined and delimited from hydrophobic ones (Rydlo et al., 2006; Najafian and Babji, 2012).

Regarding the mechanisms of action, it has been established that peptides interact with the cytoplasmic membranes of microorganisms, although the differences existing in membrane composition have implications in the mode of action and the specificity of the antibacterial compounds (Floris et al., 2003). Jenssen et al. (2006) described, as essential for peptides to exert their antimicrobial activity, the electrostatic interaction between the peptide and positively charged and anionic lipids on the surface of the microorganism, as well as the hydrophobicity and flexibility of the peptide to allow interaction with the microbial membrane. Peptides are classified into two groups depending on if the peptide acts on the cytoplasmic membrane or not. Bechinger and Lohner (2006) reported that the peptides acting on the bacterial membrane produce pores that cause membrane permeabilization and disruption of the electrochemical gradient, thereby affecting cellular respiration and allowing the flow of water and ions across the membrane, leading to swelling and celular lysis.

Several models have been proposed to explain the formation of pores: (i) barrel stavel (formation of a pore though the binding of amphipatic α-helices; the hydrophobic surface of the peptide interacts with a lipid core of the membrane and the hydrophilic surface of the peptide is oriented inside, Reddy et al., 2004); (ii) toroidal pore (peptides are included in the membrane in a perpendicular way, with the hydrophilic zones associated with phospholipids, while the hydrophobic zones are associated with the lipid core, Jenssen et al., 2006); (iii) the carpet model (peptides are located on the membrane surface so the hydrophilic surface coming in contact with the phospholipids or water molecules deformate the membrane curvature, Reddy et al., 2004); and (iv) aggregated model (peptides placed in the membrane like a micelle-like group of peptides and lipids, with no special orientation, Jenssen et al., 2006). The peptides without activity on the bacterial membrane are able to enter into the bacterial cells without causing membrane permeabilization, altering processes (inhibition of protein and nucleic acid synthesis, enzyme actitivy, etc.) that result in bacterial death (Jenssen et al., 2006).

In relation to the antifungal activity, the mechanisms of action of peptides (Jenssen et al., 2006) were described to be related to fungal cell lysis or due to interferences in fungal cell wall synthesis, membrane permeabilizaton, attack of intracellular organelles, deformation of the cell membrane, and so on. Espitia et al. (2012) reviewed the models proposed to explain the antifungal activity: (i) peptides that act through cellular lysis (characterized by their amphipathic nature; some of them are bound to the membrane surface, damaging the structure but they may not pass through it); (ii) peptides that pass into the membrane and interact with intracellular targets (interfering with synthesis of the cell wall or essential cellular components as chitin or glucan); and (iii) pore-formimg peptides (aggregated in a selective way to form pores of different sizes, allowing the passage of ions and other solutes) (De Lucca and Walsh, 1999).

The origin of the antimicrobial peptides is diverse and they have been described from mammalian, fish, avian, amphibian, insects, and so on (Rydlo et al., 2006; Kim and Wijesekara, 2010). In connection with those of marine origin, the antimicrobial activity has been described in hydrolysates derived by-products from various sources. Thus, in protein fractions derived from tuna and squid skin gelatins (1–10 kDa and <1 kDa), Gómez-Guillén et al. (2010) reported antimicrobial activity in peptide fractions, being especially sensitive to Gram-positive *Lactobacillus acidophilus* and *Bifidobacterium animalis* ssp *lactis*, and Gram-negative *Shewanella putrefaciens* and *Photobacterium phosphoreum*. These authors mentioned the reduced molecular weight in the peptide fractions (related to the elimination of aggregates), better exposure of the amino acid residues and their charges, as well as structure acquisition, among the factors facilitating the interaction of the peptide with bacterial membranes, and therefore their activity.

In mollusc and crustaceans, the antimicrobial activity of mussel (*Mytilus edulis*) protein hydrolysates is probably related to the higher cysteine residues and contents of hydrophobic amino acids (Dong et al., 2012). The resulting fraction exhibited activity against Gram-positive (*Staphylococcus aureus, Bacillus subtilis*, among others) and Gram-negative (e.g., *Escherichia coli, Pseudomonas aeruginosa*, and *Proteus vulgaris*). The digestion of oyster muscle with papain and bromelin produced an antimicrobial peptide (CgPep33) rich in cysteine (Liu et al., 2008). This peptide inhibited the growth of bacteria (*E. coli* and *P. aeruginosa*, Gram-negative and *S. aureus* and *B. subtilis*, Gram-positive) and fungi (*Botrytis cinerea* and *Penicillium expansum*). The peptide from American lobster (*Homarus americanus*) presented bacteriostatic activity against Gram-negative bacteria (*Vibrio* sp. *Halomonas* sp., and *Enterobacter aerogenes*) and parasites (Battison et al., 2008). Purified fractions of *Exocoetus volitans* backbone (Naqash and Nazeer, 2011) did not inhibit the growth of *E. aerogenes* and *S. aureus*, and slightly inhibited *Salmonella typhi*. Of the isolated fractions, fraction IIIb (with glutamic acid, lysine, glycine, and threonine as major amino acids), showed a moderate effect on killing *Vibrio cholerae*.

The antimicrobial activity of peptides derived from crab has been described in the literature. Noga et al. (1996) found that the hemolymph of blue crab (*Callinectus sapidus*) inhibited the Gram-negative bacteria (*Aeromonas hydrophila, V. parahemolyticus, V. alginolyticus, V. vulnificus*, and several strains of *E. coli*). The arasin 1 (rich in proline and arginine), derived from spider crab (*Hyas araneus)* inhibited the Gram-positive bacteria *Corynebacterium glutamicum* (Stensvag et al., 2008). Ravichandran et al. (2010) determined the antimicrobial activity of hemolymph and hemocytes of six brachyuran crabs (*Hyas araneus, Podopthalmus vigil, Dromia dehani, Charybdis helleri, Portunus sanguinolentus*, and *Portunus pelagicus*) against 16 pathogenic bacterial strains. *S. aureus* and *S. typhi* were susceptible to all samples tested. The maximum zone of inhibition corresponded to the hemolymph of *H. araneus* against *Shigella flexineri*, while the highest zone of inhibition was exhibited by the hemolymph and hemocytes on *V. cholerae* (with exceptions, Gram-positive bacteria was more susceptible to the antibacterial effect than Gram-negative bacteria). The analyses of the active fractions revealed that lipids are responsible for the antimicrobial activity in the hemolymph, showing fatty acids (oleic and linoleic acids) and fatty esters as the major components, and glycerides and glycolipids as the minor constituents. In another study, an antimicrobial peptide (Sushi I) from the horseshoe crab (*Cacinoscorpius rotundicauda*) induced the lysis of Gram-negative bacteria (Leptihn et al., 2009), with the target sites of the peptide being the outer and inner membranes and not the cytosolic space. These authors suggested the process as follows: binding (mainly by charged residues in the peptide); peptide association (peptide concentration increases evidence by changing in diffusive behavior); membrane disruption (the lipopolysaccharide is not released) and lysis (leakage of cytosolic content through membrane defects).

The applications of antimicrobial peptides are diverse. Among the biomedical applications—in addition to the activity against Gram-positive and Gram-negative bacteria, molds, and yeasts, the peptides are involved in host defense mechanisms, showing activity against viruses, parasites, nematodes, and tumor cells. They also have been considered useful in the therapy of stomach ulcers, as agents to stop sexually transmitted diseases, agents enhancing the activity of antibiotics, and so forth (Rydlo et al., 2006; Rajanbabu and Chen, 2011). In relation to food preservation, the activity of these fractions against several microorganisms makes them potential food additives, beyond their use as a protein or lipid source. Taking part in active packaging, peptides can interact with the product as well as the headspace inside the packaging and reduce or inhibit the growth of microorganisms (Soares et al., 2009).

The incorporation of bioactive peptides as functional ingredients would require some steps regarding isolation and characterization, and the assessment of safety and economical viability, until the procedure could be considered optimized and bioactive peptides could be incorporated and sold as functional products (Di Bernardini et al., 2011).

In terms of obtaining bioactive molecules, no bioactive peptides have been produced from microbial fermentation of muscle proteins, despite peptides having been obtained from other protein sources such as milk and the like (Ryan et al., 2011).

14.7.2.4　Other Biological Activities

Recently, several studies have shown the possible role of bioactive peptides obtained by controlled enzymatic hydrolysis from seafood by-products as anticancer agents by reducing cell proliferation on different human cancer cell lines. FPHs from blue whiting, cod, plaice, and salmon have been reported to be significant growth inhibitors on two breast cancer lines (Picot et al., 2006). Peptide fractions from shrimp waste proteins were found to inhibit growth of both colon and liver cancer cells (Kannan et al., 2011).

The antiproliferative activity of protein hydrolysates has been highly related to their antioxidant activity. Thus, Esperase squid gelatin hydrolysate showed a significant antioxidant activity and also the highest cytotoxic and antiproliferative effect on both breast and glioma cancer cell lines (Alemán et al., 2011a). The peptide fraction from flying fish (*Exocoetus volitans*) backbone protein hydrolysate with pepsin that exerted the highest antioxidant activity also showed a significant antiproliferative effect on human liver cancer lines (Naqash and Nazeer, 2011). In the same way, peptide fractions with high antioxidant activity isolated from muscle protein hydrolysates of flying fish, Japanese threadfin bream (*Nemipterus japonicus*), and loach exhibited strong antiproliferative activity in a dose-dependent manner on different human cancer lines (Naqash and Nazeer, 2010; You et al., 2011b).

Gastrin and cholecystokinin (CKK) are small intestinal hormones that belong to the secretagogue family. These secretagogue molecules exhibit a wide range of activities, ranging from the stimulation of protein synthesis throughout the entire body and gastric acid secretion to the control of intestinal mobility and the secretion of digestive enzymes (Johnson et al., 1978). Involvement of CCK-8 in the satiety mechanisms that control food intake in humans is well known (Bray, 2000), and peptides acting as agonists on gastrin and CCK receptors could be of interest as satietogenic ingredients in functional foods. Gastrin/CCK-like peptides have been obtained by controlled proteolysis of different seafood protein by-products, such as shrimp waste, heads, muscle, and backbones of cod, head, and the viscera of sardine, skins of North Atlantic lean fish (Cancre et al., 1999; Ravallec-Plé et al., 2000, 2001; Ravallec-Plé and Van Wormhoudt, 2003; Slizyte et al., 2009; Picot et al., 2010).

The calcitonin gen-related peptide (CGRP) is a 37-residue neuropeptide that exerts a wide variety of biological functions such as vasodilation, positive inotropic and chronotropic effect on heart, regulation of gastric acid secretion, and induction of saiety (Lenz et al., 1984; Franco-Cereceda et al., 1987; Gupta et al., 2006). The presence of CGRP-like peptides has been reported in several seafood protein by-products hydrolysates, such as cod hydrolysates from heads, stomach, viscera, and backbones (Fouchereau-Peron et al., 1999; Slizyte et al., 2009), sardine hydrolysates from heads and viscera (Ravallec-Plé et al., 2001; Rousseau et al., 2001), siki hydrolysates of industrial origin obtained from heads (Martínez-Álvarez et al., 2007), skin hydrolysates of North Atlantic lean fish (Picot et al., 2010) and saithe muscle hydrolysates (Martínez-Álvarez et al., 2008). These peptides have shown both agonist and antagonist properties *in vitro*, when their effect on the stimulation of cAMP production in rat liver membranes was tested. Thus, the CGRP-like molecules identified in the sardine hydrolysate obtained by Rousseau et al. (2001) induced an inhibition of the CGRP-stimulated adenylate cyclase activity (Rousseau et al., 2001), whereas CGRP-like molecules derived from siki heads exerted an agonistic effect (Martínez-Álvarez et al., 2007).

14.8 CONCLUSIONS

Fish processing industry generates large amounts of wastes. Sometimes some of these residues are used as fertilizer or as feed for animal/aquaculture, although in many other cases are discarded, resulting in leak of proteinaceous material and increment of environmental pollution. In recent years, there has been an increasing interest in finding an added value to the abundant fishing industry waste in order to seek sources of novel active compounds, which have great potential for use in food (as a source of proteins and/or additives), in addition to their use in nutraceuticals or pharmaceutical applications. Even though the use of these compounds is widespread, their utilization in the food area is restricted in some countries, as is the case of chitosan. In other cases, for example, fish hydrolysates, are not subjected to restricted use in foods in general, since they could be considered safe products. Owing to the diversity of origin, the potential functionality and their applications, studies in-depth of these compounds and their sustainable management, deserve to be of concern by the scientific community.

ACKNOWLEDGMENTS

The authors acknowledge Ministerio de Ciencia e Innovación (projects I3-2006701141, AGL2008-02135/ALI, AGL2011-27607) and CYTED 309ACO382 for their financial support.

REFERENCES

Aewsiri, T., Benjakul, S., Visessanguan, W., and Tanaka, M. 2008. Chemical compositions and functional properties of gelatin from pre-cooked tuna fin. *International Journal of Food Science and Technology*, 43(4), 685–693.

Aewsiri, T., Benjakul, S., and Visessanguan, W. 2009. Functional properties of gelatin from cuttlefish (*Sepia pharaonis*) skin as affected by bleaching using hydrogen peroxide. *Food Chemistry*, 115(1), 243–249.

Ahhmed, A.M. and Muguruma, M. 2010. A review of meat protein hydrolysates and hypertension. *Meat Science*, 86, 110–118.

Ahn, C.-B., Lee, K.-H., and Je, J.-Y. 2010. Enzymatic production of bioactive protein hydrolysates from tuna liver: Effects of enzymes and molecular weight on bioactivity. *International Journal of Food Science and Technology*, 45(3), 562–568.

Alder-Nissen, J. 1986. Methods in food protein hydrolysis. In: J. Adler-Nissen (Ed.), *Enzymic Hydrolysis of Food Proteins* (pp. 110–169). New York: Elsevier Applied Science Publishers.

Alemán, A., Giménez, B., Montero, P., and Gómez-Guillén, M.C. 2011c. Antioxidant activity of several marine skin gelatins. *LWT—Food Science and Technology*, 44, 407–413.

Alemán, A., Giménez, B., Pérez-Santín, E., Gómez-Guillén, M.C., and Montero, P. 2011b. Contribution of Leu and Hyp residues to antioxidant and ACE-inhibitory activities of peptides sequences isolated from squid gelatin hydrolysate. *Food Chemistry*, 125, 334–341.

Alemán, A., Pérez-Santín, E., Bordenave-Juchereau, S., Arnaudin, I., Gómez-Guillén, M.C., and Montero, P. 2011a. Antioxidant, ACE inhibitory and anticancer activities of squid gelatin hydrolysates. *Food Research International*, 44(4), 1044–1051.

Amit, K.R., Jini, R., Swapna, C.H., Sachindra, N.M., Bhaskar, N., and Baskaran, V. 2011. Application of native latic acid bacteria (LAB) for fermentative recovery of lipids and proteins for fish processing wastes: Bioactivities of fermentation products. *Journal of Aquatic Food Product Technology*, 20, 32–44.

Amit, K.R., Swapna, C.H., Bhaskar, N., Halami, P.M., and Sachindra, N.M. 2010. Effect on fermentation ensilaging on recovery of oil frem fresh water fish viscera. *Enzyme and Microbial Technology*, 46, 9–13.

Aranaz, I., Harris, R., and Heras, A. 2010. Chitosan amphiphilic derivatives. Chemistry and applications. *Current Organic Chemistry*, 14, 308–330.

Ariyoshi, Y. 1993. Angiotensin-converting enzyme inhibitors derived from food proteins. *Trends in Food Science and Technology*, 4, 139–144.

Arvanitoyannis, I.S. 2002. Formation and properties of collagen and gelatin films and coatings. Ch. 11. In: A. Gennadios (Ed.), *Protein-Based Films and Coatings* (pp. 275–304). Boca Ratón, FL: CRC Press.

Asghar, A. and Henrickson, R.L. 1982. Chemical, biochemical, functional, and nutritional characteristics of collagen in food systems. In: C.O. Chischester, E.M. Mark, and G.F. Stewart (Eds.), *Advances in Food Research* (Vol. 28) (pp. 232–372). London: Academic Press.

Atkinson, A.B. and Robertson, J.I.S. 1979. Captopril in the treatment of clinical hypertension and cardiac failure. *Lancet*, 2, 836–839.

Avena-Bustillos, R.J., Olsen, C.W., Olson, D.A., Chiou, B., Yee, E., Bechtel, P.J., and McHugh, T.H. 2006. Water vapor permeability of mammalian and fish gelatin films. *Journal of Food Science*, 71(4), 202–207.

Ayensa, M.G., An, H., Gómez-Guillén, M.C., Montero, P., and Borderías, A.J. 1999. Partial protease activity characterization of squid (*Todaropsis eblanae*) mantle. *Food Science and Technology International*, 5(5), 391–396.

Ayensa, M.G., Montero, P., Borderías, A.J., and Hurtado, J.L. 2002. Influence of some protease inhibitors on gelation of squid muscle. *Journal of Food Science*, 67(5), 1636–1641.

Bao, S., Xu, S., and Wang, Z. 2009. Antioxidant activity and properties of gelatin films incorporated with tea polyphenol-loaded chitosan nanoparticles. *Journal of the Science of Food and Agriculture*, 89(15), 2692–2700.

Barkia, A., Bougatef, A., Khaled, B., and Nasri, M. 2010. Antioxidant activities of *Sardinelle* heads and/or viscera protein hydrolysates prepared by enzymatic treatment. *Journal of Food Biochemistry*, 34(1), 303–320.

Battison, A.L., Summerfiled, R., and Patrzykat, A. 2008. Isolation and characterization of two antimicrobial peptides from haemocytes of the American lobster *Homarus americanus*. *Fish and Shellfish Immunology*, 25, 181–187.

Bechinger, B. and Lohner, K. 2006. Detergent-like actions of linear amphipatic cationic antimicrobial peptides. *BBA Biomembranes*, 1758(9), 1529–1539.

Benjakul, S., Oungbho, K., Visessanguan, W., Thiansilakul, Y., and Roytrakul, S. 2009. Characteristics of gelatin from the skins of bigeye snapper, *Priacanthus tayenus* and *Priacanthus macracanthus*. *Food Chemistry*, 116(2), 445–451.

Binsi, P.K., Shamasundar, B.A., Dileep, A.O., Badii, F., and Howell, N.K. 2009. Rheological and functional properties of gelatin from the skin of Bigeye snapper (*Priacanthus hamrur*) fish: Influence of gelatin on the gel-forming ability of fish mince. *Food Hydrocolloids*, 23(1), 132–145.

Borderías, J., Montero, P., and Martí de Castro, M.A. 1996. Gelling of hake (*Merluccius australis*) sawdust [Gelificación de serrín de merluza (*Merluccius australis*). *Food Science and Technology International*, 2(5), 293–299.

Borderías, A.J., Sánchez-Alonso, I., and Pérez-Mateos, M. 2005. New applications of fibres in foods: Addition to fishery products. *Trends in Food Science & Technology*, 16(10), 458–465.

Bougatef, A., Nedjar-Arroume, N., Manni, L., Ravallec, R., Barkia, A., Guillochon, D., and Nasri, M. 2010. Purification and identification of novel antioxidant peptides from enzymatic hydrolysates of sardinelle (*Sardinella surita*) by-products proteins. *Food Chemistry*, 118, 559–565.

Bougatef, A., Nedjar-Arroume, N., Ravallec-Plé, R., Leroy, Y., Guillochon, D., Barkia, A., and Nasri, M. 2008. Angiotensin I-converting enzyme (ACE) inhibitory activities of sardinlle (*Sardinella aurita*) by-products protein hydrolysates obtained by treatment with microbial and visceral fish serine proteases. *Food Chemistry*, 111, 350–356.

Bower, C.K., Avena-Bustillos, R.J., Hietala, K.A., Bilbao-Sainz, C., Olsen, C.W., McHugh, T.H. 2010. Dehydration of pollock skin prior to gelatin production. *Journal of Food Science*, 75(4), C317–C321.

Bray, G.A. 2000. Afferent signals regulating food intake. *Proceedings of the Nutrition Society*, 59, 373–384.

Byun, H.G. and Kim, S.K. 2001. Purification and characterization of angiotensin I converting enzyme (ACE) inhibitory peptides from Alaska pollack (*Theragra chalcogramma*) skin. *Process Biochemistry*, 36, 1155–1162.

Byun, H.G., Park, P.J., Sung, N.J., and Kim, S.K. 2003. Purification and characterization of a serine proteinase from the tuna puloric caeca. *Journal of Food Biochemistry*, 26, 479–494.

Cancre, I., Ravallec, R., Van Wormhoudt, A., Stenberg, E., Gildberg, A., and Le Gal, Y. 1999. Secretagogues and growth factors in fish and crustacean protein hydrolysates. *Marine Biotechnology*, 1(5), 489–494.

Cao, W., Zhang, C., Hong, P., and Ji, H. 2009. Optimising the free radical scavenging activity of shrimp protein hydrolysate produced with Alcalase using response surface methodology. *International Journal of Food Science and Technology*, 44(8), 1602–1608.

Cao, W., Zhang, C., Hong, P., Ji, H., and Hao, J. 2010. Purification and identification of an ACE inhibitory peptide from the peptic hydrolysate of *Acetes chinensis* and its antihypertensive effects in spontaneously hypertensive rats. *International Journal of Food Science and Technology*, 45, 959–965.

Careche, M., Borderías, A.J., and Sánchez-Alonso, I. 2004. Method of producing a functional protein concentrate from cephalopod muscle and product thus obtained, which is used in the production of analog products and other novel products. (PCT 24/06/04) Patent No. W2004/052117.

Carvalho, R.A., Sobral, P.J.A., Thomazine, M., Habitante, A.M.Q.B., Giménez, B., Gómez-Guillén, M.C., and Montero, P. 2008. Development of edible films based on differently processed Atlantic halibut (*Hippoglossus hippoglossus*) skin gelatin. *Food Hydrocolloids*, 22(6), 1117–1123.

Chen, H., Lin, C., and Kang, H. 2009. Maturation effects in fish gelatin and HPMC composite gels. *Food Hydrocolloids*, 23(7), 1756–1761.

Chen, H.M., Muramoto, K., and Yamauchi, F. 1995. Structural analysis of antioxidative peptides from soybean β-conglycinin. *Journal of Agriculture and Food Chemistry*, 43, 574–578.

Chen, Q., Xuan, G., Fu, M., He, G., Wang, W., Zhang, H., and Ruan, H. 2007. Effect of angiotensin I-converting enzyme inhibitory peptide from rice dregs protein on antihypertensive activity in spontaneously hypertensive rats. *Asian Pacific Journal of Clinical Nutrition*, 16(1), 281–285.

Chen, Y.M., Chung, Y.C., Wang, L.W., Chen, K.T., and Li, S.Y. 2002. Antibacterial properties of chitosan in waterborne pathogen. *Journal of Environmental Science. Health A*, 37, 1379–1390.

Cheung, H.S., Wang, F.L., Odentti, M.A., Sabo, E.F., and Cushman, D.W. 1980. Binding of peptide substrate and inhibitors of angiotensin-converting enzyme. *Journal of Biological Chemistry*, 255, 401–407.

Chiu, Y.T., Chiu, C.P., Chien, J.T., Ho, G.H., Yang, J., and Chen, B.H. 2007. Encapsulation of lycopene extract from tomato pulp waste with gelatin and poly(γ-glutamic acid) as carrier. *Journal of Agricultural and Food Chemistry*, 55(13), 5123–5130.

Cho, S.M., Gu, Y.S., and Kim, S.B. 2005. Extracting optimization and physical properties of yellowfin tuna (*Thunnus albacares*) skin gelatin compared to mammalian gelatins. *Food Hydrocolloids*, 19(2), 221–229.

Chobert, J.M., Sitohy, M.Z., and Whitaker, J.R. 1988. Solubility and Emulsifying properties of casein modified enzymatically by *Staphylococcus aureus* V8 protease. *Journal of Agricultural and Food Chemistry*, 36, 883–892.

Choi, S. and Regenstein, J.M. 2000. Physicochemical and sensory characteristics of fish gelatin. *Journal of Food Science*, 65(2), 194–199.

Chung, Y.C., Su, Y.P., Chen, C.C., Jia, G., Wang, H.L., Wu, J.C.G., and Lin, J.G. 2004. Relationship between antibacterial activity of chitosans and surface characteristics of cell wall. *Acta Pharmacologica Clinica*, 25, 932–936.

Cinq-Mars, C.D. and Li-Chan, E.C.Y. 2007. Optimizing angiotensin I-converting enzyme inhibitory activity of pacific hake (*Merluccius productus*) fillet hydrolysate using response surface methodology and ultrafiltration. *Journal of Agriculture and Food Chemistry*, 55, 9380–9388.

Contreras, M., Carrón, R., Montero, M.J., Ramos, M., and Recio, I. 2009. Novel casein-derived peptides with antihypertensive activity. *International Dairy Journal*, 19, 566–573.

Dallan, P.R.M. and da Luz Moreira, P. 2007. Effects of chitosan solution concentration and incorporation of chitin and glycerol on dense chitosan membrane properties. *Journal of Biomedical Materials Research Part B: Applied Biomaterials*, 80(2), 394–405.

Dávalos, A., Miguel, M., Bartolomé, B., and López-Fandiño, R. 2004. Antioxidant activity of peptides derived from egg white proteins by enzymatic hydrolysis. *Journal of Food Protection*, 67, 1939–1944.

De Lucca, A.J. and Walsh, T.J. 1999: Antifungal peptides: Novel therapeutic compounds against emerging pathogens. *Antimicrobial Agents Chemotherapy*, 43(1), 1–11.

Dekkers, E., Raghavan, S., Kristinsson, H.G., and Marshall, M.R. 2011. Oxidative stability of *Mahi mahi* red muscle dipped in tilapia protein hydrolysates. *Food Chemistry,* 124(2), 640–645.

Dermirbas, A., Pehlivan, E., and Altun, T. 2006. Potential evolution of R Turkish agricultural residues as biogas, bio-char and bio-oil sources. *International Journal of Hydrogen Energy*, 31, 613–620.

Devlieghere, F., Vermeulen, A., and Debevere, J. 2004. Chitosan: Antimicrobial activity, interactions with food components and applicability as a coating on fruit and vegetables. *Food Microbiology*, 21, 703–714.

Di Bernardini, R., Harnedy, P., Bolton, D., Kerry, J., O'Neill, E., Mullen, A.M., and Hayes, M. 2011. Antioxidant and antimicrobial peptidic hydrolysates from muscle protein sources and by-products. *Food Chemistry*, 124, 1296–1007.

Diniz, F.M. and Martin, A.M. 1996. Use of response surface methodology to describe the combined effects of pH, temperature and E/S ratio on the hydrolysis of dogfish (*Squalus acanthias*) muscle. *International Journal of Food Science and Technology*, 31, 419.

Djabourov, M., Lechaire, J., and Gaill, F. 1993. Structure and rheology of gelatin and collagen gels. *Biorheology*, 30(3–4), 191–205.

Dong, S., Song, H., Zhao, Y., Liu, Z., Wei, B., and Zeng, M. 2012. The preparation and antimicrobial activity of peptide fractions from blue mussel (*Mytilus edulis*) protein hydrolysate. *Applied Mechanisms and Materials*, 485, 340–347.

Dong, S., Zeng, M., Wang, D., Liu, Z., Zhao, Y., Yang, H. 2008. Antioxidant and biochemical properties of protein hydrolysates prepared from Silver carp (*Hypopththalmichthys molitrix*). *Food Chemistry*, 107, 1485–1493.

Dragnes, B.T., Stormo, S.K., Larsen, R., Ernstsen, H.H., and Elvevoll, E.O. 2009. Utilisation of fish industry residuals: Screening the taurine concentration and angiotensin converting enzyme inhibition potential in cod and salmon. *Journal of Food Composition and Analysis*, 22, 714–717.

Espitia, P.J.P., Soares, N.F.F., Coimbra, J.S.R., Andrade, N.J., Cruz, R.S., and Medeiros, E.A.A. 2012. Bioactive peptides: Synthesis, properties and applications in the packaging and preservation of food. *Comprehensive Reviews in Food Science and Technology,* 11(2), 187–204.

Fahmi, A., Morimura, S., Guo, H.C., Shigematsu, T., Kida, K., and Uemura, Y. 2004. Production of angiotensin I converting enzyme inhibitory peptides from sea bream scales. *Process Biochemistry*, 39, 1195–1200.

Food and Agriculture Organization [FAO]. 2010. *The State of Word Fisheries and Aquaculture 2010*. Rome, Italy: Food and Agriculture Organization of the United Nations, Electronic Publishing Policy and Support Branch.

Fernández-Díaz, M.D., Montero, P., and Gómez-Guillén, M.C. 2003. Effect of freezing fish skins on molecular and rheological properties of extracted gelatin. *Food Hydrocolloids*, 17(3), 281–286.

Ferraro, V., Cruz, I.B., Jorge, R.F., Malcata, F.X., Pintado, M.E., and Castro, P.M.L. 2010. Valorisation of natural extracts from marine source focused on marine by-products: A review. *Food Reseach International*, 43, 2221–2233.

Fitz-Gerald, R.J., Murray, B.A., and Walsh, D.J. 2004. Hypotensive peptides from milk proteins. *Journal of Nutrition*, 134, 980S–988S.

Floris, R., Recio, I., Berkhout, B., and Visser, S. 2003. Antibacterial and antiviral effects of milk proteins and derivatives thereof. *Current Pharmaceutical Design*, 9, 1257–1273.

Foh, M.B.K., Amadou, I., Foh, B.M., Kamara, M.T., Xia, W. 2010. Functionality and antioxidant properties of tilapia (*Oreochromis niloticus*) as influenced by the degree of hydrolysis. *International Journal of Molecular Science*, 11, 1851–1869.

Fouchereau-Peron, M., Duvail, L., Michel, C., Gildberg, A., Batista, I., and Le Gal, Y. 1999. Isolation of an acid fraction from a fish protein hydrolysate with a calcitonin-gene-related-peptide-like biological activity. *Biotechnology and Appied Biochemistry*, 29, 87–92.

Franco-Cereceda, A., Gennari, C., Nami, R., Agnusdei, D., Pernow, J., Lundberg, J.M., and Fischer, J.A. 1987. Cardiovascular effects of calcitonin gene related peptide I and II in man. *Circulation Research*, 60, 393–397.

Fujita, H. and Yoshikawa, M. 1999. LKPNM: A prodrug type ACE inhibitory peptide derived from fish protein. *Immunopharmacology*, 44, 123–127.

Fujita, H., Yokoyama, K., and Yoshikawa, M. 2000. Classification and antihypertensive activity of angiotensin I-converting enzyme inhibitory peptides derived from food proteins. *Journal of Food Science*, 65, 564–569.

Gallaher, D., Gallaher, C., Mahrt, G., Carr, T., Hollingshead, C., Hesslink, R., and Wise, J. 2002. A glucomannan and chitosan fiber supplement decreases plasma cholesterol and increases cholesterol excretion in overweight normocholesterolemic humans. *Journal of American College of Nutrition*, 21(5), 428–433.

Gauthier, S.F., Paquin, P., Pouliot, Y., and Turgeon, S. 1993. Surface activity and related functional properties of peptides obtained from whey proteins. *Journal of Dairy Science*, 76(1), 321–328.

Gbogouri, G.A., Linder, M., Fanni, J., and Parmentier, M. 2004. Influence of hydrolysis degree on the functional properties of salmon byproducts hydrolysates. *Journal of Food Science*, 69(8), C615–C622.

Giménez, B., Alemán, A., Montero, M.P., and Gómez-Guillén, M.C. 2009b. Antioxidant and functional properties of gelatin hydrolysates obtained from skin of sole and squid. *Food Chemistry*, 114, 976–983.

Giménez, B., Gómez-Estaca, J., Alemán, A., Gómez-Guillén, M.C., and Montero, M.P. 2009a. Physico-chemical and film forming properties of giant squid (*Dosidicus gigas*) gelatin. *Food Hydrocolloids*, 23(3), 585–592.

Giménez, B., Gómez-Guillén, M.C., and Montero, P. 2005b. The role of salt washing of fish skins in chemical and rheological properties of gelatin extracted. *Food Hydrocolloids*, 19(6), 951–957.

Giménez, B., Gómez-Guillén, M.C., and Montero, P. 2005c. Storage of dried fish skins on quality characteristics of extracted gelatin. *Food Hydrocolloids*, 19, 958–963.

Giménez, B., Turnay, J., Lizarbe, M.A., Montero, P., and Gómez-Guillén, M.C. 2005a. Use of lactic acid for extraction of fish skin gelatin. *Food Hydrocolloids*, 19(6), 941–950.

Gómez-Estaca, J., Gómez-Guillén, M.C., Fernández-Martín, F., and Montero, P. 2011. Effects of gelatin origin, bovine-hide and tuna-skin, on the properties of compound gelatin–chitosan films. *Food Hydrocolloids*, 25(6), 1461–1469.

Gómez-Guillén, M.C. and Montero, P. 2001. Extraction of gelatin from megrim (*Lepidorhombus boscii*) skins with several organic acids. *Journal of Food Science*, 66(2), 213–216.

Gómez-Guillén, M.C., Borderías, J., and Montero, P. 1997. Salt, non muscle proteins and hydrocolloids affecting rigidity changes during gelation of giant squid (*Dosidicus gigas*). *Journal of Agricultural and Food Chemistry*, 45(3), 616–621.

Gómez-Guillén, M.C., Giménez, B., and Montero, P. 2005. Extraction of gelatin from fish skins by high pressure treatment. *Food Hydrocolloids*, 19(5), 923–928.

Gómez-Guillén, M.C., Hurtado, J.L., and Montero, P. 2002a. Autolysis and protease inhibition effects on dynamic viscoelastic properties during thermal gelation of squid muscle. *Journal of Food Science*, 67(7), 2491–2496.

Gómez-Guillén, M.C., Giménez, B., López-Caballero, M.E., and Montero, P. 2011. Functional and bioactive properties of collagen and gelatin from alternative sources: A review. *Food Hydrocolloids*, 25(8), 1813–1827.

Gómez-Guillén, M.C., López-Caballero, M.E., López de Lacey, A., Alemán, A., Giménez, B., and Montero, P. 2010. Antioxidant and antimicrobial peptide fractions from squid and tuna skin gelatin. In: E. Le Bihan, and N. Koueta (Eds.), *Sea By-Products as a Real Material: New Ways of Application* (Chapter 7, pp. 89–115). Kerala, India: Transworld Research Network Signpost.

Gómez-Guillén, M.C., Pérez-Mateos, M., Gómez-Estaca, J., López-Caballero, E., Giménez, B., and Montero, P. 2009. Fish gelatin: A renewable material for the development of active biodegradable films. *Trends in Food Science and Technology*, 20, 3–16.

Gómez-Guillén, M.C., Turnay, J., Fernández-Díaz, M.D., Ulmo, N., Lizarbe, M.A., and Montero, P. 2002b. Structural and physical properties of gelatin extracted from different marine species: A comparative study. *Food Hydrocolloids*, 16(1), 25–34.

Gómez-Guillén, M.C., Borderías, J., and Montero, P. 1996. Rheological properties of gels made from high- and low-quality sardine (*Sardina pilchardus*) mince with added non-muscle proteins. *Journal of Agriculture and Food Chemistry*, 44, 746–750.

Gómez-Ruiz, J.A., Ramos, M., and Recio, I. 2004. Identification and formation of angiotensin-converting enzyme inhibitory peptides in Manchego cheese by high performance liquid chromatography-tandem mass spectrometry. *Journal of Chromatography A*, 1054, 269–277.

Goy, R., de Britto, D., and Asis, O. 2009. A review of the antimicrobial activity of chitosan. *Polymer Science and Technology*, 19(3), 241–247.

Gudmundsson, M. and Hafsteinsson, H. 1997. Gelatin from cod skins as affected by chemical treatments. *Journal of Food Science*, 62(1), 37–39 + 47.

Guerard, F., Sumaya-Martinez, M.T., Laroque, D., Chabeaud, A., and Dufossé, L. 2007. Optimization of free radical scavenging activity by response surface methodology in the hydrolysis of shrimp processing discards. *Process Biochemistry*, 42(11), 1486–1491.

Guinesi, L.S. and Cavalheiro, E.T.G. 2006. The use of DSC curves to determine the acetylation degree of chitin/chitosan samples. *Thermochimica Acta*, 444, 128–133.

Gupta, S., Mehrotra, S., Villalón, C.M., Garrelds, I.M., De Vries, R., Van Kats, J.P., Sharma, H.S., Saxena, P.R., Maassen Van Den Brink, A. 2006. Characterisation of CGRP receptors in human and porcine isolated coronary arteries: Evidence for CGRP receptor heterogeneity. *European Journal of Pharmacology*, 530(1–2), 107–116.

Hadwiger, L.A., Kendra, D.F., Fristensky, B.W., and Wagoner, W. 1986. Chitosan both activates genes in plants and inhibits RNA synthesis in fungi. In: Muzzarelli, R.A.A., Jeuniaux, C., Gooday, G.W. (Eds.), *Chitin in Nature and Technology* (pp. 209–214). New York: Plenum Press.

Harada, O., Kuwata, M., and Yamamoto, T. 2007. Extraction of gelatin from sardine scales by pressurized hot water. *Nippon Shokuhin Kagaku Kogaku Kaishi*, 54(6), 261–265.

Hardinge-Lyme, N. 2001. Chitosan-containing liquid compositions and methods for their preparation and use. E-nutraceuticals, Inc.; New York, NY; filed 1999 July 29. U.S. Patent 6,323,189 B1.

Hashiguchi, T., Kobayashi, T., Fongmoon, D., Shetty, A.K., Mizumoto, S., Miyamoto, N., Nakamura, T., Yamada, S., and Sugahara, K. 2011. Demonstration of the hepatocyte growth factor signalling pathway in the vitro neuritogenic activity of chondroitin ulfate from ray fish cartilage. *Biochimica et Biophysica Acta*, 1810, 406–413.

Haug, I.J., Draget, K.I., and Smidsrød, O. 2004. Physical and rheological properties of fish gelatin compared to mammalian gelatin. *Food Hydrocolloids*, 18(2), 203–213.

He, H.-L., Chen, X.-L., Sun, C.-Y., Zhang, Y.-Z., and Zhou, B.-C. 2006. Analysis of novel angiotensin-I-converting enzyme inhibitory peptides from protease-hydrolyzed marine shrimp *Acetes chinensis*. *Journal of Peptide Science*, 12, 726–733.

He, H.-L., Wu, H., Chen, X.-L., Shi, M., Zhang, X.-Y., Sun, C.-Y., Zhang, Y.-Z., and Zhou, B.-C. 2008. Pilot and plant scaled production of ACE inhibitory hydrolysates from *Acetes chinensis* and its *in vivo* antihypertensive effect. *Bioresource Technology*, 99, 5956–5959.

Helander, I.M., Nurmiaho-Lassila, E.-L., Ahvenainen, R., Rhoades, J., and Roller, S. 2001. Chitosan disrupts the barrier properties of the outer membrane of Gram-negative bacteria. *International Journal of Food Microbiology*, 71, 235–244.

Hernández-Ledesma, B., Dávalos, A., Bartolomé, B., and Amigo, L. 2005. Preparation of antioxidant enzymatic hydrolysates from α-lactalbumin and β-lactoglobulin. Identification of active peptides by HPLC-MS/MS. *Journal of Agricultural and Food Chemistry*, 53(3), 588–593.

Herpandi, N.H., Rosma, A., and Wan Nadiah, W.A. 2011. The tuna fishing industry. A new outlook of fish protein hydrolysates. *Comprehensive Reviews in the Food Science and Food Safety*, 10, 195–207.

Hou, H., Li, B. Zhao, X., Zhuang, Y., Ren, G., Yan, M., Cai, Y., Zhang, X., and Chen, L. 2009. The effect of Pacific cod (*Gadus macrocephalus*) skin gelatin polypeptides on UV radiation induced skin photoaging in ICR mice. *Food Chemistry*, 115(3), 945–950.

Huang, L., Chen, Y., and Morrissey, M.T. 1997. Coagulation of fish proteins from frozen fish mince wash water by ohmic heating. *Journal of Food Process Engineering*, 20, 285–300.

Huidobro, A., Montero, P., and Borderías, A.J. 1998. Emulsifying properties of an ultra-filtered protein from minced wash water. *Food Chemistry*, 61(3), 339–343.

Hultin, H.O. and Kelleher, S.D. 1999. Process for isolating a protein composition from a muscle source and protein composition. Patent US6005073.

Hultin, H.O. and Kelleher, S.D. 2000. High efficiency alkaline protein extraction. Patent US6136959.

Ichimura, T., Yamanaka, A., Otsuka, T., Yamashita, E., and Maruyama, S. 2009. Antihypertensive effect of enzymatic hydrolysate of collagen and Gly–Pro in spontaneously hypertensive rats. *Bioscience, Biotechnology and Biochemistry*, 73, 2317–2319.

Ikoma, T., Kobayashi, H., Tanaka, J., Walsh, D., and Mann, S. 2003. Physical properties of type I collagen extracted from fish scales of *Pagrus major* and *Oreochromis niloticas*. *International Journal of Biological Macromolecules*, 32, 199–204.

Jamilah, B. and Harvinder, K.G. 2002. Properties of gelatins from skins of fish black tilapia (*Oreochromis mossanbicus*) and red tilapia (*Oreochromis nilotica*). *Food Chemistry*, 77, 81–84.

Jang, A. and Lee, M. 2005. Purification and identification of angiotensin converting enzyme inhibitory peptides from beef hydrolysates. *Meat Science*, 69, 653–661.

Je, J.-Y., Park, P.-J., Kwon, J.Y., and Kim, S.-K. 2004. A novel angiotensin I converting enzyme inhibitory peptide from Alaska Pollack (*Theragra chalcogramma*) frame protein hydrolysate. *Journal of Agriculture and Food Chemistry*, 52(26), 7842–7845.

Je, J.Y., Qian, Z.J., Byun, H.G., and Kim, S.K. 2007. Purification and characterization of an antioxidant peptide obtained from tuna backbone protein by enzymatic hydrolysis. *Process Biochemistry*, 42, 840–846.

Je, J.Y., Qian, Z.J., Lee, S.H., Byun, H.G., and Kim, S.K. 2008. Purification and antioxidant properties of bigeye tuna (*Thunnus obesus*) dark muscle peptide on free radical-mediated oxidative systems. *Journal of Medicinal Food*, 11(4), 629–637.

Jellouli, K., Balti, R., Bougatef, A., Hmidet, N., Barkia, A., Nasri, M. 2011. Chemical composition and characteristics of skin gelatin from grey triggerfish (*Balistes capriscus*). *LWT—Food Science and Technology*, 44 (9), 1965–1970.

Jenssen, H., Hamill, P., and Hancock, R.E.W. 2006. Peptide antimicrobial agents. *Clinical Microbiology Reviews*, 19(3), 491–511.

Jeon, Y., Byun, H., and Kim, S. 1999. Improvement of functional properties of cod frame protein hydrolysates using ultrafiltration membranes. *Process Biochemistry*, 35(5), 471–478.

Jeon, Y.J., Shahidi, F., and Kim, S.K. 2000. Preparation of chitin and chitosan oligomers and their applications in physiological functional foods. *Food Review International*, 16(2), 159–176.

Johnson, L.R.E., Copeland, E.M., and Dudrick, S.J. 1978. Luminal gastrin stimulates growth of the dial intestine. *Scandinavian Journal of Gastroenterology*, 13(49), 95.

Johnston-Banks, F.A. 1990. Gelatin. In: P. Harris (Ed.), *Food Gels* (pp. 233–289). London: Elsevier Applied Science Publishers.

Jongjareonrak, A., Rawdkuen, S., Chaijan, M., Benjakul, S., Osako, K., and Tanaka, M. 2010. Chemical compositions and characterisation of skin gelatin from farmed giant catfish (*Pangasianodon gigas*). *LWT—Food Science and Technology*, 43, 161–165.

Jun, S.-Y., Park, P.-J., Jung, W.-K., and Kim, S.-K. 2004. Purification and characterization of an antioxidative peptide from enzymatic hydrolysate of yellowfin sole (*Limanda aspera*) frame protein. *European Food Research and Technology*, 219, 20–26.

Jung, W.-K. and Kim, S.-K. 2007. Calcium-binding peptide derived from pepsinolytic hydrolysates of hoki (*Johnius belengerii*) frame. *European Food Research and Technology*, 224, 763–767.

Jung, W.-K., Mendis, E., Je, J.-Y., Park, P.-J., Son, B.W., Kim, H.C., Choi, Y.K., and Kim, S.K. 2006. Angiotensin I-converting enzyme inhibitory peptide from yellowfin sole (*Limanda aspera*) frame protein and its antihypertensive effect in spontaneously hypertensive rats. *Food Chemistry*, 94, 26–32.

Jung, W.K., Park, P.J., Byun, H.G., Moon, S.H., and Kim, S.K. 2005. Preparation of hoki (*Johnius belengerii*) bone oligo phospho peptide with a high affinity to calcium by carnivorous intestine crude proteinase. *Food Chemistry*, 91, 333–340.

Kamil, J.Y.V.A. and Jeon, Y.J. 2002. Chitosan as an edible invisible film for quality preservation of herring and Atlantic cod. *Journal of Agricultural and Food Chemistry*, 50(18), 5167–5178.

Kannan, A., Hettiarachchy, N.S., Marshall, M., Raghavan, S., and Kristinsson, H. 2011. Shrimp shell peptide hydrolysate inhibit human cancer cell proliferation. *Journal of the Science of Food and Agriculture*, 91, 1920–1924.

Karim, A.A. and Bhat, R. 2009. Fish gelatin: Properties, challenges, and prospects as an alternative to mammalian gelatins. *Food Hydrocolloids*, 23(3), 563–576.

Kasankala, L.M., Xue, Y., Weilong, Y., Hong, S.D., and He, Q. 2007. Optimization of gelatine extraction from grass carp (*Catenopharyngodon idella*) fish skin by response surface methodology. *Bioresource Technology*, 98(17), 3338–3343.

Katayama, K., Fuchu, H., Sakata, A., Kawahara, S., Yamauchi, K., Kawamura, Y., and Muguruma, M. 2003. Angiotensin I-converting enzyme inhibitory activities of porcine skeletal muscle proteins following enzyme digestion. *Asian-Australian Journal of Animal Science*, 16, 417–424.

Khantaphant, S., Benjakul, S., and Kishimura, H. 2011. Antioxidative and ACE inhibitory activities of protein hydrolysates from the muscle of brownstripe red snapper prepared using pyloric caeca and commercial proteases. *Process Biochemistry*, 46, 318–327.

Kim, A.-K. and Mendis, E. 2006. Bioactive compounds from marine processing by-products—A review. *Food Research International*, 39, 383–393.

Kim, H.J., Yoon, M.S., Park, K.H., Shin, J.H., Heu, M.S., and Kim, J.-S. 2010. Processing optimization of gelatin from rockfish skin based on yield. *Fisheries and Aquatic Science*, 13(1), 1–11.

Kim, J.S. and Park, J.W. 2004. Characterization of acid-soluble collagen from pacific whiting surimi processing byproducts. *Journal of Food Science*, 69(8), 637–642.

Kim, J.S. and Park, J.W. 2005. Partially purified collagen from refiner discharge of pacific whiting surimi processing. *Journal of Food Science*, 70(8), 511–516.

Kim, S.K. and Wijesekara, I. 2010. Development and biological activities of marine-derived bioactive peptides: A review. *Journal of Functional Foods*, 2, 1–9.

Kim, S.K., Mendis, E., and Shahidi, F. 2008. Marine fisheries by-products as potential nutraceuticals: An overview. In: C. Barrow and F. Shahidi (Eds.), *Marine Nutraceuticals and Functional Foods* (pp. 1–22). Boca Raton, FL: CRC Press.

Kim, S., Kim, Y., Byun, H., Nam, K., Joo, D., and Shahidi, F. 2001a. Isolation and characterization of antioxidative peptides from gelatin hydrolysate of Alaska pollack skin. *Journal of Agricultural and Food Chemistry,* 49(4), 1984–1989.

Kim, S.K., Byun, H.G., Park, P.J., and Shahidi, F. (2001b). Angiotensin I converting enzyme inhibitory peptides purified from bovine skin gelatin hydrolysate. *Journal of Agricultural and Food Chemistry,* 49(6), 2992–2997.

Kim, S.K., Jeon, Y.J., Lee, B.J., and Lee, C.K. 1996. Purification and characterization of the gelatin from the bone of cod (*Gadus macrocephalus*). *Korean Journal of Life Science,* 6, 14–26.

Kim, S.-Y., Je, J.-Y., and Kim, S.-K. 2007. Purification and characterization of antioxidant peptide from hoki (*Johnius belengerii*) frame protein by gastrointestinal digestion. *Journal of Nutritional Biochemistry,* 18, 31–38.

Kim, S-K., and Wijesekara, I. 2010. Development and biological activities of marine-derived bioactive peptides: A review. *Journal of Functional Foods,* 2, 1–9.

Klompong, V., Benjakul, S., Kantachote, D., and Shahidi, F. 2007. Antioxidative activity and functional properties of protein hydrolysate of yellow stripe trevally (*Selariodes leptolepis*) as influenced by the degree of hydrolysis and enzyme type. *Food Chemistry,* 102, 1317–1327.

Kolodziejska, I., Kaczorowski, K., Piotrowska, B., and Sadowska, M. 2004. Modification of the properties of gelatin from skins of Baltic cod (*Gadus morhua*) with transglutaminase. *Food Chemistry,* 86(2), 203–209.

Kolodziejska, I., Skierka, E., Sadowska, M., Kołodziejski, W., and Niecikowska, C. 2008. Effect of extracting time and temperature on yield of gelatin from different fish offal. *Food Chemistry,* 107(2), 700–706.

Kong, X., Zhou, H., and Qian, H. 2007. Enzymatic preparation and functional properties of wheat gluten hydrolysates. *Food Chemistry,* 101, 615–620.

Korhonen, H. and Pihlanto, A. 2003. Food-derived bioactive peptides-opportunities for designing future foods. *Current Pharmaceutical Design,* 9, 1297–1308.

Kristinsson, H. and Hultin, H.O. 2003. Effect of low and high pH treatment on the functional properties of cod muscle proteins. *Journal of Agricultural and Food Chemistry,* 51(17), 5103–5110.

Kristinsson, H.G. 2007. Aquatic food protein hydrolysates. In: F. Shahidi (ed.), *Maximising the Value of Marine By-Products* (pp. 229–248). Cambridge: Woodhead Publishing Ltd.

Kristinsson, H.G. and Rasco, B.A. 2000a. Biochemical and functional properties of Atlantic salmon (*Salmo salar*) muscle proteins hydrolyzed with various alkaline proteases. *Journal of Agricultural and Food Chemistry,* 48, 657–666.

Kristinsson, H.G. and Rasco, B.A. 2000b. Fish protein hydrolysates: Production, biochemical, and functional properties. *Critical Reviews in Food Science and Nutrition,* 40, 43–81.

Kumar, M.N.V. 2000. A review of chitin and chitosan applications. *Reactive and Functional Polymers,* 46(1), 1–27.

Kumar, M.N.V.R., Muzzarelli, R.A.A., Muzzarelli, C., Sashiwa, H., and Domb, A.J. 2004. Chitosan chemistry and pharmaceutical perspectives. *Chemical Reviews,* 104(12), 6017–6084.

Kurita, K. 2006. Chitin and chitosan: Functional biopolymers from marine crustaceans. *Marine Biotechnology,* 8, 203–226.

Ledward, D.A. 1986. Gelation of gelatin. In: J.R. Mitchell, and D.A. Ledward (Eds.), *Functional Properties of Food Macromolecules* (pp. 171–201). London: Elsevier Applied Science Publishers.

Lee, S.-H., Qian, Z.-J., and Kim, S.-K. 2010. A novel angiotensin I converting enzyme inhibitory epptide from tuna frame protein hydrolysate and its antihypertensive effect in spontaneously hypertensive rats. *Food Chemistry,* 118(1), 96–102.

Lenz, H.J., Mortrud, M.T., Vale, W.W., Rivier, J.E., and Brown, M.R. 1984. Calcitonin gene related peptide acts within the central nervous system to inhibit gastric acid secretion. *Regulatory Peptides,* 9, 271–277.

Leptihn, S., Har, J.Y., Chen, J.C., Ho, B., Wohland, T., and Ding, J.L. 2009. Single molecule resolution of the antimicrobial action of quantum dot-labeled sushi peptide on live bacteria. *BMC Biology*, 7, 22–34.

Leroi, F. 2010. Occurrence and role of lactic acid bacteria in seafood products. *Food Microbiology*, 27, 698–709.

Leuenberger, B.H. 1991. Investigation of viscosity and gelation properties of different mammalian and fish gelatins. *Food Hydrocolloids*, 5(4), 353–361.

Li, C.M., Zhong, Z.H., Wan, Q.H., Zhao, H., Gu, H.F., and Xiong, S.B. 2008. Preparation and thermal stability of collagen from scales of grass carp (*Ctenopharyngodon idellus*). *European Food Research and Technology*, 227(5), 1467–1473.

Li, X.Y., Chen, X.G., Cha, D.S., Park, H.J., and Liu, C.S. 2009. Microencapsulation of a probiotic bacteria with alginate gelatin and its properties. *Journal of Microencapsulation*, 26(4), 315–324.

Li, Y., Rodrigues, J., and Tomás, H. 2012. Injectable and biodegradable hydrogels: Gelation, biodegradation and biomedical applications. *Chemical Society Reviews*, 41, 2193–2221.

Lian, W., Hsiao, H., and Chou, C. 2002. Survival of bifidobacteria after spray-drying. *International Journal of Food Microbiology*, 74(1–2), 79–86.

Liceaga-Gesualdo, A.M. and Li-Chan, E.C.Y. 1999. Functional properties of fish protein hydrolysates from herring (*Clupea harengus*). *Journal of Food Science*, 64(6), 1000–1004.

Lin, L. and Li, B. 2006. Radical scavenging properties of protein hydrolysates from Jumbo flying squid (*Dosidicus eschrichitii* Steenstrup) skin gelatin. *Journal of the Science of Food and Agriculture*, 86(14), 2290–2295.

Liu, Z.Y., Dong, S.Y., Xu, J., Zeng, M.Y., Song, H.X., and Zhao, Y.H. 2008. Production of cysteine-rich antimicrobial peptide by digestion of oyster (*Crassostrea gigas*) with alcalase and bromelin. *Food Control*, 19, 231–235.

López-Caballero, M.E., Gómez-Guillén, M.C., Pérez-Mateos, M., and Montero, P. 2005. A chitosan–gelatin blend as a coating for fish patties. *Food Hydrocolloids*, 19, 303–311.

Maezaki, Y., Tsuji, K., Nakagawa, Y., and Akimoto, M. 1993. Hypocholesterolemic effect of chitosan in adult males. *Bioscience, Biotechnology and Biochemistry*, 57, 1439–1444.

Mahmoud, M.I. 1994. Physicochemical and functional properties of protein hydrolysates in nutritional products. *Food Technology*, 48(10), 89–95.

Malde, M.K., Graff, I.E., Siljander-Rasi, H., Venäläinen, E., Julshamn, K., Pedersen, J.I., and Valaja, J. 2010. Fish-bones—A highly available calcium source for growing pigs. *Journal of Animal Physiology and Animal Nutrition*, 94, e66–e76.

Marcuse, R. 1960. Antioxidative effect of amino-acids. *Nature*, 186, 886–887.

Márquez-Ríos, E., Moran-Palacio, E.F., Lugo-Sanchez, M.E., Ocano-Higuera, V.M., and Pacheco-Aguilar, R. 2007. Postmortem biochemical behavior of giant squid (*Dosidicus gigas*) mantle muscle stored in ice and its relation with quality parameters. *Journal of Food Science*, 72(7), C356–C362.

Martínez-Álvarez, O., Guimas, L., Delannoy, C., and Fouchereau-Peron, M. 2007. Occurrence of a CGRP-like molecule in siki (*Centroscymnus coelolepsis*) hydrolysate of industrial origin. *Journal of Agriculture and Food Chemistry*, 55, 5469–5475.

Martínez-Álvarez, O., Guimas, L., Delannoy, C., and Fouchereau-Peron, M. 2008. Use of a commercial protease and yeasts to obtain CGRP-like molecules from saithe protein. *Journal of Agricultural and Food Chemistry*, 56, 7853–7859.

Megías, C., Pedroche, J., Yust, M.M., Girón-Calle, J., Alaiz, M., Millán, F., and Vioque, J. 2008. Production of copper-chelating peptides after hydrolysis of sunflower proteins with pepsin and pancreatin. *Food Science and Technology*, 41, 1973–1977.

Meisel, H. 2003. Casokinins as bioactive peptides in the primary structure of casein. In: Sawatzki, G., Renner, B. (Eds.), *New Perspectives in Infant Nutrition* (pp. 153–159). New York: Thieme, Stuttgart.

Mendis, E., Rajapakse, N., and Kim, S.K. 2005b. Antioxidant properties of a radicals scavenging peptide purified from enzymatically prepared fish skin gelatin hydrolysate. *Journal of Agriculture and Food Chemistry*, 53, 581–587.

Mendis, E., Rajapakse, N., Byun, H., and Kim, S. 2005a. Investigation of jumbo squid (*Dosidicus gigas*) skin gelatin peptides for their *in vitro* antioxidant effects. *Life Sciences*, 77(17), 2166–2178.

Miguel, M., Recio, I., Gómez-Ruiz, J.A., Ramos, M., and López-Fandiño, R. 2004. Angiotensin I-converting enzyme inhibitory activity of peptides derived from egg white proteins by enzymatic hydrolysis. *Journal of Food Protection*, 67(9), 1914–1920.

Mito, K., Fujii, M., Kuwahara, M., Matsumura, N., Shimizu, T., Sugano, S., and Karaki, H. 1996. Antihypertensive effect of angiotensin I-converting enzyme inhibitory peptides from hemoglobin. *European Journal of Pharmacology*, 304, 93–98.

Miyoshi, S., Ishikawa, H., Kaneko, T., Fukui, F., Tanaka, H., and Maruyama, S. 1991. Structures and activity of angiotensin-converting enzyme inhibitors in an α-zein hydrolysate. *Journal of Agriculture and Biology Chemistry*, 55(5), 1313–1318.

Mohtar, N.F., Perera, C., and Quek, S. 2010. Optimisation of gelatine extraction from hoki (*Macruronus novaezelandiae*) skins and measurement of gel strength and SDS–PAGE. *Food Chemistry*, 122(1), 307–313.

Montero, P. and Gómez-Guillén, M.C. 1998. Recovery and functionality of wash water protein from krill processing. *Journal of Agriculture and Food Chemistry*, 46, 3300–3304.

Montero, P. and Pérez-Mateos, M. 2001. Mince gels with hydrocolloids and salts: Composition/function relationships and discrimination of functionality by multivariate analysis. *European Food Research Technology*, 213, 338–342.

Montero, P. Giménez, B., Pérez-Mateos, M., and Gómez-Guillén, M.C. 2005. Oxidation stability of muscle with quercetin and rosemary during thermal and high-pressure gelation. *Food Chemistry*, 93 17–23.

Montero, P., Borderías, J., Turnay, J., and Leyzarbe, M.A. 1990. Characterization of hake (*Merluccius merluccius* L.) and trout (*Salmo irideus* Gibb) collagen. *Journal of Agricultural and Food Chemistry*, 38(3), 604–609.

Moreno, H.M., Carballo, J., and Borderias, J. 2010. Gelation of fish muscle using microbial transglutaminase and the effect of sodium chloride and pH levels. *Journal of Muscle Foods*, 21(3), 433–450.

Morrissey, M.T., Lin, J., and Ismond, A. 2005. Waste management and by-product utilization. In: J.W. Park (Ed.), *Surimi and Surimi Seafood* (2nd ed.). (pp. 279–323). Boca Raton, FL: CRC Press, Taylor & Francis Group.

Murado, M.A., Fraguas, J., Montemayor, M.I., Vázquez, J.A., and González, P. 2010a. Preparation of highly purified chondroitin sulphate from skate (*Raja clavata*) cartilage by-products. Process optimization including a new procedure of alkaline hydroalcoholic hydrolysis. *Biochemical Engineering Journal*, 49, 126–132.

Murado, M.A., Montemayor, M.I, Cabo, M.L.,Vázquez, J.A., and González, M.P. 2010b. Optimization of extraction and purification process of hyaluronic acid from fish eyeball. *Food Bioproducts Proccesing*, 90(3), 491–498.

Murray, B.A. and FitzGerald, R.J. 2007. Angiotensin converting enzyme inhibitory peptides derived from food proteins: Biochemistry, bioactivity and production. *Current Pharmaceutical Design*, 13, 773–791.

Muyonga, J.H., Cole, C.G.B., and Duodu, K.G. 2004a. Characterisation of acid soluble collagen from skins of young and adult Nile perch (*Lates niloticus*). *Food Chemistry*, 85(1), 81–89.

Muyonga, J.H., Cole, C.G.B., and Duodu, K.G. 2004b. Fourier transform infrared (FTIR) spectroscopic study of acid soluble collagen and gelatin from skins and bones of young and adult Nile perch (*Lates niloticus*). *Food Chemistry*, 86(3), 325–332.

Muzzarelli, R.A.A. 1997. Human enzymatic activities related to the therapeutic administration of chitin derivatives. *Cellular and Molecular Life Sciences*, 53, 131–140.

Nagai, T., Nagashima, T., Abe, A., and Suzuki, N. 2006. Antioxidative activities and angiotensin I-converting enzyme inhibition of extracts prepared from chum salmon (*Oncorhynchus keta*) cartilage and skin. *International Journal of Food Properties*, 9(4), 813–822.

Najafian, L. and Babji, A.S. 2012. A review of fish-derived antioxidant and antimicrobial peptides: Their production, assessment, and applications. *Peptides*, 33, 178–185.

Nalinanon, S., Benjakul, B., Kishimura, H., and Sahidi, F. 2011. Functionality and antioxidant properties of protein hydrolysates from muscle of ornate threadfin treated with pepsin from skipjack tuna. *Food Chemistry*, 124, 1354–1326.

Nalinanon, S., Benjakul, S., Visessanguan, W., and Kishimura, H. 2008. Tuna pepsin: Characteristics and its use for collagen extraction from the skin of threadfin bream (*Nemipterus* spp.). *Journal of Food Science*, 73(5), C413–C419.

Nam, K.A., You, S.G., and Kim, S.M. 2008. Molecular and physical characteristics of squid (*Toradores pacificus*) skin collagens and biological properties of their enzymatic hydrolysates. *Journal of Food Science*, 73(4), 249–255.

Naqash, S.Y. and Nazeer, R.A. 2010. Antioxidant activity of hydrolysates and peptides fractions of *Nemipterus japonicas* and *Exocoetus volitans* muscle. *Journal of Aquatic Food Product Technology*, 19(3–4), 180–192.

Naqash, S.Y. and Nazeer, R.A. 2011. Evaluation of bioactive properties of peptide isolated from *Exocoetus volitans* backbone. *International Journal of Food Science & Technology*, 46, 37–43.

Niki, H., Kato, T., Deya, E., and Igarashi, S. 1985. Recovery of protein from effluent of fish meat in producing surimi and utilization of recovered protein. [Surimi seizo no okeru gyoniku mizusarashieki karano tanpakushitsu no kaishu to sono riyo.] *Nippon Suisan Gakkaishi*, 51(6), 959–964.

Ninan, G., Joseph, J., and Abubacker, Z. 2010. Physical, mechanical, and barrier properties of carp and mammalian skin gelatin films. *Journal of Food Science*, 75(9), E620–E626.

Noga, E.J., Arrol, T.A., and Fan, Z. 1996. Specific and some psysicochemical characterisitics of the antibacterial activity from blue crab *Callinectus sapidus*. *Fish and Shellfish Immunology*, 6, 403–413.

Nomura, Y., Sakai, H., Ishii, Y., and Shirai, K. 1996. Preparation and some properties of type I collagen from fish scales. *Bioscience Biotechnology and Biochemistry*, 60, 2092–2094.

Norland, R.E. 1990. Fish gelatin. In M.N. Voight, and J.K. Botta, *Advances in Fisheries Technology and Biotechnology for Increased Profitability* (pp. 325–333). Lancaster: Technomic Publishing Co.

Norziah, M.H., Al-Hassan, A., Khairulnizam, A.B., Mordi, M.N., Norita, M. 2009. Characterization of fish gelatin from surimi processing wastes: Thermal analysis and effect of transglutaminase on gel properties. *Food Hydrocolloids*, 23(6), 1610–1616.

Odentti, M.A. 1977. Design of specific inhibitors of angiotensin converting enzyme: New class of orally active antihypertensive agents. *Science*, 196, 441–444.

Odentti, M.A. and Cushman, D.W. 1982. Enzymes of the renin–angiotensin system and their inhibitors. *Annual Review of Biochemistry*, 51, 283–308.

Odentti, M.A., Rubin, B., and Cushman, D.W. 1982. Enzyme of the renin–angiotensin system and their inhibitors. *Annual Review of Biochemistry*, 51, 283–308.

Ogawa, M., Portier, R.J., Moody, M.W., Bell, J., Schexnayder, M.A., and Losso, J.N. 2004. Biochemical properties of bone and scale collagens isolated from the subtropical fish black drum (*Pogonia cromis*) and sheepshead seabream (*Archosargus probatocephalus*). *Food Chemistry*, 88(4), 495–501.

Oh, J.K., Drumright, R., Siegwart, D.J., and Matyjaszewski, K. 2008. The development of microgels/nanogels for drug delivery applications. *Progress in Polymer Science*, 33, 448–477.

Pacheco-Aguilar, R., Mazorra-Manzano, M.A., and Ramírez-Suárez, J.C. 2008. Functional properties of fish protein hydrolysates from Pacific whiting (*Merluccius productus*) muscle produced by a commercial protease. *Food Chemistry*, 109, 782–789.

Papineau, A., Hoover, D., Knorr, D., and Farkas, D. 1991. Antimicrobial effect of water-soluble chitosans with high hydrostatic pressure. *Food Biotechnology,* 5(1), 45–57.

Park, C.H., Kim, H.J., Kang, K.T., Park, J.W., and Kim, J.S. 2009. Fractionation and angiotensin I-converting enzyme (ACE) inhibitory activity of gelatin hydrolysates from byproducts of Alaska pollack surimi. *Fisheries and Aquatic Science*, 12(2), 79–85.

Park, J. 2005. Ingredient technology for surimi and surimi seafood. Chapter 14. In: J.W. Park (Ed.), *Surimi and Surimi Seafood*. Boca Raton, FL: Taylor & Francis. ISBN 0-8247-2649-9.

Park, P.J., Jung, W.K., Nam, K.S., Shahidi, F., and Kim, S.K. 2001. Purification and characterization of antioxidative peptides from protein hydrolysate of lecithin-free egg yolk. *Journal of American Oil and Chemists' Society*, 78, 651–656.

Patchett, A.A., Harris, E., Tristram, E.W., Wyvratt, M.J., Wu, M.T., Taub, D., Peterson, E.R. et al. 1980. A new class of angiotensin-converting enzyme inhibitors. *Nature*, 298, 280–283.

Patil, D. and Nag, A. 2011. Production of PUFA concentrates from poultry and fish processing waste. *Journal of American Oils Chemistry Society*, 88, 589–593.

Pérez-Mateos, M. and Montero, P. 2000. Contribution of hydrocolloids to gelling properties of blue whiting muscle. *European Food Reseach Technology*, 210, 383–390.

Pérez-Mateos, M., Montero, P., and Gómez-Guillén, M.C. 2009. Formulation and stability of biodegradable films made from cod gelatin and sunflower oil blends. *Food Hydrocolloids*, 22(4), 53–61.

Phanturat, P., Benjakul, S., Visessanguan, W., and Roytrakul, S. 2010. Use of pyloric caeca extract from bigeye snapper (*Priacanthus macracanthus*) for the production of gelatin hydrolysate with antioxidative activity. *LWT-Food Science and Technology*, 43(1), 86–97.

Picot, L., Bordenave, S., Didelot, S., Fruitier-Arnaudin, I., Sannier, F., Thorkelsson, G, Bergé, J.P., Guérard, F., Chabeaud, A., and Piot, J.M. 2006. Antiproliferative activity of fish protein hydrolysates on human breast cancer cell lines. *Process Biochemistry*, 41, 1217–1222.

Picot, L., Ravallec, R., Martine, F.-P., Vandanjon, L., Jaouen, P., Chaplain-Derouiniot, M., Guérard, F. et al. 2010. Impact of ultrafiltration and nanofiltration of an industrial fish protein hydrolysate on its bioactive properties. *Journal of the Science of Food and Agriculture*, 90, 1819–1826.

Pietrowski, B.N., Tahergorabi, R., and Jaczynski, J. 2012. Dynamic rheology and thermal transitions of surimi seafood enhanced with ω-3-rich oils. *Food Hydrocolloids*, 27(2), 384–389.

Pihlanto, A., Akkanen, S., and Korhonen, H. 2008. ACE-inhibitory and antioxidant properties of potato (*Solanum tuberosum*). *Food Chemistry*, 109, 104–112.

Pranoto, Y., Lee, C.M., and Park, H.J. 2007. Characterizations of fish gelatin films added with gellan and κ-carrageenan. *LWT—Food Science and Technology*, 40(5), 766–774.

Qian, Z.-J., Je, J.-Y., and Kim, S.-K. 2007. Antihypertensive effect of angiotensin I converting enzyme-inhibitory peptide from hydrolysates of Bigeye tuna dark muscle, *Thunnus obesus*. *Journal of Agriculture and Food Chemistry*, 55, 8398–8403.

Qian, Z.J., Jung, W.K., and Kim, S.K. 2008. Free radical scavenging activity of a novel antioxidative peptide purified from hydrolysate of bullfrog skin, *Rana catesbeiana Shaw*. *Bioresource Technology*, 99, 1690–1698.

Qin, Y. and Agboh, O.C. 1997. Chtin and chitosan fibers. *Polymers for Advances Technologies*, 8, 355–365.

Quaglia, G.B. and Orban, E. 1987. Enzymic solubilisation of proteins of sardine (*Sardina pilchardus*) by commercial proteases. *Journal of the Science of Food and Agriculture*, 38, 263.

Quaglia, G.B. and Orban, E. 1990. Influence of enzymatic hydrolysis on structure and emulsifying properties of sardine (*Sardina pilchardus*) protein hydrolysates. *Journal of Food Science*, 55(6), 1571.

Quirós, A., Ramos, M., Muguerza, B., Delgado, M.A., Miguel, M., Aleixandre, A., and Recio, I. 2007. Identification of novel antihypertensive peptides in milk fermented with *Enterococcus faecalis*. *International Dairy Journal*, 17, 33–41.

Raghavan, S. and Kristinsson, H.G. 2007. Conformational and rheological changes in catfish myosin as affected by different acids during acid-induced unfolding and refolding. *Journal of Agricultural and Food Chemistry*, 55(10), 4144–4153. *Food Hydrocolloids*, 27, 384–389.

Raghavan, S. and Kristinsson, H.G. 2008b. Antioxidative efficacy of alkali-treated tilapia protein hydrolysates: A comparative study of five enzymes. *Journal of Agricultural and Food Chemistry*, 56, 1434–1441.

Raghavan, S. and Kristinsson, H.G. 2008a. Conformational and rheological changes in catfish myosin during alkali-induced unfolding and refolding. *Food Chemistry*, 107(1), 385–398.

Rahman, M.S. and Al-Mahrouqi, A.I. 2009. Instrumental texture profile analysis of gelatin gel extracted from grouper skin and commercial (bovine and porcine) gelatin gels. *International Journal of Food Science and Nutrition*, 60(7), 229–242.

Rajanbabu, V. and Chen, J.-Y. 2011. Applications of antimicrobial peptides from fish and perspectives for future. *Peptides*, 32(2), 415–420.

Rajapakse, N., Jung, W.-K., Mendis, E., Moon, S.-H., and Kim, S.-K. 2005a. A novel anticoagulant purified from fish protein hydrolysate inhibits factor XIIa and platelet aggregation. *Life Sciences*, 76, 2607–2619.

Rajapakse, N., Mendis, E., Byun, H.G., and Kim, S.K. 2005b. Purification and *in vitro* antioxidative effects of giant squid muscle peptides on free radical-mediated oxidative systems. *Journal of Nutritional Biochemistry*, 16, 562–569.

Ravallec-Plé, R. and Van Wormhoudt, A. 2003. Secretagogue activities in cod (*Gadhus morhua*) and shrimp (*Penaeus aztecus*) extracts and alcalase hydrolysates determined in AR4-2J pancreatic tumor cells. *Comparative Biochemistry and Physiology*, 134, 669–679.

Ravallec-Plé, R., Charlot, C., Pires, C., Braga, V., Batista, I., Van Wormhoudt, A., Gal, Y.L., Fouchereau-Péron, M. 2001. The presence of bioactive peptides in hydrolysates prepared from processing waste of sardine (*Sardina pilchardus*). *Journal of the Science of Food and Agriculture*, 81, 1120–1125.

Ravallec-Plé, R., Gilmartin, L., Van Wormhoudt, A., and Le Gal, Y. 2000. Influence of the hydrolysis process on the biological activities of protein hydrolysates from cod (*Gadus morhua*) muscle. *Journal of the Science of Food and Agriculture*, 80, 2176–2180.

Ravichandran, S., Wahidulla, S., D´Souza, L., and Rameshkumar, G. 2010. Antimicrobial lipids from the hemolymph of Branchyuran crabs. *Applied Biochemistry and Biotechnology*, 162, 1039–1051.

Reddy, K.V.R., Yedery, R.D., and Aranha, C. 2004. Antimicrobial peptides: Premises and promises. *International Journal of Antimicrobial Agents*, 24(6), 536–547.

Ren, J.Y., Zhao, M.M., Shi, J., Wang, J.S., Jiang, Y.M., Cui, C., Kakuda, Y., and Xue, J. 2008. Purification and identification of antioxidant peptides from grass carp muscle hydrolysates by consecutive chromatography and electrospray ionization-mass spectrometry. *Food Chemistry*, 108, 727–736.

Revilla, E., Maria, C.S., Miramontes, E., Bautista, J., García-Martínez, A., Cremades, O., Cert, R., and Parrado, J. 2009. Nutraceutical composition, antioxidant activity and hypocholesterolemic effect of a water-soluble enzymatic extract from rice bran. *Food Research International*, 42, 387–393.

Rinaudo, M. 2006. Chitin and chitosan: Properties and applications. *Progress in Polymer Science*, 31(7), 603–632.

Rousseau, M., Batista, I., Le Gal, Y., and Fouchereau-Peron, M. 2001. Purification of a functional competitive antagonist for calcitonin gene related peptide action from sardine hydrolysate. *Electronic Journal of Biotechnology*, 4(1), 25–32.

Ryan, J.T., Ross, R.P., Bolton, D., Fitzgerald, G.F., and Stanton, C. 2011. Bioactive peptides from muscle sources: Meat and fish. *Nutrients*, 3, 765–791.

Rydlo, T., Milt, J., and Mor, A. 2006. Eukaryotic antimicrobial peptides: Promises and premises in food safety. *Journal of Food Science*, 71(9), R125–R135.

Sakanaka, S. and Tachibana, Y. 2006. Active oxygen scavenging activity of egg-yolk protein hydrolysates and their effects on lipid oxidation in beef and tuna homogenates. *Food Chemistry*, 95, 243–249.

Samaranayaka, A.G.P. and Li-Chan, E.C.Y. 2008. Autolysis-assisted production of fish protein hydrolysates with antioxidant properties from Pacific hake (*Merluccius productus*). *Food Chemistry*, 107, 768–776.

Sánchez-Brambilla, G.Y., Alvarez-Manilla, G., Soto-Cordova, F., Lyon, B.G., and Pacheco-Aguilar, R. 2004. Identification and characterization of the off-flavor in mantle muscle of jumbo squid (*Dosidicus gigas*) from the Gulf of California. *Journal of Aquatic Food Product Technology*, 13(1), 55–67.

Santos, Jr. D.S., Goulet, P.J.G., Pieczonka, N.P.W., Oliveira, Jr., and Aroca, R. 2004. Gold nanoparticle embedded, self-sustained chitosan films as substrates for surface-enhanced Raman scattering. *Langmuir*, 20, 10273–10277.

Sathivel, S., Bechtel, P.J., Babbitt, J., Smiley, S., Crapo, C., Reppond, K.D., and Prinyawiwatkul, W. 2003. Biochemical and functional properties of herring (*Clupea harengus*) byproduct hydrolysates. *Journal of Food Science*, 68(7), 2196–2200.

Sathivel, S., Smiley, S., Prinyawiwatkul, W., and Bechtel, J. 2005. Functional and nutritional properties of red salmon (*Oncorhynchus nerka*) enzymatic hydrolysates. *Journal of Food Science*, 70(6), 401–406.

Schrieber, R. and Gareis, H. 2007. Gelatin handbook. In: *Theory and Industrial Practice*. Weinheim, Germany: Wiley-VCH Verlag GmbH & Co. KGaA.

Shahidi, F. and Abuzaytoun, R. 2005. Chitin, chitosan, and co-products: Chemistry, production, applications, and health effects. *Advances in Food and Nutrition Research*, 49, 93–135.

Shahidi, F., Arachchi, J.K.V., and Jeon, Y. 1999. Food applications of chitin and chitosan. *Trends in Food Science and Technology*, 10(2), 37–51.

Shahidi, F., Metusalach, J., and Brown, J.A. 1998. Carotenoid pigments in seafoods and aquaculture. *Critical Reviews in Food Science*, 38(1), 1–67.

Shahidi, F., Xiao-Quing, H., and Synowiecki, J. 1995. Production and characteristics of protein hydrolysates from capelin (*Mallotus villosus*). *Food Chemistry*, 53, 285–293.

Shimizu, K., Sato, M., Zhang, Y., Kouguchi, T., Takahata, Y., Morimatsu, F., and Shimizu, M. 2010. The bioavailable octapeptide Gly-Ala-Hyp-Gly-Leu-Hyp-Gly-Pro stimulates nitric oxide synthesis in vascular endothelial cells. *Journal of Agriculture and Food Chemistry*, 58, 6960–6965.

Simpson, B.K., Nayeri, G., Yaylayan, V., and Ashie, I.N.A. 1998. Enzymatic hydrolysis of shrimp meat. *Food Chemistry*, 61, 131–138.

Sing, D.K. and Ray, A.R. 2000. Biomedical applications of chitin, chitosan and their derivatives. *Polymer Reviews*, 40(1), 69–80.

Sini, K.T., Santosh, S., and Mathew, T.-P. 2007. Study of the production of chitin and chitosan from shrimp shell by using *Bacillus subtilis* fermentation. *Carbohydrate Reseach*, 342, 2423–2429.

Slizyte, R., Mozuraityte, R., Martínez-Álvarez, O., Falch, E., Fouchereau-Peron, M., and Rustad, T. 2009. Functional, bioactive and antioxidative properties of hydrolysates obtained from cod (*Gadus morhua*) backbones. *Process Biochemistry*, 44, 668–677.

Soares, N.F.F., Pires, A.C.S., Camilloto, G.P., Santiago-Silva, P., Espetia, P.J.P., and Silva, W.A. 2009. Active intelligent packaging for milk and milk products. In: Coimbra, J.S.R., Texeira, J.A. (Eds.). *Engineering Aspects of Milk and Dairy Products* (pp. 155–174). New York: CRC Press Taylor & Francis Group.Songchotikunpan, P., Tattiyakul, J., and Supaphol, P. 2008. Extraction and electrospinning of gelatin from fish skin. *International Journal of Biological Macromolecules*, 42(3), 247–255.

Souissi, N., Bougatef, A., Triki-Ellouz, Y., and Nasri, M. 2007. Biochemical and functional properties of sardinella (*Sardinella aurita*) by-product hydrolysates. *Food Technology and Biotechnology*, 45(2), 187–194.

Stainsby, G. 1987. Gelatin gels. In: A.M. Pearson, T.R. Dutson, and A.J. Bailey (Eds.), *Advances in Meat Research, Collagen as a Food* (Vol. 4) (pp. 209–222). New York: Van Nostrand Reinhold Company Inc.

Stensvag, K., Haug, T., Sperstad, S.V., and Rekdal, O. 2008. Arasin 1, a proline-arginine rich antimicrobial peptide isolated from the spider crab *Hyas araneus*. *Development and Comparative Immunology*, 32, 275–285.

Stine, J.J., Pedersen, L., Smiley, S., and Bechtel, P.J. 2012. Recovery and utilization of protein derived from surimi wash-water. *Journal of Food Quality*, 35, 43–50.

Strauss, G. and Gibson, S.M. 2004. Plant phenolics as cross-linkers of gelatin gels and gelatin-based coacervates for use as food ingredients. *Food Hydrocolloids*, 18(1), 81–89.

Sudarshan, N.R., Hoover, D.G., and Knorr, D. 1992. Antibacterial action of chitosan. *Food Biotechnology*, 6, 257–272.

Suetsuna, K., Ukeda, H., and Ochi, H. 2000. Isolation and characterization of free radical scavenging activities peptides derived from casein. *Journal of Nutritional Biochemistry*, 11, 128–131.

Sugahara, K., Mikami, T., Uyama, S., Mizuguchi, K., Nomura, H., and Kitagawa, H. 2003. Recent advances in the structural biology of chondroitin sulfate and dermatan sulfate. *Current Opinion Structural Biology*, 13, 612–620.

Suh, J.K. and Matthew, H.W. 2000. Application of chitosan-based polysaccharide biomaterials in cartilage tissue engineering: A review. *Biomaterials*, 21(24), 2589–2598.

Swapna, C.H., Bijinu, B., Amit, K.R., and Bhaskar, N. 2011. Simultaneous recovery of lipids and proteins by enzymatic hydrolysis of fish industry waste using different commercial proteases. *Applied Biochemistry and Biotechnology*, 164, 115–124.

Taheri, A., Abedian Kenari, A.M., Gildberg, A., and Behnam, S. 2009. Extraction and physicochemical characterization of greater lizardfish (*Saurida tumbil*) skin and bone gelatin. *Journal of Food Science*, 74(3), E160–E165.

Theodore, A.E., Raghavan, S., and Kristinsson, H.G. 2008. Antioxidative activity of protein hydrolysates prepared from alkaline-aided channel catfish protein isolates. *Journal of Agricultural and Food Chemistry*, 56, 7459–7466.

Thiansilakul, Y., Benjakul, S., and Shahidi, F. 2007. Compositions, functional properties and antioxidative activity of protein hydrolysates prepared from round scad (*Decapterus maruadsi*). *Food Chemistry*, 103, 1385–1394.

Tsai, J.S., Lin, T.C., Chen, J.L., and Pan, B.S. 2006. The inhibitory effects of freshwater clam (*Corbicula fluminea*, Muller) muscle protein hydrolysates on angiotensin I converting enzyme. *Process Biochemistry*, 41, 2276–2281.

Turgeon, S.L., Gauthier, S.F., Molle, D., and Leonil, J. 1992. Interfacial properties of tryptic peptides of b-lactoglobulin. *Journal of Agricultural and Food Chemistry*, 40, 669–675.

Vázquez, J.A., Nogueira, M., Durán, A., Prieto, M.A., Rodríguez-Amado, I., Rial, D., González, M.P., and Murado, M.A. 2011. Preparation of marine silage of swordfish, ray and shark viscera waste by lactic acid bacteria. *Journal of Food Engineering*, 103, 442–448.

Wainewright, F.W. 1977. Physical tests for gelatin and gelatin products. In: Ward, A.G., Couts, A. (Eds). *The Science and Technology of Gelatin* (pp. 507–534). New York: Academic Press.

Wang, S.M., Huang, Q.Z., and Wang, Q.S. 2005. Study on the synergetic degradation of chitosan with ultraviolet light and hydrogen peroxide. *Carbohydrate Research*, 340, 1143–1147.

Wang, Y. and Regenstein, J.M. 2009. Effect of EDTA, HCl, and citric acid on Ca salt removal from Asian (silver) carp scales prior to gelatin extraction. *Journal of Food Science*, 74(6), C426–C431.

Wang, Y., Yang, H., and Regenstein, J.M. 2008. Characterization of fish gelatin at nanoscale using atomic force microscopy. *Food Biophysics*, 3, 269–272.

Wangtueai, S. and Noomhorm, A. 2009. Processing optimization and characterization of gelatin from lizardfish (*Saurida* spp.) scales. *LWT—Food Science and Technology*, 42, 825–834.

Wasswa, J., Tang, J., Gu, X., and Yuan, X. 2007. Influence of the extent of enzymatic hydrolysis on the functional properties of protein hydrolysate from grass carp (*Ctenopharyngodon idella*) skin. *Food Chemistry*, 104(4), 1698–1704.

Weissbrodt, J. and Kunz, B. 2007. Influence of hydrocolloid interactions on their encapsulation properties using spray-drying. *Minerva Biotecnologica*, 19(1), 27–32.

Wu, H.C., Chen, H.M., and Shiau, C.Y. 2003. Free amino acids and peptides as related to antioxidant properties in protein hydrolysates of mackerel (*Scomber austriasicus*). *Food Research International*, 36, 949–957.

Wu, J. and Ding, X. 2001. Hypotensive and physiological effect of angiotensin converting enzyme inhibitory peptides derived from soy protein on spontaneously hypertensive rats. *Journal of Agriculture and Food Chemistry*, 49, 501–506.

Xu, X., Jha, A.K., Harrington, D.A., Farach-Carson, M.C., and Jia, X. 2012. Hyaluronic acid-based hydrogels: From a natural polysaccharide to complex networks. *Soft Materials*, 8, 3280–3294.

Yang, H., Wang, Y., Regenstein, J.M., and Rouse, D.B. 2007. Nanostructural characterization of catfish skin gelatin using atomic force microscopy. *Journal of Food Science*, 72, C430–C440.

Yang, J., Ho, H., Chu, Y., and Chow, C. 2008. Characteristic and antioxidant activity of retorted gelatin hydrolysates from cobia (*Rachycentron canadum*) skin. *Food Chemistry*, 110(1), 128–136.

Yeo, Y., Bellas, E., Firestone, W., Langer, R., and Kohane, D.S. 2005. Complex coacervates for thermally sensitive controlled release of flavor compounds. *Journal of Agricultural and Food Chemistry*, 53(19), 7518–7525.

Yongsawatdigul, J., Park, J.W., and Kolbe, E. 1997. Texture degradation kinetics of gels made from Pacific whiting surimi. *Journal of Food Process Engineering*, 20(6), 433–452.

You, L., Zhao, M., Liu, R.H., and Regenstein, J.M. 2011b. Antioxidant and antiproliferative activities of loach (*Misgrunus anguillicaudatus*) peptides prepared by papain digestion. *Journal of Agricultural and Food Chemistry*, 59, 7948–7953.

You, L., Zhao, M., Regenstein, J.M., and Ren, J. 2011a. *In vitro* antioxidant activity and *in vivo* anti-fatigue effect of loach (*Misgurnus anguillicaudatus*) peptides prepared by papain digestion. *Food Chemistry*, 124, 188–194.

Yust, M.M., Pedroche, J., Girón-Calle, J., Alaiz, M., Millán, F., and Vioque, M. 2003. Production of ACE inhibitory peptides by digestion of chickpea legumin with Alcalase. *Food Chemistry*, 81, 363–369.

Zhang, F., Xu, S., and Wang, Z. 2011. Pre-treatment optimization and properties of gelatin from freshwater fish scales. *Food and Bioproducts Processing*, 89(3), 185–193.

Zhang, J., Zhang, J.K., Song, Y., and He, B.K. 2005. Preparation of Cibacron Blue F3GA attached chitosan microspheres and their adsorption properties for bovine serum album. *Chemical Journal of Chinese Universities*, 26, 2363–2368.

Zhang, S., Wang, Y., Herring, J.L., and Oh, J. 2007. Characterization of edible film fabricated with channel catfish (*Ictalurus punctatus*) gelatin extract using selected pretreatment methods. *Journal of Food Science*, 72(9), 498–503.

Zhao, Y., Li, B., Liu, Z., Dong, S., Zhao, X., and Zeng, M. 2007. Antihypertensive effect and purification of an ACE inhibitory peptide from sea cucumber gelatin hydrolysate. *Process Biochemistry*, 42, 1586–1591.

Zhong, S., Ma, C., Lin, Y.C., and Luo, Y. 2011. Antioxidant properties of peptide fractions from silver carp (*Hypophthalmichthys molitrix*) processing by-product protein hydrolysates evaluated by electron spin resonance spectrometry. *Food Chemistry*, 126(4), 1636–1642.

Zhou, P. and Regenstein, J.M. 2007. Comparison of water gel desserts from fish skin and pork gelatins using instrumental measurements. *Journal of Food Science*, 72(4), C196–C201.

Zhou, P., Mulvaney, S.J., and Regenstein, J.M. 2006. Properties of Alaska pollock skin gelatin: A comparison with tilapia and pork skin gelatins. *Journal of Food Science*, 71, C313–C321.

Zhuang, Y., Hou, H., Zhao, X., Zhang, Z., and Li, B. 2009. Effects of collagen and collagen hydrolysate from jellyfish (*Rhopilema esculentum*) on mice skin photoaging induced by UV irradiation. *Journal of Food Science*, 74(6), H183–H188.

Zohuriaan-Mehr, M.J., Pourjavadi, A., Salimi, H., and Kurdtabar, M. 2009. Protein- and homo poly(amino acid)-based hydrogels with super-swelling properties. *Polymers for Advanced Technologies*, 20(8), 655–671.

Zúñiga, R.N. and Aguilera, J.M. 2009. Structure-fracture relationships in gas-filled gelatin gels. *Food Hydrocolloids*, 23(5), 1351–1357.

15 Coffee

Solange I. Mussatto and José A. Teixeira

CONTENTS

15.1 INTRODUCTION

Large amounts of by-products are constantly generated during the processing of coffee beans. Such by-products are generated in different stages, including the processing of the beans (by wet or dry process), their roasting and brewing, and basically comprise the husks, pulp, parchment, silverskin, and spent grounds. Owing to the generation in large amounts, it is of great interest for the coffee industry to find alternatives for their valorization due to both economical and environmental reasons, since they present a toxic character when disposed directly in the environment, and also because the management of industrial residues and by-products promotes the sustainable development of a country's economy.

Recently, some attempts have been made to use these by-products in the food industry since they contain antioxidant compounds that could be incorporated into other food products. Coffee by-products are also rich in sugars and minerals, and could be used as low-cost nutrient and/or carbon source for fermentative processes. Another valuable alternative for their reuse would be as biomass to produce energy, through direct burning or after conversion to ethanol or biodiesel. Coffee by-products can also be used to adsorb compounds and for the manufacturing of industrially relevant products such as organic acids and enzymes. In this chapter, the characteristics of these industrial by-products will be presented, and the most recent advances for their valorization will be discussed.

15.2 COFFEE FRUIT: PRODUCTION AND IMPORTANCE

Coffee fruit (also called berry or cherry) is the raw fruit produced from the plant of the botanical genus *Coffea*, of which there are more than 70 species. However, most of the coffee beverage consumed around the world is elaborated from the variety *Coffea arabica* (Arabica) that is considered the noblest among all coffee plants and also the variety with better sensory quality (Bertrand et al. 2003). Arabica coffee corresponds to approximately 75% of the worldwide coffee production, while the variety *Coffea canephora* (Robusta), which is more acidic, corresponds to the remaining 25%.

Approximately 60 countries produce coffee and for some of them, coffee is the main agricultural export product. The 10 largest coffee-producing countries are presented in Table 15.1. These countries are responsible for approximately 85% of the total worldwide production. Brazil, Vietnam, and Indonesia are, respectively, the first, second, and third largest producers, responsible for more than half of the worldwide production of coffee (USDA 2012). Arabica coffee is the variety cultivated in larger amount in these 10 largest coffee-producing countries, except in Vietnam, Indonesia, and India, which cultivate the Robusta variety predominantly.

The importance of coffee is mainly due to the coffee brew prepared from the roasted and ground beans. Coffee brew has been consumed for over 1000 years and today it is the most consumed drink in the world. Owing to the presence of caffeine, coffee brew is known as a stimulant; however, the number of chemical compounds identified in this beverage is large, and several of them have many beneficial health properties. Chlorogenic acid, for example, is an abundant phenolic acid in coffee and is reported to have a potent antioxidant activity as well as hepatoprotective, hypoglycemic, antibacterial, antiviral, anti-inflammatory, and anticarcinogenic activities (Farah and Donangelo 2006; Shan et al. 2009). Carbohydrates are also present in

TABLE 15.1
Ten Largest Coffee-Producing Countries (Values in Thousand 60-Kilogram Bags)

Country	Production 2010/2011	Arabica Coffee	Robusta Coffee
Brazil	54,500	41,800	12,700
Vietnam	18,735	585	18,150
Indonesia	9325	1375	7950
Colombia	8525	8525	–
India	5040	1575	3465
Ethiopia	4400	4400	–
Peru	4100	4100	–
Mexico	4000	3800	200
Honduras	3900	3900	–
Guatemala	3810	3800	10
Others	20,017	10,884	9133
Total	136,352	84,744	51,608

coffee and are reported to have important biological activities such as the ability to lower the colon cancer risk (Arya and Rao 2007).

15.3 COFFEE BEANS PROCESSING AND BY-PRODUCTS GENERATION

Coffee fruits are usually harvested when the bear fruit turns red, which normally occurs after 5 years of coffee tree plantation. This fruit is composed of two beans covered by a thin parchment, a silverskin, and further surrounded by a pulp and a husk (outer skin). Figure 15.1 illustrates a coffee bean and its structures.

The processing of the cherries initiates with their conversion into green coffee beans, and starts with the removal of both the pulp and husk, which can be done by either dry or wet methods. The dry method consists of drying the harvested coffee fruits (naturally in the sun or artificially in furnaces) and then the dried husk, pulp, parchment, and sometimes the silverskin are mechanically removed. Optionally, the silverskin can be removed by a polishing machine. In contrast, the wet process involves the use of considerable amounts of water to remove the pulp and husk, and also includes a microbial fermentation step that is able to remove any mucilage still attached to the beans. During the fermentation, the mucilage is degraded by enzymes that can be derived from both the coffee tissues and microorganisms found on the fruit skins (Belitz et al. 2009). The microorganisms that act in this step have a large influence on the final quality of the coffee beans, since the production of microbial volatile compounds results in coffee with richer aroma quality. When compared to the dry method, coffee processed by the wet method is generally considered to have a better aroma and, therefore, higher acceptance (Bytof et al. 2005). In spite of the differences between the dry and wet processing methods, both methodologies generate the coffee husk, pulp, parchment, and sometimes the silverskin as by-products. After this step, the green coffee beans are obtained, and the grains with imperfections are not used in the subsequent processing steps. The appearance of selected and rejected (with imperfections) green coffee beans are shown in Figure 15.2.

The coffee beans roasting is also an important step in coffee processing, since specific organoleptic properties (flavors, aromas, and color) are developed during

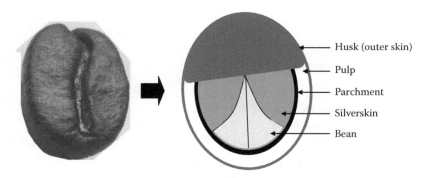

FIGURE 15.1   Appearance and structures of a coffee bean (longitudinal cross section).

FIGURE 15.2   Green coffee beans: selected (left) and rejected—with imperfections (right).

this stage, which contribute to the quality of the beans as well as of the beverage that will be produced from them (Hernández et al. 2008). This process is currently performed at 260°C for about 5 min, and leads to several changes in the chemical composition of coffee beans, as can be seen in Table 15.2. In addition, roasting promotes moisture loss and changes in the color, volume, mass, form, pH, density, and volatile composition of the green coffee beans. Figure 15.3 shows the coffee bean's appearance before and after the roasting process. Coffee silverskin (the tegument that covers the beans) is the main by-product obtained after roasting.

The roasted beans are then packed as whole beans or after grinding, as illustrated in Figure 15.4. Packaging is usually done in hermetic packages under vacuum or under inert gas to allow storage for long periods.

TABLE 15.2
Chemical Composition of Green and Roasted Coffee Beans (Values in % Dry Weight)

Constituent	Green Coffee	Roasted Coffee
Hemicellulose	23.0	24.0
Protein	11.6	3.1
Cellulose	12.7	13.2
Fat	11.4	11.9
Lignin	5.6	5.8
Chlorogenic acid	7.6	3.5
Sucrose	7.3	0.3
Caffeine	1.2	1.3
Trigonelline	1.1	0.7
Reducing sugars	0.7	0.5
Unknown	14.0	31.7

Source: Adapted from Arya, M., and Rao, L.J.M. 2007. *Crit. Rev. Food Sci. Nutr.* 47:51–67.

FIGURE 15.3 Green (left) and roasted (right) coffee beans.

An additional step of processing can be required when the objective is to produce instant coffee. In this case, an extraction procedure using water at 175°C under pressurized conditions is performed after the roasting and grinding of the beans to extract soluble solids and volatile compounds that provide aroma and flavor. A subsequent step of extraction, evaporation or freeze-concentration, is used to increase the soluble materials' concentration of the extract. The concentrated extracts are then dried to produce instant coffee. The residual solid material obtained after the aqueous extraction, which is called spent coffee grounds, is the main by-product generated during the instant coffee preparation, and is obtained in very significant amounts, with a worldwide annual production of 6,000,000 tons (Mussatto et al. 2011a).

FIGURE 15.4 Whole and ground coffee beans.

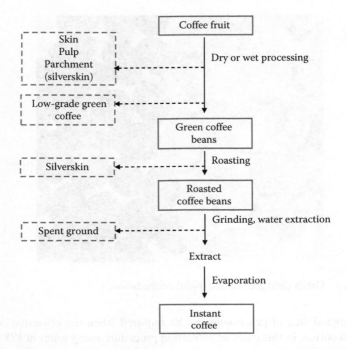

FIGURE 15.5 Schematic representation of the by-product generation during the coffee fruit processing.

A schematic representation of the by-products generation during the coffee fruit processing is shown in Figure 15.5.

15.4 COFFEE BY-PRODUCTS: CHARACTERISTICS AND APPLICATIONS

The generation of by-products is very common in any productive sector. The coffee industry, particularly, is responsible for the generation of a large amount of by-products. It is estimated that approximately 50% of the coffee fruit is not used for the production of commercialized coffee, being discarded during the processing step to obtain roasted coffee. The spent coffee grounds are also a by-product generated in significant amounts in the instant coffee industry (about 650 kg per each ton of green coffee, and about 2 kg in the wet form to each 1 kg of soluble coffee produced) (Mussatto et al. 2011a). Owing to the large amounts in which these by-products are constantly generated, there has been a strong wish to find applications for their reuse, for both economical and environmental reasons. In the last decade, this subject has been the focus of several studies involving biotechnological, chemical, and physical processes. Coffee by-products can be used, for example, in fermentative processes as substrates for microbial growth to produce a variety of valuable compounds, including enzymes, organic acids, and biofuels. Nonfermentative applications include the extraction of bioactive molecules such as antioxidant phenolic compounds. The next sections summarize the main studies and potential applications proposed to reuse coffee by-products.

15.4.1 Low-Grade Green Coffee

Low-grade green coffee is the term used to define the coffee beans that present imperfections, which may include beans of black or dark brown color, immature beans, beans damaged by insects, and so on, as shown in Figure 15.2. Imperfect coffee beans are usually obtained in significant amounts, comprising up to 20% of the coffee production.

Extracts produced from this by-product presented interesting biological activities, such as strong radical-scavenging, antioxidant, and antitumor activity (Ramalakshmi et al. 2008; 2009), which could be a consequence of the presence of specific compounds in their composition, such as caffeine, trigonelline, and chlorogenic acids. These findings open up important possibilities for the reuse of this by-product in the food industry, for example, since natural antioxidant compounds are preferred than the synthetic ones. Antioxidant and antitumor compounds also have a number of applications in the pharmaceutical area. However, the incorporation of these active molecules into food and pharmaceutical systems require safety studies to determine the toxicological and carry-through effects and dosage level.

15.4.2 Coffee Husks

Coffee husks are mainly composed of the coffee berry outer skin, but are usually obtained together with the pulp and the parchment after the processing step. However, the wet coffee processing produces a slightly different by-product, due to the pressing of the fruit through a screen in water; this procedure leaves part of the pulp, the mucilage, and the parchment attached to the seeds (Belitz et al. 2009). Figure 15.6 illustrates the coffee husk appearance. This material represents about 12% of the berry on a dry-weight basis and in terms of generation, approximately 0.18 ton are produced from each ton of coffee fruit processed (Adams and Dougan 1981). However, during

FIGURE 15.6 Coffee husks (outer skin).

the coffee cherry processing, the husks are usually separated with fractions of peel and pulp, and the mixture of these by-products may comprise nearly 45% of the coffee cherry, constituting one of the main by-products of this industry.

Coffee husks contain approximately 24.5% cellulose, 29.7% hemicelluloses, 23.7% lignin, and 6.2% ashes (Bekalo and Reinhardt 2010). Owing to the high amount of cellulose and hemicellulose fractions in the composition, coffee husks can be considered a material similar to wood and then, it could be used in several applications where wood are usually employed. In a recent study, for example, coffee husks were used for the production of particle boards and the produced coffee husk-wood board showed suitable characteristics for use in structural and nonstructural panel products. It was proposed to use coffee husks in substitution of up to 50% of wood for the production of particle board products (Bekalo and Reinhardt 2010).

Owing to the presence of antiphysiological and antinutritional compounds in their composition, such as tannins and caffeine, coffee husks are unsuitable to be directly used as animal feed or as the substrate for bioconversion processes. In these cases, a previous detoxification of this material would be required to remove (partially or entirely) such compounds, making these applications possible. Several studies have been performed with the objective of detoxifying coffee husks among which, fungal cultivation under solid-state fermentation (SSF) conditions has been considered a promising alternative. Fungal strains from the genera *Rhizopus*, *Phanerochaete*, and *Aspergillus*, in particular *Aspergillus* sp., showed good prospects for the detoxification of this material (Brand et al. 2000).

In fact, coffee husks are a good material for the application in SSF processes for the cultivation of edible fungi as well as for the production of different valuable compounds, due to their low cost and nutritional composition that allows the growth of microorganisms. For example, the production of citric acid by *Aspergillus niger* (Shankaranand and Lonsane 1994) and gibberellins (plant growth regulators) by *Gibberella fujikuroi* and *Fusarium moniliforme* (Machado et al. 2002) under SSF conditions using coffee husks as a carbon source has been reported. In the second case, it was verified that the production of gibberellic acid was higher under SSF conditions than under submerged fermentation conditions. Coffee husks can also be used as the substrate in SSF for the production of aroma compounds. The use of steam-treated coffee husk supplemented with glucose for the cultivation of *Ceratocystis fimbriata* allowed producing strong pineapple and banana aroma compounds (Soares et al. 2000a,b). Coffee husks is also a suitable raw material for the production of different enzymes, including tannase (Battestin and Macedo 2007), xylanase (Murthy and Naidu 2012a), α-amylase (Murthy et al. 2009), and protease (Murthy and Naidu 2010).

The use of coffee husks for the production of biofuels has also been reported. This material presented an excellent potential for the use in ethanol production by fermentation with *Saccharomyces cerevisiae* since it was able to provide the ethanol production results satisfactorily comparable to others reported for residues such as corn stalks, barley straw, and hydrolyzed wheat stillage (Gouvea et al. 2009). Coffee husks were also used as a raw material for biomethanation. To date, the use of coffee husks for this purpose has been limited due to their acidic pH (4.3) and the presence of antinutritional/phytotoxic factors (polyphenols) in their composition. However, the cultivation of the fungus *Mycotypha* sp. in coffee husks lowered the acidity of this material and increased the methane production (Jayachandra et al. 2011).

Some nonfermentative applications for coffee husks have also been evaluated and include their use as biosorbents for the removal of dye or heavy metals from contaminated waters (Oliveira et al. 2008a,b). Coffee husks can also be used as a source of bioactive and functional compounds (Murthy and Naidu 2012b) as well as phytochemicals, which are compounds of interest for the food and pharmaceutical industries. Fresh coffee husks composed of the outer skin and the pulp are a potential source of anthocyanins (compounds responsible for the red/blue coloration in many fruits), mainly cyanidin-3-rutinoside, which is a pigment with an important application in the food industry (Prata and Oliveira 2007).

15.4.3 COFFEE PULP

Coffee pulp is an important by-product generated during the coffee berry processing since it represents 29% of the whole berry on a dry-weight basis, and is produced in approximately 0.5 ton per one ton of coffee fruit processed (Bressani 1978). Coffee pulp is composed of carbohydrates, proteins, and minerals (especially potassium), and it also contains appreciable amounts of tannins, polyphenols, and caffeine. The polyphenols in Arabica coffee pulp can be divided into four major classes, namely, flavan-3-ols, hydroxycinnamic acids, flavonols, and anthocyanidins, and include, among other compounds, chlorogenic acid, epicatechin, catechin, rutin, protocatechuic acid, and ferulic acid (Ramírez-Coronel et al. 2004; Ramírez-Martínez 1988).

The elimination of this coffee by-product is a problem due to the large amount generated and because the direct disposal to the nature causes environmental problems. Therefore, it is of great interest to find alternatives for the coffee pulp reuse. Along with the coffee husks, the direct use of the coffee pulp for animal feed is restricted due to the presence of undesirable compounds in its composition, such as polyphenols, tannins, and caffeine (Orozco et al. 2008). The presence of these compounds has also hindered the use of coffee pulp as a substrate for the bioconversion processes. However, removal (total or at least partial) of these antiphysiological and antinutritional constituents could favor the coffee pulp reutilization for such purposes. In this sense, several biological treatments, including the use of yeasts, filamentous fungi, or bacteria, have been proposed to improve the nutritional value of coffee pulp (Orozco et al. 2008; Ulloa-Rojas et al. 2002).

Coffee pulp may also be extracted with solvents (water or organic solvents) to obtain caffeine and/or other value-added compounds. Other approaches aiming to reuse coffee pulp include the pyrolysis of the pulp impregnated with phosphoric acid to produce materials with a well-developed pore structure and high adsorption capacity (Irawaty et al. 2004), production of biogas, and production of extracts for soft drinks, jams, and other food applications (Bressani 1978).

15.4.4 COFFEE PARCHMENT

Coffee parchment (also referred to as coffee hulls) is a fibrous endocarp that covers the seeds. This material is composed of approximately 46% cellulose, 28%

hemicellulose, 34% lignin, and 0.6% ashes, with potassium and calcium being the main mineral presents (Bekalo and Reinhardt 2010). In spite of presenting a chemical composition of interest for bioprocesses, there is no study reporting this kind of application for coffee parchment. In fact, studies using this coffee by-product as a raw material to obtain valuable compounds are scarce.

Owing to the rich content of cellulose and hemicellulose, coffee parchment could be used in mixture with coffee husks for the production of particleboard (Bekalo and Reinhardt 2010). The pyrolysis of this coffee by-product produces hydrogen-rich fuel gas. Microwave-assisted pyrolysis produces much larger gas yields than conventional pyrolysis in a furnace. In addition, the gases produced in the microwave contain elevated proportions of H_2 and CO, while the gases produced in the furnace are richer in CO_2 (Domínguez et al. 2007). Coffee parchment was also demonstrated to be a good precursor for the production of activated carbons since the carbons produced from this material presented adsorption capacity comparable to those of commercial activated carbons (Brum et al. 2008).

15.4.5 Coffee Silverskin

Coffee silverskin, as shown in Figure 15.7, is the tegument that covers the coffee beans, which is separated during the roasting process. This coffee by-product is composed of cellulose (17.8%), hemicellulose (13.1%), protein (15.2%), extractives (15%), and ashes (4.7%), among other components. Glucose, xylose, galactose, mannose, and arabinose are the monosaccharides present in cellulose and hemicellulose fractions, glucose being the most abundant (Mussatto et al. 2011a). Coffee silverskin also presents high antioxidant capacity, which can be due to the presence of phenolic compounds in coffee beans, as well as other compounds formed by the Maillard reaction (such as melanoidins) (Borrelli et al. 2004).

There are few studies focused in finding valuable applications for this coffee by-product, and its main application in several countries is as fuel through burning. However, some studies have been focused in finding potential uses for coffee

FIGURE 15.7   Coffee silverskin.

silverskin. The high content of soluble dietary fiber (60%) and marked antioxidant activity, with low amount of fats and reducing carbohydrates, make it suitable for the use as a functional ingredient (Borrelli et al. 2004). In addition, aqueous extracts produced from coffee silverskin present a hyaluronidase-inhibiting effect, suggesting that this material is promising for its use in the development of functional foods and antiallergens (Furasawa et al. 2011).

In the biotechnological area, coffee silverskin has been evaluated as a raw material for the solid-state and submerged fermentation processes to obtain phenolic compounds or to produce enzymes, ethanol, or other value-added compounds. Fungal strains from the genus *Aspergillus, Mucor, Penicillium,* and *Neurospora* (particularly *A. niger* AA20, *Mucor* sp., *Penicillium purpurogenum,* and *Neurospora crassa*) have the ability to grow and release phenolic compounds from coffee silverskin under the solid-state cultivation conditions. Besides the interest of the phenolic compounds in the food and pharmaceutical industries, cultivation of these fungal strains in coffee silverskin is considered as an alternative for the biological detoxification of this material, being beneficial for its subsequent disposal to the environment (Machado et al. 2012). Another study under SSF conditions revealed the possibility of producing α-amylase by the cultivation of *N. crassa* CFR 308 over coffee silverskin (Murthy et al. 2009). This coffee by-product was also demonstrated to be an excellent material for the use as support and nutrient source during the fructooligosaccharides and β-fructofuranosidase production by *Aspergillus japonicus,* under SSF conditions (Mussatto and Teixeira 2010). In contrast, to be used in submerged fermentation systems, coffee silverskin must be first hydrolyzed with diluted sulfuric acid and the hydrolysate then obtained, containing about 20 g/L total sugars (xylose, arabinose, galactose, and mannose), can be used as the cultivation medium in fermentative processes. Some yeast strains, including *S. cerevisiae, Pichia stipitis,* and *Kluyveromyces fragilis,* are able to grow when cultivated in this hydrolysate (Mussatto et al. 2012).

15.4.6 Spent Coffee Grounds

Spent coffee grounds are the residual solid materials obtained after the treatment of raw coffee powder with hot water or steam for the instant coffee preparation. Figure 15.8 shows the appearance of this coffee by-product. This material is obtained in large amounts, it is estimated the generation of 650 kg of spent coffee grounds from each one ton of green coffee used for the preparation of instant coffee (Mussatto et al. 2011a). This value is still more significant when taking into account that from the worldwide coffee production, about 50% is used for soluble coffee preparation (Ramalakshmi et al. 2009).

Spent coffee grounds are mainly composed of sugars (45.3%, with mannose and galactose being the most abundant) and proteins (13.6%) (Mussatto et al. 2011b). Once obtained, this material contains elevated moisture (in the range of 80–85%), which combined with the high sugars and protein contents constitutes a suitable environment for the growth of fungal strains that are able to deteriorate this material promptly. However, uncontrolled fungal growth leads to the loss of carbon and protein sources present in this material, which could be used for other more valuable

FIGURE 15.8   Spent coffee grounds.

applications, and the discharge of this material into the environment and sanitary landfill should be avoided due to its toxic character and the presence of organic matter. To avoid these problems, coffee industries around the world are searching for alternatives to reuse the spent coffee grounds. In some cases, this material is used as a fuel in industrial boilers of the same industry, due to its high calorific power (approximately 5000 kcal/kg) (Silva et al. 1998).

Some recent studies report more valuable alternatives to reuse this coffee by-product, mainly in the biotechnological area. Spent coffee grounds can be used, for example, as the substrate (without any nutritional supplementation) for the cultivation of the edible fungus *Flammulina velutipes* under SSF conditions (Leifa et al. 2001). This material also proved to be a good substrate to support the growth of fungal strains from the genera *Penicillium, Aspergillus, Neurospora*, and *Mucor* (Machado et al. 2012; Murthy and Naidu 2012a). Besides growing in spent coffee grounds, some fungal strains from these genera were also able to release phenolic compounds from this material when cultivated under solid-state conditions. This is an additional advantage because phenolic compounds are largely utilized in the food and pharmaceutical industries.

The possibility of using spent coffee grounds in submerged fermentation systems has also been evaluated. In a recent study, a distilled beverage was produced by submitting this material to an extraction with water, followed by fermentation of the produced liquid fraction supplemented with sugars, and subsequently distillation of the fermented broth. The produced spirit presented characteristics (flavor and presence of volatile compounds) and organoleptic properties acceptable for consumption, which were different when compared to spirits commercially found (Sampaio 2010). Fuel ethanol has also been produced by submerged fermentation using spent coffee grounds as raw material. In this case, a sugar-rich hydrolysate produced by dilute acid hydrolysis of spent coffee grounds is used as the fermentation medium for ethanol production (Mussatto et al. 2012; Sampaio 2010). Different yeasts are able to convert the sugars present in this hydrolysate to ethanol, among which *S. cerevisiae* is able to provide elevated conversion yield (Mussatto et al. 2012).

Spent coffee grounds have also been evaluated as a raw material for the production of other fuels, including biodiesel and fuel pellets. This material is composed of approximately 10–15% oil depending on the coffee species (Arabica or Robusta), which can be extracted and subsequently converted with high efficiency (close to 100%) into biodiesel by transesterification processes (Burton et al. 2010; Kondamudi et al. 2008). The solid waste residue from the extraction process could be further used for the production of fuel pellets (Kondamudi et al. 2008) that would be an interesting alternative from the point of view of the application of the biorefinery concept.

Other valuable application for spent coffee grounds is to obtain extracts, since the extracts produced from this material present high antioxidant capacity and anti-tumor activity, being of interest for the application in several industrial fields (Bravo et al. 2011; Mussatto et al. 2011c; Ramalakshmi et al. 2009). This material is also an inexpensive and easily available adsorbent for the removal of cationic dyes from wastewater (Franca et al. 2009).

15.5 CONCLUSIONS AND PERSPECTIVES

Coffee is a product largely consumed around the world and, as a consequence of this big market, the generation of large amounts of by-products is inevitable. Valorization of these by-products is of interest for environmental and economic reasons. After specific pretreatments with chemical, physical, and/or biological agents, coffee by-products may provide, for example, value-added natural antioxidants, sugars, lipids, proteins, pigments, among several other compounds of interest to the pharmaceutical, cosmetic, and food industries. In addition, several other valuable compounds, including enzymes, fuels, beverages, and others, can be produced by fermentation using coffee by-products as raw materials. To date, the use of coffee by-products for the production of valuable compounds is scarce and there are also a limited number of studies considering these by-products as raw material for industrial processes. However, there is a great interest in increasing the research as well as the real applications of these industrial by-products in the near future, not only for environmental concerns but also because coffee by-products have a low cost and present a chemical composition that allows their application in different areas. Cellulose, for example, is a homopolymer of glucose units usually found in these by-products and that can be used for the production of pulp and paper or to obtain glucose, which can be subsequently used as a carbon source for the production of a variety of fuels and chemicals, including ethanol, butanol, hydrogen, glycerol, hydroxymethylfurfural, and organic acids, among others. Hemicellulose is another polymeric structure present in the composition of coffee by-products and that can be used for different purposes. Sugars present in this structure can be used, for example, for the production of chemicals such as xylitol, mannitol, arabitol, furfural, and ethanol. All these potential alternatives for the application of coffee by-products, and the interest in finding alternatives to avoid their disposal into the environment and at the same time to valorize them, create a positive expectance of expansion of this research area in the near future.

REFERENCES

Adams, M.R., and Dougan, J. 1981. Biological management of coffee processing. *Trop. Sci.* 123:178–196.

Arya, M., and Rao, L.J.M. 2007. An impression of coffee carbohydrates. *Crit. Rev. Food Sci. Nutr.* 47:51–67.

Battestin, V., and Macedo, G.A. 2007. Tannase production by *Paecilomyces variotii. Bioresour. Technol.* 98:1832–1837.

Bekalo, S.A., and Reinhardt, H.-W. 2010. Fibers of coffee husk and hulls for the production of particleboard. *Mater. Struct.* 43:1049–1060.

Belitz, H.-D., Grosch, W., and Schieberle, P. 2009. Coffee, tea, cocoa. In *Food Chemistry* (4th edition), ed. H.-D. Belitz, W. Grosch, and P. Schieberle, 938–970. Germany: Springer.

Bertrand, B., Guyot, B., Anthony, F., and Lashermes, P. 2003. Impact of the *Coffea canephora* gene introgression on beverage quality of *C. arabica. Theor. Appl. Genet.* 107:387–394.

Borrelli, R.C., Esposito, F., Napolitano, A., Ritieni, A., and Fogliano, V. 2004. Characterization of a new potential functional ingredient: Coffee silverskin. *J. Agric. Food Chem.* 52:1338–1343.

Brand, D., Pandey, A., Roussos, S., and Soccol, C.R. 2000. Biological detoxification of coffee husk by filamentous fungi using a solid state fermentation system. *Enzyme Microb. Technol.* 27:127–133.

Bravo, J., Monente, C., Juániz, I., Paz De Pena, M., and Cid, C. 2013. Influence of extraction process on antioxidant capacity of spent coffee. *Food Res. Int.* 50:610–616.

Bressani, R. 1978. Potential uses of coffee-berry by-products. In *Coffee Pulp. Composition, Technology, and Utilization*, ed. J.E. Braham, and R. Bressani, 5–10. Canada: INCAP.

Brum, S.S., Bianchi, M.L., Silva, V.L., Gonçalves, M., Guerreiro, M.C., and Oliveira, L.C.A. 2008. Preparation and characterization of activated carbon produced from coffee waste. *Quim. Nova* 31:1048–1052.

Burton, R., Fan, X., and Austic, G. 2010. Evaluation of two-step reaction and enzyme catalysis approaches for biodiesel production from spent coffee grounds. *Int. J. Green Energy* 7:530–536.

Bytof, G., Knopp, S.-E., Schieberle, P., Teutsch, I., and Selmar, D. 2005. Influence of processing on the generation of γ-aminobutyric acid in green coffee beans. *Eur. Food Res. Technol.* 220:245–250.

Domínguez, A., Menéndez, J.A., Fernández, Y. et al. 2007. Conventional and microwave induced pyrolysis of coffee hulls for the production of a hydrogen rich fuel gas. *J. Anal. Appl. Pyrolysis* 79:128–135.

Farah, A., and Donangelo, C.M. 2006. Phenolic compounds in coffee. *Braz. J. Plant Physiol.* 18:23–36.

Franca, A.S., Oliveira, L.S., and Ferreira, M.E. 2009. Kinetics and equilibrium studies of methylene blue adsorption by spent coffee grounds. *Desalination* 249:267–272.

Furasawa, M., Narita, Y., Iwai, K., Fukunaga, T., and Nakagiri, O. 2011. Inhibitory effect of a hot water extract of coffee "silverskin" on hyaluronidase. *Biosci. Biotechnol. Biochem.* 75:1205–1207.

Gouvea, B.M., Torres, C., Franca, A.S., Oliveira, L.S., and Oliveira, E.S. 2009. Feasibility of ethanol production from coffee husks. *Biotechnol. Lett.* 31:1315–1319.

Hernández, J.A., Heyd, B., and Trystram, G. 2008. On-line assessment of brightness and surface kinetics during coffee roasting. *J. Food Eng.* 87:314–322.

Irawaty, W., Hindarso, H., Felycia, E.S., Mulyono, Y., and Kurniawan, H. 2004. Utilization of Indonesian coffee pulp to make an activated carbon. The 10th APCCHE Congress, Kitakyushu, Japan, October 17–21.

Jayachandra, T., Venugopal, C., and Anu Appaiah, K.A. 2011. Utilization of phytotoxic agro waste- coffee cherry husk through pretreatment by the ascomycetes fungi *Mycotypha* for biomethanation. *Energy Sustainable Dev.* 15:104–108.

Kondamudi, N., Mohapatra, S.K., and Misra, M. 2008. Spent coffee grounds as a versatile source of green energy. *J Agric. Food Chem.* 56:11757–11760.

Leifa, F., Pandey, A., and Soccol, C.R. 2001. Production of *Flammulina velutipes* on coffee husk and coffee spent-ground. *Braz. Arch. Biol. Technol.* 44:205–212.

Machado, C.M.M., Soccol, C.R., Oliveira, B.H., and Pandey, A. 2002. Gibberellic acid production by solid-state fermentation in coffee husk. *Appl. Biochem. Biotechnol.* 101–103:179–191.

Machado, E.M.S., Rodriguez-Jasso, R.M., Teixeira, J.A., and Mussatto, S.I. 2012. Growth of fungal strains on coffee industry residues with removal of polyphenolic compounds. *Biochem. Eng. J.* 60:87–90.

Murthy, P.S., and Naidu, M.M. 2010. Protease production by *Aspergillus oryzae* in solid-state fermentation utilizing coffee by-products. *World Appl. Sci. J.* 8:199–205.

Murthy, P.S., and Naidu, M.M. 2012a. Production and application of xylanase from *Penicillium* sp. utilizing coffee by-products. *Food Bioprocess Technol.* 5:657–664.

Murthy, P.S., and Naidu, M.M. 2012b. Recovery of phenolic antioxidants and functional compounds from coffee industry by-products. *Food Bioprocess Technol.* 5:897–903.

Murthy, P.S., Naidu, M.M., and Srinivas, P. 2009. Production of α-amylase under solid-state fermentation utilizing coffee waste. *J. Chem. Technol. Biotechnol.* 84:1246–1249.

Mussatto, S.I., Ballesteros, L.F., Martins, S., and Teixeira, J.A. 2011c. Extraction of antioxidant phenolic compounds from spent coffee grounds. *Sep. Purif. Technol.* 83:173–179.

Mussatto, S.I., Machado, E.M.S., Carneiro, L.M., and Teixeira, J.A. 2012. Sugars metabolism and ethanol production by different yeast strains from coffee industry wastes hydrolysates. *Appl. Energy* 92:763–768.

Mussatto, S.I., Machado, E.M.S., Martins, S., Teixeira, J.A. 2011a. Production, composition, and application of coffee and its industrial residues. *Food Bioprocess Technol.* 4:661–672.

Mussatto, S.I., Carneiro, L.M., Silva, J.P.A., Roberto, I.C., and Teixeira, J.A. 2011b. A study on chemical constituents and sugars extraction from spent coffee grounds. *Carbohydr. Polym.* 83:368–374.

Mussatto, S.I., and Teixeira, J.A. 2010. Increase in the fructooligosaccharides yield and productivity by solid-state fermentation with *Aspergillus japonicus* using agro-industrial residues as support and nutrient source. *Biochem. Eng. J.* 53:154–157.

Oliveira, L.S., Franca, A.S., Alves, T.M., and Rocha, S. D. F. 2008a. Evaluation of untreated coffee husks as potential biosorbents for treatment of dye contaminated waters. *J. Hazard. Mat.* 155:507–512.

Oliveira, W.E., Franca, A.S., Oliveira, L.S., and Rocha, S.D. 2008b. Untreated coffee husks as biosorbents for the removal of heavy metals from aqueous solutions. *J. Hazard. Mat.* 152:1073–1081.

Orozco, A.L., Pérez, M.I., Guevara, O. et al. 2008. Biotechnological enhancement of coffee pulp residues by solid-state fermentation with *Streptomyces*. Py-GC/MS analysis. *J. Anal. Appl. Pyrolysis* 81:247–252.

Prata, E.R.B.A., and Oliveira, L.S. 2007. Fresh coffee husks as potential sources of anthocyanins. *LWT - Food Sci. Technol.* 40:1555–1560.

Ramalakshmi, K., Kubra, I.R., and Rao, L.J.M. 2008. Antioxidant potential of low-grade coffee beans. *Food Res. Int.* 41:96–103.

Ramalakshmi, K., Rao, L.J.M., Takano-Ishikawa, Y., and Goto, M. 2009. Bioactivities of low-grade green coffee and spent coffee in different *in vitro* model systems. *Food Chem.* 115:79–85.

Ramírez-Martínez, J.R. 1988. Phenolic compounds in coffee pulp: quantitative determination by HPLC. *J. Sci. Food Agric.* 43:135–144.

Ramírez-Coronel, M.A., Marnet, N., Kumar Kolli, V S., Roussos, S., Guyot, S., and Augur, C. 2004. Characterization and estimation of proanthocyanidins and other phenolics in coffee pulp (*Coffea arabica*) by thiolysis-high-performance liquid chromatography. *J. Agric. Food Chem.* 52:1344–1349.

Sampaio, A.R.M. 2010. Desenvolvimento de tecnologias para produção de etanol a partir do hidrolisado da borra de café. MSc Thesis, Department of Biological Engineering, University of Minho, Braga, Portugal.

Shan, J., Fu, J., Zhao, Z., et al. 2009. Chlorogenic acid inhibits lipopolysaccharide-induced cyclooxygenase-2 expression in RAW264.7 cells through suppressing NF-kB and JNK/AP-1 activation. *Int. Immunopharmacol.* 9:1042–1048.

Shankaranand, V.S., and Lonsane, B.K. 1994. Coffee husk: an inexpensive substrate for production of citric acid by *Aspergillus niger* in a solid-state fermentation system. *World J. Microbiol. Biotechnol.* 10:165–168.

Silva, M.A., Nebra, S.A., Machado Silva, M.J., and Sanchez, C.G. 1998. The use of biomass residues in the Brazilian soluble coffee industry. *Biomass Bioenergy* 14:457–467.

Soares, M., Christen, P., Pandey, A., and Soccol, C.R. 2000a. Fruit flavour production by *Ceratocystis fimbriata* grown on coffee husk in solid-state fermentation. *Process Biochem.* 35:857–861.

Soares, M., Christen, P., Pandey, A., Raimbault, M., and Soccol, C.R. 2000b. A novel approach for the production of natural aroma compounds using agro-industrial residue. *Bioprocess. Eng.* 23:695–699.

Ulloa-Rojas, J.B., Verreth, J.A.J., van Weerd, J.H., and Huisman, E.A. 2002. Effect of different chemical treatments on nutritional and antinutritional properties of coffee pulp. *Anim. Feed Sci. Technol.* 99:195–204.

USDA—United States Department of Agriculture. 2012. Coffee: World Markets and Trade. Circular series, December 2011. http://www.fas.usda.gov/coffee_arc.asp (accessed May 12, 2012).

16 Beer

Solange I. Mussatto, Nuno G.T. Meneses,
and José A. Teixeira

CONTENTS

16.1 INTRODUCTION

The generation of by-products is inevitable in any industrial process, and could not be different during the process for beer elaboration. In fact, a large variety of by-products and residues are generated by the brewing industry, which include products derived from the raw materials used for fermentation as well as broken bottles, crushed cans, metal lids, plastics, and so on. Some of them, mainly those derived from the raw materials (namely, the spent grains, surplus yeast, and spent hops, generated from the barley malt, yeast, and hop, respectively), are obtained in large amounts and, considering that they are rich in sugars, proteins, and minerals, have great potential to be reused as raw materials for other industrial processes. However, the industrial applications for these three brewing by-products are still limited, and they are basically sold to farmers for use as cattle feed, or simply as a landfill.

Some attempts have been made to use these by-products in the food industry for the production of valuable compounds such as xylitol, arabitol, and lactic acid, as well as the substrate for the culture of microorganisms, or simply as raw material for the extraction of compounds such as sugars, proteins, vitamins, acids, and antioxidants. They have also been considered as raw materials for the production of biofuels. In this chapter, the characteristics and most recent advances on the valorization of brewer's spent grains (BSGs), surplus yeast, and spent hops are summarized and discussed.

429

16.2 BREWING PROCESS AND BY-PRODUCTS GENERATION

A typical brewing process starts by adding the milled malted barley and water in a mash tun. After mixing these components, the temperature is slowly increased to ca. 78°C to promote the enzymatic hydrolysis of malt constituents and to solubilize their breakdown products. With the temperature increase, malt starch is mainly converted to maltose, and maltotriose (fermentable sugars), and dextrins (nonfermentable sugars), while the proteins are partially degraded to polypeptides and amino acids. At the end of this mashing process, a sweet liquid called wort is produced, and the nondegraded part of the malted barley grain (the spent grains) is allowed to settle to form a bed in the bottom of the mash tun. The sweet wort is then filtered through this spent grains bed and is subsequently transferred to the brewing kettle, where it is boiled. The spent grains are obtained as a by-product of this filtration process (Mussatto et al. 2006a). This by-product can be constituted by the nondegraded part of the malted barley grains only, or by a mixture of this material with adjuncts, that is, unmalted cereals such as corn (maize), rice, wheat, oats, rye, or sorghum that can be used to replace part of the barley malt (usually 15–20%) with the objective of producing beers of distinct qualities or to reduce the production costs.

In the brewing kettle, the wort is supplemented with hops and is boiled for least 1 h, aiming at transferring the bitter and aromatic hop components to wort. Hops can be used in the natural form or in the form of powder, pellets, or extract. The hop components confer bitter taste, flavor, and foam stability to the final beer, thus proving to be of great importance for the beer quality. At the end of the boiling stage, the medium is cooled and the liquid extract is separated from the residual solid material (spent hops), which is mainly obtained when natural hops are used during boiling. A fraction of the hop components usually ends up in the trub, a precipitate formed after wort boiling constituted by insoluble hop materials, condensation products of hop polyphenols and wort proteins, and isomerized hop acids adsorbed onto trub solids (Huige 2006).

The cooled hopped wort is then transferred to the fermentation vessel where it is inoculated with the yeast that will convert the fermentable sugars to ethanol and carbon dioxide. During this process, the yeast biomass is significantly increased (three- to sixfold) and the surplus yeast is separated at the end of the fermentation (Keukeleire 2000).

A schematic representation of the brewing process and the points where the three main by-products (spent grains, spent hops, and surplus yeast) are generated is shown in Figure 16.1. Besides these three by-products, other residual materials are produced in the brewing industry, which include (1) broken bottles, crushed cans and metal lids, and plastic and cardboard packaging, which are segregated and sold to recycling companies; (2) diatomaceous earth used for beer clarification, which is dried and subsequently referred to as inert landfill; and (3) sludge, which is generated in considerable amounts in the water treatment plant and is managed as waste.

The next sections deal with the characteristics and potential alternatives that have been evaluated to use the three main brewing by-products.

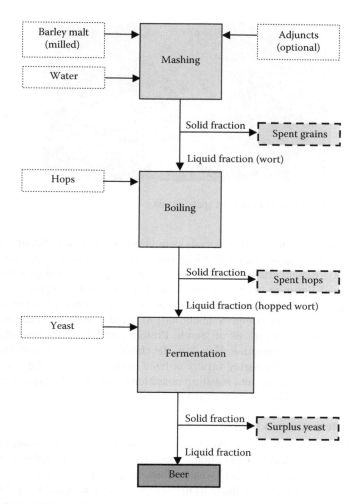

FIGURE 16.1 Schematic representation of the brewing process and points where the main by-products are generated.

16.3 BREWING BY-PRODUCTS: CHARACTERISTICS AND APPLICATIONS

16.3.1 BREWER'S SPENT GRAINS

16.3.1.1 Chemical Composition and Characteristics

BSG is the main by-product generated during the brewing process as it is derived from the most important raw material used in this process—the barley malt—and therefore, it is obtained in a very large amount. In terms of structure, BSG consist of a heterogeneous matrix comprising the barley grain husk in mixture with the pericarp and fragments of endosperm. The appearance of this brewing by-product is shown in Figure 16.2.

FIGURE 16.2 Appearance of brewer's spent grains.

BSG presents a fibrous structure mostly composed of cellulose (a linear homopol-ymer of repeated D-glucose units), hemicellulose (a heteropolysaccharide composed by pentoses (xylose and arabinose), hexoses (mannose, glucose, and galactose), and acetic acid), and lignin (a heterogeneous long-chain polymer composed mostly of phenyl propane units). The chemical composition of this material is presented in Table 16.1. Carbohydrates comprise about half of the dry mass and consist of ara-binoxylans, β-glucan, and traces of starch. Protein and lignin are also important fractions in the BSG composition. However, the chemical composition of this mate-rial varies according to the barley variety utilized in the brewing process as well as its harvest time, the malting and mashing conditions, and with the type of adjuncts (other cereal grains) added in the brewing process.

16.3.1.2 Alternatives to Reuse

Although BSG is rich in sugars, proteins, and minerals, which are compounds of interest for several valuable applications, this material has been traditionally sup-plied to local farmers to be used as cattle feed. In fact, BSG is a good substrate material for use as animal feed due to its high contents of protein and fiber (Mussatto 2009). However, there is a need to find alternatives for the valorization of this by-product due to its rich chemical composition and because it is produced in large amounts every year. For these reasons, when compared to the other two by-products obtained during the brewing process (spent hops and surplus yeast), a larger number of researchers have been aiming to reuse BSG.

BSG has been evaluated as a raw material for chemical and biotechnologi-cal processes, but the number of studies on biotechnological applications is much more significant. Since BSG is rich in polysaccharides and protein, and also has high moisture content (about 80–85% w/w), it is susceptible to microbial growth and degrades in a few days. Owing to this characteristic, several studies have been per-formed to use BSG as a raw material for the cultivation of microorganisms, including bacteria such as *Bifidobacterium adolescentis* 94BIM and *Lactobacillus* sp. (Novik et al. 2007), actinobacteria, especially *Streptomyces* (Szponar et al. 2003), and sev-eral fungal strains such as *Pleurotus ostreatus* (Gregori et al. 2008), *Penicillium janczewskii* (Terrasan et al. 2010), *Penicillium brasilianum* (Panagiotou et al. 2006),

TABLE 16.1
Chemical Composition of Brewer's Spent Grains

Component (% Dry Weight Basis)	Reference[a]				
	1	2	3	4	5
Cellulose	21.7	16.8	25.4	21.9	25.3
Hemicellulose	19.3	28.4	21.8	29.6	41.9
Xylan	13.6	19.9	NR	20.6	NR
Arabinan	5.6	8.5	NR	9.0	NR
Lignin	19.4	27.8	11.9	21.7	16.9
Acetyl groups	NR	1.3	NR	1.1	NR
Proteins	24.7	15.3	24.0	24.6	NR
Ash	4.2	4.6	2.4	1.2	4.6
Extractives	NR	5.8	NR	NR	9.5
Minerals (mg/kg Dry Weight Basis)					
Phosphorus	6000.0	5186.0	NR	NR	NR
Potassium	600.0	258.1	NR	NR	NR
Calcium	3600.0	3515.0	NR	NR	NR
Magnesium	1900.0	1958.0	NR	NR	NR
Sulfur	2900.0	1980.0	NR	NR	NR
Iron	154.9	193.4	NR	NR	NR
Manganese	40.9	51.4	NR	NR	NR
Boron	3.2	NR	NR	NR	NR
Copper	11.4	18.0	NR	NR	NR
Zinc	82.1	178.0	NR	NR	NR
Molybdenum	1.4	NR	NR	NR	NR
Sodium	137.1	309.3	NR	NR	NR
Aluminum	81.2	36.0	NR	NR	NR
Barium	8.6	13.6	NR	NR	NR
Strontium	10.4	12.7	NR	NR	NR
Chromium	<0.5	5.9	NR	NR	NR
Tin	<1.3	NR	NR	NR	NR
Lead	<1.6	NR	NR	NR	NR
Cobalt	17.8	NR	NR	NR	NR
Iodine	11.0	NR	NR	NR	NR
Cadmium	<0.2	NR	NR	NR	NR
Nickel	<0.5	NR	NR	NR	NR
Selenium	<1.6	NR	NR	NR	NR
Gallium	<1.5	NR	NR	NR	NR
Silicon	NR	10,740.0	NR	NR	NR

Note: NR, nonreported.

[a] 1, From Meneses (2011); 2, From Mussatto and Roberto (2006); 3, From Kanauchi et al. (2001); 4, From Carvalheiro et al. (2004); 5, From Silva et al. (2004).

among others. In several cases, the production of important enzymes was observed when cultivating microorganisms in BSG, as, for example, xylanolytic enzymes by *Penicillium janczewskii* (Terrasan et al. 2010) and *Streptomyces avermitilis* (Bartolomé et al. 2003), protease by *Aspergillus awamori* (Bhumibhamon 1978), feruloyl esterases by *Talaromyces stipitatus, Humicola grisea* var. *thermoidea* (Mandalari et al. 2008), and *Streptomyces avermitilis* (Bartolomé et al. 2003), α-amylase by *Aspergillus oryzae* (Bogar et al. 2002; Francis et al. 2003), *Bacillus licheniformis* (Okita et al. 1985), and *Bacillus subtilis* (Duvnjak et al. 1983), and cellulases by *Trichoderma reesei* (Sim and Oh 1990).

As it was mentioned before, BSG is constituted by several polysaccharides. Such compounds can be degraded into their corresponding constituents by means of a hydrolytic procedure that can be hydrothermal, enzymatic, or acidic. During the hydrolysis, cellulose is converted to glucose, while the hemicellulose fraction is converted to xylose, arabinose, mannose, galactose, and acetic and hydroxycinnamic (ferulic and *p*-coumaric) acids (Mussatto and Roberto 2004). All these compounds have an industrial interest, mainly in the food industry. Ferulic acid, for example, has potential applications as a natural antioxidant, food preservative/antimicrobial agent, anti-inflammatory agent, photoprotectant, and food flavor precursor, while *p*-coumaric acid exhibits chemoprotective and antioxidant properties (Bartolomé et al. 2002; Faulds et al. 2002; Mussatto et al. 2007a). The compounds released from the BSG structure during the hydrolysis process can be purified for use as such, or can be used as substrate and/or nutritional factors in fermentative processes to obtain different products of industrial interest such as ethanol (Meneses 2011; Laws and Waites 1986), xylitol (Carvalheiro et al. 2005; Mussatto and Roberto 2008), lactic acid (Cruz et al. 2007; Mussatto et al. 2007b), and pullulan (Roukas 1999). In all these cases, the use of a low-cost raw material such as BSG significantly contributes to minimize the production costs.

BSG can also be used in fermentation processes as the support material for cell immobilization. In this case, BSG must be first submitted to a pretreatment with acid and/or base to hydrolyze the residual starchy endosperm and promote some delignification in its structure. The irregular shape and nonhomogeneity structure then obtained provide active sites that are readily colonized by yeasts. Figure 16.3 shows the modification in the structure of BSG after the pretreatment stage for cell immobilization. Other advantages of using BSG as the support for immobilization include its low cost, high cell loading capacity, low mass transfer limitations, stability, rigidity, reusability, availability, nontoxicity, and food grade (Almeida et al. 2003; Dragone et al. 2007). The reuse of BSG as the support for cell immobilization during the fermentation stage of the brewing process could be an interesting alternative to reuse this by-product in the same industry; however, the technology for beer production with immobilized cells have not been much explored on an industrial level. Another possibility to reuse BSG in the same brewing industry is by using the extract obtained from its pressing as an antifoaming agent in the fermenter. Besides, to be effective for such a purpose, the addition of this extract improves hop utilization, enhances yeast performance, and does not affect the properties of the final beer (Roberts 1976; Mussatto 2009). Additionally, it is of great interest to reuse the BSG extract since it has a high biological oxygen demand and constitutes an undesirable effluent.

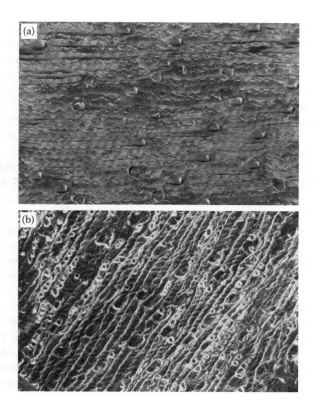

FIGURE 16.3 Structure of brewer's spent grains before (a) and after pretreatment (b) for cells immobilization.

BSG can also be used in chemical processes, for example, as an adsorbent for removing volatile organic compound emissions or organic material from effluents. This material presents a significant potential as a bioadsorbent for application in the remediation of metal-contaminated wastewater streams (especially those containing Cu (II) ions) (Lu and Gibb 2008). The reactive functional groups such as hydroxyl, amine, and carboxyl that can be activated in BSG are responsible for the metal ion binding (Li et al. 2009), while the sorption capacity for cadmium and lead can be enhanced by submitting this material to a pretreatment with 0.5 M NaOH (Low et al. 2000). Other possible applications of this brewing by-product are as raw material for the production of activated carbons (Mussatto et al. 2010), cellulose pulps that could be used in the manufacture of some types of papers (Mussatto et al. 2006b, 2008), or to obtain building materials, due to its low ash and high fiber contents (Aliyu and Bala 2011).

16.3.2 Surplus Yeast

16.3.2.1 Chemical Composition and Characteristics

Two classes of yeasts can be used in the brewing process: the bottom-fermenting and the top-fermenting yeasts. As a function of the yeast strain and consequently of the conditions used for fermentation, the produced beer is divided into two categories, namely, lager and ale. Lager yeasts, usually *Saccharomyces pastorianus* or *S. carlsbergensis*, runs the fermentation at cool temperatures (between 8°C and 15°C), and form a cloudy mass at the bottom of the vessel at the end of the process (Bamforth 2003). The beers produced with these yeasts are called bottom fermented and are the most widespread beer types throughout the world (>90%). Otherwise, ale beers are usually produced with strains of *S. cerevisiae* at temperatures varying between 16°C and 25°C (Ferreira et al. 2010). In both the cases, the yeast biomass is separated at the end of the process being called as surplus yeast.

Yeast cells contain approximately 80% water and are also rich in protein (40–50% cell dry weight), carbohydrates (30–45% cell dry weight), minerals (5–10% cell dry weight), RNA (6–8% cell dry weight), lipids (up to 5% cell dry weight), and vitamins. The most abundant element is carbon, which accounts for approximately 50% of the cell dry weight. Other major elemental components include oxygen (30–35%), nitrogen (5%), hydrogen (5%), and phosphorus (1%). The protein content present several bound amino acids, namely, arginine, cystine, glycine, histidine, isoleucine, leucine, lysine, methionine, phenylalanine, threonine, tryptophan, tyrosine, and valine, among which, leucine, lysine, and tyrosine are the most abundant. Vitamins in yeast cells may include biotin, choline, folic acid, niacin, pantothenic acid, riboflavin, thiamin, pyridoxine, and vitamin B-6. Minerals comprise a variety of elements, specially potassium and phosphorus (Briggs et al. 2004; Huige 2006; Lewis and Young 1995). The chemical composition of some brewer's surplus yeast is shown in Table 16.2; however, the composition of each class of macromolecules within a given cell varies as a function of physiological condition and phase in growth cycle (Briggs et al. 2004).

16.3.2.2 Alternatives to Reuse

The surplus yeast, namely, *Saccharomyces* yeast biomass, is the second major by-product generated in the brewing industry, after spent grains. Currently, part of this excess is sold to food industries; however, it is still underutilized, being largely used for animal feed (especially for pigs and ruminants) (Ferreira et al. 2010). Brewer's yeast is also used as replacement for fish meal (Oliva-Teles and Gonçalves 2001), being able to replace 50% of fishmeal protein with no negative effects in fish performance.

The brewer's yeast is a by-product of large interest for the food industry since it is an inexpensive nitrogen source generally recognized as safe and has good nutritional characteristics (Chae et al. 2001). However, the inclusion of yeast in food products is limited by the amount of nucleic acid present in its composition, primarily ribonucleic acid (RNA), because humans metabolize RNA to uric acid, which can lead to gout (Huige 2006). One of the most common applications of yeast cells in the food industry is to obtain yeast extract, which is a mixture of amino acids, peptides, nucleotides, and other soluble components of yeast cells, produced by breaking down of yeast cells using enzymes. The brewer's yeast can also be used, after enzymatic treatment, as flavors, flavor enhancers, or flavor potentiating in a wide variety of foods, including meat products, sauces and gravies, soups, chips, and crackers. Other applications of brewer's yeast in the food industry are as vitamin supplements in health food (Chae et al. 2001; Huige 2006), and as raw material to produce yeast protein concentrates. Brewer's yeast products can be found in the form of powders,

TABLE 16.2

Chemical Composition of Surplus Yeast

Components (% Dry Weight Basis)	Reference[a]		
	1	2	3
Protein	48	NR	50
Lipid	NR	1	NR
Ash	7	8	7
Crude fiber[b]	3	NR	NR
Carbohydrates	NR	36	42
Minerals in Ash (%)			
Calcium	0.12	NR	NR
Chlorine	0.12	NR	NR
Iron	0.01	NR	NR
Magnesium	0.24	NR	NR
Phosphorus	1.43	NR	NR
Potassium	1.71	NR	NR
Sodium	0.09	NR	NR
Sulfur	0.38	NR	NR
Vitamins (mg/100 g)			
Niacin	NR	NR	50
Thiamin	NR	NR	15
Panthotenate	NR	NR	10
Riboflavin	NR	NR	7
Folic acid	NR	NR	4
Pyridoxine	NR	NR	3
Biotin	NR	NR	0.2

Note: NR, nonreported.

[a] 1, From Huige (2006); 2, From Lamoolphak et al. (2006); 3, From Lewis and Young (1995).

[b] Glucans, mannans, and polymeric hexosamines.

flakes, tablets, and in the liquid form. These products can be sprinkled on food, used as a seasoning or mixed with milk, juices, soups, and gravies (Ferreira et al. 2010).

Owing to the composition rich in compounds of industrial interest, several studies have been focused in finding valuable applications for the surplus yeast. The brewer's yeast can be used, for example, to obtain proteins, vitamins, amino acids, cytochromes, DNA, and RNA, which are compounds with a number of industrial applications. Such compounds can be recovered from the yeast by processes such as autolysis, plasmolysis in organic salt solution or nonpolar organic solvent, acid- or alkali-catalyzed hydrolysis, enzymatic hydrolysis, or hydrothermal decomposition (Lamoolphak et al. 2006). β-Glucan, a hydrocolloid of great interest to the pharmaceutical and food industries, can also be extracted from brewer's yeast. This compound has the potential to improve the functional properties of food products, being used as a thickening, water-holding, or oil-binding agent, and emulsifying or foaming stabilizer (Romero and Gomez-Basauri 2003; Thammakiti et al. 2004). Brewer's yeast is also a good source of trivalent chromium, which has been studied extensively for its medicinal properties, namely, due to the ability to potentiate insulin activity (Cefalu and Hu 2004; Ding et al. 2000).

Besides being useful as raw material to obtain valuable compounds, brewer's yeast has also proved to be of value for the application in fermentative processes. For example, the use of brewer's yeast autolysate during the fermentation of vegetable juices by *Lactobacillus acidophilus* is able to increase the number of lactic acid bacteria, reducing the fermentation time, with the additional advantage of enriching the vegetable juices with amino acids, vitamins, minerals, and antioxidants (Rakin et al. 2007). The use of yeast extract as a source of nutrients in the formulation of microbiological media is also well known and largely used.

The brewer's yeast can also be reutilized in the environmental area. Dead cells of *Saccharomyces* yeast are an efficient biosorbent of metal ions (particularly lead, zinc, copper, and nickel), and this application is considered a low-cost technology for detoxifying metal-bearing effluents. It must be emphasized that heat-killed cells have a higher ability to remove heavy metals than live cells, thus being more suitable for bioremediation applications (Machado et al. 2009, 2008; Zouboulis et al. 2001).

16.3.3 SPENT HOPS

16.3.3.1 Chemical Composition and Characteristics

The hop plant, *Humulus lupulus*, is a hardy, dioecious, climbing plant that belongs to the hemp family and is essentially used for brewing. It is rich in bitter constituents (α-acids (humulones), β-acids (lupulones), soft and hard resins) and ethereal oils that confer bittering and aroma components to beer, which are of great importance from the viewpoint of the brewing technology and for the taste of the beer (Esslinger and Narziss 2005). However, only 15% of the hops constituents are solubilized during the wort boiling, that is, 85% of this raw material will become spent hops (Huige 2006). The lupulones, for example, are insoluble at the normal pH value of wort and are largely removed with spent hops and trub. Phenolic components from hops, including *p*-hydroxycoumaric, gallic, ferulic, protocatechinic, and caffeic acids, catechins, flavones, and anthocyanidines, among others, are also precipitated with proteins during wort boiling (Esslinger and Narziss 2005).

Spent hop is a material with high contents of nitrogen-free extract, fibers, and proteins. Crude fiber is composed of sugars such as rhamnose, arabinose, xylose, mannose, galactose, and glucose, among which, glucose and xylose are the most abundant. Pectic sugars, arabinose, rhamnose, galactose, and uronic acid account for 46% of the polysaccharides in spent hops (Oosterveld et al. 2002). Mono- and multifunctional aliphatic carboxylic acids in this by-product include oxalic, glucaric, gluconic, threonic, glyceric, glycolic, lactic, and acetic acids (Fischer and Bipp 2005). The chemical composition of spent hops is summarized in Table 16.3.

Part of the hop components ends up in the trub fraction (the precipitate formed after the wort boiling), and this amount is still more significant when hop is used

TABLE 16.3
Chemical Composition of Spent Hops

Component (% Dry Matter Basis)	Reference[a]	
	1	2
Protein	23.0	22.4
Lipid	4.5	NR
Ash	6.5	6.0
Crude fiber	26.0	23.6
Nitrogen-free extract	40.0	NR

[a] 1, From Huige (2006); 2, From Briggs et al. (1981).

in the form of powder, pellets, or extract. Hot trub consists of protein (40–70%), bitter substances (7–15%) organic compounds such as polyphenols and mineral substances (20–30%), while the cold trub consists of protein (50%), polyphenols (15–25%), and carbohydrates of high molecular mass (20–30%) (Esslinger and Narziss 2005).

16.3.3.2 Alternatives to Reuse

The main application of spent hops is as a soil conditioner or a fertilizer, due to its high nitrogen content, as shown in Table 16.3. Although the spent hops usually contain residual bitter substances, the presence of these components does not hinder their use as a fertilizer. On the other hand, the presence of such components is not desirable for application as feed supplement. Animals unwillingly eat bitter fodder and they are discouraged by sedative–hypnotic properties of 2-methyl-3-buten-2-ol, which is the product of bitter acid degradation. Some microorganisms have the ability to degrade hop bitter acids, which could be an environmentally friendly alternative to overcome this problem (Huszcza et al. 2008).

Some studies have been performed to add value to spent hops. A possible alternative to reuse this material is as raw material to obtain flavors, saccharides, or organic acids through the oxidation or hydrolysis processes (Fischer and Bipp 2005; Laufenberg et al. 2003; Oosterveld et al. 2002; Vanderhaegen et al. 2003). All these compounds have an industrial interest. Among them, the hop acids, particularly, have potential application as natural antibacterial agents, being safe to control bacteria in fermentation processes (Ruckle and Senn 2006). Additionally, β-acids in alkali metal salt form are soluble and stable when dissolved in propylene glycol (1,1-propanediol), and therefore, they are especially useful for use as antibacterial agents in food products, as agents in water treatment, and in appropriate pharmaceutical and cosmetic applications (Wilson et al. 2008).

Pectins represent a large part of the polysaccharides in spent hops and can be recovered by the acid extraction of this material. Pectins are widely used as an ingredient (gelling and thickening agent) in the food industry and their recovery constitutes an interesting alternative to valorize spent hops. Residual hops resins can also be recovered by extraction with acetone, obtaining an unsaturated drying oil that can be used for paints (Huige 2006).

Studies on the valorization of the trub are scarce. It was proposed in a recent study to add this material to the pitching wort, since trub components increase yeast vitality and fermentation performance. This positive effect of the trub is associated with several components present in its composition such as lipids, unsaturated long-chain fatty acids, and zinc (Kühbeck et al. 2007).

16.4 CONCLUSIONS AND PERSPECTIVES

A variety of by-products are generated during the process for beer elaboration, among which the spent grains, surplus yeast, and spent hops are those produced in the largest amounts. Up till now, spent grains and spent hops are practically unused, being supplied to farmers for use as animal feed. The yeast biomass has some industrial applications, but a surplus still exists. Interesting and valuable alternatives for reuse of these by-products have been proposed, which would make possible their application in food and pharmaceutical products. The valorization of these industrial by-products is of great interest as they are generated in large amounts around the world and could serve as raw materials for different industrial processes. In brief, the valorization of these by-products would be beneficial for the brewing industry that could sell them obtaining an additional financial gain, for other industries that could use them as low-cost raw materials for other processes, and would be also beneficial to the environment, since their disposal to the nature, which could promote environmental problems, would be avoided. For all the reasons mentioned before, an increase in the reutilization of the brewing by-products is expected in a near future.

REFERENCES

Aliyu, S. and Bala, M. 2011. Brewer's spent grain: A review of its potentials and applications. *Afr. J. Biotechnol.* 10:324–331.

Almeida, C., Brányik, T., Moradas-Ferreira, P., and Teixeira, J. 2003. Continuous production of pectinase by immobilized yeast cells on spent grains. *J. Biosci. Bioeng.* 96:513–518.

Bamforth, C. 2003. *Beer: Tap into the Art and Science of Brewing*, 2nd edn. New York: Oxford University Press, Inc.

Bartolomé, B., Gómez-Cordovés, C., Sancho, A.I. et al. 2003. Growth and release of hydroxycinnamic acids from brewer's spent grain by *Streptomyces avermitilis* CECT 3339. *Enzyme Microb. Technol.* 32:140–144.

Bartolomé, B., Santos, M., Jimenez, J.J., del Nozal, M.J., and Gómez-Cordovés, C. 2002. Pentoses and hydroxycinnamic acids in brewers' spent grain. *J. Cereal Sci.* 36:51–58.

Bhumibhamon, O. 1978. Production of acid protease and carbohydrate degrading enzyme by *Aspergillus awamori. Thai. J. Agr. Sci.* 11:209–222.

Bogar, B., Szakacs, G., Tengerdy, R.P., Linden, J.C., and Pandey, A. 2002. Production of α-amylase with *Aspergillus oryzae* on spent brewing grain by solid substrate fermentation. *Appl. Biochem. Biotechnol.* 102–103:453–461.

Briggs, D.E., Boulton, C.A., Brookes, P.A., and Stevens, R. 2004. *Brewing: Science and Practice*. Cambridge: Woodhead Publishing.

Briggs, D.E., Hough, J.S., Stevens, R., and Young, T.W. 1981. *Malting and Brewing Science, Vol 1, Malt and Sweet Wort*, 2nd edn. London: Chapman and Hall.

Carvalheiro, F., Duarte, L.C., Lopes, S., Parajó, J.C., Pereira, H., and Gírio, F.M. 2005. Evaluation of the detoxification of brewery's spent grain hydrolysate for xylitol production by *Debaryomyces hansenii* CCMI 941. *Process Biochem.* 40:1215–1223.

Carvalheiro, F., Esteves, M.P., Parajó, J.C., Pereira, H., and Gírio, F.M. 2004. Production of oligosaccharides by autohydrolysis of brewery's spent grain. *Bioresource Technol.* 91:93–100.

Cefalu, W.T. and Hu, F.B. 2004. Role of chromium in human health and in diabetes. *Diabetes Care*, 27:2741–2751.

Chae, H.J., Joo, H., and In, M.-J. 2001. Utilization of brewer's yeast cells for the production of food-grade yeast extract. Part 1: Effects of different enzymatic treatments on solid and protein recovery and flavor characteristics. *Bioresource Technol.* 76:253–258.

Cruz, J.M., Moldes, A.B., Bustos, G., Torrado, A., and Domínguez, J.M. 2007. Integral utilisation of barley husk for the production of food additives. *J. Sci. Food Agr.* 87:1000–1008.

Ding, W.J., Qian, Q.F., Hou, X.L., Feng, W.Y., and Chai, Z.F. 2000. Determination of chromium combined with DNA, RNA and proteins in chromium-rich brewer's yeast by NAA. *J. Radioanal. Nucl. Chem.* 244:259–262.

Dragone, G., Mussatto, S.I., and Almeida e Silva, J.B. 2007. High Gravity brewing by continuous process using immobilised yeast: Effect of wort original gravity on fermentation performance. *J. Inst. Brew.* 113:391–398.

Duvnjak, Z., Budimir, A., and Suskovic, J. 1983. Effect of spent grains from beer production on production of α-amylase by *Bacillus subtilis* 21+. *Prehram. Technol. Rev.* 21:97–101.

Esslinger, H.M. and Narziss, L. 2005. *Beer.* Weinheim: Wiley-VCH Verlag GmbH & Co.

Faulds, C., Sancho, A., and Bartolomé, B. 2002. Mono- and dimeric ferulic acid release from brewer's spent grain by fungal feruloyl esterases. *Appl. Microbiol. Biotechnol.* 60:489–494.

Fischer, K. and Bipp, H.-P. 2005. Generation of organic acids and monosaccharides by hydrolytic and oxidative transformation of food processing residues. *Bioresource Technol.* 96:831–842.

Ferreira, I.M.P.L.V.O, Pinho, O., Vieira, E., and Tavarela, J.G. 2010. Brewer's *Saccharomyces* yeast biomass: Characteristics and potential applications. *Trends Food Sci. Technol.* 21:77–84.

Francis, F., Sabu, A., Nampoothiri, K.M. et al. 2003. Use of response surface methodology for optimizing process parameters for the production of α-amylase by *Aspergillus oryzae*. *Biochem. Eng. J.* 15:107–115.

Gregori, A., Svagelj, M., Pahor, B., Berovic, M., and Pohleven, F. 2008. The use of spent brewery grains for *Pleurotus ostreatus* cultivation and enzyme production. *New Biotechnol.* 25:157–161.

Huige, N.J. 2006. Brewery by-products and effluents. In: *Handbook of Brewing,* 2nd edn, eds. F.G. Priest and G.G. Stewart. Boca Raton: CRC Press.

Huszcza, E., Bartmanska, A., Anioł, M., Maczka, W., Zolnierczyk, A., and Wawrzenczyk, C. 2008. Degradation of hop bitter acids by fungi. *Waste Manage.* 28:1406–1410.

Kanauchi, O., Mitsuyama, K., and Araki, Y. 2001. Development of a functional germinated barley foodstuff from brewers' spent grain for the treatment of ulcerative colitis. *J. Am. Soc. Brew. Chem.* 59:59–62.

Keukeleire, D.D. 2000. Fundamentals of beer and hop chemistry. *Quim. Nova* 23:108–112.

Kühbeck, F., Müller, M., Back, W., Kurz, T., and Krottenthaler, M. 2007. Effect of hot trub and particle addition on fermentation performance of *Saccharomyces cerevisiae*. *Enzyme Microb. Technol.* 41:711–720.

Lamoolphak, W., Goto, M., Sasaki, M. et al. 2006. Hydrothermal decomposition of yeast cells for production of proteins and amino acids. *J. Hazard. Mater.* 137:1643–1648.

Laufenberg, G., Kunz, B., and Nystroem, M. 2003. Transformation of vegetable waste into value added products: (A) the upgrading concept; (B) practical implementations. *Bioresource Technol.* 87:167–198.

Laws, D.R.J. and Waites, M.J. 1986. Utilization of Spent Grains. Patent number 85–305109 169068, Brewing Research Foundation, UK.

Lewis, M.J. and Young, T.W. 1995. *Brewing*. London: Chapman and Hall.

Li, Q., Chai, L., Yang, Z., and Wang, Q. 2009. Kinetics and thermodynamics of Pb(II) adsorption onto modified spent grain from aqueous solutions. *Appl. Surf. Sci.* 255:4298–4303.

Low, K.S., Lee, C.K., and Liew, S.C. 2000. Sorption of cadmium and lead from aqueous solutions by spent grain. *Process Biochem.* 36:59–64.

Lu, S. and Gibb, S.W. 2008. Copper removal from wastewater using spent grain as biosorbent. *Bioresource Technol.* 99:1509–1516.

Machado, M.D., Janssens, S., Soares, H.M., and Soares, E.V. 2009. Removal of heavy metals using a brewer's yeast strain of *Saccharomyces cerevisiae*: Advantages of using dead biomass. *J. Appl. Microbiol.* 106:1792–1804.

Machado, M.D., Santos, M.S., Gouveia, C., Soares, H.M., and Soares, E.V. 2008. Removal of heavy metals using a brewer's yeast strain of *Saccharomyces cerevisiae*: The flocculation as a separation process. *Bioresource Technol.* 99:2107–2115.

Mandalari, G., Bisignano, G., Lo Curto, R.B., Waldron, K.W., and Faulds, C.B. 2008. Production of feruloyl esterases and xylanases by *Talaromyces stipitatus* and *Humicola grisea* var. thermoidea on industrial food processing by-products. *Bioresource Technol.* 99:5130–5133.

Meneses, N. 2011. Produção de bebida destilada e etanol combustível a partir de resíduos da indústria cervejeira. Master dissertation, University of Minho, Braga, Portugal.

Mussatto, S.I. 2009. Biotechnological potential of brewing industry by-products. In: *Biotechnology for Agro-Industrial Residues Utilisation. 1 edn., Vol 1.* eds. P.S. Nigam, and A. Pandey. Netherlands: Springer, pp. 313–326.

Mussatto, S.I., Dragone, G., and Roberto, I.C. 2006a. Brewer's spent grain: Generation, characteristics and potential applications. *J. Cereal Sci.* 43:1–14.

Mussatto, S.I., Dragone, G., and Roberto, I.C. 2007a. Ferulic and *p*-coumaric acids extraction by alkaline hydrolysis of brewer's spent grain. *Ind. Crops Prod.* 25:231–237.

Mussatto, S.I., Dragone, G., Rocha, G.J.M., and Roberto, I.C. 2006b. Optimum operating conditions for brewer's spent grain soda pulping. *Carbohydr. Polym.* 64:22–28.

Mussatto, S.I., Fernandes, M., Dragone, G., Mancilha, I.M., and Roberto, I.C. 2007b. Brewer's spent grain as raw material for lactic acid production by *Lactobacillus delbrueckii*. *Biotechnol. Lett.* 29:1973–1976.

Mussatto, S.I., Fernandes, M., Rocha, G.J.M., Orfão, J.J.M., Teixeira, J.A., and Roberto, I.C. 2010. Production, characterization and application of activated carbon from brewer's spent grain lignin. *Bioresource Technol.* 101:2450–2457.

Mussatto, S.I. and Roberto, I.C. 2004. Alternatives for detoxification of dilute-acid lignocellulosic hydrolyzates for use in fermentative processes: A review. *Bioresource Technol* 93:1–10.

Mussatto, S.I. and Roberto, I.C. 2006. Chemical characterization and liberation of pentose sugars from brewer's spent grain. *J. Chem. Technol. Biotechnol.* 81:268–274.

Mussatto, S.I. and Roberto, I.C. 2008. Establishment of the optimum initial xylose concentration and nutritional supplementation of brewer's spent grain hydrolysate for xylitol production by *Candida guilliermondii*. *Process Biochem.* 43:540–546.

Mussatto, S.I., Rocha, G.J.M., and Roberto, I.C. 2008. Hydrogen peroxide bleaching of cellulose pulps obtained from brewer's spent grain. *Cellulose* 15:641–649.

Novik, G.I., Wawrzynczyk, J., Norrlow, O., and Szwajcer-Dey, E. 2007. Fractions of barley spent grain as media for growth of probiotic bacteria. *Microbiol.* 76:804–808.

Okita, H., Yamashita, H., and Yabuuchi, S. 1985. Production of microbial enzymes using brewers' spent grain. *Hakko Kogaku Kaishi—J. Ferment. Technol.* 63:55–60.

Oliva-Teles, A. and Gonçalves, P. 2001. Partial replacement of fishmeal by brewers yeast (*Saccaromyces cerevisae*) in diets for sea bass (*Dicentrarchus labrax*) juveniles. *Aquaculture,* 202:269–278.

Oosterveld, A. Voragen, A.G.J., and Schols, H.A. 2002. Characterization of hop pectins shows the presence of an arabinogalactan-protein. *Carbohydr. Polym.* 49:407–413.

Panagiotou, G., Granouillet, P., and Olsson, L. 2006. Production and partial characterization of arabinoxylan-degrading enzymes by *Penicillium brasilianum* under solid-state fermentation. *Appl. Microbiol. Biotechnol.* 72:1117–1124.

Rakin, M., Vukasinovic, M., Siler-Marinkovic, S., and Maksimovic, M. 2007. Contribution of lactic acid fermentation to improved nutritive quality vegetable juices enriched with brewer's yeast autolysate. *Food Chem.* 100:599–602.

Roberts, R.T. 1976. Use of an extract of spent grains as an antifoaming agent in fermentors. *J. Inst. Brew.* 82:96.

Romero, R. and Gomez-Basauri, J. 2003. Yeast and yeast products, past, present and future: From flavors to nutrition and health. Nutritional Biotechnology in the Food and Feed Industries. In: *Proceedings of Alltech's 19th International Symposium,* eds. T.P. Lyons, and K.A. Jacques, Loughborough, Leics: Nottingham University Press, pp. 365–378.

Roukas, T. 1999. Pullulan production from brewery wastes by *Aureobasidium pullulans. World J. Microbiol. Biotechnol.* 15:447–450.

Ruckle, L. and Senn, T. 2006. Hop acids can efficiently replace antibiotics in ethanol production. *Int. Sugar J.* 108:139–147.

Silva, J.P., Sousa, S., Rodrigues, J. et al. 2004. Adsorption of acid orange 7 dye in aqueous solutions by spent brewery grains. *Sep. Purif. Technol.* 40:309–315.

Sim, T.S. and Oh, J.C.S. 1990. Spent brewery grains as substrate for the production of cellulases by *Trichoderma reesei* QM9414. *J. Ind. Microbiol. Biotechnol.* 5:153–158.

Szponar, B., Pawlik, K.J., Gamian, A., and Dey, E.S. 2003. Protein fraction of barley spent grain as a new simple medium for growth and sporulation of soil actinobacteria. *Biotechnol. Lett.* 25:1717–1721.

Thammakiti, S., Suphantharika, M., Phaesuwan, T., and Verduyn, C. 2004. Preparation of spent brewer's yeast β-glucans for potential applications in the food industry. *Int. J. Food Sci. Technol.* 39:21–29.

Terrasan, C.R.F., Temer, B., Duarte, M.C.T., and Carmona, E.C. 2010. Production of xylanolytic enzymes by *Penicillium janczewskii. Bioresource Technol.* 101:4139–4143.

Vanderhaegen, B., Neven, H., Coghe, S., Verstrepen, K.J., Derdelinckx, G., and Verachtert, H. 2003. Bioflavoring and beer refermentation. *Appl. Microbiol. Biotechnol.* 62:140–150.

Wilson, R.J.H., Smith, R.J., and Haas, G. 2008. Application for hop acids as anti-microbial agents. Patent number US 7,361,374 B2.

Zouboulis, A.I., Matis, K.A., and Lazaridis, N.K. 2001. Removal of metal ions from simulated wastewater by *Saccharomyces* yeast biomass: Combining biosorption and flotation processes. *Sep. Sci. Technol.* 36:349–365.

Index